Materials Science and Technologies

Manufacturing Technology Research

nova
science publishers

www.novapublishers.com

Materials Science and Technologies

Biocomposites: Advances in Research and Applications
James R. Bush (Editor)
2023. ISBN: 979-8-89113-267-2 (Softcover)
2023. ISBN: 979-8-89113-313-6 (eBook)

Ferrite Materials and Technologies
Ravi Panwar, PhD (Editor)
Dharmendra Singh (Editor)
2023. ISBN: 979-8-89113-086-9 (Hardcover)

Perovskite Solar Cells: From Materials Science to Device Engineering
Thembinkosi Donald Malevu, PhD (Editor)
2023. ISBN: 979-8-89113-130-9 (Softcover)
2023. ISBN: 979-8-89113-188-0 (eBook)

More information about this series can be found at
https://novapublishers.com/product-category/series/materials-science-and-technologies/

Manufacturing Technology Research

Computer-Aided Design: Advances in Research and Applications
Dimitrios Tzetzis, PhD (Editor)
Panagiotis Kyratsis, PhD (Editor)
2023. ISBN: 979-8-88697-609-0 (Softcover)
2023. ISBN: 979-8-89113-147-7 (EPUB)

Additive Manufacturing and Reverse Engineering: Research and Manufacturing of Complex Parts
Betim Shabani (Author)
Vladimir Dukovski (Author)
2021. ISBN: 978-1-53619-718-1 (eBook)

Industry 4.0: Principles, Effects and Challenges
Yilmaz Uygun (Editor)
2020. ISBN: 978-1-53618-331-3 (Hardcover)
2020. ISBN: 978-1-53618-423-5 (eBook)

More information about this series can be found at https://novapublishers.com/product-category/series/manufacturing-technology-research/

Gamze D. Tetik
Gizem Celep
and Fulya Yilmaz
Editors

Textile Insights and Future Aspects

nova
science publishers
www.novapublishers.com

Library of Congress Cataloging-in-Publication Data

ISBN: 979-8-89113-289-4

Published by Nova Science Publishers, Inc. † New York

Contents

Preface

Textile history is almost as old as human history. Human being has felt the need to cover up from the first moment of his existence, and throughout the history, the need for dressing has been met with the use of different materials and techniques. Along with meeting the basic needs, the need for different types of clothing and diversity in clothing increased and new textile products were developed with the differentiation of cultural levels and developing technology over time. The term "textile" is originated from the Latin Word "texere" that means "woven, weave a fabric or piece of cloth". Whatever it is called in history, textile has been endured as a science that humanity needed since the beginning of civilization. Moreover, it updated itself with the developments in engineering and technology by innovative approaches to meet the demands of humanity. The birth of textile as an engineering field corresponds to the 1750s that the Industrial Revolution occurred. While manual fabrication of textile materials appears in before this date, the transition to the industrial production scales created textile engineering field and enabled the fabrication of various textile materials that will suffice the world population which gradually increases. Nowadays, the global world population has been reached over 8 billion people and this situation result with more needs for resources and raw materials. As textiles are one of the main necessities for human beings, the technological improvements, developments in science, and digitalization in processes make textile industry more common in various areas. In addition to new technological textile production, textile-based innovative solutions have made textiles to take one step further by working with multidisciplinary areas. Over time, while the numerical quantity of range of textile products applications increase, it also diversifies qualitatively. This is reflected in every stage of textile engineering that serves manufacturing of diversified textile products. Textile science tries to respond consumer demands with new applications in the field.

The diversity in textile raw materials affected product performance and paved the way for progress in quality, comfort, and technical fields. New product developments in textiles have enabled the use of much lighter, technical, and multi-functional textiles that respond to various stimuli in many fields such as the military, health, construction, and automotive sectors. While the developments in textile technology meet human needs and consumer demands, they trigger the search for new resources. Besides that, the development of new materials allowed the development of new technologies for the processing of these textile materials. Studies in the field of textile product development show themselves with significant growth, digitalization applications, and innovations in the textile and apparel industry.

Textile technology consists of many stages, from fiber production to the forming of the garment. These stages include fiber formation, yarn production, knitting/weaving production, fabric coloring and finishing processes, and apparel. In every step, new developments including textile materials, textile production and processes have been arisen day by day. Together with the natural and man-made fibers production, new types of fibers have been produced especially for their sustainability and also functional features.

Since the newly developed technologies are utilized in the textile and the apparel industry, textile production rates are faster, and it is possible to obtain products with higher quality and low defects. Textile processes have been also changed by developing functional properties for textiles such as flame retardancy, liquid repellency, anti-bacterial activity etc. Smart textiles, nanofibers technology, and plasma technology are also continuously promoting technologies. Apparel manufacturing is also progressing by the usage of newly adopted computer technologies like 3D scanning, robotics in different processes.

In forthcoming world, it will be possible to see much more technological use in the textile industry as in every industrial field. In this book, it is aimed to figure out the general textile improvements in various sub-disciplines of textiles. It mainly focused on the topics that will be much more popular as future expectation. Technical fabrics, smart textiles, nanofibers, sustainable finishing technologies, composites, functional textiles, sustainability in clothing industry, and 3D garment designing are the trending topics for future aspect. In this book, 33 expert scientists studying textile engineering and sub-disciplines participate and contribute with 23 original chapters having various perspectives. Each chapter aims to provide scientific and original contributions to the literature by addressing the innovative aspects of their sub-disciplines in the field of textile engineering. The most important purpose of this book is to present up-to-date knowledge of textile engineering and to shed a light on other studies in this field.

Chapter 1

Recent Progress and Innovative Applications in Knitted Spacer Fabric Technology

Gözde Ertekin*, PhD
and Arzu Marmarali, PhD
Department of Textile Engineering, Ege University, Izmir, Türkey

Abstract

Spacer fabrics are three-dimensional (3D) textiles that have two fabric surfaces connected together with spacer yarns or layers. Due to their 3D composition, spacer fabrics have special requirements that conventional textiles are unable to meet. Compression, breathability, high elastic recovery, high water vapor transfer properties, etc. are the main characteristics of spacer fabrics. Because of the ability to use a number of various materials, good performance properties, and 3D construction, spacer textiles have found applications in the fields of medical textiles, automotive textiles, composites, protective textiles, sportswear, etc. Additionally, these materials are utilized in different technical application fields through various techniques, including coating and lamination. The technologies of nonwoven, weaving, and knitting can be used to manufacture spacer textiles, and both weft and warp knitting techniques are the most commonly used techniques for the production of spacer fabrics. Weft-knitted spacer fabrics can be manufactured on both the electronic V-bed flat knitting machine and the double jersey circular knitting machine using a variety of fiber types. In this technique, two separate fabric layers are knitted with two sets of needles and are connected by tuck loops. Flat knitting machines have a strong potential for the production of spacer fabrics in preform shapes. These preforms could be used for the manufacturing of composites that have less waste and shorter production times. On the other hand, thanks to the many technical advantages warp knitting technology offers various spacer fabrics in different structures and properties. Spacer fabrics knitted with both techniques are frequently used for medical textiles, protective textiles, automotive textiles, composites, sports textiles, and equipment. In this chapter, recent advances in the field of spacer-knitted fabrics produced with both weft and warp knitting techniques, as well as innovative application areas and approaches where these structures are used, will be investigated.

Keywords: weft knitted spacer fabric, warp knitted spacer fabric, innovative applications

---

* Corresponding Author's Email: gozde.damci@ege.edu.tr.

In: Textile Insights and Future Aspects
Editors: Gamze D. Tetik, Gizem Celep and Fulya Yilmaz
ISBN: 979-8-89113-289-4

Introduction

Creating 3D structures in a single knitting process has been increasingly preferred over using additional joining procedures in order to expand the usage and value of knitted textiles in industrial and technical applications. Several 3D fabrics have been knitted with the use of advanced weft and warp knitting technologies to have remarkable mechanical, physical, and thermal characteristics, including lightweight, excellent dimensional stability, and high stiffness and strength (Hu, 2008). Flat and circular weft knitting, as well as warp knitting technology, can be used in order to produce 3D knitted fabrics. The most common 3D knitted structures are spacer structures, directionally oriented structures, net-shaped structures, and tubular structures (Chen, 2015). Spacer fabrics are 3D structures, which have two fabric surfaces connected together with spacer yarns or layers. Due to their 3D composition, spacer fabrics have special requirements that conventional textiles are unable to meet. Compression, breathability, high elastic recovery, high water vapor transfer properties, etc. are the main characteristics of spacer fabrics. Due to the ability to use a number of various materials, good performance properties, and 3D construction, spacer textiles have found applications in the fields of medical textiles, automotive textiles, composites, protective textiles, sportswear, etc. Additionally, these materials are utilized in unique technical application fields through various techniques, including coating and lamination. It is possible to produce spacer fabrics using weaving, nonwoven, and knitting technologies. Both weft and warp knitting techniques are the most commonly applied techniques for the production of spacer fabrics. Weft-knitted spacer fabrics can be produced using either flat or circular knitting machines with two sets of needles (Patnaik & Patnaik, 2019; Ertekin, 2014; Liu, 2013; Crina et al. 2013; Ciobanu, 2011; Ertekin, 2009; Badawi, 2008; Bruer et al., 2005).

This chapter reviews recent advances in the field of spacer-knitted fabrics produced with both weft and warp knitting techniques, as well as innovative application areas and approaches where these structures are used.

Spacer Fabric Production in Flat Weft Knitting Machines

Thanks to advanced technologies in computerized flat knitting machines, it is possible to manufacture spacer fabrics in which spacer yarns or fabric layers connect two outer layers. Two fabric layers are knitted on the front and back needle beds individually and connected by tucks on both needle beds (Liu & Hu, 2011). The thickness of the flat knitted spacer fabric is determined by the distance between the two needle beds, which can be varied in a range of 2-10 mm (Ertekin, 2009; Bruer et al., 2005). A cross-sectional view and its needle diagram of a flat knitted spacer fabric are shown in Figure 1. Additionally, in flat knitting technology, it is possible to connect two outer surfaces knitted in two needle beds using fabric layers. In this technique, spacer fabrics can be knit with or without loop transfer, and the fabric thickness is independent of the distance between the front and back needle beds (Chen, 2015).

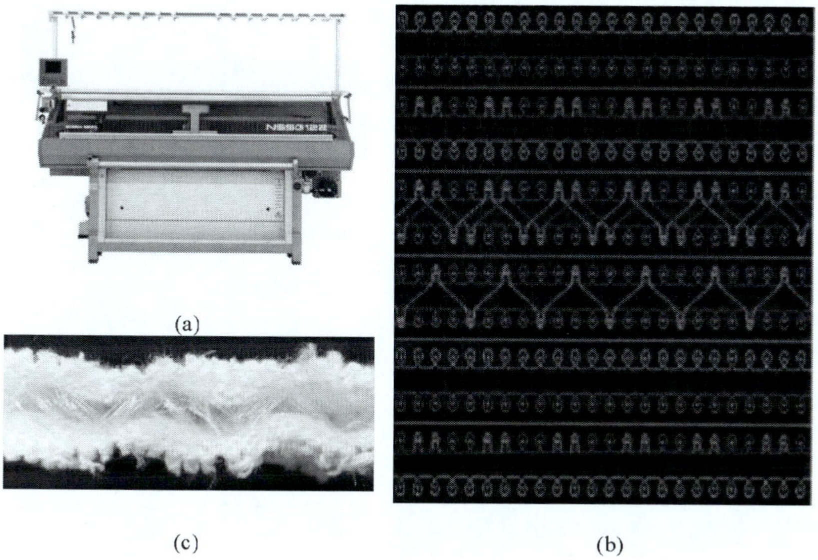

Figure 1. a) One of the computerized flat knitting machines (Shima Seiki) for spacer fabric, b) Knitting notation, c) Cross-sectional view of flat knitted spacer fabric (Ertekin, 2017).

Spacer Fabric Production in Circular Weft Knitting Machines

Double bed (cylinder and dial) circular knitting machines are used to produce the spacer fabrics. The connection of the two outer fabrics to each other is made with spacer yarns. For each course of the fabric, at least three different yarns must be used, including yarn for cylinder needles, yarn for dial needles, and spacer yarn. Monofilament yarn is generally used to connect the fabric layers as seen in Figure 2 (Ertekin & Marmaralı, 2011; Bruer et al., 2005).

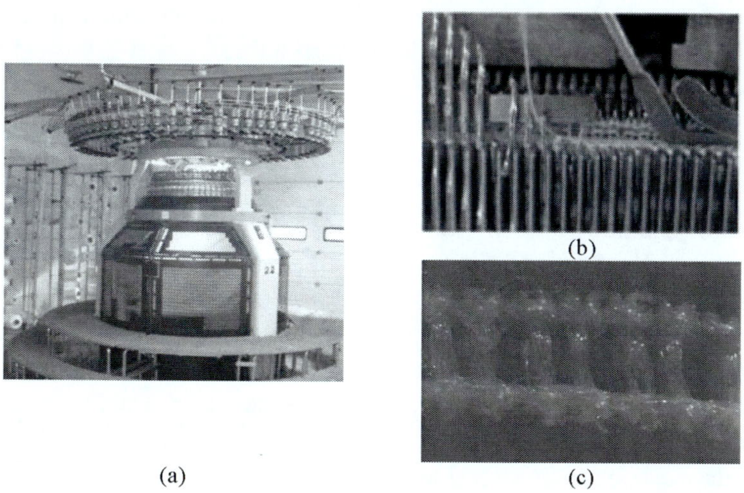

Figure 2. (a) One of the circular knitting machines (Mayer & Cie) for the spacer fabric, (b) Knitting zone, (c) Cross-sectional view of the circular knitted spacer fabric (Ertekin, 2009).

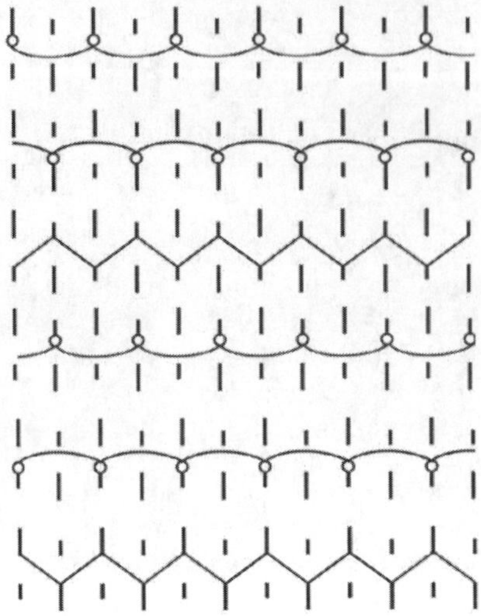

Figure 3. Needle diagram of the circular knitted spacer fabric (Ertekin, 2009).

There are two methods to produce circular knit spacer textiles: (a) Macking tucks on dial and cylinder needles on the same feeder (Figure 3) and (b) knitting/plating on dial needles while knitting on cylinder needles (Ertekin & Marmaralı, 2011). Jacquard technology can be used to produce various patterns on the fabric's outer layers. Adjusting height of the dial, the distance between the two layers of fabric can be changed from 1.5 to 5.5 mm (Chen, 2015).

Spacer Fabric Production in Warp Knitting Machines

Warp knitting technology is the most commonly used technique compared to other types of spacer fabrics, with their higher productivity and a wider range of structural variants. Double needle bar raschel machines are used to produce warp-knitted spacer fabrics (Figure 4). The machine must have at least four guide bars; generally five or seven guide bars are preferred. For the production of spacer fabrics, guide bars 1 and 2 feed the front needle bar, while guide bars 5 and 6 feed the back needle bar to knit the outer layers. The spacer yarns are laid around both needle bars by guide bars 3 and 4. As a result, various spacer fabrics can be produced using different yarns and structures in the front and back needle bars. It is possible to create mesh or closed outer layer structures with various patterns, using various threading and lapping movements of each guide bar (Chen, 2015; Ertekin, 2014). In recent years, warp-knitted jacquard spacer fabrics have gained widespread use due to their physical properties such as lightness, heat conduction, buffering, compression resistance, moisture permeability, and air permeability. According to the application area, two outer layers of warp-knitted spacer fabrics may be the same or different, have a mesh structure on both faces or just one face, and may even have different mesh sizes on each side. The demand for different patterns and functions in these fabrics is also increasing (Wen & Liu, 2022). They are widely used in clothing, footwear, medical care, construction engineering, safety protection, and other fields. Generally, multi-filament yarns are always used to knit outer layers, while monofilament yarns are preferred as spacer yarns. The characteristics of the spacer yarn influence the properties of the final product. Additionally, the inclination angle of the spacer yarn can be adjusted in different geometrical configurations, such as vertical, diagonal, etc. by altering the underlapping movements and the threading order. The thickness of the spacer fabric can be adjusted in a range of 1.5 to 20.0 mm by changing the distance between the needle bars. In addition, a wide range of spacer fabrics with thicknesses ranging from 20 mm to 65 mm can be produced by using the HighDistance® machine of Karl Mayer. After the knitting process, the spacer fabrics need to be heat-seat in order to obtain and maintain the desired form (Patnaik & Patnaik, 2019; Ertekin & Marmaralı, 2016; Chen, 2015; Ertekin, 2014; Liu et al., 2012a; 2012b; 2014).

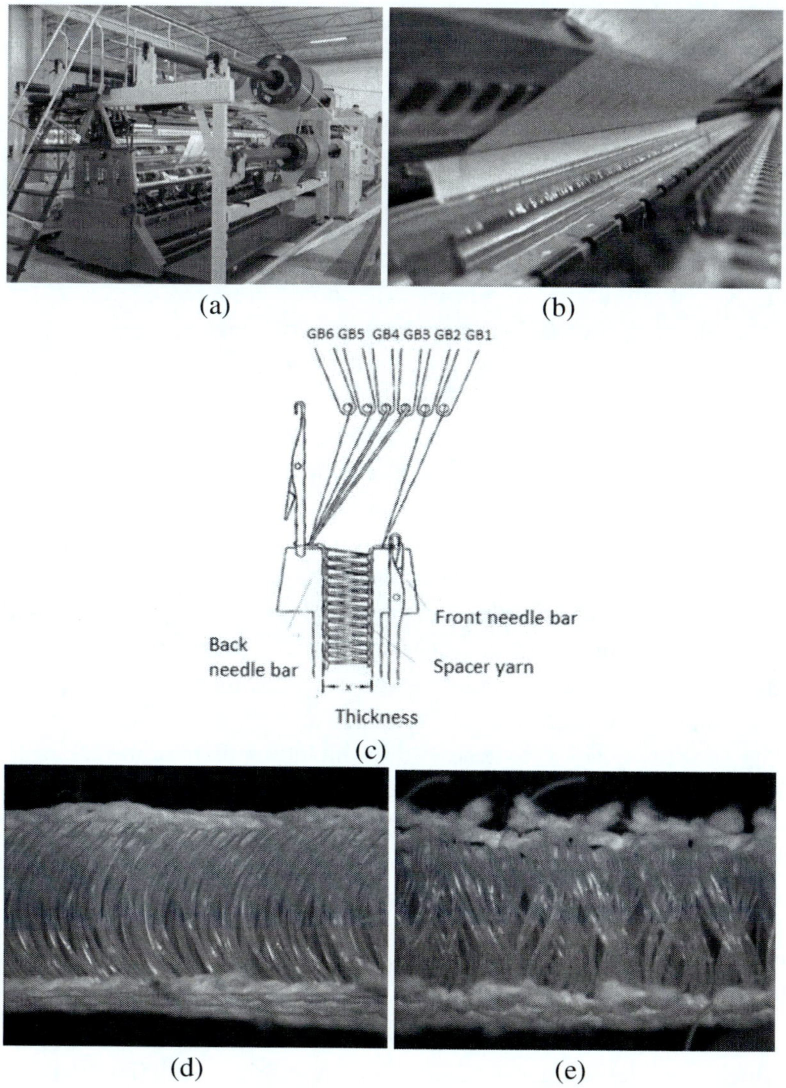

Figure 4. (a) One of the double needle bar raschel warp knitting machine (Karl Mayer), (b) and (c) Knitting zones of the double needle bar raschel machine, cross-sectional view from the (d) course direction and (e)wale direction (Ertekin, 2014).

Characteristics and Applications

Spacer knitted textiles have a wide range of application areas like cushioning materials, protective textiles, well-being and comfort, medical textiles, insole applications, aerogel applications, composite reinforcements, sound and noise absorption, vibration isolation, and sensor technology, etc. Several studies conducted on recent progress and innovative application areas of spacer knitted fabrics within the last few years are summarized below:

Cushioning Materials

Spacer knitted fabrics are promising for cushioning purposes in protective clothing and equipment for body protection against impacts due to their high impact response and energy absorption properties (Ertekin & Marmaralı, 2017). The cushioning characteristic of these fabrics is significantly influenced by their compression properties. The shorter distance from the spacer yarn, the coarser monofilaments, and the higher density of the spacer yarn are characteristics of weft-knitted spacer fabrics. Therefore, they are proper to use as impact protectors with their better compression resistance and absorption energy (Zhao et al., 2018). The compression resistance, energy absorption, and dissipation capacity of the weft-knitted spacer fabrics can be increased with the number of miss or tuck stitches in successive courses of the surface layer and higher inclination angle in spacer yarns. On the other hand, in this case, their thickness, resilience, and compressibility values decrease. Due to their higher energy absorption, improved resiliency, and compressibility, spacer textiles with close surface structures are more suitable for cushioning applications (Asayesh & Amini, 2021a; 2021b).

Rudy and Wardiningsih (2021) investigated the maximum impact force and force attenuation capacity of the weft-knitted spacer fabrics produced using varying diameters of nylon monofilament yarns. They reported that larger diameter spacer yarns resulted in lower impact force and higher force attenuation capacity, while greater force attenuation can be achieved by using multiple layers of spacer fabric. According to Chen et al. (2021), fabric thickness, mass per unit area, and surface structures, including plain or mesh construction, have significant effects on the heat and mass transfer properties of polyester warp-knitted spacer fabrics, which are widely used in sportswear, functional clothing, and protective clothing. As volume density increases, water vapor permeability of spacer fabrics decreases, while water absorption ratio increases. Additionally, spacer fabrics with one-sided mesh, and one-sided close surface structures provide better water vapor and air transmission properties.

Amirifard and Asayesh (2022) aimed to enhance the compression characteristics of weft-knitted spacer fabrics by incorporating elastic yarns into the surface layers of the fabric with a variety of tensions and alignments, and they found that utilizing higher elastic yarn tension and fewer missed stitches in the surface layers of spacer fabrics can effectively improve their cushioning properties. Different compression behaviors can be seen with the incorporation of an elastic inlay in spacer fabrics (Yu et al., 2022a; 2022b; 2021). Spacer fabrics can be obtained in different thicknesses, densities, and compression behaviors while retaining air permeability and lower fabric weight by using different inlay patterns and elastic yarns (Yu et al., 2022a). However, the creation of wearable cushioning products that can adapt to body contours presents a significant challenge. Yu et al. (2022b) proposed a method to control the curvature of the weft-knitted spacer fabric to achieve a high degree of body conformity. They reported that the feeding rate of elastic yarn has a significant effect on the fabric curvature and produces a thicker, more curved, lower weight and density fabric with lower feeding rates of elastic yarn.

The cushioning characteristic of the warp-knitted spacer fabrics is influenced by the fineness of the spacer filament. Compressive stress and indentation force deflection increase with increasing fineness of the spacer filament. A warp-knitted spacer fabric with 0.25 mm spacer filament fineness is suggested for mattresses with both high comfort and pressure resistance. It is recommended that the compression depth of a warp-knitted spacer fabric should not exceed 40% for optimal compression in mattresses (Qu et al., 2022).

The compression behavior of warp-knitted spacer fabrics measured using a pair of flat or spherical plates can be characterized by different models, such as the genetic algorithm–optimized viscoelastic model (Yousefpour et al., 2018), the finite element method (Datta et al., 2019a; Sun et al., 2019), regression analysis and the K-means cluster method (Li et al., 2018). For the evaluation of the compression characteristic of spacer fabrics in real conditions, in a study, Datta et al. (2019b) characterized warp-knitted spacer fabric by analyzing the reaction force in spherical compression of the fabric, along with its thermal resistance, air, and moisture transfer properties. They suggested warp-knitted spacer fabric as an effective component of sports bras to control breast movement while preserving the comfort of sportswomen.

Well-Being and Comfort

Two outer layers of the spacer fabric are held apart by spacer yarns, creating a layer of ventilated air that allows heat and moisture to escape. Thermoregulation properties, which are especially important for applications that involve the human body, are influenced by the types of material and characteristics of the surface layers (Buzaite & Mikucioniene, 2022; Chen, 2015). The porosity, air permeability, and thermoregulation properties of warp-knitted spacer fabrics were evaluated based on the linear density of the polyester filaments by using the experimental design of Box and Behnken. According to this evaluation, there is a high correlation between the linear density of the filament and the air permeability. The face layer has a significant impact on all the responses, while the bottom layer has a negligible effect (Rajan et al., 2021; Rajan & Das, 2020). The thermal comfort properties of 3D warp-knitted spacer fabrics were investigated as potential replacements for polyurethane foams in car seats and back supports by Arumugam et al. (2018; 2019a). Fabric parameters such as raw material and type of spacer yarn, density, thickness, and tightness of the surface layer of spacer fabrics significantly affect the thermal conductivity and water vapor permeability. Spacer fabrics with a hexagonal net structure had high air permeability, water vapor permeability, and good thermal conductivity with a more open surface structure than fabrics with locknit surface structures. These results have important implications for the usage of spacer fabrics for car seats and back supports. In a study, in comparison with standard liner fabrics, warp-knitted spacer fabric is suggested due to its higher energy absorption, breathability, and evaporative heat transfer properties as the inner padding layer of a motorcycle helmet is a crucial component to provide comfort and stability while wearing the helmet (Kumar et al., 2022).

Medical Applications

Weft-knitted spacer fabrics are an excellent substitute for wound dressings due to their higher air permeability, water vapor permeability, absorbency, and thermal conductivity. Additionally, an increase in the inclination angle of spacer yarn leads to an increase in air permeability, water vapor permeability, and absorbency and a decrease in thermal conductivity and compressibility (Asayesh et al., 2019). Warp-knitted spacer fabrics have also been used as a hip protectors due to their desirable properties, such as high air and water vapor permeability, good compressibility, and lightweight. Smaller hexagonal mesh size, open/closed surface structure, lower thickness, and latex coating can be suggested to improve the durability, comfort, and lifetime of warp-knitted spacer fabrics appropriate for use as hip protectors (Ertekin & Marmaralı, 2018). Another application area of warp-knitted spacer fabric in medical textiles is tissue engineering. The findings of a study revealed that heat setting is an effective approach for modifying the properties of 3D bioabsorbable poly (lactic acid) spacer fabric's scaffolds, thickness, porosity, pore size, and stiffness, to create an environment suitable for tissue regeneration. The precise control and tunability of spacer fabric properties make it a promising candidate for patient-specific applications for regenerating different tissues (Caronna et al., 2022). Moreover, Schäfer et al. (2020) investigated the possibilities of elasticity- and porosity-adjustable textile-reinforced hydrogels as a flexible platform for soft tissue engineering. With its adaptable mechanics, the textile-reinforced hydrogel system presents a promising framework for the fabrication of a range of soft tissues, such as cartilage, tendon, and muscle.

Insole Applications

Insoles are used to offer comfort during walking and to provide resistance to ground reaction forces. Li et al. (2022) proposed a novel structure of weft-knitted spacer fabric incorporating inlays for insoles that can absorb shock, resist pressure, and enhance thermal comfort through heat dissipation. The compression resistance of the fabric increases by enhancing the density of the inlay and the spacer yarn, leading to a decrease in impact forces. Moreover, the higher density of the spacer yarn provides improved air permeability but decreases thermal resistance. Special surface treatments, such as plasma treatment, a well-known method for surface modification, can enhance the comfort characteristics of the materials. Based on fabric thickness and gas processing duration, Rajan and Sundaresan (2020) employed atmospheric plasma oxygen and argon to examine the thermal comfort characteristics of polyester warp-knitted spacer fabric. With atmospheric plasma treatment, air permeability and thermal conductivity declined while water vapor permeability and thermal resistance increased. Plasma treatments can improve the thermal comfort properties of polyester warp-knitted spacer fabrics, with argon plasma treatment being more effective than oxygen plasma treatment.

The weft-knitted spacer fabric has potential use in orthopedic foot insole applications. The insole made of weft-knitted spacer fabric with superelastic nitinol monofilaments exhibited a reduction in peak plantar pressure, improved breathability, and uniform pressure distribution rather than commercial diabetic foot insoles, and maintained the cushioning properties, pressure distribution, and endurance even after one month. For this reason, the NiTi insole is promising as a tool to reduce ulceration and pressure sores in diabetic feet (Hamedi & Salimi, 2022; Hamedi et al., 2020).

Aerogel Applications

The silica aerogel coating significantly improves the hydrophobic and oleophilic properties of 3D weft-knitted spacer fabrics and these fabrics are considered suitable for industrial applications such as oil spill cleanup requiring hydrophobic and oleophilic characteristics. Several parameters such as concentration rate, fabric structure (Islam et al., 2022a), temperature, oil type, and water medium (Islam et al., 2022b) were investigated to determine the effect of these parameters on the hydrophobic and oleophilic properties of silica aerogel treated spacer fabrics. This treatment enhances the hydrophobicity and oleophilic characteristics of fabrics with increasing water contact angles and oil sorption capacities. Due to higher pore diameter, specific surface area, surface roughness, the percentage silica add-on, water contact angle, and lower surface energy, these fabrics have higher oil sorption efficiency (Islam et al., 2022c; 2022d; 2023a). Weft-knitted spacer fabrics coated with silica aerogels+poly(vinyl-alcohol) sols exhibited higher thermal resistance, lower thermal conductivity, and better thermal stability, which could be used as eco-friendly and multifunctional oil sorbent materials for industrial applications (Islam et al., 2023b). Moreover, silica-aerogel coated three-layered weft-knitted spacer fabrics with higher thermal resistance and porosity, as well as higher air permeability, and lower density could have potential applications in thermal insulation for heating and cooling (Islam et al., 2019).

Composite Reinforcements

Flexible composites reinforced with spacer-knitted fabrics have gained increasing popularity in the last few years. The compression behavior and energy-absorption efficiency of polyurethane (PU)-matrix composites by impregnating warp-knitted spacer fabrics with flexible PU foam were analyzed via experimental and theoretical methods (using finite element analysis (Chen, 2021a), and the compression meso-mechanics theoretical model based on the Winkler elastic foundation beam theory (Chen, 2021b). In a study, Jiang et al. (2022) stated that increasing the filling rate of silicone rubber improved the compressive and impact properties of flexible composites reinforced with warp-knitted spacer fabrics. The structural parameters affecting the impact properties of flexible composites were classified in descending order, such as fabric thickness, spacer inclination angle, spacer density, and mesh density.

Another example of flexible composites reinforced with warp-knitted spacer fabrics is fabric impregnated with shear thickening fluid that improves the low-velocity impact and compressive characteristics of the composites. Shear thickening fluid-filled warp-knitted spacer fabric composites may serve as damping or energy-absorptive materials for human body protection (Gu et al., 2023; Xu et al., 2022).

Auxetic Warp Knitted Spacer Fabric

Warp-knitting technology is utilized method to create 3D auxetic fabrics, which possess negative Poisson ratios. Thanks to this technology, it is possible to produce auxetic warp-knitted fabrics with two auxetic geometries such as a re-entrant hexagonal network and rotating square solids through a single knitting process avoiding the need for additional post-compression and heat fixation processes (Chang et al., 2022). Gao et al. (2020) determined the auxetic property of 3D warp-knitted spacer fabrics with the reconstruction of a 3D image obtained by synchrotron radiation X-ray computed tomography. This model provides a foundation for further numerical analysis using finite element analysis. Additionally, Chang and Ma (2019) stated that fabrics with negative Poisson ratios exhibited relatively better energy absorption when their yarn loading capacity was equivalent.

Sound and Noise Absorption

3D knitted spacer fabrics are intriguing due to their voluminous, soft, and highly porous nature, which makes them suitable for sound absorption applications. Both weft- and warp-knitted spacer fabrics can be used as sound absorbers in buildings, automobile upholstery, auditoriums, and other places. The sound absorption coefficient of knitted spacer fabrics can be enhanced with increasing thickness, massivity, and high airflow resistance. The weft-knitted spacer structure with covered elastomeric yarns demonstrated great potential for noise reduction and control engineering (Fu et al., 2022; Arumugam et al., 2019b). Davoudabadi Farahani et al. (2022; 2023) studied the effect of the inclination angle of spacer yarns at two different spacing and layering (single-layered and three-layered) on the sound absorption properties of warp-knitted spacer fabrics enhanced with polyacrylonitrile nanofiber. The sound absorption capacity at all frequencies increases with the increase in spacing and inclination angle, layering, and nanofiber enhancement.

Additionally, the sound absorption performance of warp knitted spacer fabrics was modeled and predicted by means of artificial neural network system (Davoudabadi Farahani et al., 2023), statistical evaluation (Arumugam et al., 2019b), and empirical models of Delany and Bazley, Garai and Pompoli, and Dunn and Davern (Abedkarimi et al., 2020).

Vibration Isolation

Each spacer yarn inside the spacer fabrics acts as an Euler rod, which can reduce the vibration of the weft-knitted spacer fabrics. These fabrics can be used as functional materials for personal protective equipment against vibration in various applications, such as antivibration gloves, wheelchair cushions, etc. Various factors including monofilament diameter, the distance between the surface layers, and excitation acceleration level have a significant effect on the vibration behavior of the fabric. Spacer fabrics with smaller monofilament diameters and longer distances should be preferred for vibration isolation (Chen et al., 2023). The damping properties (vibration indexes, such as damping ratio and coefficient, logarithmic decrease, and free vibration period), vibration transmission coefficient, and natural frequency values of spacer fabrics can be analyzed by using a single-degree-of-freedom model. The results obtained from the theoretical analysis support the development of better designed spacer fabrics with improved vibration transmission properties (Chen et al., 2018a; 2018b; Taghvaie et al., 2020).

Sensor Technology

Due to possible uses for wearable power supplies and self-powered sensors, the integration of newly developed triboelectric nanogenerator (TENG) science with conventional textile technology has received significant attention in recent years. However, the limited electrical output of textile TENGs has been a major challenge in their development. To address this issue, due to its higher compression elasticity and enhanced power output, a scalable and washable warp-knitted spacer fabric has been developed. Potential uses of this fabric include self-powered sensors, wearable micro/nano power sources, and human-machine interfaces (Wang et al., 2021). Dou et al. (2022) fabricated highly stretchable and compressible strain sensors from carbon nanotubes and weft-knitted polypyrrole-deposited spacer fabrics. With the ability to stretch up to 450% and compress up to 60%, it displayed good electrical conductivity and reliable strain-sensing capability under various stretching deformations.

Conclusion

3D knitted textiles are already widely employed as technical textiles in a variety of applications including automotive textiles, personal protective clothing, sports textiles, foundation garments, swimwear pads, buffer clothing, and medical textiles. In technical textiles, the usage of weft-knitted spacer textiles is limited compared to warp-knitted spacer fabrics. Therefore, the most important goal for weft-knitted spacer fabrics in the future is to widen their applications to include the automotive sector, the medical textile, composite reinforcement, smart textiles, etc. New kinds of material, coating techniques, and finishing treatments can be used to enhance the characteristics of warp-knitted spacer fabric while also exploiting new application areas. In order to achieve this goal, novel fiber materials such as high-performance fibers, optical fibers, biodegradable fibers, and shape memory fibers are becoming essential. The combination of spacer fabric knitting with new materials, processes, and treatments has become a significant trend for the future.

Disclaimer

None.

References

Abedkarimi, R., Hasani, H., Soltani, P., & Talebi, Z. (2020). Experimental and computational analysis of acoustic characteristics of warp-knitted spacer fabrics. *The Journal of the Textile Institute*, *111*(4), 491-498. https://doi.org/10.1080/00405000.2019.1648140.

Amirifard, M., & Asayesh, A. (2022). The effect of elastic yarn tension and alignment on the tensile and compression characteristics of weft-knitted spacer fabrics for cushioning applications. *The Journal of the Textile Institute*, 1-12. https://doi.org/10.1080/00405000.2022.2131351.

Arumugam, V., Mishra, R., Militky, J., & Tomkova, B. (2019b). Noise attenuation performance of warp knitted spacer fabrics. *Textile Research Journal*, *89*(3), 281-293. https://doi.org/10.1177/0040517517743687.

Arumugam, V., Mishra, R., Militky, J., Davies, L., & Slater, S. (2018). Thermal and water vapor transmission through porous warp knitted 3D spacer fabrics for car upholstery applications. *The Journal of the Textile Institute*, *109*(3), 345-357. https://doi.org/10.1080/00405000.2017.1347023.

Arumugam, V., Mishra, R., Militky, J., Kremenakova, D., Tomkova, B., & Venkatraman, M. (2019a). Thermo-physiological properties of 3D warp knitted spacer fabrics for car seat application. *Indian Journal of Fibre & Textile Research*, *44*(4), 475-485.

Asayesh, A., & Amini, M. (2021a). Analysis of the compression performance of weft-knitted spacer fabrics for protective applications in view of the surface layer structure. *Fibers and Polymers*, *22*, 3469-3478. https://doi.org/10.1007/s12221-021-0248-y.

Asayesh, A., & Amini, M. (2021b). The effect of fabric structure on the compression behavior of weft-knitted spacer fabrics for cushioning applications. *The Journal of the Textile Institute*, *112*(10), 1568-1579. https://doi.org/10.1080/00405000.2020.1829330.

Asayesh, A., Ehsanpour, S., & Latifi, M. (2019). Prototyping and analyzing physical properties of Weft knitted spacer fabrics as a substitute for wound dressings. *The Journal of the Textile Institute*, *110*(9), 1246-1256. https://doi.org/10.1080/00405000.2018.1557358.

Badawi, M. S. S. (2008). *Development of the weaving machine and 3D woven spacer fabric structures for lightweight composites materials.* [Doctoral dissertation]. Technische Universität Dresden.

Bruer, S. M., Powell, N., & Smith, G. (2005). Three-dimensionally knit spacer fabrics: a review of production techniques and applications. *Journal of Textile and Apparel, Technology and Management*, *4*(4), 1-31.

Buzaite, V., & Mikucioniene, D. (2022). Effect of inner layer structures of weft-knitted spacer fabrics on thermal insulation and air permeability. *Textile Research Journal*, *92*(15-16), 2862-2872. https://doi.org/10.1177/00405175211021452.

Caronna, F., Glimpel, N., Paar, G. P., Gries, T., Blaeser, A., Do, K., Dolan, E. B., & Ronan, W. (2022). Manufacturing, characterization, and degradation of a poly (lactic acid) warp-knitted spacer fabric scaffold as a candidate for tissue engineering applications. *Biomaterials Science*, *10*(14), 3793-3807. https://doi.org/10.1039/D1BM02027G.

Chang, Y., & Ma, P. (2019). Energy absorption and Poisson's ratio of warp-knitted spacer fabrics under uniaxial tension. *Textile Research Journal*, *89*(6), 903-913. https://doi.org/10.1177/0040517518758005.

Chang, Y., Liu, Y., Shuaiquan, Z., & Hong, H. (2022). Design and manufacture of three-dimensional auxetic warp-knitted spacer fabrics based on re-entrant and rotating geometries. *Textile Research Journal*, *92*(3-4), 467-478. https://doi.org/10.1177/00405175211037204.

Chen, C., Chen, J., Sun, F., Yang, H., & Du, Z. (2018a). Analysis of the damping property of warp-knitted spacer fabrics under damped free vibration. *Textile Research Journal*, *88*(7), 790-799. https://doi.org/10.1177/0040517517690628.

Chen, C., Chen, J., Sun, F., Yang, H., Lv, Z., Zhou, Q., Du, Z., & Yu, W. (2018b). Study of the vibration transmission property of warp-knitted spacer fabrics under forced sinusoidal excitation vibration. *Textile Research Journal*, *88*(8), 922-931. https://doi.org/10.1177/0040517517690630.

Chen, F., Wang, J., Gao, S., Ning, X., Yan, P., & Tian, M. (2023). An experimental study on the vibration behavior and the physical properties of weft-knitted spacer fabrics manufactured using flat knitting technology. *Textile Research Journal*, *93*(3-4), 615-631. https://doi.org/10.1177/00405175221124929.

Chen, Q., Shou, D., Zheng, R., Tang, K. P. M., Fu, B., Zhang, X., & Ma, P. (2021). Moisture and thermal transport properties of different polyester warp-knitted spacer fabric for protective application. *AUTEX Research Journal*, *21*(2), 182-191. https://doi.org/10.2478/aut-2020-0013.

Chen, S. (2021a). Low-velocity impact properties of polyurethane composites reinforced with warp-knit spacer fabric. *AATCC Journal of Research*, *8*(2), 38-47. https://doi.org/10.14504/ajr.8.2.5.

Chen, S. (2021b). The compression meso-mechanics theoretical model and experimental verification of Polyure-thane-based warp-knit spacer fabric composites. *AATCC Journal of Research*, *8*(5), 30-38. https://doi.org/10.14504/ajr.8.5.4.

Chen, X. (Ed.). (2015). *Advances in 3D Textiles*. Elsevier.

Ciobanu, L. (2011). Development of 3D knitted fabrics for advanced composite materials. In B. Attaf (Ed.), *Advances in composite materials-ecodesign and analysis* (pp. 161-192). IntechOpen. https://doi.org/10.5772/14876.

Crina, B., Blaga, M., Luminita, V., & Mishra, R. (2013). Comfort properties of functional weft knitted spacer fabrics. *Textile and Apparel, 23*(3), 220-227.

Datta, M. K., Behera, B. K., & Goyal, A. (2019a). Prediction and analysis of compression behaviour of warp-knitted spacer fabric with cylindrical surface. *Journal of Industrial Textiles, 48*(9), 1489-1504. https://doi.org/10.1177/1528083718769936.

Datta, M. K., Behera, B. K., & Goyal, A. (2019b). Characterization of warp knitted spacer fabric for application in sports bra. *Fibers and Polymers, 20*, 1983-1991. https://doi.org/10.1007/s12221-019-9039-0.

Davoudabadi Farahani, M., Jamshidi Avanaki, M., & Jeddi, A. A. (2022). Sound absorption of warp knitted spacer fabrics based on knit structure and nanofiber enhancement. *Journal of Industrial Textiles, 51*(10), 1539-1557. https://doi.org/10.1177/1528083720903411.

Davoudabadi Farahani, M., Jeddi, A. A. A., & Hasanzadeh, M. (2023). Predicting the sound absorption performance of warp-knitted spacer fabrics via an artificial neural network system. *Fibers and Polymers, 24*(4), 1491-1501. https://doi.org/10.1007/s12221-023-00151-6.

Dou, L., Zeng, Z., Cheng, D., Li, S., Ke, W., & Cai, G. (2022). Weft-knitted spacer fabric for highly stretchable–compressible strain sensor, supercapacitor, and joule heater. *Nanomaterials, 12*(20), 3684. https://doi.org/10.3390/nano12203684.

Ertekin, G. (2009). A research about general properties of spacer fabrics produced in circular knitting machines. [Master dissertation]. Ege University.

Ertekin, G. (2014). A research about performance and comfort properties of mechanical impact protection textile products. [Doctoral dissertation]. Ege University.

Ertekin, G. (2017). Analysis of thermo-physiological comfort and moisture management properties of flat knitted spacer fabrics. *Textile and Apparel, 27*(3), 241-250.

Ertekin, G., & Marmarali, A. (2011). Heat, air and water vapor transfer properties of circular knitted spacer fabrics. *Textile and Apparel, 21*(4), 369-373.

Ertekin, G., & Marmaralı, A. (2016). The effect of heat-setting conditions on the performance characteristics of warp knitted spacer fabrics. *Journal of Engineered Fibers and Fabrics, 11*(3), 64-71. https://doi.org/10.1177/155892501601100309.

Ertekin, G., & Marmaralı, A. (2017). Impact resistance behaviour of silicone coated warp knitted spacer fabrics used for protective clothing. *The Journal of the Textile Institute, 108*(12), 2123-2131. https://doi.org/10.1080/00405000.2017.1315796.

Ertekin, G., & Marmaralı, A. (2018, December). An evaluation of usability of warp knitted spacer fabrics as hip protector. *IOP Conference Series: Materials Science and Engineering, 459*(1), 012038. IOP Publishing. https://doi.org/10.1088/1757-899X/459/1/012038.

Fu, S., Zeng, P., Zhou, L., Tang, X., & Liu, Y. (2022). Sound absorption coefficient analysis and verification of weft-knitted spacer fabrics for noise reduction application. *Textile Research Journal, 92*(23-24), 4541-4550. https://doi.org/10.1177/00405175221084277.

Gao, Y., Ma, P., Li, K., & Xu, J. (2020). Characterization of a 3D auxetic warp-knitted spacer fabric by synchrotron radiation x-ray computed tomography. *Fibers and Polymers, 21*, 930-933. https://doi.org/10.1007/s12221-020-9763-5.

Gu, L., Du, Z., Liu, G., & Ma, M. (2023). Mechanical properties of shear thickening fluid filled warp-knitted spacer fabrics. *Journal of Industrial Textiles, 53*, 15280837231159939. https://doi.org/10.1177/15280837231159939.

Hamedi, M., & Salimi, P. (2022). Reducing foot plantar pressure using superelastic nitinol monofilaments as spacer yarns in a novel weft knitted spacer fabric insole. *Journal of Industrial Textiles, 52*, 15280837221127314. https://doi.org/10.1177/15280837221127314.

Hamedi, M., Salimi, P., & Jamshidi, N. (2020). Improving cushioning properties of a 3D weft knitted spacer fabric in a novel design with NiTi monofilaments. *Journal of Industrial Textiles, 49*(10), 1389-1410. https://doi.org/10.1177/1528083718817560.

Hu, J. (2008). *3-D fibrous assemblies: Properties, applications and modelling of three-dimensional textile structures*. Elsevier.

Islam, S. R., Alassod, A., Naveed, T., Dawit, H., Ahmed, K., & Jiang, J. (2022a). The study of hydrophobicity and oleophilicity of 3D weft-knitted spacer fabrics integrated with silica aerogels. *Journal of Industrial Textiles*, *51*(5_suppl), 8804S-8825S. https://doi.org/10.1177/15280837211029048.

Islam, S. R., Patoary, M. K., Estifanos, H. D., Lugoloobi, I., Yousif, A. H., Jiang, J., & Shao, H. (2022b). Hydrophobic and oleophilic 3D weft-knitted spacer fabrics coated by silica aerogels with five different concentrations. *Journal of Industrial Textiles*, *52*, 15280837221118063. https://doi.org/10.1177/15280837221118063.

Islam, S. R., Patoary, M. K., Farooq, A., Naveed, T., Ahmed, K., Shao, H., & Jiang, J. (2022c). 3D Weft-knitted spacer fabrics (WKSFs) coated with silica aerogels as oil intercepting sorbents for use in static and dynamic water tests. *Industrial Crops and Products*, *186*, 115169. https://doi.org/10.1016/j.indcrop.2022.115169.

Islam, S. R., Patoary, M. K., Yousif, A. H., Chaudary, A., Estifanos, H. D., Naveed, T., Jiand, J., & Shao, H. (2023a). SiO_2 aerogels (SAs) coating on the surface of 3D weft-knitted spacer fabrics (WKSFs) used as sorbent in oil spill cleanup. *Journal of Water Process Engineering*, *51*, 103451. https://doi.org/10.1016/j.jwpe.2022.103451.

Islam, S. R., Wang, W., Junjie, P., Jiang, J., Shao, H., & Chen, N. (2023b). Study on thermal, mechanical, and wettability properties of three-dimensional weft-knitted spacer fabrics with various silica aerogels coating. *Textile Research Journal*, 00405175231161777. https://doi.org/10.1177/00405175231161777.

Islam, S. R., Yousif, A. H., Estifanos, H. D., Lugoloobi, I., Alassod, A., Wendi, W., Jiang, J., & Shao, H. (2022d). Using various concentrations of SiO_2 aerogel for oil wicking, spreading, and interception tests of 3D weft-knitted spacer fabrics. *The Journal of the Textile Institute*, 1-11. https://doi.org/10.1080/00405000.2022.2110027.

Islam, S. R., Yu, W., & Naveed, T. (2019). Influence of silica aerogels on fabric structural feature for thermal isolation properties of weft-knitted spacer fabrics. *Journal of Engineered Fibers and Fabrics*, *14*, 1558925019866446. https://doi.org/10.1177/1558925019866446.

Jiang, S., Ding, M., Chen, W., & Wang, J. (2022). Preparation and Properties of Flexible Cushioning Composites Based on Silicone Rubber and Warp—Knitted Spacer Fabric. *Silicon*, *15*, 1-16. https://doi.org/10.1007/s12633-022-02267-5.

Kumar, N. M., Thilagavathi, G., & Periasamy, S. (2022). Development and characterization of warp knitted spacer fabrics for helmet comfort liner application. *Journal of Industrial Textiles*, *51*(2_suppl), 2053S-2070S. https://doi.org/10.1177/1528083720939215.

Li, M., Wu, Y., Chen, C., & Du, Z. (2018). Characterization of plantar press-comfort performance of warp-knitted spacer fabrics. *Textile Research Journal*, *88*(4), 357-366. https://doi.org/10.1177/0040517516679152.

Li, N. W., Yick, K. L., Yu, A., & Ning, S. (2022). Mechanical and thermal behaviours of weft-knitted spacer fabric structure with inlays for insole applications. *Polymers*, *14*(3), 619. https://doi.org/10.3390/polym14030619.

Liu, Y. (2013). *A study of warp-knitted spacer fabrics as cushioning materials for human body protection.* [Doctoral dissertation]. The Hong Kong Polytechnic University.

Liu, Y., & Hu, H. (2011). Compression property and air permeability of weft-knitted spacer fabrics. *The Journal of the Textile Institute*, *102*(4), 366-372. https://doi.org/10.1080/00405001003771200.

Liu, Y., Au, W. M., & Hu, H. (2014). Protective properties of warp-knitted spacer fabrics under impact in hemispherical form. Part I: Impact behavior analysis of a typical spacer fabric. *Textile Research Journal*, *84*(4), 422-434. https://doi.org/10.1177/0040517513495941.

Liu, Y., Hu, H., Long, H., & Zhao, L. (2012a). Impact compressive behavior of warp-knitted spacer fabrics for protective applications. *Textile Research Journal*, *82*(8), 773-788. https://doi.org/10.1177/0040517511433147.

Liu, Y., Hu, H., Zhao, L., & Long, H. (2012b). Compression behavior of warp-knitted spacer fabrics for cushioning applications. *Textile Research Journal*, *82*(1), 11-20. https://doi.org/10.1177/0040517511416283.

Patnaik, A., & Patnaik, S. (Eds.). (2019). *Fibres to smart textiles: Advances in manufacturing, technologies, and applications*. CRC Press.

Qu, M., Xia, F., Fang, F., Huang, L., & Ma, P. (2022). Influence of spacer filament fineness on indentation characterizations of warp-knitted spacer fabrics for mattresses. *Journal of Engineered Fibers and Fabrics, 17*, 15589250221131884. https://doi.org/10.1177/15589250221131884.

Rajan, T. P., & Das, S. (2020). Evaluation of air permeability behaviour of warp knitted spacer fabrics. *Indian Journal of Fibre & Textile Research, 45*(1), 32-39. https://doi.org/10.56042/ijftr.v45i1.20205.

Rajan, T. P., & Sundaresan, S. (2020). Thermal comfort properties of plasma-treated warp-knitted spacer fabric for the shoe insole. *Journal of Industrial Textiles, 49*(9), 1218-1232. https://doi.org/10.1177/1528083718811084.

Rajan, T. P., Kandhavadivu, P., & Periyasamy, A. P. (2021). Influence of porosity and yarn linear density on the thermal behaviour of polyester warp knitted spacer fabrics. *Fibers and Polymers, 22*(11), 3212-3221. https://doi.org/10.1007/s12221-021-0707-5.

Rudy, R., & Wardiningsih, W. (2021). Force attenuation capacity of weft-knitted spacer fabric in low-velocity impact. *International Journal of Clothing Science and Technology, 33*(6), 942-952.

Schäfer, B., Emonts, C., Glimpel, N., Ruhl, T., Obrecht, A. S., Jockenhoevel, S., Gries, T., Beier, J.P, & Blaeser, A. (2020). Warp-knitted spacer fabrics: a versatile platform to generate fiber-reinforced hydrogels for 3D tissue engineering. *Materials, 13*(16), 3518. https://doi.org/10.3390/ma13163518.

Sun, F., Ma, M., Pan, X., Liu, G., Yang, S., Zheng, D., Li, M., & Du, Z. (2019). Simulation of plate compression behavior of warp-knitted spacer fabrics based on geometry and property parameters. *Textile Research Journal, 89*(6), 1051-1064. https://doi.org/10.1177/0040517518760758.

Taghvaie, M., Payvandy, P., & Jalili, M. M. (2020). Introduction and optimization of a novel nonlinear model for the free vibration of warp-knitted spacer fabrics. *Fibers and Polymers, 21*(8), 1849-1856. https://doi.org/10.1007/s12221-020-9780-4.

Wang, Q., Peng, X., Zu, Y., Jiang, L., & Dong, K. (2021). Scalable and washable 3D warp-knitted spacer power fabrics for energy harvesting and pressure sensing. *Journal of Physics D: Applied Physics, 54*(42), 424006. https://doi.org/10.1088/1361-6463/ac181c.

Wen, Z., & Liu, S. (2022). Process and design of jacquard warp knitted spacer fabric. *Fibres & Textiles in Eastern Europe, 1*(151), 67-70. https://doi.org/10.5604/01.3001.0015.6464.

Xu, W., Yan, B., Hu, D., & Ma, P. (2022). Preparation of auxetic warp-knitted spacer fabric impregnated with shear thickening fluid for low-velocity impact resistance. *Journal of Industrial Textiles, 51*(4_suppl), 7188S-7204S. https://doi.org/10.1177/1528083720927013.

Yousefpour, M., Ahmadi, M. S., & Payvandy, P. (2018). Comparison of GA-optimized viscoelastic models for the characterization of compression behavior of warp-knitted spacer fabrics. *Autex Research Journal, 18*(3), 209-215. https://doi.org/10.1515/aut-2017-0027.

Yu, A., Sukigara, S., & Shirakihara, M. (2021). Effect of silicone inlaid materials on reinforcing compressive strength of weft-knitted spacer fabric for cushioning applications. *Polymers, 13*(21), 3645. https://doi.org/10.3390/polym13213645.

Yu, A., Sukigara, S., & Takeuchi, S. (2022a). Effect of inlaid elastic yarns and inlay pattern on physical properties and compression behaviour of weft-knitted spacer fabric. *Journal of Industrial Textiles, 51*(2_suppl), 2688S-2708S. https://doi.org/10.1177/1528083720947740.

Yu, A., Sukigara, S., & Yick, K. L. (2022b). Curvature control of weft-knitted spacer fabric through elastic inlay. *Textile Research Journal, 92*(19-20), 3826-3837. https://doi.org/10.1177/00405175221097098.

Zhao, T., Long, H., Yang, T., & Liu, Y. (2018). Cushioning properties of weft-knitted spacer fabrics. *Textile Research Journal, 88*(14), 1628-1640. https://doi.org/10.1177/0040517517705630.

Chapter 2

Phase Change Material Integrated Fiber and Yarn-Based Thermal Adaptive Textiles

Sennur Alay Aksoy*, PhD
and Demet Yılmaz, PhD
Department of Textile Engineering, Süleyman Demirel University, Isparta, Türkey

Abstract

The textile and apparel industry is one of the most important industrial branches in Türkiye, with its employment and production amounts as well as its export performance. Today, the representatives of the textile industry in Türkiye have to catch up with new trends in order to maintain their presence in world markets and increase their contribution to the economy of the country. For example, they should have a vision that can closely monitor and adapt to changes and opportunities in issues such as adaptation to sustainable production technologies, development of functional and smart products with added value, and branding. Functional and smart textiles are textiles that have qualities that meet the expectations of consumers or the needs of the end-use area, unlike the basic features expected from conventional textile products. For example, textiles with functional properties such as flame retardant, antibacterial activity, water repellency, and high chemical and mechanical resistance can be developed with the use of different methods or materials. Considering consumer expectations, the greatest expectation in terms of apparel textile products is the development of textile structures that can offer enhanced thermal comfort performance. Textiles with thermal comfort, which is defined as adaptive or dynamic comfort, unlike a conventional garment with a passive thermal insulation feature, contain smart materials with a heat regulation system that will keep the microclimate temperature constant even in changing environmental conditions and/or physical activity situations. Passive insulation is the insulation associated with passive air provided by conventional fibrous textile materials that can be retained in the structure depending on the construction. The type of fiber is also of secondary importance to passive insulation. Smart materials such as phase change materials (PCMs), shape memory polymers, and temperature sensitive polymers are materials that can change their physical properties by being temperature sensitive, and they have different structures and properties. They can change the properties of textile materials, such as water vapor and air permeability, sensitive to temperature or they can provide a buffering effect sensitive to temperature

* Corresponding Author's Email: sennuralay@sdu.edu.tr.

In: Textile Insights and Future Aspects
Editors: Gamze D. Tetik, Gizem Celep and Fulya Yilmaz
ISBN: 979-8-89113-289-4

changes with their heat energy storage/dissipation properties. For example, a temperature-sensitive polymer can give the fabric variable porosity and water vapor permeability depending on the variable temperature, while phase change materials (PCMs) can provide heating, cooling, and temperature regulation properties with their heat energy storage/dissipation properties.

Phase change materials (PCMs) store and release high-capacity latent heat energy as their physical state changes under nearly isothermal conditions. In the textile industry, PCMs are used to produce temperature-regulating fibers and smart textiles, which have applications in various fields such as military, aerospace, sportswear, home textiles, automotive, conditioning, medicine, and health. Various methods are currently used to incorporate PCMs into textile materials, such as melt spinning, wet fiber spinning, impregnation, exhaustion, coating, and lamination. Each method has some disadvantages, such as loss of thermal conductivity, softness, flexibility, breathability, or moisture-carrying capacity, and poor thermal properties in final products. In this study, in addition to known existing methods, newly developed techniques or methods will be introduced, and studies in the literature and commercial product examples will also be included.

Keywords: phase change material, temperature regulation, microcapsule, functional yarn, nanofiber

Introduction

Today, the textile industry, beyond classical clothing products, can be designed and produced in a way that undertakes a specific function and meets the needs of different sectors. Representatives of the textile industry, which produces products/materials for different fields in a wide range of sectors, make great efforts on the production of new technologies, raw materials, and auxiliary chemicals in order to develop textile products with quality and performance that will meet different expectations and needs. Examples of efforts in this direction are the development of bicomponent fibers/yarns, nanofibers, sandwich fabrics, and smart materials. In particular, smart materials attracting researchers' attention are sensitive to environmental variables or stimuli, which have different cross-sections and can contain different polymeric structures. The ability of the textile industry to produce high-tech, functional, high-value-added, and innovative products has become a necessity not only for the company to improve its competitiveness in the market but also for the country's economy. In this context, the development of materials and technology for the production of unique smart textiles that are sensitive to environmental stimuli should also be considered an opportunity.

Flexible textile-based temperature, humidity, etc. sensors (Frydrysiak, 2020; Wendler et al., 2020; Ozek et al., 2021; Yu et al., 2021; Wang et al., 2022a; Li et al., 2022; Liu et al., 2022a; Tözüm et al., 2022), adaptive thermal textiles (Tabor et al., 2020; Demirbağ Genç & Alay Aksoy, 2019, 2022; Korkmaz Memiş & Kaplan, 2022; Wang et al., 2022b; Demirbağ Genç et al., 2023), and nanofiber structures that can offer stimulus-sensitive controlled release performance (Štular et al., 2019; Zhang et al., 2020; Chatterjee & Hui, 2021; Schoeller et al., 2021) can be given as examples of these types of textiles. Stimulus-sensitive textiles are textiles that can change their color, shape, and physical properties by being sensitive to many different stimuli such as heat, light, and electric current (Jocic, 2016). It is possible to use different types of materials for textile production capable of adaptive thermal management with stimulus sensitive and shape memory polymers, temperature-sensitive polymers and phase change materials (PCMs) are among such materials.

Phase-change materials (PCMs) can store and release high-capacity latent heat energy as their physical state changes under almost isothermal conditions. In addition to these features, they can be used for cooling and heating applications in different fields and industries (Benmoussa et al., 2016; Alay Aksoy et al., 2017; Noyan et al., 2018) due to their important advantages such as minimal volume change, slight temperature fluctuation, and good chemical stability. In the textile industry, PCMs enable the production of advanced textiles that offer improved thermal comfort for the human body, due to their absorption and release mechanisms of thermal energy.

Today, in addition to all expectations from clothing systems such as protection, performance, functionality, and aesthetics, the efficiency of thermal management performance that can adapt to variable environmental conditions has become an indispensable requirement, especially for sportswear, workwear, medical, and other protective clothing. Thermophysiological clothing comfort is an important component of clothing comfort and is a concept related to heat and moisture transfer between body and clothing environments. While in thermal equilibrium, the body core temperature is typically around 37°C, and the skin temperature must be in the range of 33-34°C for continued comfort. While a change in core body temperature in the range of 0.5–1°C is acceptable, a body temperature above 37.5–38.3°C causes hyperthermia and a body temperature below 35°C results in hypothermia. Both situations are a potential threat to human life. Providing thermal comfort is extremely important in terms of issues such as energy saving, efficiency, and economy, as well as protecting human physical and psychological health. There is a close relationship between maintaining the thermal comfort of the human body and heating, ventilation, and air conditioning (HVAC) systems in buildings. On the other hand, the lack of thermal comfort negatively affects industrial productivity. Regulating the thermal comfort of the human body, building heating, which constitutes approximately 40% of total building energy consumption, has a significant effect on the saving of energy consumption of HVAC systems. For example, maintaining the thermal comfort of the human body within the 2°C extended heating or cooling setpoint range can be achieved with approximately 20% savings in HVAC energy consumption (Peng & Cui, 2020; Tabor et al., 2020).

Body heat balance is a function of metabolic heat production and environmental conditions. The body works to keep its temperature constant through various heat transfer mechanisms. Garment systems can support or inhibit body heat transfer, depending on the various properties of the textiles, including the finishing additives, as well as the structure of the yarn and fabric. The comfort performance of standard garments developed with conventional textile materials and production techniques is passive and insufficient to maintain body temperature balance in dynamic conditions with environmental and metabolic temperature changes (Memiş & Kaplan, 2020). Individuals often add or remove layers of clothing several times to accommodate unstable thermal comfort conditions. For this reason, there is a need for textile structures that can adapt to changes in body and environmental conditions and manage the heat transfer between the body and the environment. These structures are called textiles with an active comfort regulation or thermal adaptive textiles. In recent years, textile manufacturers and materials scientists have focused on the development of textiles with active thermal comfort regulation functions.

Innovations in materials science and textile technology make it possible to design advanced textiles with dynamic or active thermal comfort. For enhanced thermal comfort, using various techniques such as advanced material science, fiber engineering, and finishing techniques, various products have been designed and developed under brands such as Omni-heat™ (Columbia), CoolMax® (Dupont), AeroReact (Nike), Dri-FIT (Nike), Verycool™ (Yonex), HeatGear® (Under Armor), ForMotion (Adidas), Gore-Tex, etc. (Peng & Cui, 2020; Tabor et al., 2020). A significant part of the efforts to develop garments that offer active thermal comfort functionality has focused on adapting smart polymeric materials such as stimuli-responsive polymers and phase change materials (PCMs) to textile structures (Sarier & Onder, 2012; Demirbağ & Aksoy, 2016; Chaudhuri & Wu, 2020; Genç & Aksoy, 2022a; 2022b). Today, there are examples of PCM-added textile products in different forms such as thermoregulated denim fabric (Outlast Europe GmbH, Heidenheim and UCO Sportswear NV, Ghent/Belgium), cooling vest (TST Sweden Ab), fibers, and coating materials (OutlastVR) (Mattila, 2006; Paul, 2019).

In this study, the methods used for the application of PCMs directly or in microencapsulated form to textiles are mentioned. The methods used for PCM application to textile products in the form of natural or synthetic fibers and yarns are explained, and the findings of ongoing studies are included. In particular, in addition to the known existing methods, newly developed techniques or methods are introduced, and studies in the literature and commercial product examples are also included.

PCMs Application Methods in Textile

In the textile industry, PCMs enable the production of smart textiles that offer improved thermal comfort and dynamic or adaptive thermoregulation function, due to their thermal energy absorption and release mechanisms. Textile materials containing PCM not only provide dynamic insulation properties in clothing fabrics but also provide heating or cooling effects under different environmental conditions, in case of sudden temperature changes. For this reason, the use of PCMs in the textile industry to improve the thermal comfort of clothing and to prevent effects such as sweating and sudden changes in body temperature, which adversely affect the performance of users, such as professional athletes, and physical workers, is gaining importance. While the advantages of PCMs in terms of daily wear are not serious, they can extend the survival time of the user by means of their heat exchange features in extreme situations. Thus, PCMs offer advantages in medical, protective, and technical products in life-threatening situations or performance improvement, and in products for energy saving (such as home textile products) with its temperature regulation feature. Many efforts to fabricate textile products containing PCMs are focused on the development of PCM application methods for textiles. There are various methods for incorporating PCMs into textiles and the methods can be summarized into two groups. They are:

1. Application in fabric form with different methods (such as coating, impregnation, shrinking, foaming, etc.).
2. Incorporation into the fiber structure during manmade fiber (such as melt spinning, wet spinning, etc.) and yarn production

Each method has some drawbacks such as reduced thermal conductivity, softness, flexibility, breathability or moisture transport capability, and non-durable thermal properties to external effects on the final products. In this study, in addition to the known existing methods, newly developed techniques or methods will be introduced to fabricate PCMs doped fibers and yarns.

Incorporation into the Fiber Structure during Manmade Fiber

The incorporation of PCM within the filament to fabricate PCM fiber during the manmade fiber spinning process is a very attractive method to improve the lifetime of the thermo-regulating effect of the woven or knitted fabrics produced from such filament (Zhang et al., 2006; Gao et al., 2009; Li et al., 2014; Luo et al., 2018; Xiang et al., 2019; Ahn et al., 2021; Liu et al., 2022b). In the literature, various technologies were used to fabricate PCM fibers (Yang et al., 2022). However, some technologies, including electrospinning, melting spinning, and solution spinning, have gained more importance.

In the production of manmade PCM fiber with heat regulation feature, PCMs are either incorporated directly into the fiber matrix structure by mixing it into the molten polymer or polymer solution (McCann, 2006; Ke et al., 2014) or the hollow structure of the hollow fibers is filled with PCM (Vigo & Frost, 1983; Tao, 2008; Luo et al., 2018; Xiang et al., 2019). The applications are simple, and produced fibers have high heat storage capacity. However, PCMs easily move away from the fiber structure. In the production of heat-regulating fiber, the third option is the inclusion of PCMs in the fiber structure after microencapsulation (Bryant & Colvin, 1988; Leskovsek et al., 2004; Zhang et al., 2005; Zhang et al., 2006; Jiang et al., 2008; Gao et al., 2009). In this case, microencapsulated PCMs are mixed into the polymer melt or solution, and fiber spinning is performed. The application of microcapsules to the fiber ensures the permanence, that is, the heat storage/release properties of the fibers do not decrease even after very long use. In addition, the microcapsule application method based on fiber spinning does not negatively affect the properties of fabrics such as handle, color, drape, and softness. However, industrial fiber production with this method is difficult and requires advanced technology and know-how. Despite the advantage of permanence of the fiber spinning application method, there are many important problems, such as negatively affected fiber mechanical properties with higher microcapsule application, limited and low microcapsule usage opportunity (around 5-10%), difficult fiber spinning due to aggregation of microcapsules, insufficient stability to fiber production conditions, unsuitable for all kinds of fiber, and high production cost (Yang et al., 2022). In addition, the particle size of the produced microcapsules is the limiting parameter, and the size of microcapsules generally needs to be smaller than 10 µm on average (Cox, 1998; Hartmann, 2004; Zhang et al., 2005; Sarıer & Önder, 2012; Yang et al., 2022).

Other PCM Fiber and Yarn Production Methods

Nanofibers

Most of the research on PCM fiber fabrication has focused on nanofiber and nanofibrous surface fabrication. A broadly used method for producing nanofibers, electrospinning has attracted extensive attention in the research and development of PCM nanofibers. Electrospinning technology is the widely used method to produce nanofibers and is a simple, convenient, economical, and versatile technique for fabricating nano-sized fibers. Many types of polymers and/or polymer blends, drug-impregnated polymers, and polymers loaded with nanoparticles, or functional molecules have been used in electrospinning. Due to their small diameter, high surface area, and complex pore structures, electro-spun PCM nanofibers have attracted interest for the production of thermoregulating and thermal energy storage materials (Babapoor et al., 2016; Babapoor et al., 2017). They can be used in the production of enhanced efficiency thermal regulating garments, electronic components, etc., due to their ultrafine size, huge specific surface areas with high porosity networks, being lightweight and allowing direct use in various composites (Semnani Rahbar et al., 2016; Babapoor et al., 2017; Noyan et al., 2018).

The nanofibers can be obtained with either single- or double-nozzle electrospinning methods. Single or coaxial electrospinning gave an opportunity to develop thermoregulating hybrid materials with improved thermal, mechanical, and physical features. Electrospinning played an important role in improving the application field of PCMs, and various types of PCMs nanofibers were prepared by electrospinning (Semnani Rahbar et al., 2016; Noyan et al., 2018; Darzi et al., 2019). The double-nozzle technique is known as the coaxial method and is an advanced electrospinning method used for the fabrication of core-shell structured nanofibers. In this technique, core and shell solutions flow through a coaxial capillary to obtain the core-sheath structured nanofibers. Depending on the used materials, the electrospun core-sheath structured nanofibers have a great potential application in various end uses such as microfluidics, photonics, and energy storage systems (Babapoor et al., 2017).

In the method of single-nozzle electrospinning, the host polymer mixed or grafted with PCMs is electrospun to form fibers with PCM dispersed in the polymer matrix. Single-nozzle electrospinning is the most preferred method because of its easy applicability. Its outstanding problem is the insufficient encapsulation of PCM in the shell and partial leakage of PCM through the fibers. Recently the coaxial electrospinning technique has attracted attention. In this technique, simultaneous electrospinning of two or more different materials is carried out to obtain core-shell structured nanofibers. The method is used to effectively encapsulate both hydrophilic and oleophilic PCMs by various polymer types and offers many advantages such as better encapsulation efficiency, less PCMs leakage, and higher mechanical properties of the PCM nanofibers (Noyan et al., 2018). Besides coaxial electrospinning, emulsion electrospinning is another method for preparing core-sheath PCM fibers. In the emulsion electrospinning method, the insoluble core component is dispersed into the easy-to-fiber-forming sheath polymer component using an emulsifier. The uniform spinning emulsion obtained later is electrospun as a nanofiber form using the uniaxial spinneret equipment (Quan et al., 2022). It was found that the single-nozzle electrospinning technology provided nano-sized PCM fibers, while the coaxial electrospinning technology and the emulsion electrospinning technology enhanced thermal stability. Several research studies have recently been conducted on the fabrication of nano PCM fibers with various polymer shells and PCMs cores by electrospinning in recent years (Cai et al., 2014; Noyan et al., 2018; Darzi et al., 2019; Ismar & Sarac, 2019; Ahn et al., 2021).

PCM Yarns with Conventional Textile Processes

To obtain the fabrics from PCM fibers, it is required to produce the PCM yarns. Unlike the investigation of PCM fibers, how to make PCM yarns (PCM-containing yarns) using conventional industrial processes is another point to develop (Yang et al., 2022). Conventional textile processes will provide the opportunity to remove the limitations for mass production and quality parameters such as textile wear comfort and mechanical properties. For this reason, studies aimed at the production of PCM-added yarn with traditional textile production processes have been carried out in recent years.

Apart from the fiber/yarn and fabric application methods available in the literature, the patent titled "process for depositing microcapsules into multifilament yarn and the products produced" by Berlinger (2005) aimed to apply microcapsules to multifilament yarns (EP 1 609 896 B1). In the patent, it was stated some of the methods that microcapsules are sprayed on the surface of the filament bundle in the form of continuous slivers and tows. In these methods, microcapsules can be applied to materials in yarn form by "spraying or pulverization" and "coating" methods. In the spraying method, the microcapsules are applied by spraying them onto the dry continuous tows or strips with a gun-like apparatus. In the method, microcapsules can be applied to more surfaces of the fiber bundle and cannot be applied sufficiently to the inner part. In the coating method, a solution containing adhesive and microcapsules is applied to the yarn surface on a coating machine. The researcher stated that microcapsules remain on the yarn surface and hence the fabric handle becomes harder. Therefore, spraying and coating methods require post-treatment processes to improve fabric properties such as the handle. To overcome these problems, the researcher suggested the requirement of online processes during the classical textile processes. Starting from these points, a method was developed based on the principle of spraying microcapsules on the separated filaments in untwisted form and then attaching the microcapsules between the fibers by twisting (EP 1 609 896 B1). In the method, microcapsules were applied to POY (partially oriented) filament yarns during the texturing process. After the false twist texturing process in an air-jet false-twisting machine, the yarn was untwisted with an air-jet, the filaments were separated from each other, and an aqueous microcapsule suspension was sprayed onto the untwisted open filament bundle. The obtained filaments were wound on each other again and thus the microcapsules were compressed in the multifilament yarn structure. It was stated that there are microcapsules with a diameter of 0.1-200 μm located on the filament surfaces. Although the developed method allowed microcapsule integration during the texturing process, there was no commercial example of the method or detailed study results on the method. In addition, many questions related to the application remained unclear, such as how the solution was delivered during the air-jet texturing process or whether the solution dispersed after application.

Marsal (2012) worked on microcapsule applications on roving and ring-spinning machines. For the application on the roving machine, a solution consisting of 10% microcapsules and 10% resin was prepared. The solution was applied to the roving samples for 30 minutes at 70 ºC in a bath, and then the samples were dried by radio frequency. It was observed that the uniformity of the roving samples deteriorated after the application and this case caused considerably high yarn breaks, and hence noil (0.3%) during the spinning process. Therefore, roving microcapsule applications resulted in problems in spinning. Besides the inadequate quality of the yarn, another drawback of the method in front of industrialization was stated as the requirement of a high bath ratio for dosing in industrial dyeing equipment, and this caused excessive microcapsule consumption and high production cost.

In the following part of the study, microcapsule application trials were performed on the ring-spinning machine, and various apparatuses were developed. However, it was determined that the yarn quality was insufficient. In the last part of the study, experiments were conducted on core-spun yarn production. In this method, a stainless steel bath was firmly attached to a suitable area at the front of the ring-spinning machine above the drafting system for each spindle. In other words, a bath is placed on the feed rollers, where the core yarn, such as elastane, is fed. The yarn used as the core material was fed into the bath located in the drafting system. The core yarn that passed through the bath with microcapsule solution was wrapped with staple fibers during ring-spinning, and the microcapsule applied core-spun yarn was obtained. The yarn images indicated the presence of microcapsules on the yarn surface. In order to improve resistance to washing, drying was performed between 60 °C and 120 °C by using 1, 2, and 3 ratios of resin. In the study, it was determined that the 2% resin concentration and the drying temperature 100 °C were the optimum working parameters for industrial application. After the first washing process, it was determined that a significant part of the microcapsules had been removed. However, it was observed that small-sized microcapsules remained after 15 washes. In the study, it was stated that bathroom maintenance is easy, and the method is suitable for all kinds of fibers. The study concluded that the core-spun application method is suitable for industrial application, and it is possible to place a bath on ring-spinning machines with more than 1600 spindles. In the method, which can be considered as one of the alternative microcapsule application methods, it should be examined whether the microcapsule given to the core of the yarn is sufficient to provide the expected properties. The yarn and fabric properties should be evaluated. During the whole yarn spinning process, the bath should be filled continuously in order to apply the same amount of microcapsule to the core yarn. The core yarn should be tied easily after the core yarn breakage or after doffing.

In the literature, there are some studies based on the usage of PCMs without microencapsulation in yarn production. Lin et al. (2016) used PCM roving to produce far-infrared/anion-releasing elastic warp-knitted fabrics. The core-spun yarn production method was used to produce composite yarns. In the core-spun yarn production, PCM roving and bamboo roving or both rovings were used as a shell material, while bamboo roving and stainless steel was used as a core material. The yarns produced with different core-shell material combinations were used as a weft yarn for the elastic warp-knitted fabrics.

Yu et al. (2022a) worked on the fabrication of friction-spun composite yarns with a phase change function. First, a cotton core yarn was coated by PCM (the polyethylene glycol) through the sizing method and then PCM applied cotton core yarn was fed to the friction spinning machine and wrapped with different sheath materials (cotton, wool, and polyester sliver). It was found that the enthalpy of the produced yarn reached up to 108.72 J/g for the cores-pun yarn composed of polyester sheath fiber produced at 8000 rpm.

In another work, Yu et al. (2022b) used the friction spinning method to fabricate the core-shell structured yarn for the fabrics with phase change and electrothermal functions. First, cotton/stainless steel wire was produced by friction spinning, and then this produced yarn was coated by polyethylene glycol as PCM via the sizing method. This PCM yarn was used as a core component for friction spinning and was covered with a cotton sliver. The composite yarn was weaved with 2-ply cotton yarns to form an electrothermal phase change textile. At the end of the study, the described method was stated as a new technique for textile-based electrothermal heating elements.

Yılmaz et al. (2022) suggested an alternative microcapsule application method to the available methods such as fiber spinning, impregnation & drying, coating, etc. widely used in practice in order to eliminate their significant disadvantages. The method is based on the application of microcapsules into the staple fiber bundle on a ring-spinning machine. The focus is on short staple ring spinning because it is a production technique that dominates the yarn market. It was found that the microcapsules settled densely through the fiber surface and integrated into the yarn structure by twisting the motion of the yarn. The durability of the microcapsules was provided after the winding and knitting processes in spite of the friction effect that occurred in these processes. When the performance of an alternative application method with the impregnation method, and it was not determined a considerable difference in intensity and distribution character of microcapsules in both fabric samples. Both fabrics displayed similar microcapsule durability effects against repeated washings, even though a crosslinker was used in the impregnation method. The fabrics exhibited similar thermoregulation properties. In the study, common application processes (such as impregnation, coating, etc.) and alternative application methods were evaluated from an ecological point of view, and it was reported that low water and microcapsule/auxiliary material consumptions and no need for a fabric drying process in an alternative method could enable ecological yarn production compared to known application methods as well as minimizing the losses in fabric attitude and comfort properties.

Miao et al. (2023) proposed a liquid flow spinning (LFS) method to overcome the commercial PCMs drawbacks such as high rigidity, no-breathability, easy leakage, and abrasion and to increase their potential applications. In the paraffin-coated yarns (PCYs) production method, a core yarn was first obtained by injecting molten paraffin into a silicone tube under negative pressure of a syringe, and then the staple fibers were evenly wrapped onto the core on a friction spinning frame. It was determined that the obtained PCYs showed enhanced mechanical properties due to the core-sheath yarn structure and the excellent temperature-regulating effect due to the high melting enthalpy of PCYs. In particular, compared to commercial PCMs-based textiles, PCYF was superior in thermal protection performance (9 °C lower).

Nanofiber-Based Yarns

In a typical electrospun nanofiber web, fibers are generally collected in a random state, and uneven fiber form occurs. This uneven structure finds usage areas such as filtration, composite reinforcement, protective textiles, and biomedical applications. However, the inadequate mechanical properties of nanofiber webs prepared by electrospinning and centrifugal spinning technologies result from the relatively low fiber orientation and limit their applications in many fields (Yang et al., 2022). Thus, the electrospun fibers should be converted to continuous aligned fiber bundles or yarns having required mechanical stability for various end use areas. For the conversion of nanofibers into textile surfaces, the textile industry is switching its attention to the production of yarns from nanofibers.

In the current literature, there are patents and research on the production of twisted and untwisted yarns from nanofibers. In nano-yarn production, the twist inserted into a fiber bundle can improve the mechanical properties. Semnani Rahbar et al. (2016) prepared the thermal regulating form stable PA 6 nanofibre yarns containing microencapsulated phase change materials (MPCMs) via twisting electrospinning technology. In the study, it was determined that MPCM was settled on the nanofiber surface, exhibiting a random arrangement without clustering. In addition, the melting enthalpy values of the composite nanofiber yarns increased to more than 80% of the heat storage capacity of MPCMs.

Chen et al. (2020) designed phase change nanofiber hollow yarns in terms of the thermal energy storage property of PCMs. In the study, it was used a special twisting setup in the electrospinning process to prepare hybrid yarns containing PCMs. In the study, polyvinyl alcohol (PVA) polymer was used as the core material of the nanofibers having PCM in the shell layer and PVA polymer was later removed by hot water to get hollow PCM fibers. The PVA removing process caused a reduction trend in heat storage capacity, but still, considerable enthalpy values were preserved.

Ke et al. (2022) fabricated polyacrylonitrile (PAN)/PEG/carbon nanotube (CNT)/cotton composite yarn by impregnating PEG polymers into CNT/cotton yarn and coating electrospun PAN around its surface. It was found that the resulting compound yarn presented thermal energy storage and electrical conductivity properties and good electrothermal heat conversion and storage capabilities.

Yılmaz et al. (2023) worked on the fabrication of core-shell structured PCM-loaded polylactic acid (PLA)/cotton biocomposite yarn. PCM nanocapsules synthesized with poly (methyl methacrylate-co-methacrylic acid) shells were incorporated into cotton fibers at different application conditions using their developed method. The PCM-loaded yarn was used as a core component in the modified electrospinning set-up and covered by the PLA nanofibres with/without PCM nanocapsules to benefit from the outstanding characteristics of nanofibers. Thermal measurements indicated that PCMs loaded cotton core and core-shell structured hybrid yarns exhibited a temperature difference of at least 2 °C which was required for thermoregulation effect compared to the reference yarns. It was determined that PCM nanocapsule-loaded yarns heated up less in high-temperature environments and provided a thermoregulation cooling effect compared to unloaded yarns. At the end of the study, the research suggested that the developed hybrid yarns combined the properties of cotton fibers and PCMs in a single structure and had the potential to be used as thermal adaptive textiles products based on natural fiber yarns.

Conclusion

Phase change materials are materials that can absorb and release latent heat energy as they pass from one phase to the other without showing a significant volume change at an almost constant temperature. These materials are used for different purposes such as temperature regulation, and heating-cooling in the textile field. Today, interest in textile products containing phase change materials is increasing day by day both in academic and industrial fields. In this study, studies on PCMs, microencapsulated PCMs, proper PCMs for textile products, and particularly, their application on textile products in the form of fiber and yarn are included. Among phase change materials, in particular, organic-based materials are widely used. Organic-based PCMs are primarily preferred in the field of textiles due to their non-toxicity, high heat storage capacity, and suitable phase transition temperatures for body temperature. However, efforts to improve these materials, which generally transition from solid to liquid phase, are continuing due to their negative properties, such as flowing away from the liquid phase and low thermal conductivity. A large part of these efforts has focused on encapsulation technology. In addition to physical or chemical micro/nanoencapsulation methods for encapsulation, electrospin is of interest. Micro/nanoencapsulation is an effective method in terms of creating a surface area that will increase the thermal conductivity of PCMs, as well as being a solution for the liquid phase flow problems of solid-liquid PCMs. In the first studies on microencapsulation, organic polymers were used as wall-forming materials. However, in recent years, it has been observed that inorganic and organic-inorganic hybrid wall materials are used in PCM microencapsulation as an alternative to organic wall materials due to the low thermal conductivity and insufficient mechanical resistance properties of organic materials. In this context, inorganic and hybrid capsule walls offer the opportunity to improve the thermal response of PCMs with their high thermal conductivity potentials. In addition, hybrid capsule walls can also provide thermal energy storage for the capsules, as well as functional properties such as antibacterial activity, flame retardancy, and UV protection.

Manmade fiber and yarn production technologies and fabric finishing methods are widely used for the application of PCMs to textile materials. These methods are used in the industrial field as well as being the subject of R&D activities. Commercial product samples are available. However, these methods have some limitations of their own. For this reason, it is among the current research topics on the development of alternative methods/techniques in the titles of performance, ecological, and industrialization, apart from the methods known in the relevant technological fields. For example, it is observed that research studies are concentrated on the use of co-axial electrospinning in addition to conventional single-spinning, and the production of nano yarns to improve the performance of PCMs nanofibers, which are limited in use due to their poor mechanical properties. Considering this point, in this study, some studies on thermoregulated fiber and yarn production with PCM additives and developed production methods were examined.

As a result, it has been determined that studies on the development of PCM-based products have focused on the creation of a hybrid wall structure in the field of microcapsule production and the development of PCM textiles in yarn form. In addition, considering the advantages of the combination of nano/microstructures, it is thought that it will be beneficial to increase the efforts for the development of nanofiber/conventional fiber composite structures in studies on PCM yarn production.

Disclaimer

None.

References

Ahn, Y. H., DeWitt, S. J., McGuire, S., & Lively, R. P. (2021). Incorporation of phase change materials into fibers for sustainable thermal energy storage. *Industrial & Engineering Chemistry Research*, *60*(8), 3374-3384. https://doi.org/10.1021/acs.iecr.0c06140.

Alay Aksoy, S., Alkan, C., Tözüm, M. S., Demirbağ, S., Altun Anayurt, R., & Ulcay, Y. (2017). Preparation and textile application of poly(methyl methacrylate-co-methacrylic acid)/n-octadecane and n-eicosane microcapsules. *The Journal of the Textile Institute*, *108*(1), 30–41. https://doi.org/10.1080/00405000.2015.1133128.

Babapoor, A., Karimi, G., & Khorram, M. (2016). Fabrication and characterization of nanofiber-nanoparticle-composites with phase change materials by electrospinning. *Applied Thermal Engineering*, *99*, 1225-1235. https://doi.org/10.1016/j.applthermaleng.2016.02.026.

Babapoor, A., Karimi, G., Golestaneh, S. I., & Mezjin, M. A. (2017). Coaxial electro-spun PEG/PA6 composite fibers: Fabrication and characterization. *Applied Thermal Engineering*, *118*, 398-407. https://doi.org/10.1016/j.applthermaleng.2017.02.119.

Benmoussa, D., Molnar, K., Hannache, H., & Cherkaoui, O. (2016). Development of thermo-regulating fabric using microcapsules of phase change material. *Molecular Crystals and Liquid Crystals*, *627*(1), 163–169. https://doi.org/10.1080/15421406.2015.1137385.

Berlinger, M. (2005). *Process for depositing microcapsules into multifilament yarn and the products*. U.S. Patent, 1/138,859 (EP 1 609 896 B1).

Bryant, Y., & Colvin, D. (1988). *Fiber with reversible enhanced thermal storage, properties and fabrics made therefrom*. US Patent, 4756958.

Cai, Y., Zong, X., Zhang, J., Du, J., Dong, Z., Wei, Q., Zhao, Y., Chen, Q., & Fong, H. (2014). The improvement of thermal stability and conductivity via incorporation of carbon nanofibers into electrospun ultrafine composite fibers of lauric acid/polyamide 6 phase change materials for thermal energy storage. *International Journal of Green Energy*, *11*, 861–875. https://doi.org/10.1080/15435075.2013.829068.

Chatterjee, S., & Hui, P. C. L. (2021). Review of applications and future prospects of stimuli-responsive hydrogel based on thermo-responsive biopolymers in drug delivery systems. *Polymers*, *13*(13), 2086. https://doi.org/10.3390/polym13132086.

Chaudhuri, S., & Wu, C. M. (2020). Switchable wettability of Poly(NIPAAm-co-HEMA-co-NMA) coated PET fabric for moisture management. *Polymers*, *12*, 100. https://doi.org/10.3390/polym12010100.

Chen, W., Fu, M., & Weng, W. (2020). Electrospinning of continuous nanofiber hollow yarns for thermal storage and insulation by a multi-step twisting method. *Textile Research Journal*, *90*, 1045–1056. https://doi.org/10.1177/0040517519886023.

Cox, R. (1998). Synopsis of the new thermal regulating fiber Outlast. *Chemical Fibers International*, *48*, 475–479.

Darzi, M. E., Golestaneh, S. I., Kamali, M., & Karimi, G. (2019). Thermal and electrical performance analysis of coelectrospun-electrosprayed PCM nanofiber composites in the presence of graphene and carbon fiber powder. *Renewable Energy*, *135*, 719–728. https://doi.org/10.1016/j.renene.2018.12.028.

Demirbağ Genç, S., & Alay-Aksoy, S. (2022). Production of CS-g-PNIPAM copolymer and stimuli responsive and antibacterial cotton fabric. *International Journal of Clothing Science and Technology*, *34*(6), 852-868. https://doi.org/10.1108/IJCST-08-2021-0105.

Demirbağ Genç, S., Alay Aksoy, S., & Alkan, C. (2023). A smart cotton fabric with adaptive moisture management through temperature sensitive poly (2-hydroxyethyl-6-(vinyl amino)hexanoate) finishing. *Cellulose*, *30*, 2467–2481. https://doi.org/10.1007/s10570-022-04934-7.

Demirbağ, S., & Aksoy S. A. (2016). Encapsulation of phase change materials by complex coacervation to improve thermal performances and flame retardant properties of the cotton fabrics. *Fibers and Polymers*, *17*, 408-417. https://doi.org/10.1016/j.tca.2007.11.007.

Demirbağ, S., & Alay Aksoy, S. (2019). Fabrication of thermoresponsive cotton graft PNIPAA fabric. *The Journal of the Textile Institute*, *110*(2), 171-178. https://doi.org/10.1080/00405000.2018.1525843.

Frydrysiak, M. (2020). Comparison of textile resistive humidity sensors made by sputtering, printing and embroidery techniques. *Fibres & Textiles in Eastern Europe*, *5*(143), 91-96. https://doi.org/10.5604/01.3001.0014.2391.

Gao, X., Han, N., Zhang, X., & Yu, W. Y. (2009). Melt-processable acrylonitrile–methyl acrylate copolymers and melt-spun fibers containing MicroPCMs. *Journal of Materials Science*, *44*, 5877–5884. https://doi.org/10.1007/s10853-009-3830-z.

Genç, S. D, & Aksoy, S. A. (2022a). Synthesis of poly (N-isopropylacrylamide) polymer for fabrication of thermo-responsive cotton fabric. *Indian Journal of Fibre & Textile Research (IJFTR)*, *46*(4), 385-397. https://doi.org/10.56042/ijftr.v46i4.40797.

Genç, S. D., & Aksoy, S. A. (2022b). Development of temperature and pH responsive smart cotton fabrics by P (NIPAM–co-MAM) copolymer finishing. *Textile and Apparel*, *32*(3), 193-207. https://doi.org/10.32710/tekstilvekonfeksiyon.979303.

Hartmann, M. H. (2004). *Stable phase change materials for use in temperature regulating synthetic fibers, fabrics and textiles*. US Patent, 6,689,466.

Ismar, E., & Sarac, A. S. (2019). Electrospun polyacrylonitrile–lauric acid composite nanofiber webs as a thermal energy storage material. *Journal of Engineered Fibers and Fabrics*, *14*, 1558925018824890. https://doi.org/10.1177/1558925018824890.

Jiang, M., Song, X., Ye, G., & Xu, J. (2008). Preperation of PVA/parafin thermal regulating fiber by in situ microencapsulation. *Composite Science and Technology*, *68*, 2231-2237. https://doi.org/10.1016/j.compscitech.2008.04.004.

Jocic, D. (2016). Polymer-based smart coatings for comfort in clothing. *Tekstilec*, *59*, 107–114. https://doi.org/10.14502/Tekstilec2016.59.107-114.

Ke, G, Jin X, Cai G, Li W, & Xu A. (2022). A novel composite cotton yarn with phase change and electrical conductivity functions. *Journal of Industrial Textiles*, *51*, 554S-568S. https://doi.org/10.1177/15280837211003166.

Ke, H., Pang, Z., Xu, Y., Chen, X., Fu, J., Cai, Y., Huang, F., & Wei, Q. (2014). Graphene oxide ımproved thermal and mechanical properties of electrospun methyl stearate/polyacrylonitrile form-stable phase change composite nanofibers. *Journal of Thermal Analysis and Calorimetry*, *117*, 109-122. https://doi.org/10.1007/s10973-014-3669-7.

Korkmaz Memiş, N., & Kaplan, S. (2022). Smart polyester fabric with comfort regulation by temperature and moisture responsive shape memory nanocomposite treatment. *Journal of Industrial Textiles*, *51*(5_suppl), 7920S-7941S. https://doi.org/10.1177/15280837209756.

Leskovˇsek, M., Jedrinoviˊ, C. G., & Stankoviˇc Elesini, U. (2004). Properties of polypropylene fibres with incorporated microcapsules. *Acta Chimica Slovenica*, *51*(4), 99-715.

Li, P., Sun, Z., Wang, R., Gong, Y., Zhou, Y., Wang, Y., Liu, X., Zhou, X., Ouyang, J., Chen M., Hou, C., Chen, M., & Tao, G. (2022). Flexible thermochromic fabrics enabling dynamic colored display. *Frontiers of Optoelectronics*, *15*(1), 40. https://doi.org/10.1007/s12200-022-00042-3.

Li, W., Ma, Y. J., Tang, X. F., Jiang, N., Zhang, R., Han, N., & Zhang, X. X. (2014). Composition and characterization of thermoregulated fiber containing acrylic-based copolymer microencapsulated phase-change materials (MicroPCMs). *Industrial & Engineering Chemistry Research*, *53*(13), 5413-5420. https://doi.org/10.1021/ie404174a.

Lin, J. H., Huang, Y. T., Li, T. T., Lin, C. M., & Lou, C. W. (2016). Bamboo charcoal/ phase change material/stainless steel ring-spun complex yarn and its far-infrared/anion-releasing elastic warp-knitted fabric: Fabrication and functional evaluation. *Journal of Industrial Textiles*, *46*, 624–642. https://doi.org/10.1177/1528083715595007.

Liu, X., Miao, J., Fan, Q., Zhang, W., Zuo, X., Tian, M., Zhu, S., Zhang, X., & Qu, L. (2022a). Recent progress on smart fiber and textile based wearable strain sensors: Materials, fabrications and applications. *Advanced Fiber Materials*, *4*(3), 361-389. https://doi.org/10.1007/s42765-021-00126-3.

Liu, Z., Hu, D., Yao, J., Wang, Y., Zhang, G., Křemenáková, D., Militky, J., Wiener, J., Li, L., & Zhu, G. (2022b). Fabrication and performance of phase change thermoregulated fiber from bicomponent melt spinning. *Polymers*, *14*(9), 1895. https://doi.org/10.3390/polym14091895.

Luo, D., Wei, F., Shao, H., Xiang, L., Yang, J., Cui, Z., Qin, S., & Yu, J. (2018). Shape stabilization, thermal energy storage behavior and thermal conductivity enhancement of flexible paraffin/MWCNTs/PP hollow fiber membrane composite phase change materials. *Journal of Materials Science*, *53*(22), 15500-15513. https://doi.org/10.1007/s10853-018-2722-5.

Marsal Amenós, F. (2012). New method for applying microcapsules in the spinning process. *Annals of the University of Oradea, Fascicle of Textiles*, Leatherwork, 98.

Mattila, H. R. (2006). *Intelligent Textiles and Clothing*. Woodhead Publishing Ltd.

McCann, J. T., Marquez, M., & Xia, Y. (2006). Melt coaxial electrospinning: A versatile method for the encapsulation of solid materials and fabrication of phase change nanofibers. *Nano Letters*, 6(12), 2868-2872. https://doi.org/10.1021/nl0620839.

Memiş, N. K., & Kaplan, S. (2020). Wool fabric having thermal comfort management function via shape memory polyurethane finishing. *The Journal of the Textile Institute*, *111*(5), 734-744. https://doi.org/10.1080/00405000.2019.1661936.

Miao, Y., Fu, C., Li, L., Tang, W., Shang, L., Chen, D., Xu, A., Wu, Z., Wu, M., Jia, L., Xu, W., Su, B. & Xia, Z. (2023). Liquid flow spinning mass-manufactured paraffin cored yarn for thermal management and ultra-high protection. *Composites Science and Technology*, *235*, 109972. https://doi.org/10.1016/j.compscitech.2023.109972.

Noyan, E. C., Onder, E., Sarier, N., & Arat, R. (2018). Development of heat storing poly(acrylonitrile) nanofibers by coaxial electrospinning. *Thermochimica Acta*, *662*, 135-148. https://doi.org/10.1016/j.tca.2018.02.008.

Ozek, E. A., Tanyeli, S., & Yapici, M. K. F. (2021) Flexible graphene textile temperature sensing RFID coils based on spray printing. *IEEE Sensors Journal*, *21*(23), 26382-26388. https://doi.org/10.1109/JSEN.2021.3075902.

Paul, R. (2019). *High performance technical textiles*. John Wiley & Sons Ltd.

Peng, Y., & Cui, Y. (2020). Advanced textiles for personal thermal management and energy. *Joule*, *4*(4), 724–742. https://doi.org/10.1016/j.joule.2020.02.011.

Quan, Z., Xu, Y., Rong, H., Yang, W., Yang, Y., Wei, G., Ji, D., & Qin, X. (2022). Preparation of oil-in-water core-sheath nanofibers through emulsion electrospinning for phase change temperature regulation. *Polymer*, *256*, 125252. https://doi.org/10.1016/j.polymer.2022.125252.

Sarier, N., & Onder, E. (2012). Organic phase change materials and their textile applications: an overview. *Thermochimica Acta*, *540*, 7–60. https://doi.org/10.1016/j.tca.2012.04.013.

Schoeller, J., Itel, F., Wuertz-Kozak, K., Gaiser, S., Luisier, N., Hegemann, D., Ferguson, S. J., Fortunato, G., & Rossi, R. M. (2021). pH-responsive chitosan/alginate polyelectrolyte complexes on electrospun PLGA nanofibers for controlled drug release. *Nanomaterials*, *11*(7), 1850. https://doi.org/10.3390/nano11071850.

Semnani Rahbar, R., Maleki, H., & Kalantari, B. (2016). Fabrication of electrospun nanofibre yarn based on nylon 6/microencapsulated phase change materials. *Journal of Experimental Nanoscience*, *11*, 1402–1415. https://doi.org/10.1080/17458080.2016.1233582.

Štular, D., Kruse, M., Župunski, V., Kreinest, L., Medved, J., Gries, T., Blaeser, A., Jerman, I., Simončič, B., & Tomšič, B. (2019). Smart stimuli-responsive polylactic acid-hydrogel fibers produced via electrospinning. *Fibers and Polymers*, *20*, 1857-1868. https://doi.org/10.1007/s12221-019-9157-8.

Tabor, J., Chatterjee, K., & Ghosh, T. K. (2020). Smart textile-based personal thermal comfort systems: Current status and potential solutions. *Advanced Materials Technologies*, *5*(5), 1901155. https://doi.org/10.1002/admt.201901155.

Tao, X. (2008). *Smart Fibres, Fabrics and Clothing*. Woodhead Publishing Limited.

Tözüm, M. S., Alkan, C., & Aksoy, S. A. (2022). Developing of thermal energy storing visual textile temperature indicators based on reversible color change. *Journal of Industrial Textiles*, *51*(2_suppl), 1964-1988. https://doi.org/10.1177/1528083720980831.

Vigo, T. L., & Frost, C. M. (1983). Temperature adaptable hollow fibers containing polyethylene glycols. *Journal of Industrial Textiles*, *12*(4), 243–254. https://doi.org/10.1177/152808378301200405.

Wang, C., Gong, X., Li, J., Chen, Y., Li, B., Zhang, L., & Fu, S. (2022a). Ultrahigh-sensitivity thermochromic smart fabrics and flexible temperature sensors based on intramolecular proton-coupled electron transfer. *Chemical Engineering Journal*, *446*, 136444. https://doi.org/10.1016/j.cej.2022.136444.

Wang, Y., Yu, X., Liu, R., Zhi, C., Liu, Y., Fan, W., & Meng, J. (2022b). Shape memory active thermal-moisture management textiles. *Composites Part A: Applied Science and Manufacturing*, *107037*. https://doi.org/10.1016/j.compositesa.2022.107037.

Wendler, J., Maraite, D., Nocke, A., & Cherif, C. (2020). Novel textile moisture sensors based on multi-layered braiding constructions. *Textile Research Journal*, *90*(3-4), 469-477. https://doi.org/10.1177/0040517519868172.

Xiang, L., Luo, D., Yang, J., Sun, X., Qi, Y., & Qin, S. (2019). Preparation and comparison of properties of three phase change energy storage materials with hollow fiber membrane as the supporting carrier. *Polymers*, *11*(8), 343. https://doi.org/10.3390/polym11081343.

Yang, K., Venkataraman, M., Zhang, X., Wiener, J., Zhu, G., Yao, J., & Militky, J. (2022). Review: incorporation of organic PCMs into textiles. *Journal of Materials Science*, *57*, 798–847. https://doi.org/10.1007/s10853-021-06641-3.

Yılmaz, D., Alay Aksoy, S., Maleki, H., Rahbar, R. S., Söğüt, B., & Azimi, S. S. (2023). Fabrication of PCM-loaded polylactic acid (PLA)/cotton biocomposite yarn with thermoregulation function. *Cellulose*, *30*, 3995–4009. https://doi.org/10.1007/s10570-023-05101-2.

Yılmaz, D., Özen, Ö., Özkayalar, S., Alay Aksoy, S., & Söğüt, B. (2022). An innovative method for microcapsule based functional yarn production. *Journal of Natural Fibers*, *19*(14), 8051-8064. https://doi.org/10.1080/15440478.2021.1958435.

Yu, L., Ke, G., Wang, Y., & Tang, J. (2022a). Fabrication and characteristics of polyethylene glyco/cotton friction spun composite yarn. *Journal of Energy Storage*, *48*, 103978. https://doi.org/10.1016/j.est.2022.103978.

Yu, L., Qin, X., Ke, G., Tang, J., & Wang, Y. (2022b). Fabrication of cotton/polyethylene glycol/stainless steel wire friction spun composite yarn and its application in electrothermal phase change textile. *Composites Communications*, *32*, 101152. https://doi.org/10.1016/j.coco.2022.101152.

Yu, M., Jin, J., Wang, X., Yu, X., Zhan, D., & Gao, J. (2021). Development and design of flexible sensors used in pressure-monitoring sports pants for human knee joints. *IEEE Sensors Journal*, *21*(22), 25400-25408. https://doi.org/10.1109/JSEN.2021.3087005.

Zhang, J., Li, J., Xu, C., Xi, Z., Ren, Y., Song, Q., & Lu, S. (2020). Novel pH-sensitive drug-loaded electrospun nanofibers based on regenerated keratin for local tumor chemotherapy. *Textile Research Journal*, *90*(19-20), 2336-2349. https://doi.org/10.1177/0040517520919920.

Zhang, X. X., Wang, X. C., Tao, X. M., & Yick, K. L. (2005). Energy storage polymer/MicroPCMs blended chips and thermo-regulated fibers. *Journal of Material Science*, *40*(14), 3729-3734. https://doi.org/10.1007/s10853-005-3314-8.

Zhang, X. X., Wang, X. C, Tao, X. M., & Yick, K. L. (2006). Structures and properties of wet spun thermo-regulated polyacrylonitrile-vinylidene chloride fibers. *Textile Research Journal*, *76*, 351–359. https://doi.org/10.1177/0040517506061959.

Chapter 3

The Development of Heater Fabrics on Circular Knitting Machines

Arzu Marmarali[1,*], PhD
and Utku Varol[2], MD
[1]Department of Textile Engineering, Ege University, Izmir, Türkey
[2]Sun Tekstil San. ve Tic. A.S., Torbalı, Izmir, Türkey

Abstract

Conductive yarns or fabrics are increasingly being used in protective applications, such as monitoring physiological parameters or active wearing of warm clothing in the health, sports, and military fields. In recent years, various studies have been carried out on the performance and production of conductive yarns and fabrics to develop heating products for special conditions or daily applications. It is clear that with the use of heating textiles, it is possible to control body temperature and ensure comfort conditions for humans in extremely cold conditions. Heating textiles can be obtained by either adding an electronic mechanism to manufactured fabric or using conductive yarns during the production of fabrics. The first method has some disadvantages, such as more weight, a more rigid structure, and physiological and thermoregulatory discomfort. Recently, conductive yarns, which have the characteristics of textile yarns, are generally integrated directly into fabrics during production. This chapter aims to develop and compare two different electro-textile-based heating fabric structures that can be used to protect the body functions of people in extremely cold environmental conditions. Heater fabrics were produced using the conductive yarn in certain courses of interlock and three-thread fleece fabrics. It is possible to reach the targeted heating level by changing the settlement of the conductive yarns in the fabrics. The dimensional values, heating performance, and thermal comfort properties of developed fabric structures were measured and evaluated. When the whole characteristics of the two types of fabrics are compared, it is seen that the interlock fabric structure developed in this chapter has higher conductive heating performances besides lower fabric density and weight, higher thickness, higher air permeability, thermal resistance, and water vapor resistance values. The heating fabric developed within the study is expected to contribute to state-of-the-art technology with high conductive heating performance besides being suitable for mass production and easy assembly of the circuit.

---

* Corresponding Author's Email: arzu.marmarali@ege.edu.tr.

In: Textile Insights and Future Aspects
Editors: Gamze D. Tetik, Gizem Celep and Fulya Yilmaz
ISBN: 979-8-89113-289-4

Keywords: cold weather clothing, conductive yarn, electrically heated fabric, personal heating garment (PHG), thermal comfort

Introduction

Heating is a basic human need and has a major effect on our standard of living. Generally, humans feel comfortable within the temperature range of 18–25 °C. Even though the existence of a self-regulating system for core body temperature in the human body, additional heating can be necessary for human survival in extreme weather conditions. Failure to protect the human body in very cold environmental conditions can lead to injuries caused by cold and health problems such as dehydration, numbness, shivering, frostbite, cold-related limb loss, and hypothermia (Wang et al., 2010). Although conventional textile products provide protection against environmental conditions, which is one of the basic needs of humans, they cannot show the expected performance in extreme climate conditions. As garments become smarter, it is possible to protect vital organs against cold stress both in outdoor and indoor environments. Heating textiles have a wide range of end uses, including medical applications such as orthopedic supports, electrotherapy, clinical cover to maintain patients internal heat levels, strain sensors, movement tracking devices, etc. Additionally, thermotherapy has received great attention today and has become an efficient physiotherapy method due to its effects on relieving pain and revitalizing muscles for injured skin/joints/tissues (Zhang et al., 2017; Repon et al., 2022).

Textile products help maintain body temperature by creating an insulation layer between the body and the environment in the cold. The function of clothing is to reduce heat transfer to the environment, especially by limiting convective and radiant heat loss. The body constantly releases humidity, either as insensible perspiration or sweat. At low temperatures, this humidity may condense within the textile layers and negatively affect thermal insulation and, as a consequence, the thermal and tactile comfort of the wearer (Rossi, 2009).

Personal heating fabrics, which can be used for thermoregulation or physiotherapy, are examined under four headings according to their working principles. These principles are phase change materials, chemical reactions, fluids, and electric current. Wang et al. (2010) reviewed the advantages and disadvantages of personal heating garments and explained that uncontrolled temperature is a vital disadvantage for chemical heating garments. Additionally, they pointed out that electrical heating garments with precise temperature-controlled units and flexible heating elements have a promising future. For this reason, the usage of heating textiles with metallic wires as a shield in low-temperature environments is increasing (Padleckienė et al., 2020). Such apparel helps to control the body temperature and prolong comfortable conditions. An electrically heated glove patented in 1910 was one of the first e-textiles, and the heating effect was created by an electric current run through the conductive fibers. Different devices that have heating elements were created and patented between the 1930s and 1970s, including various heated blankets, electrically heated socks, and heated jackets. The growth of e-textiles in the later part of the 20th century rapidly increased thanks to a series of developments in material science and electronics, expanding the possibilities of embedding electronics within clothing (Hughes-Riley et al., 2018). Various applications of e-textiles have been used in a variety of fields, such as heated bed linings, footwear, furniture, sportswear, leisurewear, medical applications, and automotive furniture (Adamhasan, 2008).

Smart heating textiles should have three main features: ensuring freedom of movement, compatibility with body temperature, and actively keeping body temperature stable besides lightweight, low-cost, and nontoxic features. In the beginning, heated textiles were produced mainly in two ways: The first was to integrate electronic components into the previously produced fabric, and the other was to produce a textile surface using a technical yarn with electronic features (Hamdani et al., 2016). Nowadays, various textile manufacturing technologies and treatments such as knitting, weaving, nonwoven, spinning, breading, coating/laminating, printing, sewing, embroidery, and chemical treatment have been used for integrating a heating element into a textile material. Heating components made from non-woven textiles are of little use, as they have less electrical conductivity. On the other hand, it is determined that despite having the same dimensions, due to their structural geometries, the heating resistance of woven-based heater components is lower than knitted ones (Repon et al., 2022).

In recent years, research has been increasingly conducted to develop fabrics that provide active protection in extremely cold environments. As can be seen from the review of the available literature, generally, just heating performance was investigated in those research studies. However, this chapter also aims to investigate and develop thermal comfort properties of electro-textile-based heating fabric structures differently from previous research. Additionally, these structures will be suitable for mass production on circular knitting machines, and the circuit assembly can be made easily. Fabric samples were knitted in interlock and three-thread fleece structures using electrically conductive yarns on a circular knitting machine. The heating performance and thermal comfort properties of these fabrics were examined, and suggestions for the final product were presented.

Previous Studies

Textile-based heating systems have become one of the most important studied subjects in textiles due to the developments of both conductive yarns and computerized knitting systems. Kukkonen et al. (2001) developed modular heating products which have low power consumption, small dimensions, and low weight. To obtain the heating function, they supported the heating garments with portable power supplies. Wang and Lee (2009) presented a flexible and washable electrically heated vest (EHV) for cold environments.

In the literature, there are various studies on knitted fabrics that can reach different temperatures by embedding steel-based conductive threads or silver threads into the fabric (Kayacan et al., 2009; Mey et al., 2014; Uner & Gurcum, 2019). Tong et al. (2014) have explained that besides the linear resistance of the conducting yarns, the contact resistance between the conducting yarns plays an essential role in the total electrical resistance of the conductive knitted fabric. Wang et al. (2019) investigated the use of tricomponent elastic and conductive composite yarns for the production of heating fabrics. Kayacan et al. (2015) determined the resistance values before and after physical and chemical processes to evaluate the performance of a fabric that contains conductive yarns. Yılmaz (2015) designed an interactive woven fabric supported by carbon fiber having an electronic infrastructure. Liu et al. (2015; 2016) fabricated three different knitted heating fabrics (plain, rib, and interlock fabrics) using conductive silver yarns and investigated their thermoelectric properties. From the results, they found that the maximum equilibrium temperature on the surface of the conductive fabrics is strongly correlated with their power consumption and is not affected by the structure of the fabric. Zhang et al. (2017) developed a stretchable and conductive weft-knitted carbon fabric, which can be transformed from traditional fabrics with a simple heat treatment, for the practical applications of ultra-stretchable and wearable thermal therapy devices. Kursun et al. (2015; 2018) integrated an electronic module into the panels produced using different conductive yarns to form a wearable heating vest. Eroglu et al. (2019) informed that knitted fabrics generally exhibit better conductivity properties than woven fabrics, because of the continuity of conductive yarns in the knitted structure and its uniform effect on the entire surface. Maurya et al. (2023) showed that heavy knitted fabrics with a low loop length had a higher heating performance.

Trials for the Heater Knitted Fabric

Assembly of Power Supply and Circuit Design

Conductive yarns, the essential building element of e-textiles, transmit electrical signals through electronic assemblies embedded into this e-textile. Therefore, electric conductive yarns must have resistance to possible changes in electrical quantities besides low electrical resistance, or good electrical conductivity. The electric resistance is closely related to the structure of the conductive yarns and the way they are embedded into a textile material. A certain amount of heat is generated in the fabric after a certain time when the power at a specific voltage is applied to the conductive yarns embedded in the fabric and the temperature rises (Penava et al., 2022).

In this chapter, it is planned to obtain a portable heater fabric structure that can produce heat energy in the range of 5-16 W using the lithium polymer power supply. The technical specifications of this power supply are a power range of 10,000-20,000 mAh and DC 5-7.4V 1-2A. The circuit was created using conductive yarns and strips on fabric samples with dimensions of 20 cm by 50 cm. The parallel circuit diagram designed to reach the targeted heating power is shown in Figure 1. The black parallel horizontal lines show the conductive yarns, and the gray vertical columns show the conductive strips.

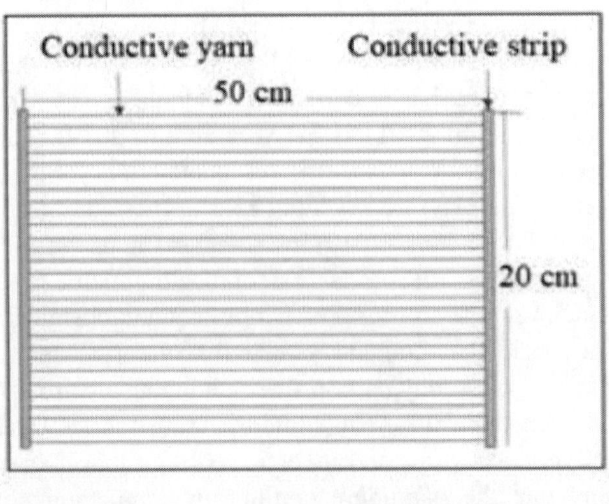

Figure 1. Heater fabric circuit diagram.

Production of Samples

The samples were produced using 150 Denier/72 filament continuous PET yarn as the ground yarn. During the preliminary trials, the resistance values of the conductive yarns supplied by various companies were compared. For this purpose, DC 5V-7.4V voltage was applied to the yarns and heating performances depending on time were obtained with the FLIR E8 thermal camera. At the end of the preliminary trials, it was decided to use 330dtex steel-coated PET yarn with a resistance value (12-15 ohm/m) considering the economic and knittability parameters besides conductivity. Narrow woven strips produced from 100% steel filaments were used to create the parallel circuit. The characteristic properties of the steel strip are 2mm thickness, 5mm width, and 14-48 Ω/m resistance value.

The knitwear fabrics produced on flat knitting machines were generally examined in the literature. Circular knitting machines are preferred in this chapter for developing the heater fabric structures, unlike previous studies. These machines have higher productivity and are preferred for mass production. Considering that the natural elasticity of the conductive yarns is very low, it was decided to choose knitting structures in which these yarns can be placed horizontally and flat without forming loops. Two different structures were designed to develop heater fabric structures suitable for mass production using a circular knitting machine.

Interlock Fabric

The interlock fabric sample was produced using a Terrot double bed circular knitting machine (diameter 38" and gauge 18). The conductive yarns were laid as horizontal weft yarn in every 20 rows between the cylinder and dial bed loops. In this way, a distance of 1 cm was created between the rows of conductive threads. To complete the circuit, the conductive strips must be placed vertically on the fabric surface, in contact with the conductive yarns. It was decided to create vertical gaps where the conductive strips could be placed to obtain a smooth fabric surface. For this, as shown in Figure 2.a, after 300 loops were knitted with both cylinder needles and dial needles, 8 loops were knitted only with dial needles. Then the conductive strip was placed into the opening approximately 1 cm wide. For high heating performance, good contact must be maintained between the conductive strip and all the rows of conductive yarns in the fabric, and the conductive strip should be fixed to the fabric. For assembly, sewing, soldering, sticking, etc. methods can be used. In the experimental setup in this chapter, the strips were fixed by using sticking tape. By increasing the surface where the conductive yarn will contact the conductive strip, more effective heating has been provided on the fabric. (Figures 2. b-c).

Three-Thread (3:1) Fleece Fabric

The three-thread fleece fabric structure was produced on the single bed circular knitting machine (diameter 38" and gauge 20) considering the circuit design. Eight courses between conductive courses were knitted using the same PET yarn for ground, binding, and fleece yarn. To create a 1 cm distance between the courses, including conductive yarns, the conductive yarn was fed as every eighth course. In this course, the conductive yarn was fed as fleece yarn, while the same PET yarn was used as ground and binding yarn (Figure 3. a). Then to complete the heating fabric circuit, the conductive strips were vertically placed by sticking tape on the fleece (back) side of the fabric (Figures 3. b-c).

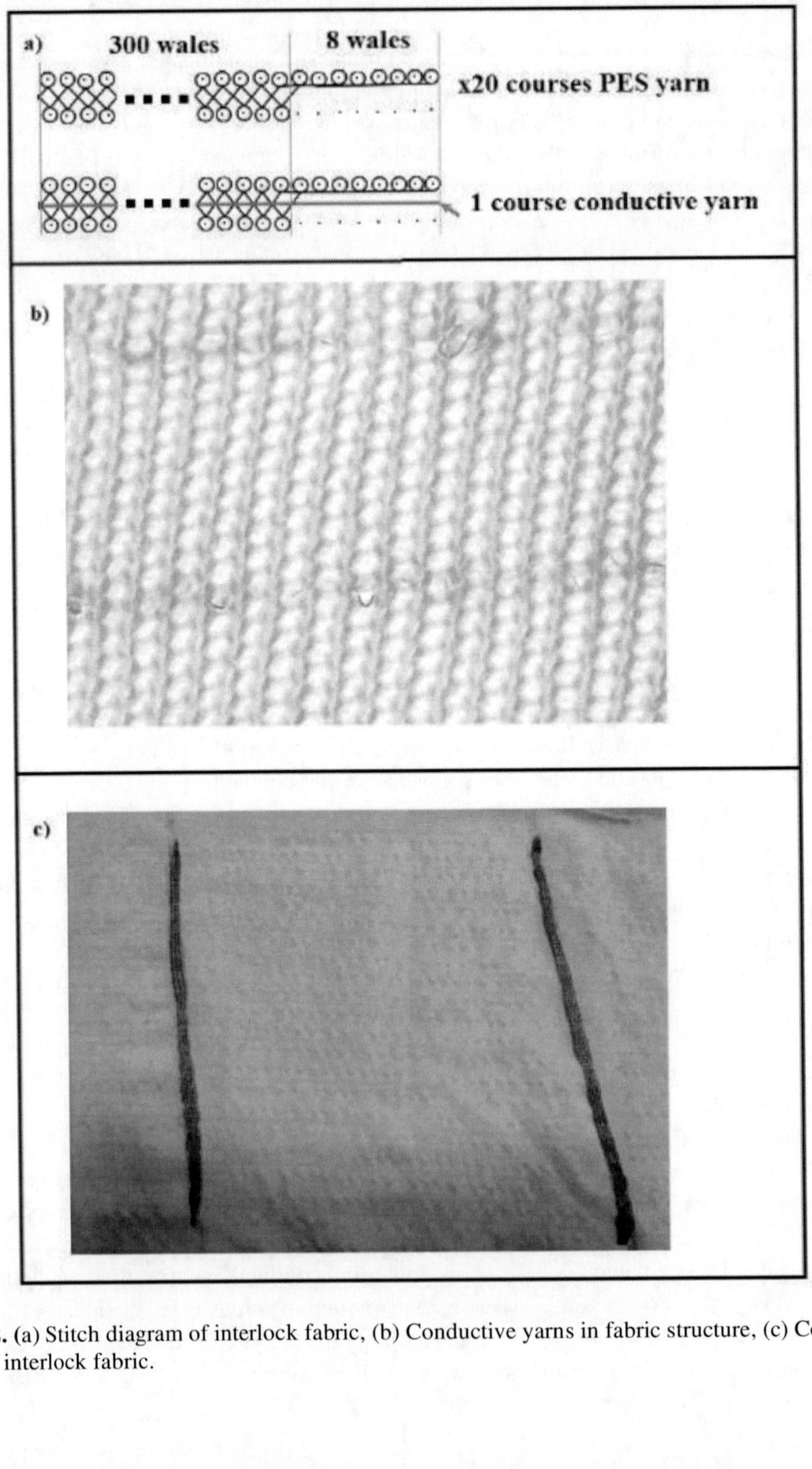

Figure 2. (a) Stitch diagram of interlock fabric, (b) Conductive yarns in fabric structure, (c) Conductive strips on interlock fabric.

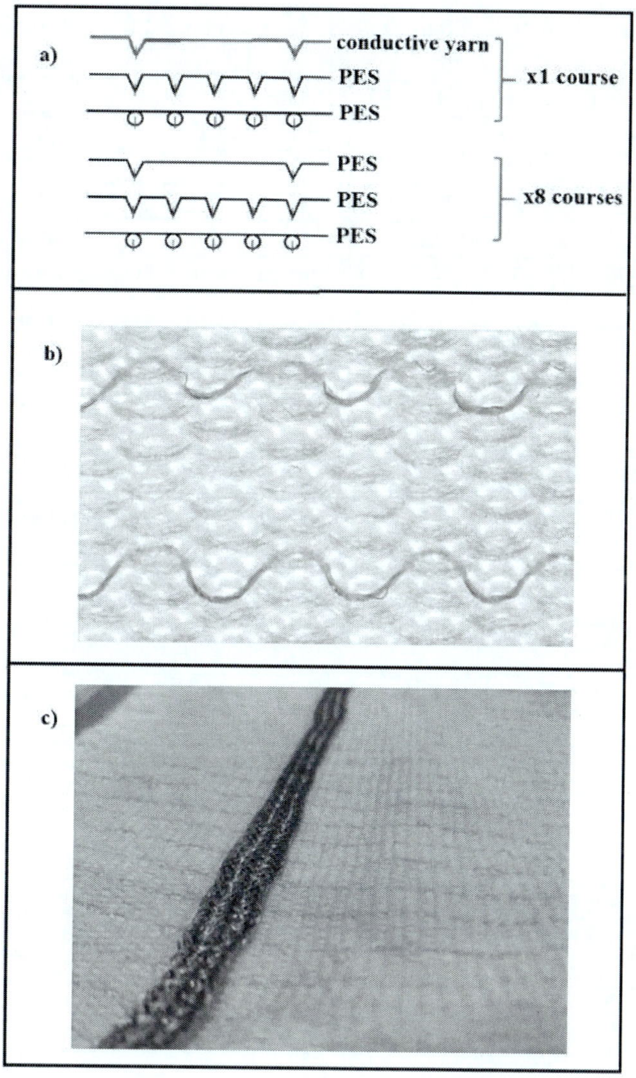

Figure 3. (a) Stitch diagram of 3:1 fleece fabric, (b) Conductive yarns in fabric structure, (c) Conductive strips on 3:1 fleece fabric.

Measurements

In addition to the heating performances of the fabrics, their structural parameters (thickness and weight), mechanical properties (bursting strength and elongation), and thermal comfort properties (air permeability, thermal resistance, and water vapor resistance) were tested. The test devices and related standards used during the measurements are given in Table 1. For the rating of the resistance to water vapor, the Ret comfort-rating system (Table 2) developed by The Hohenstein Institute was used (Reljić et al., 2016).

The fabric density parameter was calculated by Equation FD = G / T. Where FD is the fabric density (g/m^3), G is the weight of fabric (g/m^2) and T is the fabric thickness (m).

Table 1. The tested parameters, their units, and test methods

Parameter	Symbol (unit)	Related standard	Test device
Weight	G (g/m^2)	TS EN 12127	Digital scale
Thickness	T (mm)	ISO 8301:1991	SENSORA Alambeta
Fabric density	FD (kg/m^3)		Calculated
Bursting strength	BS (kPa)	ISO 13938-1	James Heal Tru Burst
Elongation in the vertical direction	E	TS EN ISO 13934-1	Zwick Z010 Proline Universal
Air permeability	AP (mm/s)	TS 391 EN ISO 9237	Textest FX 3300
Thermal resistance	Rct (m^2K/W)	ISO 8301:1991	SENSORA Alambeta
Water vapor resistance	Ret (m^2Pa/W)	ISO 11092	SENSORA Permetest
Radiant heating performance of the yarn	RHPY (°C)		Test setup
Conductive heating performance	CHP (°C)		Test setup

Table 2. Ret comfort-rating system (Reljić et al., 2016)

Rating	Ret value (m^2Pa/W)	Description
Very good	0-6	Extremely breathable and comfortable at a higher level of activity
God	7-13	Very breathable and comfortable at a moderate rate of activity
Satisfactory	14-20	Breathable, but uncomfortable at a higher rate of activity
Unsatisfactory	21-30	Slightly breathable, giving moderate comfort at a low rate of activity
Very unsatisfactory	31 +	Not breathable and uncomfortable, with a short tolerance time

The conductive method was used to determine the heat-generating capacity of the developed fabrics which measured the temperature of the heated fabric surface. For the measurements, the mechanism given in Figure 4 was prepared, and during the tests, energy was provided using an adjustable power supply. Before testing, the fabrics were conditioned for 24h under normal conditions (65% humidity, at a temperature of 23°C), and the tests were completed in the standard conditioned room.

The test mechanism includes a thermal camera and nine iButton data loggers. The iButton data loggers are equally spaced to take measurements from different parts of the fabric surface for conductive heating measurements. On the other hand, a FLIR E8 thermal camera was placed at a height of 50 cm to provide radiant measurement of the reached temperature of the yarns in the fabric structure. Thus, it was possible to measure the mean surface temperature of the fabric with the maximum temperature value reached by the conductive yarns. Additionally, time-dependent temperature changes for conductive heating measurements were observed for 20 minutes, and temperatures at 20 second intervals with the iButton data loggers were recorded as graphics.

SPSS 24.0 statistical software was used in the statistical evaluation of the measurements. First of all, the suitability of the distribution type of all the data was evaluated to the normal distribution. Spearman's correlation test was applied to measure the strength of the relationship between the two variables for data that did not fit the normal distribution.

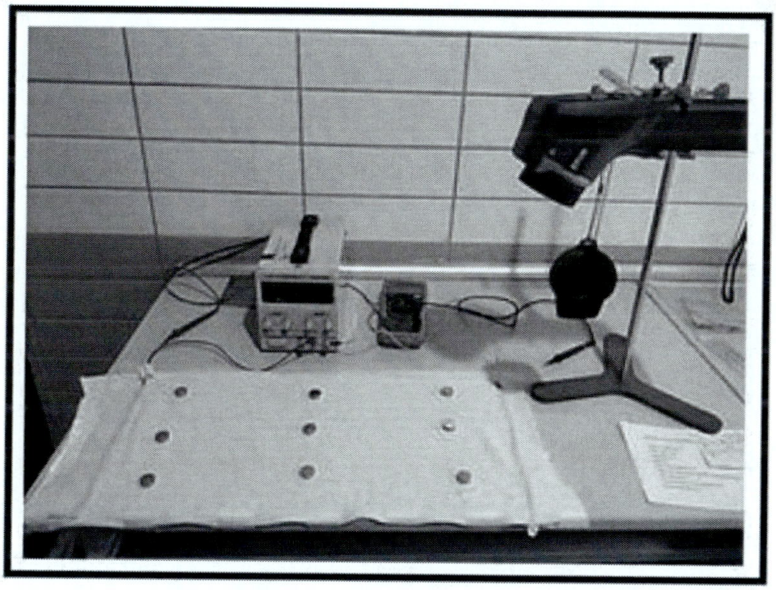

Figure 4. Test mechanism prepared to measure the heating performance of fabrics.

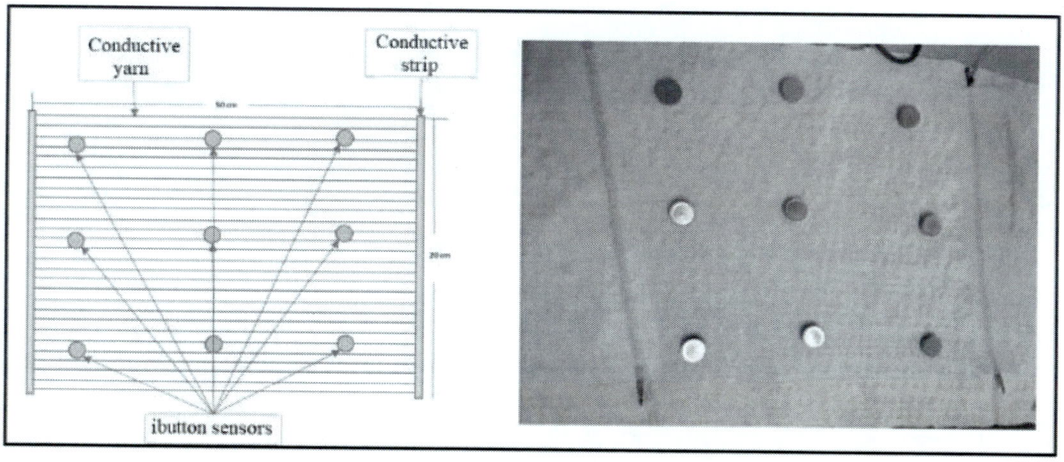

Figure 5. Placement of iButton sensors on fabrics.

Results and Evaluation

The results of the structural, mechanical, comfort properties, and heating performance of the heating fabrics (Table 3) as well as the evaluations of these measurements are included in this section.

Table 3. The measured characteristics of the samples

Fabric	G	T	FD	BS	E	AP	Rct	Ret	RHPY	CHP
	g/m²	mm	kg/m³	kPa	%	mm/s	m²K/W	m²Pa/W	°C	°C
Interlock	321	2,63	121,9	633	31	1578	0,0654	4,10	69,9	33
3:1 fleece	368	1,52	242,2	652	25	1356	0,0352	3,85	60,1	29

Spearman's correlation analyses were performed to determine the effects of the measured fabric parameters on the radiant heating and conductive heating performances of the samples. According to the test results given in Table 4, it is seen that the effect of each measured fabric parameter on the radiant heating and the conductive heating performances is the same. In addition, it has been determined that there is a statistically significant relationship between all fabric parameters examined except the bursting strength and the radiant and conductive heating values.

Table 4. Spearman's correlation test results

		G	T	FD	BS	E	AP	Rct	Ret
Radiant heating	Correlation coefficient	-.895**	.876	-.870**	-.453	.878**	.876**	.870**	.801**
	Sig (2-tailed)	0.000	0.001	0.001	0.189	0.001	0.001	0.001	0.005
	N	10	10	10	10	10	10	10	10
Conductive heating	Correlation coefficient	-.895**	.876	-.870**	-.453	.878**	.876**	.870**	.801**
	Sig (2-tailed)	0.000	0.001	0.001	0.189	0.001	0.001	0.001	0.005
	N	10	10	10	10	10	10	10	10
**. Correlation is significant at the 0.01 level (2-tailed).									
*. Correlation is significant at the 0.05 level (2-tailed).									

Structural and Mechanical Characteristics

The weight of the interlock fabric is less than of the three-thread fleece fabric, although its thickness is quite high. This is because the conductive yarns slightly protruded out from the fabric surface to the outside on both sides of the interlock fabric. The conductive yarns are laid horizontally between the front and rear bed stitches without forming loops during fabric production. Therefore, the density of interlock fabric is quite low.

The results show that the bursting strength value of the three-thread fleece fabric is higher than that of the interlock fabric due to the high fabric density value. However, on the other hand, the percentage elongation value of the interlock fabric in the longitudinal direction is higher than that of the three-thread fleece fabric. It is predicted that this is because the loops can change shape more easily under force, and the fabric can easily stretch because of the low fabric density of the interlock fabric.

Thermal Comfort Characteristics

Air permeability is an indicator of how well air passes through the fabric. One of the main factors that influence the amount of air flowing through the fabrics is the fabric density, and fabrics with high fabric density value have low air permeability (Hu et al., 2006; Oğlakcioğlu & Marmarali, 2007). According to the results, the air permeability value of the three-thread fleece fabric is lower (15%) than that of the interlock fabric due to the higher fabric density, as mentioned in the literature.

Thermal resistance is the resistance offered by the fabric against heat flow, and it is greatly influenced by fiber type, yarn properties, and fabric structure. When the results were examined, it was observed that because of its lower fabric density, the interlock fabric had a higher thermal resistance value (86%) than the fleece fabric. As stated in previous studies (Blaga et al., 2015; Havenith, 2002), fabrics that have lower fabric density have higher thermal resistance values as a result of a higher amount of stagnant air in their structure.

Water vapor resistance is the measure of the fabric's reluctance to let water vapor pass through. If the resistance to water vapor in a fabric is higher, it prevents evaporative cooling of the skin, and stored heat cannot be dissipated from the body, which may lead to uncomfortable feelings for the wearer (Kakvan et al., 2015). The results show that three-thread fleece fabric has a lower value of water vapor resistance (6.5%) than interlock fabric because of the higher fabric density as mentioned by Gorjanc et al., (2012). According to the Hohenstein Institute classification criterion, as given in Table 2, it can be concluded that both fabric samples, which are developed in this chapter, have a very good water vapor resistance rating.

Heating Performance

As mentioned in the Methods section, the heating performance measurements of the developed fabrics within the study were carried out using two different methods such as radiant heat measurement of yarns, and conductive heat measurement of fabric surfaces. As is known, radiant heat is the heat energy value that the material gives to the environment through radiation, whereas conductive heat is the heat value transferred by surface contact.

Thermal camera images were taken from both fabrics to observe the regular radiant heating on the yarn's surface placed in the fabric structure. It was observed that the regular radiant heating values on the surface of the yarns embedded in the interlock fabric and the three-thread fleece fabric were 69.9 °C and 60.1 °C, respectively.

As mentioned above, the conductive heat values depending on time were measured using iButton sensors, which were placed in different places on the fabric surface. The iButton data, for both fabrics under normal room conditions, are shown in Figures 6. a and 6. b. As it is seen, the conductive heat values of the nine iButtons were largely equal to each other for both types of fabric. This result reveals that the surface temperature distribution is uniform in both fabrics. It was observed that for both fabrics the surface temperature of the heated samples rises rapidly for the first half minute during the conductive heating performance measurements. Additionally, the samples reached their maximum heating performance after 40 seconds and maintained the same temperature for 20 minutes. The maximum conductive heating values of the surfaces of interlock fabric and three-thread fleece fabrics are 33 °C and 29 °C, respectively.

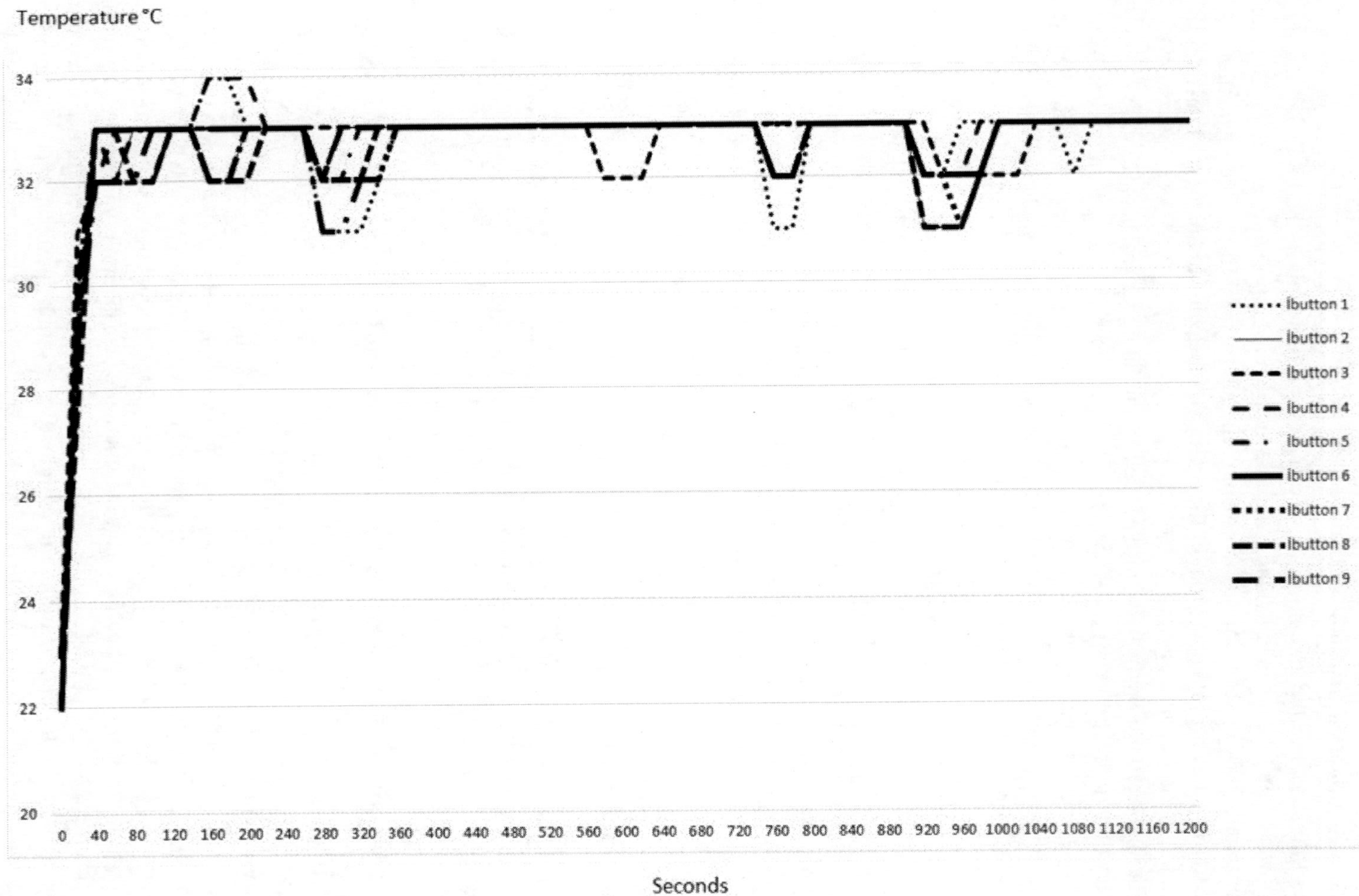

(a)

Figure 6. (Continued).

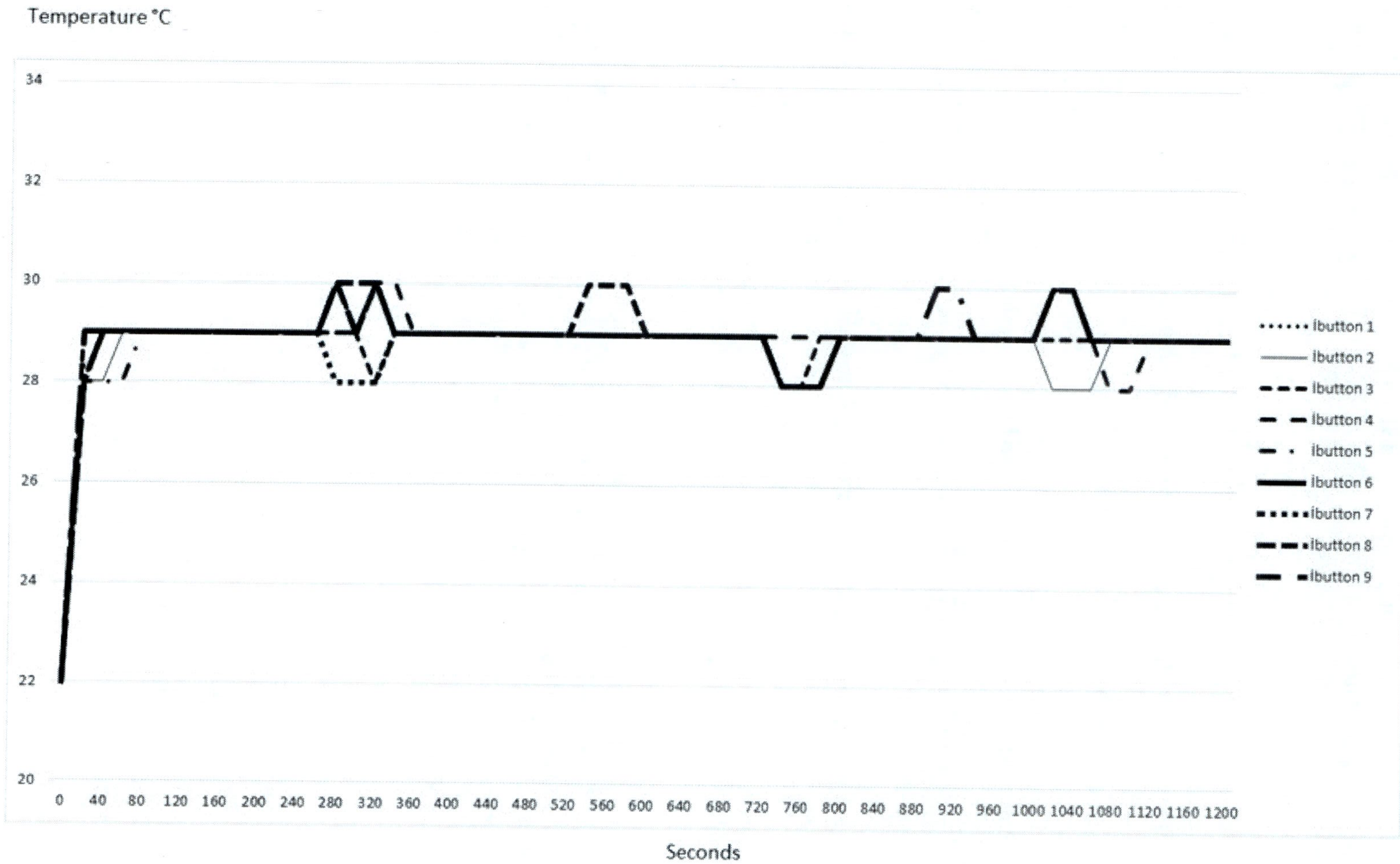

(b)

Figure 6. The heating performance graphs were performed using iButton for (a) interlock fabric, and (b) three-threaded fleece fabric.

In addition, it was determined that the conductive heating performance of the interlock fabric was 11% higher than that of the three-thread fleece fabric while all other parameters were constant. It is thought that this result is due to the low fabric density of the interlock fabric despite the low fabric weight and high thickness values. As is well known, the porosity of the fabric with low density and the amount of stagnant air in the pores are high. In this way, the thermal resistance value of the fabric is higher, and the heat obtained in this way is kept on the fabric for a longer time. Although there is no statistically significant relationship between bursting strength and pilling values and the heating performance of fabrics, it is important to consider these values in terms of evaluating product life.

As a result, it can be recommended that the use of interlock fabric, which has higher radiant and conductive heating performances, for cool climate garments. Additionally, it has a high elongation rate, showing that the clothes to be produced from this fabric are suitable for making comfortable clothes, due to better adaptation to body movements.

Conclusion

Considering the obtained results, it is seen that the interlock fabric structure developed in this chapter has lower fabric density and weight, higher thickness, more elongation, higher air permeability, thermal resistance, and water vapor resistance values. Additionally, the radiant heating value of the conductive yarn in the interlock structure and the conductive heating performance values of interlock fabric are comparatively higher than that of the fleece fabric. Especially for radiant heating, considerably high-temperature values were obtained for the yarns in both types of fabric. It was also seen that the heating power in the yarns was homogenously spread over the fabric surface and a temperature increase was achieved. Therefore, it is thought that the developed heating fabrics can be used as a middle layer in multilayered clothing assemblies or after being covered with another fabric with a high thermal resistance value efficiently. The findings of the study show that the developed knitted fabric structures can be used in multilayer heated clothing, medical applications, etc. Improving heating performance is very important for these types of products. In future studies, it is possible to improve the heating performance of knitted fabrics by changing the materials and properties of conductive yarn and ground yarns, fabric structure, machine settings, and circuit design.

Disclaimer

None.

References

Adamhasan, A. S. (2008). Examining conductivity characteristics and production of poliester/polianiline, cotton/polianiline composite fabrics. [*Master dissertation*]. Süleyman Demirel University.

Blaga, M., Ciobanu, A. R., & Marmaralı, A. (2015). Investigation of the physical and thermal comfort characteristics of knitted fabrics used for shoe linings. *Textile and Apparel*, 25(2), 111-118.

Eroglu, N. S., Canoglu, S., & Yukseloglu, S. M. (2019). Heating and conductive properties of fabrics made of metal yarns. *European Journal of Engineering and Applied Sciences*, 2(2), 13-16.

Gorjanc, D. S., Dimitrovski, K., & Bizjak, M. (2012). Thermal and water vapor resistance of the elastic and conventional cotton fabrics. *Textile Research Journal*, 82(14), 1498-1506. https://doi.org/10.1177/0040517512445337.

Hamdani, S. T. A., Fernando, A., & Maqsood, M. (2016). Thermo-mechanical behavior of stainless steel knitted structures. *Heat and Mass Transfer*, 52, 1861-1870. https://doi.org/10.1007/s00231-015-1707-z.

Havenith, G. (2002). Interaction of clothing and thermoregulation. *Exogenous Dermatology*, 1(5), 221-230. https://doi.org/10.1159/000068802.

Hu, J. Y., Li, Y. I., & Yeung, K. W. (2006). Air permeability, clothing biosensory engineering. In Y. Li & A. S. W. Wong (Eds.), *Cloth biosensory* (1st ed., pp. 252-260). Woodhead Publishing Limited.

Hughes-Riley, T., Dias, T., & Cork, C. A. (2018). A historical review of the development of electronic textiles. *Fibers*, 6(34), 2-15. https://doi.org/10.3390/fib6020034.

Kakvan, A., Najar, S. S., & Psikuta, A. (2015). Study on the effect of blend ratio on thermal comfort properties of cotton/nylon-blended fabrics with high-performance Kermel fibre. *The Journal of The Textile Institute*, 106(6), 674-682. https://doi.org/10.1080/00405000.2014.934045.

Kayacan, O., Bulgun, E., & Sahin, O. (2009). Implementation of steel-based fabric panels in a heated garment design. *Textile Research Journal*, 79(16), 1427–1437. https://doi.org/10.1177/00405175081014.

Kayacan, O., Kayacan, O., & Bulgun, E. (2015). Design methodology and performance studies of a flexible electro-textile surface. *Autex Research Journal*, 15(3), 153-157. https://doi.org/10.1515/aut-2015-0013.

Kukkonen, K., Vuorela, T., & Rantanen, J. (2001). The design and implementation of electrically heated clothing [Conference session]. *5th Int. Symposium on Wearable Computers, Zurich, Switzerland*.

Kursun Bahadir, S., & Sahin, U. K. (2018). A wearable heating system with a controllable e-textile- based thermal panel. In J. H. Ortiz (Ed.), *Wearable technologies* (1st ed., pp. 175-194). IntechOpen. https://doi.org/10.5772/intechopen.76192.

Kursun Bahadir, S., Kalaoglu, F., & Jevšnik, S. (2015). The use of hot air welding technologies for manufacturing E-textile transmission lines. *Fibers and Polymers*, 16(6). 1384-1394. https://doi.org/10.1007/s12221-015-1384-z.

Liu, H., Jin, L., & Li, C. (2015). Thermal-electronic behaviors investigation of knitted heating fabrics based on silver plating compound yarns. *Textile Research Journal*, 86(13), 1398-1412. https://doi.org/10.1177/0040517515612359.

Liu, H., Li, J., Chen, L., Liu, L., Li, Y., Li, X., Li, X., & Yang, H. (2016). Thermal-electronic behaviors investigation of knitted heating fabrics based on silver plating compound yarns. *Textile Research Journal*, 86(13) 1398–1412. https://doi.org/10.1177/0040517515612359.

Maurya, S. K., Das, A., & Kumar, B. (2023). Optimization of heating performance of the rib-knitted wearable heating pad. *Engineering Proceedings*, 30(1), 2-5. https://doi.org/10.3390/engproc2023030001.

Mey, G., Ozcelik, M., & Schwarz, A. (2014). Designing of conductive yarn knitted thermal comfortable shirt using battery operated heating system. *Textile and Apparel*, 24(1), 26-29.

Oğlakcioğlu, N., & Marmaralı, A. (2007). Thermal comfort properties of some knitted structures. *Fibres & Textiles in East Europe*, 15(5–6), 64 – 65.

Padleckienė, I., Stygienė, L., & Krauledas, S. (2020). Development and investigation of a textile heating element ensuring thermal physiological comfort. *Fibres&Textiles in East Europe*, 28(5), 56-62. https://doi.org/10.5604/01.3001.0014.2385.

Penava, Z., Penava, D. S., & Knezić, Z. (2022). Heat as a conductivity factor of electrically conductive yarns woven into the fabric. *Materials*, 15(3), 1202-1208. https://doi.org/10.3390/ma15031202.

Reljic, M., Stepanovic, J., Lazic, B., Ćirković, N., & Cerovic, D. (2016). The change of water-vapor resistance of materials used for the clothing production during exploitation. *Advanced Technologies*, 5(2), 73-78. https://doi.org/10.5937/savteh1602073R.

Repon, R., Laureckiene, G., & Mikucioniene, D. (2022). Effect of stretching on the thermal behavior of electro-conductive weft-knitted composite fabrics. *Polymers*, 14(249), 1-23. https://doi.org/10.3390/polym14020249.

Rossi, R. (2009). Comfort and thermoregulatory requirements in cold weather clothing. In J. T. Williams (Ed.), *Textiles for cold weather apparel* (1st ed., pp. 3-18), Woodhead Publishing.

Tong, J., Ding, F., & Tao, X. (2014). Temperature effect on the conductivity of knitted fabrics embedded with conducting yarns. *Textile Research Journal*, 84(17), 1849–1857. https://doi.org/10.1177/0040517514530026.

Uner, I., & Gurcum, B. H. (2019). Applications of technical embroidery and conductive yarns in e-textile design. *International Journal of Social Humanities Sciences Research*, 6(44), 3376-3388.

Wang, F., & Lee, H. (2009). Evaluation of an electrically heated vest (EHV) using a thermal manikin in cold environments. Annals of Occupational Hygiene, 54(1), 117–124. https://doi.org/10.1093/annhyg/mep073.

Wang, F., Gao, C., Kuklane, K., & Holmér, I. (2010). A review of the technology of personal heating garments. *International Journal of Occupational Safety and Ergonomics*, 16(3), 387-404. https://doi.org/10.1080/10803548.2010.11076854.

Wang, Y., Gordon, S., & Yu, W. (2019). Structural architecture of wearable materials based on tri-component elastic-conductive composite yarn: Toward a Joule heating application. *Textile Research Journal*, 89(16), 3303–3311. https://doi.org/10.1177/0040517518811937.

Yılmaz, N. (2015). *Power sourced heated jacket. [Master dissertation]. Marmara University.*

Zhang, M. C., Wang, C. Y., Liang, X. P., Yin, Z., Xia, K. L., Wang, H. M., Jian, M. Q., & Zhang, Y. Y. (2017). Weft-knitted fabric for a highly stretchable and low-voltage wearable heater. *Advanced Electronic Materials*, 3, 1-8. https://doi.org/10.1002/aelm.201700193.

Chapter 4

Laminated Breathable Fabrics Designed with Nanofiber Membranes

Çigdem Akduman[1,*], PhD
and Nida Oğlakcıoğlu[2], PhD
[1]Pamukkale University, Denizli Vocational School of Technical Sciences, Denizli, Türkey
[2]Ege University, Faculty of Engineering, Department of Textile Engineering, Izmir, Türkey

Abstract

Breathable fabrics find applications in active sportswear, protective clothing, survival clothing, outdoor clothing, and especially military clothing due to their comfortable structures. Tents, cargo transport packaging, or sleeping bags are some other application areas. There are many different approaches to designing breathable fabrics according to the end-use area, such as using microfibers, creating microporous polymer coating-based membranes, laminated films, or finally coating solid polymer films on the fabric. Laminating nanofiber membranes is a new approach for breathable fabrics. Nanofiber membranes show the ability to control airflow through the fabric without significantly affecting comfort properties such as thermal insulation or water vapor transmission. These materials can be used for a whole garment, such as jackets or trousers, or they can only be used to reduce air permeability for a certain part of a garment. This review explains the production and alternatives of breathable laminated fabrics and compares them with fabric design with novel nanofiber membranes.

Keywords: air permeability, breathability, lamination, nanofiber membranes

Introduction

Nanofiber technology offers solutions to some global issues by creating and fabricating functional nanofibers for diverse applications. Electrospinning from polymer solutions or melts is a practical and direct method to produce nanofibers (Ragaisiene et al., 2014) varying from nanometers to a few microns (Kim, 1999). The high performance of nanofibers in many

* Corresponding Author's Email: cakduman@pau.edu.tr.

In: Textile Insights and Future Aspects
Editors: Gamze D. Tetik, Gizem Celep and Fulya Yilmaz
ISBN: 979-8-89113-289-4

applications is due to their ability to form a highly porous network with nano-sized fiber diameters and large surface-to-volume ratios (Na et al., 2012; Kucukali-Ozturk et al., 2017; Gorji et al., 2018).

The unique features of nanofibers make them appropriate for medical to basic users, industrial to innovational products for capacitors, transistors, drug delivery systems, battery separators, fuel cells, sensors, acoustic applications, filtration purposes, and so on (Fadil et al., 2021; Chee et al., 2017; Wang et al., 2021; Wang & Hsiao, 2016; Son et al., 2014). Nanofibers can be produced mainly from polymers and their blends, and also from the addition of ceramic precursors, metals, and nanocomposites to the polymer solution or melt (Yousefzadeh et al., 2011; Si et al., 2014). Increasing awareness of the current and potential applications of nanofibers has encouraged the research and advancement of nanofiber integrated solutions.

Clothing that provides protection from bad weather conditions has been used for millions of years. Advanced impermeable water-breathable fabrics are created for garments that provide protection, especially in cold weather conditions from wind, rain, and loss of body heat (Gibson et al., 2014). These fabrics find application as comfort fabrics in active sports clothing, protective clothing, survival clothing, outdoor clothing and military clothing. Besides these protections, they can show different functionalities such as antibacterial, self-cleaning, the ability to capture volatile organic substances, filter airborne particles or electromagnetic shielding (Gibson et al., 2001; Kadam et al., 2021; Akduman et al., 2021; Gouda et al., 2017; Yang et al., 2016). However, comfort; the human sensory response to clothing is identified by various thermal, physiological, and mechanical parameters and is still the most important characteristic of clothing from the consumer's point of view because it is not possible to wear an uncomfortable garment for a long time even it is a necessity and serves several benefits (Das & Alagirusamy, 2011). When the temperature of our skin is about 34°C, we feel comfortable. Naturally, the physiology of our body successfully controls this temperature under normal conditions (Abbott, 2001a). For this reason, when normal conditions are exceeded, besides functionality, the diffusion of water vapor from the clothing should be maintained to prevent the increase in the body temperature, and heat loss should be prevented by providing adequate insulation (Amini et al., 2016).

The idea for the lamination of nanofiber layers with other textile materials is similar to that in current outdoor and protective clothing. Outdoor clothing generally includes a membrane or hydrophobic coating as an upper layer, which is used to achieve hydrostatic resistance, wind resistance, and water vapor permeability. Depending on the specific polymer used for nanofiber membranes and the amount of nanofibers due to the electrospinning period or nozzle number, various fabric properties, such as strength, weight, and porosity, can be obtained. Especially a wide range of porosity is possible. While, the high dense nanofiber membrane shows a nearly nonporous polymer nanofiber coating, a very smaller amount of nanofiber web results in a highly porous and fragile fibrous structure (Gibson et al., 2001). Electrospun nanofiber membranes behave like a microporous membrane or coating because their nano- to micro-sized fibers and connections resulted in high porosity with small pores. The resistance to convective gas and fluid flow is quite large and is comparable to the flow resistance of commercial membranes such as microporous polytetrafluoroethylene (PTFE) membranes, which are often used as a component of protective clothing systems.

Therefore, the integration of nanofiber membranes into conventional fabrics may open up new opportunities instead of commercial microporous membranes, such as improving comfort performance and thermal management properties (Oğlakcıoğlu et al., 2021). This study gives general information on lamination and coating, which are possible binding technologies for nanofiber membranes to the fabric, explains the methods of designing functional breathable fabrics, and finally, an overview of the research on the utility of the nanofiber layers for improved performance and breathability properties.

Lamination

Laminating is the process that joins multiple layers together and by using one of the binding techniques such as adhesives, heat, pressure, or mechanical bonding produces steady and consistent multilayer structures (Shim, 2013). The reason for the application of coating and lamination is to produce fabrics with features and functionality that the fabric cannot achieve alone or as it is. Flame, wet adhesive, hot melt, and dry heat lamination are the classification of the laminating process according to the bonding methods and the type of adhesive used.

Flame lamination is a continuous and simple laminating process. It has a high production speed and relatively low cost. A prepared thin foam sheet is used and moved continuously through an open flame. The open flame creates a sticky thin film of polymer melt on the foam surface, and then it passes through a nip roller with the fabric. The melted surface of the foam acts as an adhesive and binds to the fabric as it cools. Foams (often polyurethane foams) and/or other lofted non-woven polyesters are the most commonly applied material in the process of flame lamination. The inclusion of a thick foam layer will produce a stiffer laminated fabric (Dawes & Owen, 1972), and the process produces a solid bond layer with low porosity. Another disadvantage is that it is not an environmentally friendly process, because it gives rise to harmful emissions throughout the lamination process (Fung et al., 2002; Shim, 2013).

Wet adhesive lamination employs liquid adhesives to bind the lamination layers together. By using one of the coating techniques, such as roll, spray, knife coating, etc, adhesives are preliminarily applied to one or both surfaces (Fung et al., 2002). After that, the substrates are joined together under pressure while the bonding agent is still in the liquid state and then dried and/or cured in a dryer. The final properties of laminated fabrics are mainly dependent on the adhesives employed. The penetration depth of the adhesive affects adhesion and structural flexibility. The optimal balance should be achieved for good adhesion without losing flexibility. To achieve good adhesion, some penetration is necessary, but deep penetration may lead to tough structures and resulted in a loss of drapability, handle, tear strengths, and breathability. When solvent-based and water-based adhesives are compared, solvent-based adhesives dry quickly, but give rise to volatile organic compounds emissions, while water-based adhesives are more environmentally friendly and do not include toxic solvents, but require a drying process that increases energy consumption (Shim, 2013).

Hot melt lamination employs hot melt adhesives instead of wet adhesives. They are 100% solid materials and are applied from the melt. When hot melt adhesives are heated to specific temperatures, they melt and when they cool, they solidify to form bonds (Singha, 2012; Shim, 2013). Hot melt adhesives can be thermoplastic or reactive. When thermoplastic adhesives are employed, lamination is usually sensitive to high temperatures, steam, and water, so the adhesive type should be selected according to the application ranges. When reactive hot melt adhesives are used, the adhesive solidifies again when it cools, but also crosslinking reaction takes place. Cross-linking turns materials to thermoset; after the cross-linking, heat does not soften the adhesives again. They are highly durable, with good heat and climate tolerance, but when compared to thermoplastic adhesive is that they are rather expensive. Another advantage of hot melt over wet adhesives, they do not need a drying unit, they are environmentally friendly, free of solvent, and do not use water (Fung et al., 2002). Bonds are formed right after solidifying without drying; therefore, operating speeds are not dependent on drying, thus a high production speed can be achieved. In hot melt lamination by applying a very low coating level with little adhesive, a soft handle and flexibility can be achieved (Shim, 2013).

Dry heat lamination uses 100% solid adhesive without medium or being melted. These solid adhesives are laid on the substrate. Then, the solid adhesive is activated by heat and pressure, and the adhesion is achieved between the layers. From the environmental point of view, the advantage of this process is that high solid content does not give rise to harmful emissions. However, the melting process of the dry solid adhesive needs high energy, and the lamination layers are exposed to high temperatures, so it can be applied to substrates that are stable at high temperatures (Shim, 2013).

An alternative type of dry adhesive is a delicate, breathable nonwoven web roll or film (Fung et al., 2002). These types of dry adhesive can be directly laid between substrates in the lamination process (Jarrell, 1992). The low add-on can be obtained. It is an optimal method where powder adhesives tend to fill the gaps of open structures rather than stay on the surface. However, they have a high cost when compared to the powder adhesive. For different widths and add-on requirements, the adhesive roll should be changed (Shim, 2010).

Coating

The coating is one of the options for adding or enhancing function to a textile fabric, adding functionality to a material, and/or creating a material with certain properties (Smith, 2010). A coated fabric includes a fabric, sometimes mentioned as a base fabric or substrate, and a polymer compound (Holden, 1987). In the coating process, polymeric compounds are applied on the surface of the fabric (Singha, 2012). Coatings, including fillers and extenders, and pigments bonded with a polymer from composites (Ferrigno, 1998). Coating materials are generally liquids or solids. Gases coatings are also employed and each need individual processing methods to create a coating on the fabric. The purpose of the coating is to produce a uniform and durable coating of an adjustable and appropriate thickness with good adhesion to the fabric. The coating material is transferred to the surface of the substrate and forms a coating layer. Application of the coating material can be carried out with direct immersion of the fabric in an application bath for both sides or a couple of rollers, spraying, and direct extrusion. Processing parameters affect the mass of the applied coating material, the consistency of the coating on the surface, the adhesion between the coating layer and the fabric, and the penetration level of the coating material (Shim, 2010).

One of the main characteristics of a coating is its thickness, and the typical coating level varies from a few microns to several hundred microns, depending on the purpose it serves (Jones et al., 2017). Most coating formulations consist of binders, a vehicle, and additives (Shim, 2010; Jones et al., 2017). Binders ensure the bonding of the fabric and other components by forming a continuous film, adhering to the surface of the fabric, and controlling the adhesion properties between the coating and the base material. Binders also manage the mechanical properties of the coated material. Coating materials can be found in the form of liquid, melt, foam, continuous film, web, or powder. There are several types of coatings, and some of them are: immersion/dip coating in a bath, layer by layer, knife or blade coating, transfer coating, traditional roll or engraved roll coating, slot die or extrusion coating, powder coating, and spray coating, etc. (Smith, 2010; Fung et al., 2002).

For weather-proof clothing, earlier fabrics focused on rubber or synthetic coatings, such as PVC, PUR, or acrylate. These coatings have high water resistance but their water vapor transmission was very little (Gottwald, 1996).

Regardless of the type or purpose, coatings application has a common problem: people find the garment uncomfortable when wearing them for a long time or during the time of intense activity. It has been recommended that the clothing used during physical activity have a water vapor permeability of 6000 g/m^2 per day. Even the densest sportswear fabric without a coating certainly satisfies this requirement. The application of an impermeable coating decreases the permeability by about 100 g/m^2 per day. To provide comfortable coated clothing, this permeability level must be significantly increased (Abbott, 2001b).

Although coating and laminating processes are different processes, they share several principles, necessities, and machinery.

Both coating and lamination are important technologies to expand possibilities of textiles. These processes add beneficial functions and properties, developing real engineered composite systems (Shim, 2010). Each component synergistically adds different features and functionality to the entire material. However, while the functionality and properties of fabrics improve, their comfort properties must also be provided.

Designing Breathable Protective Fabrics

The technical design and development process for protective clothing considers the demands of the human body and adaptability to certain conditions. Protection from rain and cold is one of the most universal needs of humans. Other protection needs can be more specific and related to exposed dangers such as extreme heat, toxic or corrosive liquids, toxic fumes, potentially dangerous chemicals, viruses, bacteria, or other sources of contamination, X-ray or other electromagnetic or nuclear radiation because of the environmental and working conditions (Abbott, 2001a; Baji et al., 2020). Many threats can be countered by adding new properties and chemicals to the coatings and laminates, as mentioned above; consequently, some comfort issues are likely to occur. It should be kept in mind that there can be a negative relation between the level of protection and the comfort property (Ahn et al., 2011). Therefore, in the development of the protective clothing, the process should provide thermal comfort properties with considering durability, cleanability, aesthetic properties, cost, and availability (Gohlke & Tanner, 1976).

Over the years, many researchers have tried to define the relationship between thermal comfort and protection, and it has been seen that the thermal environment is the main factor that affect this (HSE, 2023). Although the comfort of the people mainly depends on the temperature of the environment; characteristics such as relative humidity, airflow velocity, the metabolic rate, insulation, and the water vapor transmission have also a significant impact on feeling comfortable (Ahn et al., 2011). For instance, in cold conditions, if clothing does not provide adequate insulation, the wearer may be exposed to cold stress and suffer from conditions such as frostbite or hypothermia. On the contrary, in hot weather, wearing impermeable airtight, and water vapor clothing increases humidity inside because sweat does not evaporate sufficiently. For this reason, water vapor permeability becomes the most important feature of comfort and has a key role (Amini et al., 2016). The feeling of comfort in clothing is mainly identified by heat exchange and a balanced moisture transmission, namely "breathability" between the human body and the environment (Ahn et al., 2011).

Depending on the end-use application, breathable protective fabrics can be designed using high-density wovens, using microfibers, lamination of textiles with breathable films, microporous membranes, and finally hydrophobic or hydrophilic coating of the fabric with a solid polymer film (Kannekens, 1994; Jassal & Agrawal, 2010; Ahn et al., 2011).

Because of the thin and smooth yarns of microfibers, high-density wovens provide windproof fabric with exceptional water vapor transmission when compared to laminated and coated fabrics (Kannekens, 1994). However, relatively less water repellency requires an additional water-repellent finish, as these types of fabrics are used mostly in outdoor applications. The main benefit of these fabrics is their nice handle and drape.

Breathable laminates use either a microporous film or a hydrophilic film. A microporous film has very tiny pores; they let water diffuse as vapor but not in the liquid state. A hydrophilic film comprises hydrophilic group molecules: as a result, only the vapor molecules pass through the film. These laminates are generally impermeable to water as required. High hydrostatic pressures and water vapor permeability are common features; however, the vapor permeability of the laminate may be affected by the lamination process, especially the amount of glue and type used for lamination (Kannekens, 1994).

Microporous membranes are able to transfer water vapor by their stable pore structure while preventing liquids' penetration. Penetration of liquids through textiles with barriers occurs by capillary action. This phenomenon is mainly controlled by pore size and structure, fiber chemistry, and surface treatments such as water-repellent finishes (Dolez, 2013). The pores of the microporous membranes are approximately 0.02 μm – 1 μm in diameter. The strong interaction of water molecules shapes a bigger drop diameter of about 200 μm – 5000 μm when compared to that of pores of the membrane. Therefore, they are too large to penetrate through the membrane pores. Whereas, sweat or water vapor molecules formed due to metabolic actions are smaller than 0.0003 μm and can easily pass through the pores of the polymer membrane. It takes place through the capillary transfer mechanism of water vapor transmission through the pores of microporous membranes (Gulbinienė et al., 2007).

Conversely, hydrophilic membranes are non-porous membranes and deliver water vapor through a molecular mechanism. These membranes are resistant to water molecules and generally show high hydrostatic pressure. Their permeability and morphologies depend on their chemical composition. The basis of the water vapor transmission of hydrophilic materials in this group is the chemical adsorption of water vapor through the soft segments, such as the polyether chain, whereas water droplets are repelled by the hard segments, such as the urethane chain (Ahn et al., 2011). There is a difference in water vapor pressure between two ends of the membrane, known as non-activated diffusion, which causes a concentration gradient within the membrane, which results in the passing of the water vapor molecules through the membrane (Gulbinienė et al., 2007).

In breathable coatings, three basic approaches are used to the manufacture of a water vapor permeable coating, resistant to the passage of liquid water:

1. Creating tiny pore-like holes all over the film with needles, laser beams, or a series of micrometer-sized hole-making techniques.

2. Making the coating film from a material that can be separated into thin, fiber-like structures with micrometer-sized gaps between the fibers.
3. Creating monolithic polymer membranes as coating layers that do not contain pores, in which the transmission of water vapor takes place through the activated diffusion process. Water vapor from the skin condenses and dissolves on the surface of these membranes, and then diffuses to the other side of the film. On the other hand, water vapor is released and evaporates into the environment (Abbott, 2001a).

To make breathable microporous and hydrophilic coatings, at first, the acrylate and polyurethane solutions are applied as coatings. The coagulation process creates micropores. When hydrophilic groups were included in the chemicals used, coagulation was not required. By forming hydrophilic groups, hydrophilic coatings are made. For water vapor permeable coatings, the most common coating weights are between 60 and 100 g/m^2 in the dry state. Hydrostatic pressure of 60 to 600 cm is obtained in terms of water impermeability, depending on the structure of the base substrate. Paste coatings are often lower than stable foam coatings, and typical values for water vapor permeability of paste coatings vary between 3000 and 6000 g/m^2/24h (Kannekens, 1994).

The waterproofness of the microporous coating decreases after the first wash and remains stable thereafter. It has better breathability than the hydrophilic coating. The breathability of the hydrophilic coating is influenced by the thickness of the coating and decreases with the increase of the coating thickness. While the hydrophilic coating is cheaper and simpler and the production speed is higher, it has a tougher handle. Since the hydrophilic coating does not have holes, it is less sensitive to possible deterioration (Kramar, 1998).

Considering both breathable laminate and coated fabrics, these fabrics can be basically considered into two groups: microporous materials and hydrophilic materials. Comfortable and breathable fabrics transport water vapor effectively from inside of the clothing to the environment either through micropores or hydrophilic structure through activated diffusion, as well as protect the human body from wind, rain, and cold to maintain a constantly comfortable clothing microclimate under different environmental conditions (Ahn et al., 2011). The body tries to balance the increase in temperature by secreting sweat, the sweat on the skin evaporates and needs to be removed to prevent the feeling of wetness. The clothing serves both in absorbing the heat and removing the water vapor from the skin. Depending on external conditions, clothing maintains a regulated equilibrium between temperature and humidity inside the air gap immediately surrounding the body (Gugliuzza & Drioli, 2013).

One of the most preferred technologies for breathable outdoor and protective clothing is polymeric microporous membranes. A microporous membrane or coating is defined as a thin polymeric film that has a controlled open porous morphology and millions of microscopic pores, preferably smaller than 3 μm in diameter (Hunter & Fan, 2009). Therefore, their interconnected network of tiny holes shows a breathable impermeable polymeric structure. The term "breathable" is used to describe membranes with a water vapor permeability greater than 400 g/m^2/day. Common methods to measure membrane breathability are the upright cup (ASTM E96B), inverted cup (ASTM E96BW), the desiccant inverted cup (JIS 1099), dynamic moisture permeation cell (ASTM F 2298), and upright cup methods (BS 7209) which measure the water vapor transmission in g/m^2/day, and sweating hot plate methods (ASTM F 1868, ISO 11092) which measure the energy lost through the fabric by evaporation in m^{2}Pa/W (Kramar, 1998; Gugliuzza & Drioli, 2013). The choice of testing method would have to be made according to the final usage of the clothing. For instance, some best-known commercial membranes under the trademark Gore-Tex, Sympatex, and Scholler show much better water vapor transport characteristics in a humid environment than in a dry environment (Gugliuzza & Drioli, 2013). Depending on the end use of the garment, the polymer material and membrane structure are determined, and then is the application method is decided, whether it will be coating or lamination. It can be cast or coated directly onto the fabric or formed into a membrane and then laminated to the fabric (Hunter & Fan, 2009).

Poly (tetrafluoroethylene) (PTFE) and polyurethane (PU) are the most popular polymers for the fabrication of microporous membranes, they can also be made from polyolefins, polyamides, polyester, and polyether-based copolymers (Hunter & Fan, 2009). In the 1970s, the first use of a microporous membrane was commercialized under the Gore-Tex brand to produce breathable fabric and garments. It was an important step in producing the first comfortable breathable waterproof fabrics (Holmes, 2000). The Gore-Tex microporous polymeric film is an expanded PTFE (ePTFE), having 0.001 inch thickness, 82% pore volume, and about 9 billion pores per square inch with a maximum pore size of 0.2 μm in diameter (Hunter & Fun, 2009; Gorji et al., 2017; Gohlke & Tanner, 1976). The maximum pore size of Gore-Tex is much smaller than that of a water droplet and larger than that of a water vapor molecule. This ePTFE membrane is extremely stable to chemicals, is smooth, UV resistant, water repellent, durable, and can withstand high temperatures and high water pressures. Gore-Tex is the combination of the very small pore size and the hydrophobic nature. High pore volume, a large number of pores, and the low thickness of the film provide a large open area that allow the diffusion of water vapor (Gohlke & Tanner, 1976).

Nowadays, for military use, outdoor clothing or work wear by laminating a microporous membrane between an outer waterproof fabric and an inner soft and flexible fabric, new durable and high-quality breathable fabrics are produced. Besides, just laminating a microporous membrane on an outer fabric and using an insulation material inside is also popular.

Many types of breathable protective membranes and fabrics with various chemical compositions have been proposed. Donaldson's ePTFE membrane under the trade name Tetratex, PUs under the trade names Porelle® (Porelle Membranes, 2023) and Polartech® (Polartec, 2023) which uses weatherproof technology and Permax® which is used in high-performance hydrophilic coating applications (Gorji et al., 2017; Lubrizol, 2023), monolithic polyester membranes under the trade name of Smylex (Exten, 2023), Dupont's thermoplastic polyester elastomer Hytrel (Dupont, 2023), and DSM's Arnitel which is bio-based mass-balanced thermoplastic elastomer (DSM, 2023), polyester/polyamid blend under the trade name of Tepor® (Panther, 2023), and polyester-based hydrophilic membrane under the trade name of Sympatex (Sympatex, 2023) are some examples. Besides, plasticizer-free poly (ether-b-amide) Pebax can also form naturally breathable membranes when extruded alone or in combination with otherwise non-breathable resins (Solvay, 2023).

When the coated and laminated products are compared, the coated products have been currently sold at lower prices than laminated fabrics, and by varying the number, the thickness, and the type of coated layers, different performance levels could be achieved providing high flexibility. Coated fabrics have a better handle than laminated fabrics. But the laminating production speed is higher than the coating process. All types of textile substrates can be laminated (Kramar, 1998), while there are some limitations in the coating process.

Nevertheless, innovations in materials science, nanotechnology, and biotechnology bring new materials to coating and lamination processes, expanding the possibility of highly functional textiles by providing comfort properties as well (Harifi & Montazer, 2017). In this scope, novel nanofiber coating or nanofiber lamination processes open up new opportunities to design breathable fabrics, that combine comfort and functionality.

Nanofiber Membranes for Breathable Fabrics

The electrospun nanofiber web has high porosity, excellent structural mechanical properties, high flexibility, and high air and water vapor permeability. The pore size in the electrospun nanofiber membrane is very small and depends on the diameter of the nanofibers and the thickness of the nanofiber membrane (Akduman, 2019; 2021). This micron pore size prevents the penetration of water droplets while it is large enough to allow the transfer of water vapor from the body to the environment (Amini et al., 2016). Therefore, many studies have examined the combination of nanofiber webs and membranes with textile fabrics.

Membranes made of submicron electrospun fibers are thin but thicker than webs, have very small pore sizes, and have a high surface area to volume ratio. The application of electrospun nanofibers has the advantages of high penetration resistance depending on the chemical character of the nanofibers, good breathability (evaporative cooling), low areal mass, low-pressure drop (Nurfaizey et al., 2012), and ability of functionalization (Baji et al., 2020) for protective clothing.

In order to understand, the difference between the measurements of air permeability of the textile surfaces and the pressure drop(ΔP) of the filter materials needs to be explained. Air permeability is defined as the rate of airflow passing perpendicularly through a known area under a prescribed air pressure differential between the two surfaces of a material (ISO 9237).

Air is drawn through the specimen into a closed chamber and out through an orifice that measures the flow. A monometer calculates the pressure drop on both sides. Measurement equipment converts pressure drop to airflow. Units are measured in terms of airflow (per unit area) per unit of pressure drop per unit of time, such as $l/m^2/s$ (Hutten, 2007). Higher air permeability let better windproof properties (Hes & Williams, 2011). When the air permeability of nanofiber surfaces is evaluated according to ASTM 737-96 or ISO9237, the very small pore structure shows resistance to pressurized air, in a sense, to the wind. For this reason, in protective clothing wind resistance is achieved by very tiny pores or by blocking all the pores.

On the other hand, the pressure drop of an air filter is also found by measuring its resistance to the airflow passing through the filter, given for specific face velocities. The face velocity is measured in m/s and can be calculated as the volume flow rate (m^3/s) divided by the face area (m^2). The Aerosol carrying air flow rate is settled before the measurement.

The measuring equipment gives the penetration of the aerosol with pressure drop, usually in Pascal (Pa). When air passes through the filter material with a lower pressure drop, the air to be filtered can pass through the filter more easily. This means that the filter device will use less energy. When higher air flow rates are used, penetration does not change, but the pressure drop increases. That is why air permeability for protective clothing and pressure drop for filters are used in a different manner. Low air permeability against pressurized air is expected in protective clothing, and their breathability can be expressed by their water vapor permeability, while in aerosol filtration like face masks, pressure drop should be low enough to pass air for breathing. Thus, air permeability and aerosol penetration should be in good balance infiltration. Nanofibers can be used for both applications, while thicker nanofiber membranes are preferred for protective clothing, and relatively thinner nanofiber webs are preferred for air filtration purposes.

One of the first reviews on the transport properties of porous nanofiber membranes was carried out by Gibson et al. (2001). They summarized the transport properties of nanofiber membranes for single layers, and the coating, which was applied to military chemical/biological protective clothing by means of water vapor transmission, air permeability, and aerosol particle penetration. For the water vapor diffusion test, they explained a dynamic moisture permeation cell, with known gas fluxes and water concentrations and calculated the water vapor transport through the cell. For efficiency performance, the mentioned filtration test system generates particles between 1 and 5 μm and measures the total amount of aerosols removed from the gas stream. According to the summarized data, the convective flow resistance of the polyacrylonitrile and polybenzimidazole nanofiber fiber mats was quite large compared to those of normal clothing materials and showed still high water vapor diffusion, allowing evaporative cooling to take place. They gave the comparison of polybenzimidazole and polyacrylonitrile electrospun membrane with PTFE microporous membrane and commercial Sympatex and Gore-Tex laminate water vapor diffusion resistances. Electrospun membranes showed water vapor diffusion resistance quite lower than 200-1500 s/m of membrane laminates results and similar to PTFE microporous membrane. They also analyzed the water vapor diffusion resistance and airflow resistance properties of electrospun PA 6.6 that was deposited on an activated-carbon-loaded polyurethane foam of different thicknesses. While the uncoated foam sample showed approximately 200 s/m of diffusion resistance, electrospun layers added only more than 10 s/m to the total of the foam layer. They said that direct application of nanofiber membranes to garment systems is possible for breathable fabrics instead of costly manufacturing steps of laminating and curing.

Liu et al. (2006) produced electrospun Nylon-6, Epoxy 609, and thermoplastic polyurethane with a thickness ranging from 20 to 150 μm. The electrospun membranes were attached to one side of a glass/epoxy fabric lamina before lamination, and each fabric was placed between two adjacent plies of the laminate. They only investigated the mechanical properties of the laminate and found that the integrated nanofiber layer had no significant effect on the mechanical properties of the laminates.

Lee and Obendorf (2006) reported the barrier properties of liquid penetration polypropylene (PP) electrospun webs and laminates by melt-electrospinning. They have chosen PP because of its relatively low surface energy, chemically inert character, lightweight, and low cost. After electrospinning, they laminated the PP webs onto the adhesive side of a 100% polyester nonwoven substrate at 90°C for 15 s at a pressure of 15 gf/cm^2. The majority of the produced fibers were in the micrometer range. Produced electrospun webs showed 90–100% barrier performance against various surface tensions. For the level of web thickness 0.006 g/cm^2, the air permeability was approximately 270 cm^3/s/cm^2, and for the level 0.009 g/cm^2, it was approximately 150 cm^3/s/cm^2. The increase in thickness of the electrospun webs lowered the air permeability. Lamination also reduced the final air permeability by approximately 20%. However, moisture vapor transport only decreased with increasing web thickness by 2–9%, with lamination by 6–12%. So, they showed that these electrospun laminated fibers can be used as barrier membranes by laminating and still have enough breathability comparable to typical woven work clothing.

In another study, they reported that a layered fabric system with electrospun PU fiber web was used on a spun-bonded nonwoven to develop protective textile materials as a barrier to liquid penetration. Layered fabric structures with web area density of 1.0 and 2.0 g/m^2 exhibited air permeability of greater than 100 cm^3/s/cm^2, higher than most personal protective equipment materials currently in use; moisture vapor transport was 19.90 and 19.35 g/h/m^2, respectively, and similar to nonwovens and typical woven work clothing fabrics (Lee & Obendorf, 2007).

Heikkilä et al. (2007) produced polyamide-66 nanofiber coatings on different substrates for filtration purposes. They found that the filtration efficiency significantly increased even with 0.02 g/m^2 nanofibers coating when compared to uncoated samples. When 0.5 g/m^2 nanofiber coating was applied, the capture of the submicron aerosol particles reached over 90%. They did not investigate the breathability, but they evaluated the pressure drop of the nanofiber coatings. The pressure drop increased slightly with the lowest coating weight, but with a 0.10 g/m^2 coating weight it reached 2-fold or higher, and with the highest coating weight it increased 0- to 20-fold due to microporous structures of the nanofiber coating. Reached pressure drop values were still acceptable for achieved filtration efficiency. It is important that they have suggested covering the nanofiber layer with a protection layer for the first time due to the very delicate state and tends to peel off.

Many researchers used PU nanofiber membranes for the development of water/windproof, breathable fabrics because of their tough and durable structure. Thus, Kang et al. (2007) electrospun 0.02 mm of PU nanofiber web directly onto a polyester/nylon fabric without any binding agent. For comparison, polyester/nylon fabrics were coated under different conditions of 0.02 mm thickness of PU resin membranes with a coating solution with a knife edge on the face fabric. These coatings showed a high weight increase between 20.96 and 46.72% compared to nanofiber-coated fabric (6.55%). The PU web-applied fabric showed higher air permeability. In the case of water vapor transmission, the PU nanofiber-coated fabric showed 375.80 g/h/m^2 close to the control fabric while the PU resin-coated fabrics showed 13.45 to 64.40 g/h/m^2 and qmax was the lowest for the nanofiber coated fabric, indicating greater comfort sensation. However, the water resistance value did not reach that of resin-coated fabrics (Kang et al., 2007).

Han et al. (2008) also developed a functional PU composite nanofiber for textile application using coaxial electrospinning. They used the advantages of good adhesion, better function ability of polycarbonate (PC) (shell), and good mechanical performance of PU (core). Nanofibers were electrospun from different percentages of PC shells and PU cores, and the fiber diameters and porosity of the composite nanofibers were investigated. With the increasing core percentage of PU, the thickness of the membranes varies from 0.156 to 0.240 mm. Compared to the water vapor transport of pure cotton of 5570 g/m^2/day, they still showed a fairly high transmission between 3258 and 3417 g/m^2/day while pure PU (8%wt) showed 3240 g/m^2/day. Their initial trials indicated that it was possible to coat core-shell composite nanofibers onto a cotton fabric with still significant breathability.

Although PU membranes are tough and durable, their adhesion is very low to the base substrate; thus, they are easily peeling off. Therefore, some researchers have tried to use binding agents or lamination to bond the nanofiber membranes to the substrates. Because of similar reasons and the need for durability for laundering in clothing applications, Sumin et al. (2009) emphasized the need for lamination to overcome the poor mechanical properties of nanofiber membranes. So, they laminated the mass-produced PU nanofiber web produced by Finetex which have average diameters of 660, 670, and 700 nm and pore size of 0.55, 0.57, and 0.60 μm^2 by hot melt and solvent lamination onto a nylon fabric as a substrate. They did not evaluate breathability of the laminates, but they observed from the SEM images after wash cycles that the nanofibers preserved their fibrous structure and did not peel off (Sumin et al., 2009). They evaluated the waterproofness, calcium chloride permeability, warm/cold feeling, thermal conductivity, and wash fastness test properties in their other study (Sumin et al., 2010). To determine vapor permeability with calcium chloride, they used 0.8 m/s wind velocity in a chamber, so they both evaluated water coloum for waterproofness and wind resistance. They found that nanofiber laminated fabrics had over 18000 (mmH_2O) waterproofness with a decreasing tendency after 10 washings to over 12000 which was quite sufficient for a raincoat. The nanofiber laminated fabric had about 10000 (g/m^2/day) vapor permeability which was also adequate for outdoor clothing. They used q-max, the capacity for instant heat transfer to approximate the index of warm/cool feeling. The nanofiber laminated fabric had a q-max of less than 0.2 (W/m^2), which meant that the laminated fabric had more of a warm than a cool feeling in contact with the skin. Besides, the nanofiber laminated fabric had a low thermal conductivity of under 0.0035 (W/mK) which led to good insulation.

Yoon and Lee (2011) developed different levels and structures of nanofiber layered fabrics and examined their breathability and waterproofness, compared with a densely woven fabric, a microporous PTFE membrane laminated fabric, and a hydrophilic nonporous PU coated fabric. They used a multiple nozzle system for uniform nanofiber webs in their laboratory and produced PU nanofibers with 300 to 500 nm and 5.6 and 10.2 g/m^2. A 100% polyester dense woven fabric with a thickness of 0.17 mm was used as a substrate. An aerosol adhesive was applied to improve the adhesion of the PU nanofiber webs. The mass-produced nanofiber web from Finetex had a weight of 5.2 g/m^2. The nanofiber webs were laminated on substrate fabrics using a mesh roller and a polyurethane adhesive, and two-layered and three-layered laminates were produced. Nanofiber-layered fabrics showed high water vapor transmissions between 2899 and 4300 g/m^2/day, much higher than those of PU-coated fabrics, lower air permeability than the densely woven fabric, and higher than those of the PU-coated fabrics.

Bagherzadeh et al. (2012) developed a sandwich of woven fabric systems with electrospun polyacrylonitrile (PAN) nanofiber web layers for breathable barrier textile materials. Air permeability, water vapor transmission, and water repellency properties were compared with Gore-tex. The PAN nanofibers were electrospun onto a tight woven fabric for 60, 90, and 180 min with single needle set-up, and an additional woven fabric was laminated onto the PAN nanofiber coated fabric. The microscopic pores of the nanofiber-integrated laminates were too small to allow airflow at a prescribed pressure differential and showed low air permeabilities. All samples had high water vapor permeabilities like Gore-tex and were close to the control fabric. According to the Bundesmann water repellency test, with the increasing time of electrospinning, the lower amount of water passed.

Ahn et al. (2011) produced a PU nanofiber membrane onto a water-repellent finished rip stop PA 6.6 woven fabric. By applying an adhesive PU electrospun nanofiber and PTFE membrane laminated to the surface of the nylon fabric followed by a heat fixation at 160°C. According to their results, the nanoweb laminate had a higher water vapor transmission rate but lower water resistance (3.62 m^{2}Pa/W) than the PTFE laminate. Satisfactory air permeability and water penetration results were obtained. The wearing test revealed that the nanoweb laminated clothing provided a more comfortable clothing-microclimate than the PTFE laminate.

Faccini et al. (2012) deposited PA 6 nanofibers onto nonwoven viscose fabric to design protective clothing against nanoparticulate aerosols. The nanofiber diameters were in the range of 66 to 195 nm. Hot-press lamination was used by dot PA thermoplastic resin application to improve the adhesion of the nanofiber layer. Firmly attached nanofibers were obtained, but after several torsion and compression cycles, significant deformations occurred. They only evaluated the filtration capability and observed that sodium chloride penetration was affected mainly by the thickness level of the nanofibers and provided up to 80% retention of 20 nm size particles and over 50% retention of 200 nm size.

Knizek et al. (2015) produced two- and three-layered laminates, including a PU nanofiber membrane with a mass per unit area of 5 g/m^2 for outdoor clothing. PU nanofibers were first deposited onto a nonwoven spunbond backing fabric and then laminated with another top layer by sticking at 140°C and a pressure of 1 kg/cm^2. They achieved 100% wind resistance, water vapor permeability resistance of 71.2 to 76.2 m^{2}Pa/W, and water column height of 1974 to 2944 cm. In the study of Heinisch et al. (2013), the binder was applied by deep printing, and laminated at 150°C with 1.5 kg/cm^2 pressure. Water vapor resistance increased to 0.4 m^{2}Pa/W for nanofiber applied two-layered laminate, 100% polyester woven fabric as the top layer with a bulk porosity of about 68%.

Ah Hong et al. (2015) coated PU nanofiber with waterborne PU (WPU) coatings and investigated the water vapor permeability, waterproofness, abrasion resistance and compared them with Gore-Tex XCR. The nanofiber membrane was laminated by hot melt dot lamination with an additional adhesive of 9 and 13 g/m^2, onto water repellent finish applied PA 6.6 fabrics. The WPU coating process gave a pickup rate of 6±0.5 g/m^2 with a gravure roll coater. It protected the nanofiber layer and helped counteract the weakness of the nanofiber membrane. In general, coating lowered the water vapor permeability rate of the developed laminates, but some of them showed better results than Gore-Tex. Although they had some promising results, adding a coating layer onto a nanofiber layer can take away the advantage of the porous character of nanofiber membrane and can also damage the nanofibers itself. It should be discussed that there is a need to add a layer of nanofiber to the base, while porous membranes can already be achieved with the coating.

Turaga et al. (2014) wanted to use the hydrophilic, biocompatible and nontoxic properties of PVA for producing nanofibers membranes. They evaluated breathability after heat crosslinking at 150°C for 3h. However, they did not mention the collection period or the thickness of the membrane. The cross-linked nano webs showed 1211.3 g/m^2/day of water vapor transmission.

The thermal comfort properties of hydrophobic thermoplastic PU (TPU) and hydrophilic PVA nanofiber membranes were also investigated and compared by the authors without using any binder or lamination glue (Akduman et al., 2018). Nanofibers were just electrospun onto a polyester screen fabric as a carrier. Polycarboxylic acid was used to cross-link the PVA nanofibers. TPU nanofiber diameters were ranged from 625 to 661 nm, while PVA nanofibers were ranged from 221 to 249 nm. Different membrane thicknesses were achieved by changing the electrospinning time (1, 3, and 5h). While the air permeability of the carrier fabric had 4740 l/m^2/s, both nanofiber membranes showed very low air permeabilities and decreased simultaneously with increasing the electrospinning time, 33.67 l/m^2/s for the thickest TPU membrane and 17.3 l/m^2/s for the thickest PVA membrane. Membranes that had thinner nanofiber diameter revealed a higher air permeability. Despite the high resistance to air permeability, all nanofiber membranes showed a high water vapor permeability between 669 to 700 g/m^2/day. Since there was no significant air gap between the carrier and nanofiber membranes, the thermal insulation remained low.

Ragaišienė et al. (2014) used a different method for integrating the nanofibers into the fabric structure. They inserted the nanofiber layer into the inner structure of the knitted fabric and tried to protect the nanofiber layer from damage. Roller electrospinning was used to produce 15% PA6 nanofibers. 21.5 gr/m^2 polypropylene nonwoven, cotton yarns and a knitted fabric used as deposition materials. After spinning, nanofibers-coated cotton yarns were used in a flat double bed weft knitting machine. Then these knittings were again covered with nanofibers. However, the SEM images revealed that the nanofiber coating was damaged and could not withstand knitting friction.

Serbezeanu et al. (2015) prepared hybrid membranes by electrospinning aromatic polyimide (PI) nanofibers onto Kevlar support for improved thermal performance and moisture comfort as heat and flame protective materials. Kevlar/PI membranes revealed a water vapor transmission between 797.2 and 863.4 g/m^2/day, high enough compared to non-wovens and typical woven work clothing fabrics. There were also no significant changes observed by increasing the electrospinning time because of the highly porous structure. The air permeability decreased with increasing PI web thickness relative to that of pure Kevlar. The hydrostatic pressure of the Kevlar/PI membranes was similar, 0.31 and 0.39 mH_2O, and higher than pure Kevlar. They also investigated the hybrid structure after annealing at 260°C. The air permeability was negligibly affected, the water vapor transmission remained the same, and a significant increase was observed for the water resistance.

Amini et al. (2016) fabricated PU and PU/PA 6.6 hybrid nanofiber membranes and examined them as waterproof and breathable clothing. The single nozzle and two opposite nozzle electrospinning methods were used and compared. For the two nozzle systems, both solutions were electrospun simultaneously. The air permeability of the hybrid samples was higher than that of the PU samples. In the case of water vapor transmission, PU samples were higher than hybrid samples because of the thicker nanofiber diameters with greater pores. Also, the PU membrane coating showed better resistance to acidic water permeation.

A similar two nozzle electrospinning set-up was also used by Gorji et al. (2018) for bi-functional PU and poly (2-acryloylamido-2-methylpropanesulfonic acid)-graphene oxide (PAMPS-GO) nanofibers. The hydrophobic nature of PU was at the outer face and the hydrophilic nature of PAMPS was at the inner face. Since the mechanical and dimensional stability of PAMPS was not satisfactory, they decided to improve it using PU and produced bi-functional three-layer nanofiber membranes. The average diameters of the PU and PAMPS nanofibers were 500 nm and 83 nm, respectively. With increasing the graphene oxide content, the water vapor permeability and the stability of PAMPS nanofibers increased; whereas they decreased by increasing the surface density of the nanofibers. During the water vapor transfer, PAMPS nanofibers absorbed the water, partially dissolved, and formed a barrier film across the membrane, which decreased the water vapor transmission. Therefore, it can be said that for better water vapor transmission and water impermeability, hydrophobic nanofibers seem to be better options.

Yang et al. (2016) produced hydrophobic PVDF nanofiber membranes that show fine water and windproof properties along with breathable characteristics. They integrated the surface hydrophobicity and interconnected porous structure, which was the result of tuning the DMAc/acetone percentages and the NaCl concentrations. With increasing DMAc/acetone ratio from 1/9 to 7/3, maximum pore size decreased from 3.4 to 1.6 μm and mean pore size decreased from 2.4 to 1.1 μm, hydrostatic pressure increased from 44.1 to 82.3 kPa and air permeability of the membranes changed from 24 to 8.1 mm/s, due to thinner nanofibers. Also, the water vapor transmission rate of the PVDF nanofiber membranes fell from 11.8 to 11.5 kg/m^2/day. Higher porosity provided more transmission channels and presented better breathability. The produced nanofiber membranes had a good mechanical performance with strength and toughness for possible utility in clothing applications.

Sometimes PUs seemed to be the best option for breathable nanofiber membrane applications, but waterproofness may not be enough. Thus, Zhao et al. (2017) synthesized fluorinated polyurethane containing short perfluorohexyl chains and introduced it as a hydrophobic agent into PU solution for nanofiber production. Magnesium chloride helped reduce the maximum pore size of the membranes. The prepared composite nanofiber membranes showed a hydrostatic pressure of 104 kPa, a water vapor transmission rate of 11.5 kg/m^2/day, and a tensile strength of 12.4 MPa.

Ghaffari et al. (2019) used sewing to integrate the polyacrylonitrile nanofiber membrane into the clothing assemblies. They explored the thermal insulation, waterproofness, and breathability of the nanofiber membrane combination of polyester fabric and a warp-knitted interlining by altering the porosity of the membrane to design a weightless cold weather protective clothing. The sewing and fusing were selected for joining the layers, and the effect was investigated as well as nanofibers diameter. Warp-knitted interlining was fused with nanofiber membrane at 100°C, under the pressure of 2 bar for 10 s. A single-needle lockstitch sewing machine was used to sew polyester fabrics to the nanofiber-bonded interlinings. The two-plate togmeter test method was used for thermal resistance according to BS 4745. In order to improve thermal insulation, the air gap thickness is increased, but if it is increased too much (usually ≥8 mm), heat transfer through the air layer increases as natural convection occurs. Therefore, a thin layer of nanofiber membrane can improve thermal insulation. Their results proved that the thermal insulation of the PAC nanofiber (concentration of 12%) bonded interlining as face fabric increased from 177% to 315% when a layer of polyester fabric was sewn for the third layer.

In another study, the authors (Oğlakcıoğlu et al., 2021) improved the wind-resistance and breathability properties of conventional knitted fabrics by using the TPU nanofiber coating. The functionality of cotton, viscose, modal, and lyocell single jersey fabrics was increased by nanofiber coating from the point of wind resistance characteristic. The TPU nanofibers were selected because of their high mechanical strength, elasticity, and durability. The diameters of the TPU fibers were in the range of 756 and 837 nm. A binder paste was applied to the surface of the jersey fabrics and relatively short deposition times of 10, 20, and 30 min were used. With increasing deposition time, the mean pore size of each jersey fabric decreased dramatically (84 µm to 7.7 µm for the modal fabric). For all fabric types, the pore size distribution was narrowed. The air permeability values were also significantly decreased to 287, 308, 371, and 380 l/m^2/s for cotton, viscose, modal, and lyocell, respectively, while the water vapor resistance did not change dramatically. They indicated that using a very little amount of nanofiber could be very effective in improving the windproof properties than changing the fabric structure, such as linear yarn density, fabric tightness, etc.

These are some of the studies investigating the use of nanofiber membranes in the design of breathable clothing. Their breathability and air/waterproofness are summarized. Within this scope, it is possible to use nanofibers instead of traditional membranes, but it is necessary to produce them in mass production and modify the existing equipment for the lamination processes.

Conclusion

Recently, there has been great interest in electrospun nanofibers as nanofiber membranes as an alternative to porous membranes used in protective clothing or outdoor applications. This is mainly attributed to their unique properties of small diameter and high surface area, which allow high breathability and high air resistance property, easy to be functionalized, and possibilities for specific applications. Neoshell from Polartec Inc, Nexture® from Lime (formerly Finetex), and Respilon from Respilon Group s.r.o are examples of commercial applications of nanofiber membranes. They all mass-produced nanofiber membranes, exhibit high breathability along with high air/water vapor permeability. Polartec Neoshell is an elastomeric polyurethane membrane that is used in commercial outerwear fabrics. It can tune airflow through the fabric without significantly affecting thermal insulation and water vapor transmission. These membranes may be used for either whole clothing, such as jackets or pants, or incorporated into certain areas of the garment for allowing increased ventilation (Gibson et al., 2014; Polartec, 2023). Similar to Neoshell, Nexture provides freshness through low resistance to evaporative transmission, high breathability, high air permeability, and water/windproof properties for many outdoor applications (Ghaffari et al., 2019; Yu et al., 2019, Limenano, 2023). Respilon has also developed a membrane to protect from external impacts for extreme conditions with its breathability and stretchability (Respilon, 2023).

There is no doubt in the future that this field will continue to advance toward the production of even more efficient membranes and efficient combinations with traditional textiles to improve their properties. For commercial applications, the strength and durability of laundering should be deeply investigated and improved. There are still many opportunities to evolve, including the combination of new active chemicals or biological structures into the nanofiber during spinning or by post-processing techniques (Nurfaizey et al., 2012).

Disclaimer

None.

References

Abbott, N. J., (2001a). Coated fabrics for apparel use: The problem of comfort. In D. Satas & Arthur A. Tracton (Eds.), *Coatings technology handbook* (2nd ed., pp. 825-829). Marcel Dekker, Inc.

Abbott, N. J., (2001b). Coated fabrics for protective clothing. In D. Satas & Arthur A. Tracton (Eds.), *Coatings technology handbook* (2nd ed., pp. 819-825). Marcel Dekker, Inc.

Ah Hong, K., Sook Yoo, H., & Kim, E. (2015) Effect of waterborne polyurethane coating on the durability and breathable waterproofing of electrospun nanofiber web-laminated fabrics. *Textile Research Journal, 85*(2), 160-170. https://doi.org/10.1177/0040517514542141.

Ahn, H. W., Park, C. H., & Chung, S. E. (2011). Waterproof and breathable properties of nanoweb applied clothing. *Textile Research Journal*, *81*(14), 1438-1447. https://doi.org/10.1177/0040517510392.

Akduman, C. (2021). Cellulose acetate and polyvinylidene fluoride nanofiber mats for N95 respirators. *Journal of Industrial Textiles, 50*(8), 1239-1261. https://doi.org/10.1177/1528083719858760.

Akduman, C., Demirel, V., & Tezcan, F. (2021). Filter life comparison of different levels of nanofiber coated cleanable-surface filter for gas turbine. *Journal of Applied Polymer Science, 138*(33), 50820. https://doi.org/10.1002/app.50820.

Akduman, C., Oğlakcıoğlu, N., Akcakoca Kumbasar, P. A., & Sarı, B. (2018). Investigation of thermal comfort properties of electrospun nanofiber mats. *Journal of Fashion Technology Textile Engineering, S4*(015). https://doi.org/10.4172/2329-9568.S4-015.

Akduman, Ç. (2019). PVDF electrospun nanofiber membranes for microfiltration: The effect of pore size and thickness on membrane performance. *European Journal of Science and Technology, 16*, 247-255. https://doi.org/10.31590/ejosat.556748.

Amini, G., Samiee, S., Gharehaghaji, A. A., & Hajiani, F. (2016). Fabrication of polyurethane and nylon 66 hybrid electrospun nanofiber layer for waterproof clothing applications. *Advances in Polymer Technology, 35*(4), 419-427. https://doi.org/10.1002/adv.21568.

Bagherzadeh, R., Latifi, M., Najar, S. S., Tehran, M. A., Gorji, M., & Kong, L. (2012). Transport properties of multi-layer fabric based on electrospun nanofiber mats as a breathable barrier textile material. *Textile Research Journal, 82*(1), 70-76. https://doi.org/10.1177/004051751142076.

Baji, A., Agarwal, K., & Oopath, S. V. (2020). Emerging developments in the use of electrospun fibers and membranes for protective clothing applications. *Polymers, 12*(2), 492. https://doi.org/10.3390/polym12020492.

Chee, W. K., Lim, H. N., Zainal, Z., Harrison, I., Huang, N. M., Andou, Y., Chong K. F., & Pandikumar, A. (2017). Electrospun nanofiber membranes as ultrathin flexible supercapacitors. *RSC Advances, 7*(20), 12033-12040. https://doi.org/10.1039/C7RA00406K.

Das, A., & Alagirusamy, R. (2011). Improving tactile comfort in fabrics and clothing. In G. Song (Ed.), *Improving comfort in clothing* (1st ed., pp. 216-244). Woodhead Publishing.

Dawes, V. H., & Owen, J. D. (1972). The handle and bending behaviour of fabric laminates. *Journal of the Textile Institute, 63*(8), 443-474. https://doi.org/10.1080/00405007208630369.

Dolez, P. I. (2013). Smart barrier membranes for protective clothing. In R. A. Chapman (Ed.), *Smart textiles for protection* (1st ed., pp. 148-189). Woodhead Publishing.

DSM. (2023, April 7). *Sympatex goes green with Arnitel® B-MB*. https://www.dsm.com/engineering-materials/en_US/insights/blog/sympatex-goes-green-with-arnitel-b-mb.html.

DuPont™. (2023, April 7). *Hytrel® handling and processing precautions.* https://www.distrupol.com/Hytrel_Handling_and_Processing_Precautions.pdf.

Exten. (2023, April 7). *Film and laminates, Symlex, waterproof breathable microembossed membranes.* https://panthersafety.it/en/technology/lining-and-membranes.

Faccini, M., Vaquero, C., & Amantia, D. J. J. O. N. (2012). Development of protective clothing against nanoparticle based on electrospun nanofibers. *Journal of Nanomaterials, 2012,* 1-9. https://doi.org/10.1155/2012/892894.

Fadil, F., Affandi, N. D. N., Misnon, M. I., Bonnia, N. N., Harun, A. M., & Alam, M. K. (2021). Review on electrospun nanofiber-applied products. *Polymers, 13*(13), 2087. https://doi.org/10.3390/polym13132087.

Ferrigno, T. H. (1988). Fundamental principles of formulating coatings for fabrics. *Journal of Coated Fabrics, 17*(4), 265-271. https://doi.org/10.1177/15280837880170.

Fung, W. (2002). Production Methods. In W. Fung (Ed.) *Coated and laminated textiles* (1st ed., pp. 83-148). Woodhead Publishing. https://doi.org/10.1533/9781855737518.83.

Ghaffari, S., Yousefzadeh, M., & Mousazadegan, F. (2019). Investigation of thermal comfort in nanofibrous three-layer fabric for cold weather protective clothing. *Polymer Engineering & Science, 59*(10), 2032-2040. https://doi.org/10.1002/pen.25203.

Gibson, P., Schreuder-Gibson, H., & Rivin, D. (2001). Transport properties of porous membranes based on electrospun nanofibers. *Colloids and Surfaces A: Physicochemical and Engineering Aspects, 187*, 469-481. https://doi.org/10.1016/S0927-7757(01)00616-1.

Gibson, P., Sieber, M., Bieszczad, J., Gagne, J., Fogg, D., & Fan, J. A. (2014). A design tool for clothing applications: Wind resistant fabric layers and permeable vents. *Journal of Textiles, 2014,* 1-7. https://doi.org/10.1155/2014/925320.

Gohlke, D. J., & Tanner, J. C. (1976). Gore-Tex® waterproof breathable laminates. *Journal of Coated Fabrics, 6*(1), 28-38. https://doi.org/10.1177/152808377600600104.

Gorji, M., Bagherzadeh, R., & Fashandi, H. (2017). Electrospun nanofibers in protective clothing. In Mehdi Afshari (Ed.), *Electrospun nanofibers* (1st ed., pp. 571-598). Woodhead Publishing.

Gorji, M., Karimi, M., & Nasheroahkam, S. (2018). Electrospun PU/P (AMPS-GO) nanofibrous membrane with dual-mode hydrophobic–hydrophilic properties for protective clothing applications. *Journal of Industrial Textiles, 47*(6), 1166-1184. https://doi.org/10.1177/1528083716682920.

Gottwald, L. (1996). Water vapor permeable PUR membranes for weatherproof laminates. *Journal of Coated Fabrics, 25*(3), 168-175. https://doi.org/10.1177/152808379602500302.

Gouda, M., Hebeish, A. A., & Aljaafari, A. I. (2017). New route for development of electromagnetic shielding based on cellulosic nanofibers. *Journal of Industrial Textiles, 46*(8), 1598-1615. https://doi.org/10.1177/1528083715627166.

Gugliuzza, A., & Drioli, E. (2013). A review on membrane engineering for innovation in wearable fabrics and protective textiles. *Journal of Membrane Science, 446*, 350-375. https://doi.org/10.1016/j.memsci.2013.07.014.

Gulbinienė, A., Jankauskaitė, V., Sacevičienė, V., & Mickus, K. V. (2007). Investigation of water vapour resorption/desorption of textile laminates. *Materials Science-Medziagotyra*, *13*(3), 255-261.

Han, X. J., Huang, Z. M., He, C. L., Liu, L., & Wu, Q. S. (2008). Coaxial electrospinning of PC (shell)/PU (core) composite nanofibers for textile application. *Polymer Composites, 29*(5), 579-584. https://doi.org/10.1002/pc.20194.

Harifi, T., & Montazer, M. (2017). Application of nanotechnology in sports clothing and flooring for enhanced sport activities, performance, efficiency and comfort: a review. *Journal of Industrial Textiles, 46*(5), 1147-1169. https://doi.org/10.1177/1528083715601512.

Heikkilä, P., Sipilä, A., Peltola, M., Harlin, A., & Taipale, A. (2007). Electrospun PA-66 coating on textile surfaces. *Textile Research Journal, 77*(11), 864-870. https://doi.org/10.1177/0040517507078241.

Heinisch, T., Bajzík, V., Knížek, R., & Gregušová, Z. (2013). Effect of the process of lamination microporous nanofiber membrane on the evaporative resistance of the two-layer laminate. *Advanced Materials Research*, *677*, 103–108. https://doi.org/10.4028/www.scientific.net/amr.677.103.

Hes, L., & Williams, J. (2011). Laboratory measurement of thermo-physiological comfort. In G. Song (Ed.), *Improving comfort in clothing* (1st ed., pp.114-137). Woodhead Publishing.

Holden, D. W. (1987) Coated and laminated fabrics-some future trends. *Journal of Coated Fabrics, 17*(1), 46-49. https://doi.org/10.1177/152808378701700106.

Holmes, D. A. (2000). Waterproof breathable fabrics. In A. R. Horrocks & S. C. Anand (Eds.), *Handbook of technical textiles* (1st ed., pp.12-282). Woodhead Publishing.

HSE. (2023, April 7). *Health and safety executive, thermal comfort.* https://www.hse.gov.uk/temperature/thermal/index.htm.

Hunter, L., & Fan, J. (2009). Waterproofing and breathability of fabrics and garments. In J. Fan & L. Hunter (Eds.), *Engineering apparel fabrics and garments* (1st ed., pp.283-308). Woodhead Publishing.

Hutten, I. M. (2007). Testing Nonwoven Filter Median. In I. M. Hutten (Ed.) *Handbook of nonwoven filter media* (1st ed., pp. 343-408). Elsevier. https://doi.org/10.1016/B978-0-08-098301-1.00006-X.

Jarrell, C. A (1992). New process for coating and laminating face-finished fabrics. *Journal of Coated Fabrics*, *21*(3), 212-221. https://doi.org/10.1177/152808379202100307.

Jassal, M., & Agrawal, A. K. (2010). Intelligent breathable coatings and laminates for textile applications. In W. C. Smith (Ed.), *Smart textile coatings and laminates* (1st ed., pp. 189-221). Woodhead Publishing.

Jones, F. N., Nichols, M. E., & Pappas, S. P. (2017). Polymerization and Film Formation. In, *Organic coatings: Science and technology* (4th ed., pp. 6-28). John Wiley & Sons.

Kadam, V., Truong, Y. B., Schutz, J., Kyratzis, I. L., Padhye, R., & Wang, L. (2021). Gelatin/β–Cyclodextrin Bio–Nanofibers as respiratory filter media for filtration of aerosols and volatile organic compounds at low air resistance. *Journal of Hazardous Materials, 403*, 123841. https://doi.org/10.1016/j.jhazmat.2020.123841.

Kang, Y. K., Park, C. H., Kim, J., & Kang, T. J. (2007). Application of electrospun polyurethane web to breathable water-proof fabrics. *Fibers and Polymers, 8*, 564-570. https://doi.org/10.1007/BF02875881.

Kannekens, A. (1994). Breathable coatings and laminates. *Journal of Coated Fabrics, 24*(1), 51-59. https://doi.org/10.1177/152808379402400105.

Kim, J. O. (1999). Dynamic moisture vapor transfer through textiles: Part III: Effect of film characteristics on microclimate moisture and temperature changes. *Textile Research Journal, 69*(3), 193-202. https://doi.org/10.1177/004051759906900306.

Knizek, R., Karhankova, D., & Fridrichova, L. (2015). Two and three layer lamination of nanofiber. *International Journal of Materials and Metallurgical Engineering, 9*(10), 1266-1269. https://doi.org/10.5281/zenodo.1129806.

Kramar, L. (1998). Recent and future trends for high performance fabrics providing breathability and waterproofness. *Journal of Coated Fabrics, 28*(2), 106-115. https://doi.org/10.1177/152808379802800203.

Kucukali-Ozturk, M., Ozden-Yenigun, E., Nergis, B., & Candan, C. (2017) Nanofiber-enhanced lightweight composite textiles for acoustic applications. *Journal of Industrial Textiles, 46*(7), 1498-1510. https://doi.org/10.1177/1528083715622242.

Lee, S., & Kay Obendorf, S. (2006). Developing protective textile materials as barriers to liquid penetration using melt-electrospinning. *Journal of Applied Polymer Science, 102*(4), 3430-3437. https://doi.org/10.1002/app.24258.

Lee, S., & Obendorf, S. K. (2007). Transport properties of layered fabric systems based on electrospun nanofibers. *Fibers and Polymers, 8*, 501-506. https://doi.org/10.1007/BF02875872.

Limenano (2023, April 7). *Technical textiles*. http://limenano.com/product/fiber.php.

Liu, L., Huang, Z. M., He, C. L., & Han, X. J. (2006). Mechanical performance of laminated composites incorporated with nanofibrous membranes. *Materials Science and Engineering: A, 435*, 309-317. https://doi.org/10.1016/j.msea.2006.07.064.

Lubrizol Performance Coatings. (2023, April 7). *Permax*, https://www.lubrizol.com/Coatings/Products/Permax-202.

Na, Y., Agnhage, T., & Cho, G. (2012). Sound absorption of multiple layers of nanofiber webs and the comparison of measuring methods for sound absorption coefficients. *Fibers and Polymers, 13*(10), 1348. https://doi.org/10.1007/s12221-012-1348-5.

Nurfaizey, A. H., Tucker, N., Stanger, J., & Staiger, M. P. (2012). Functional nanofibers in clothing for protection against chemical and biological hazards. In Q. Wei (Ed.), *Functional nanofibers and their applications* (1st ed., pp. 236-261). Woodhead Publishing.

Oğlakcıoğlu, N., Akduman, C., & Sarı, B. (2021). Investigation of thermal comfort properties of electrospun thermoplastic polyurethane fiber coated knitted fabrics for wind-resistant clothing. *Polymer Engineering & Science, 61*(3), 669-679. https://doi.org/10.1002/pen.25607.

Panther. (2023, April 7). *Linings and Membranes.* https://panthersafety.it/en/technology/lining-and-membranes.

Polartec®. (2023, April 7). *Power Shield®, bio-based weather protection.* https://www.polartec.com/fabrics/weather-protection/power-shield.

Porelle membranes. (2023, April 7). *Introduction to Porelle®.* https://www.porellemembranes.com/.

Ragaišienė, A., Rukuižienė, Ž., Mikučionienė, D., & Milašius, R. (2014). Insertion of electrospun nanofibres into the inner structure of textiles. *Fibres & Textiles in Eastern Europe, 6*(108), 59-62.

Respilon Ltd. (2023, April 7). *Nanofiber waterproof, windproof and breathable apparel membrane*. https://www.respilon.com/products/apparel-membranes/home/.

Serbezeanu, D., Popa, A. M., Stelzig, T., Sava, I., Rossi, R. M., & Fortunato, G. (2015). Preparation and characterization of thermally stable polyimide membranes by electrospinning for protective clothing applications. *Textile Research Journal, 85*(17), 1763-1775. https://doi.org/10.1177/0040517515576326.

Shim, E. (2010). Coating and laminating processes and techniques for textiles. In W. C. Smith (Ed.), *Smart textile coatings and laminates* (1st ed., pp. 11-45). Woodhead Publishing.

Shim, E. (2013). Bonding requirements in coating and laminating of textiles. In I. Jones and G. K. Stylios (Eds.), *Joining textiles* (1st ed., pp. 309-351). Woodhead Publishing.

Si, Y., Tang, X., Yu, J., & Ding, B. (2014). Electrospun nanofibers: solving global issues. In B. Ding & J. Yu (Eds.), *Electrospun nanofibers for energy and environmental applications*, (1st ed., pp. 3-38). Springer-Verlag Berlin Heidelberg.

Singha, K. (2012). A review on coating & lamination in textiles: processes and applications. *American Journal of Polymer Science*, 2(3), 39-49. https://doi.org/10.5923/j.ajps.20120203.04.

Smith, W. C. (2010). Overview of textile coating and lamination. In W. C. Smith (Ed.), *Smart textile coatings and laminates* (1st ed., pp. 3-9). Woodhead Publishing.

Solvay. (2023, April 4). https://www.solvay.om/en/brands/hyflon-perfluoropolymers.

Son, Y. J., Kim, W. J., & Yoo, H. S. (2014). Therapeutic applications of electrospun nanofibers for drug delivery systems. *Archives of Pharmacal Research, 37*, 69-78. https://doi:10.1007/s12272-013-0284-2.

Sumin, L., Kimura, D., Lee, K. H., Park, J. C., & Kim, I. S. (2010). The effect of laundering on the thermal and water transfer properties of mass-produced laminated nanofiber web for use in wear. *Textile Research Journal, 80*(2), 99-105. https://doi.org/10.1177/0040517508102308.

Sumin, L., Kimura, D., Yokoyama, A., Lee, K., Park, J. C., & Kim, I. (2009). The effects of laundering on the mechanical properties of mass-produced nanofiber web for use in wear. *Textile Research Journal*, *79*(12), 1085-1090. https://doi.org/10.1177/0040517508101622.

Sympatex. (2023, April 7). https://www.sympatex.com/en.

Turaga, U., Singh, V., Behrens, R., Korzeniewski, C., Jinka, S., Smith, E., Kendal R. J., & Ramkumar, S. (2014). Breathability of standalone poly (vinyl alcohol) nanofiber webs. *Industrial & Engineering Chemistry Research, 53*(17), 6951-6958. https://doi.org/10.1021/ie5005465.

Wang, X., & Hsiao, B. S. (2016). Electrospun nanofiber membranes. *Current Opinion in Chemical Engineering*, *12*, 62-81. https://doi.org/10.1016/j.coche.2016.03.001.

Wang, Y., Yokota, T., & Someya, T. (2021) Electrospun nanofiber-based soft electronics. *NPG Asia Materials, 13*(1), 22. https://doi.org/10.1038/s41427-020-00267-8.

Yang, F., Li, Y., Yu, X., Wu, G., Yin, X., Yu, J., & Ding, B. (2016). Hydrophobic polyvinylidene fluoride fibrous membranes with simultaneously water/windproof and breathable performance. *RSC Advances, 6*(90), 87820-87827. https://doi.org/10.1039/C6RA17565A.

Yoon, B., & Lee, S. (2011). Designing waterproof breathable materials based on electrospun nanofibers and assessing the performance characteristics. *Fibers and Polymers, 12*, 57-64. https://doi.org/10.1007/s12221-011-0057-9.

Yousefzadeh, M., Latifi, M., Teo, W. E., Amani-Tehran, M., & Ramakrishna, S. (2011). Producing continuous twisted yarn from well-aligned nanofibers by water vortex. *Polymer Engineering & Science*, *51*(2), 323-329. https://doi.org/10.1002/pen.21800.

Yu, X., Wu, X., Si, Y., Wang, X., Yu, J., & Ding, B. (2019). Waterproof and breathable electrospun nanofibrous membranes. *Macromolecular Rapid Communications, 40*(8), 1800931 (11-19). https://doi.org/10.1002/marc.201800931.

Zhao, J., Li, Y., Sheng, J., Wang, X., Liu, L., Yu, J., & Ding, B. (2017). Environmentally friendly and breathable fluorinated polyurethane fibrous membranes exhibiting robust waterproof performance. *ACS Applied Materials & Interfaces, 9*(34), 29302-29310. https://doi.org/10.1021/acsami.7b08885.

Chapter 5

Automotive Upholstery Fabrics, Production Technologies and Tests

Reyhan Ozcan Berber[1,2,*], PhD
and Semih Oylar[1,3], MD
[1]Martur Automotive Seating and Interiors, Bursa, Türkey
[2]Department of Textile Engineering, Ege University, İzmir, Türkey
[3]Department of Textile Engineering, Uludag University, Bursa, Türkey

Abstract

Technical textiles are defined as textile materials and products produced primarily for their technical and performance properties, rather than their aesthetic or decorative properties. One of the important sub-categories of technical textiles in the classification of usage areas is mobiletech. According to research, approximately 18% of textiles used in vehicles are automotive upholstery fabrics. The textiles used in the interior upholstery of the vehicles should have properties such as durability, dust resistance, and comfort. In general, all textile production processes are suitable for technical textile production. However, since the mechanical, chemical, and comfort properties of technical textiles are important, special production processes are also required. Automotive upholstery fabrics traditionally consist of combining three layers: upper surface fabric, lamination foam, and backing fabric. At the top surface of the 3-component structure; jacquard or dobby woven fabrics, woven velvet fabrics, warp knitted fabrics, raschel double needle bar knitted fabrics, and circular knitted fabrics called face fabric are used. PET-based air-texturized or friction-texturized yarns, which meet automotive standards such as high abrasion resistance, flame retardancy, and light fastness, are widely used in the production of face fabrics. Upper-surface fabric structures are produced to provide basic technical and aesthetic properties. After the face fabric is produced, washing is performed to remove impurities from the fabric and improve the hand. Then, the washed fabric is passed through the stenter machine, and the drying process is carried out by providing aspect stabilization, which is called the finishing process. Chemical finishing is done if necessary to meet the OEM-defined specifications. As an intermediate layer, polyurethane or polyester-based lamination foam is used between the upper surface fabric and the backing fabric. It is used to increase comfort feeling with its resilience property and to prevent wrinkling of the upper surface fabric. The backing fabric is produced using PET or PA-based yarn in warp or circular

* Corresponding Author's Email: reyhan.ozcan@marturfompak.com.

In: Textile Insights and Future Aspects
Editors: Gamze D. Tetik, Gizem Celep and Fulya Yilmaz
ISBN: 979-8-89113-289-4

knitted fabric construction. The main purpose of the backing layer is to reduce the frictional resistance that may occur from the foam-foam combination during the assembly of the seat cover to the seat foam. Various lamination methods can be used to bond the upper surface fabric, foam, and backing fabric layer. In this technique, flame is applied to both sides of the foam and the surfaces of the foam are melted to ensure adhesion. In the meantime, the upper surface fabric and the backing fabric are brought together with the foam, and adhesion is ensured. Thus, a three-layer composite structure is obtained. After this three-layered textile structure is produced, it is subjected to the cutting-sewing process and ready to be trimmed in the interior parts of the vehicle intended for use.

Keywords: mobiletech, automotive interiors fabric, foam, flame lamination, performance tests

Introduction

Technical textiles are defined as textile materials and products manufactured mainly for their technical and performance properties rather than their aesthetic or decorative properties.

When technical textile products are classified according to their application areas, they are generally divided into twelve groups as Mobiltech, Indutech, Medtech, Hometech, Clothtech, Agrotech, Buildtech, Sportech, Packtech, Geotech, Protech, and Oekotech. It may not be possible to separate textile products from each other with precise lines. Thus, some products may be included in several different classifications. One of the most significant categories in the classification of technical textiles according to their use areas is vehicle textiles.

The 2021 textile market research shows that vehicle textiles have a 16% market share in the global technical textile market. The main products in this application area are: seat belts, airbags, vehicle interior surface coating materials, automotive upholstery, cord fabrics, tires, carpets, curtains, hoses, belts, ropes, filters, battery separators, and composite structures. One of the important reasons for the automotive industry to prefer textile surfaces is that they are lightweight materials compared to alternatives. A vehicle with significantly reduced weight has the advantage of lower fuel consumption. Weight reduction is important to increase the range of electric vehicles, which have recently become more common, and to achieve relatively low greenhouse gas emissions for vehicles using petroleum-derived fuels.

Automotive upholstery has a market share of 54% in the global automotive textile market. The interiors of a vehicle can be categorized into 6 groups: cockpit, seats, doors and side coverings, headliner, trunk, and floor carpets (Dolle, 2013).

Approximately 20 kg of textile material are used in an automobile, a significant part of which is trunk/floor carpets (4.5 kg) and upholstery (3.5 kg) (Erth et al., 2005).

Automotive upholstery plays an important role in the first interaction during vehicle sales. One of the most effective instruments for the perception of the brand identity desired to be given with color, pattern, design, all of these is the upholstery that contacts the person most intensely. In addition, as it is not difficult to predict that autonomous vehicles will become widespread in the near future, it is thought that expectations from the upholstery, which is at the forefront for meeting the additional function needs that will arise with the freedom of the driver, will increase.

Automotive upholstery fabrics traditionally consist of a combination of three layers: upper surface fabric, lamination foam, and backing fabric, as shown in Figure 1.

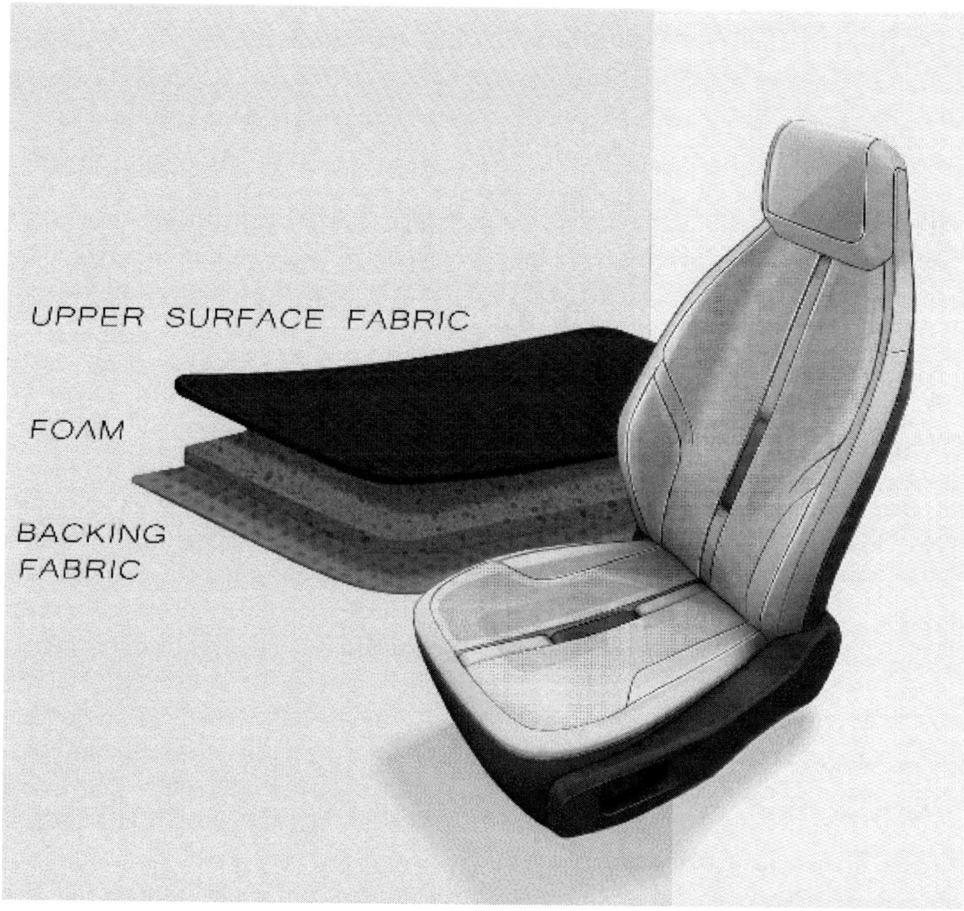

Figure 1. Automotive upholstery fabric components.

Technical Properties which Affect Design Parameters and Requirements

The automotive industry becomes the most demanding sector in the group of technical textiles. That demand brings new approaches to design automobile textiles. Regarding the utilization area of fabric as an interior, different properties are required. While the fabric meets the technical requirements, it should also follow the actual trends defined by the major automobile manufacturers (Akkus, 2019).

Because it is the most visible interior part inside of an automobile, seating fabrics are expected to meet those trends. While designing an automobile fabric, general aspects, color, structure, haptic, trimability, and cost are the main properties which affect the design phase (Powell, 2004). All those properties must be considered while selecting fabric type, material, yarn type, structure, and final treatments. Generally, the material is selected as PET, yarn type is air-texturized or friction-texturized. For any of the fabric types mentioned above, long yarn floats are not preferred because of abrasion resistance. Color is another parameter that affects fabric design. One of the most used colors for automobile interior fabrics is gray but to catch the general harmony of automobile, other effect yarns can be used to make the design richer.

Fibers and Yarns Used in Automotive Upholstery Fabric

Fibers

In order to meet the expectations of automobile upper surface fabrics, the yarns to be used should have a soft touch, high light fastness, and very high abrasion resistance (Gunay & Sezen, 2016). The abrasion properties of fabrics largely depend on the structure and type of yarn, whether it is textured or not, the fineness of the filaments, and most importantly the type of fiber (Pamuk, 2006; Fung & Hardcastle, 2001).

Automobile seat cover fabric structures are generally produced from polyester, wool, and wool/polyester blended, polyamide, acrylic, and leather or artificial leather structures (Akgun & Kamacı, 2021). Since the late 1990s, polyester has been used in 95% of seat covers in automobiles due to its high strength and modulus, high performance against abrasion, UV radiation, and heat, outstanding anti-aging performance, good shape retention, dimensional stability, and low cost (Fung & Hardcastle, 2001; Camlibel, 2018). Other properties that make polyester ideal are its high tear strength, mildew resistance, low moisture absorption making it easy to clean, excellent flexibility, and crease resistance. Another important reason for preference of pes fibers is their fast adaptation to texturing (FT/Friction texturing or ATY/Air-jet texturing) and dye finishing processes. However, its low moisture absorption feature is a disadvantage in terms of providing thermal comfort in hot weather (Pamuk, 2006; Fung & Hardcastle, 2001). Fabrics made from recycled PET such as automotive seat covers have become crucial given the benefits they provide in terms of waste recycling. Recycled PES achieves the high performance of virgin PES textile materials and can completely replace them in the vehicle when considering both knitted and plain-woven structures. Using recycled PES as automotive upholstery can reduce the environmental impact of car manufacturing. Ford is currently using REPREVE®, a 100% recycled polyester yarn made from both post-consumer and post-industrial waste, in five vehicles around the world, making it a truly global material (Cuc, 2017).

Another fiber used in the upper surface fabric of automobile upholstery is acrylic, which has high light and UV resistance. This fiber has a soft handle, but its abrasion resistance is not that good compared to other synthetic fibers (Pamuk, 2006; Fung & Hardcastle, 2001). Polypropylene is another fiber that can be an alternative. It is cheaper, more easily recyclable, and slightly lighter than polyester, but the disadvantages of this fiber are more than its advantages. While polypropylene cannot be commercially dyed, it only provides profit when the fiber is dyed, while other disadvantages of using polypropylene yarns in automobile seats are its low melting point and certain abrasion resistance. In addition, moisture absorption is lower than that of polyester (Pamuk, 2006). Although wool fibers can be used on the upper surface of automobile seat upholstery, they are used in high-level automobiles due to their high price. Wool is a hygroscopic fiber that absorbs water vapor and therefore provides better thermal comfort than polyester (Pamuk, 2006; Fung & Hardcastle, 2001).

Yarns

Fabrics are generally produced from bulked continuous filament (BCF) textured polyester yarns (Khalifa et al., 2016). High stretch with bulk, greater cover (opacity), greater breathability and absorption than regular filament yarn, more insulation, and a soft, and dryer hand. The fabric made from textured yarn is more wrinkle-resistant than regular yarn (Hada & Garg, 2015). The appearance and touch of polyester can be altered from smooth, lustrous, and flat to crimped, dull, and soft due to the thermoplastic nature of the polymer. False Twist (Friction Texturisation) and Air Jet method of texturization are preferred over other methods like stuffier box, gear crimping, etc. (Hada & Garg, 2015; Gunay & Sezen, 2013).

False Twist (Friction Texturisation) - (FT): Yarn type has bulky structure and elasticity (stretch-type ones) properties. It can be used for both knitted and woven fabrics. Additional processes such as intermingling, half texturization, and low twist can be applied to improve the strength performance of the yarn (Fung & Hardcastle, 2001).

Air jet texturized yarn (ATY-TASLAN): Structurally, friction is more rigid than textured yarn, its stretching properties are low, and it has an entangled structure due to the intermingled filaments. The use of ATY yarn in knitted fabrics is quite limited, and they are mainly used as weft and warp threads in woven fabrics (Gunay & Sezen, 2013).

Different types of yarn are used to highlight the different properties of the fabric. Polyester staple yarn, polyester monofilament, polyamide straight filament, polyamide/polyester blend, acrylic, and polyamide flocked yarn are used to give an effect to the fabric (Gunay & Sezen, 2013). For instance, FDY yarns are bright by nature and can be preferred in parts of the fabric that will not be exposed to abrasion. Another type of yarn is Chenille as a fancy yarn with beautiful, soft, and fuzzy surfaces (Khalifa et al., 2016).

The yarns commonly used on the upper surface of automobile upholstery and their usage rates in Western Europe are texturized (Intermingled, ASG, Plied Types) and air texturized yarn. Flat yarn, PES yarn from staple fiber and Twist, Fancy (effect yarns), Chenile, Flock acyrilic types, Elastane-covered yarns are rarely preferred when needed for design (Gunay & Sezen, 2013).

Automotive Upholstery Fabric Production

Upper Surface Fabric

The structure of seat covers has remained unchanged for more than 30 years (Avcu & Unal, 2018). There are three different components, connected by means of adhesive layers; the first one is the top layer material produced by weaving or knitting techniques, the second is the polyurethane (PUR) foam layer, which is followed by a backing layer made of polyester or polyamide (Shishoo, 2008).

Since the expectations of the fabric used on the upper surface and the fabric used on the backside are different, the yarn type and fabric structures used in these fabrics are different. While upper surface fabrics are produced with thicker yarns in higher weight and more special structures, lining fabrics are generally weft- or warp-knitted fabric structures produced from thinner yarns and low weight. The upper surface fabric structures that can be used in seat upholstery can be summarized as follows (Adanur, 1999; Hoeval, 1999).

- Woven: Plain Dobby and Jaquard Woven, Velvet Woven
- Circular Knitted; Plain Knit, Velvet Knit, Circular Knit Spacer
- Warp-Knitted: Tricot, Double needle bar raschel fabric, Warp knit Spacer
- Alcantara, Leather, Artificial Leather (Vinyl) (Adanur, 1999; Hoeval, 1999).

The main types of car seat fabric with typical weight ranges are flat woven fabric (200–400 g/m^2), flat woven velvet (360–450 g/m^2), warp knit tricot (generally pile surface,160–340 g/m^2), raschel double needle bar knitted (pile surface, 280–370 g/m^2), circular knits (generally pile surface, 160–230 g/m^2), Fabrics in nylon tend to be towards the lower weight range (Khalifa et al., 2016).

Woven Fabrics

They are the most widely used woven fabrics as upper surface fabric in automobile upholstery with a share of 51% all throughout the world. Woven fabrics have a very durable structure (Smith, 2013). Dobby plain woven fabrics have a share of 32%, while jacquard woven fabrics have a share of 18%. The usage rate of velvet fabrics is 0.45% (Shishoo, 2008).

Plain woven fabrics are especially used as a bolster for the seat cover. Jacquard-woven fabrics are used in the most striking part of the sofa due to its visual richness and in the flatter part due to its seating style. This area, called the main insert, is the sitting and back areas of the seat that come into contact with the person (Figure 2).

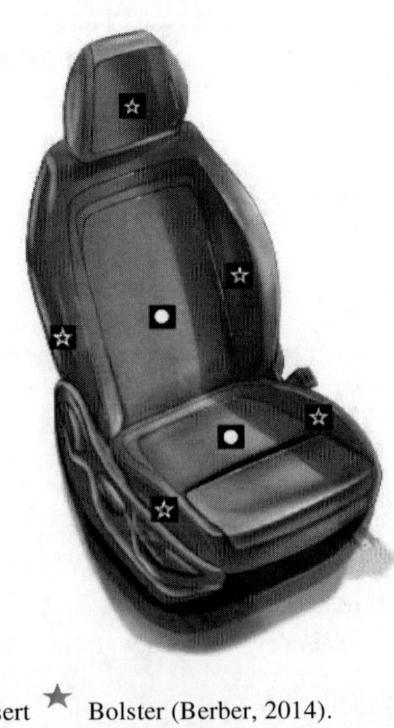

Figure 2. Seat Cover Parts ● Main Insert ★ Bolster (Berber, 2014).

Velvet woven fabrics are preferred mostly for bus seats. Velvet woven fabrics can also be used in some passenger cars launched as a special series due to their fuller and richer appearance.

Knitted Fabrics

Knitted fabrics are considered the growing trend in recent years in the automotive sector due to their soft touch, high elasticity, variety of patterns, and ease of production. An analysis of the fabrics used in the vehicles exhibited at the major automobile fairs in recent years reveals that knitted fabrics are used extensively in RSA group (Renault, Nissan, and Dacia), PSA group (Peugeot and Citroen) automobiles, and Fiat brand vehicles (Berber, 2014).

During the assembly of the cover, serious physical force is applied to fit the cover on the seat foam. Fabrics with very high dimensional stability make this process even more difficult. Knitted fabrics have an advantage in this situation due to their high elasticity and are preferred by many OEM's as the bolster part of the seat (Berber, 2014).

Warp knitting is preferred over weft knit fabrics for seat upholstery due to flatter, closer, less elastic structures such as Milanese, Raschel, simplex, and tricot.

Weft Knitted Fabrics: Circular knitted velvet fabric, one of the fabrics manufactured by weft knitting, can be preferred in all parts of the seat cover. Although plain fabrics have higher potential for use in bolster parts of the seat, they can also be used as main insert parts by enriching their appearance with various applications. Jacquard circular knitted fabrics are the most popular fabric type for the central parts of the seat cover.

Warp-Knitted Fabrics: The most popular warp-knitted fabrics used on the upper surface are plain and jacquard fabrics produced on tricot machines. The high production speed of tricot machines makes it possible to produce flat fabrics with low weight and high abrasion resistance at a low cost. Therefore, this type of flat fabric is widely used on the bolster parts of the seat. In addition, velvet fabrics produced on warp knitting special plush(pol) machines and double needle bar raschel machines are still used by some car manufacturers, although they have lost their popularity in recent years (Pamuk, 2006; Shishoo, 2008).

Spacer Fabrics: Spacer fabrics are produced on circular knit or double needle bar Raschel machines. Spacer fabrics have become an increasing trend in the automotive industry due to their sporty appearance due to their porous structure and the comfort they provide with their high air permeability (Smith, 2013). The spacer fabrics overcome the long-time problem of breathing through the seat. Multifilament yarn is used as background yarn and mono-filament yarn is used as spacer yarn (Hada & Garg, 2015). The space between the layers can vary between 1-15 mm, and according to the requirement, it provides lightness, volume, softness, high air permeability, moisture permeability, thermal control, and tear resistance in the usage area (Hada & Garg, 2015). Many studies have been carried out on its use as an alternative material to seat foam. Especially in luxury vehicles, high-distance fabrics can be used instead of foam on the seat (Smith, 2013; Hada & Garg, 2015). In addition, due to its high air permeability properties, it is used above the fans in seats with the ventilation system. Spacer fabrics are also preferred as upper surface fabric in cars such as Fiat, Skoda, Nissan, and Volkswagen (Berber, 2014).

Backing Fabric

As automobile upholstery fabrics are used generally with 3 components, backing fabric, also known as backing fabric, is selected from the fabric types of circular or warp knitting. Woven or non-woven backing fabrics are rarely used. Backing fabrics are not expected to have the same technical properties as surface (visual) fabrics. For this reason, these fabrics can be less heavy than surface fabrics (Horrocks & Anand, 2016). The material can be selected as PA or PES. Backing fabrics generally have a weight per meter square between 20-150 g/m². The main purpose of backing fabric usage is to make the trimming process easier by providing a sliding mechanism while trimming a seat cover to the backrest or cushion foam. In order to create a composite structure with balanced elongation values, when fabrics with high dimensional stability, such as woven fabrics are used as the upper surface fabric, backing fabrics with high elasticity produced in circular knitting machines are preferred. When fabrics with relatively high elongation values, such as knitted, are used as upper surface fabrics, warp knitted backing fabric structures are preferred in order to obtain a balanced final elongation at break (Berber, 2014).

Lamination Foam

Lamination foam ensures that the seams on the seam line are deep enough and that the occupants feel comfortable by covering the improper surfaces on the seat (Jerkovic et al., 2010).

It also prevents wrinkling and bagging of the top surface of the upholstery fabric and provides softness. With the addition of flame-retardant chemicals, it enables the fabrics to obtain flame retardant properties. Polyester polyurethane or polyether polyurethane-based materials are generally preferred as lamination foam (Fung & Hardcastle, 2001).

With increasing environmental regulations, the use of polyurethane foam in car seats has been questioning due to the drawbacks that have come to light in recent years, in particular the fact that as a result of the flame lamination process used, toxic fumes are produced from combustion during manufacturing (Shishoo, 2008; Avcu & Unal, 2018). Moreover, the low permeability (air and water vapor) properties of the seat fabrics due to the usage of PUR foam make the user uncomfortable. As a result, the replacement of PUR foam has been investigated with three-dimensional fabric structures (Avcu & Unal, 2018).

Final Treatments for Automotive Upholstery Fabrics

Generally, final treatments for automotive interior fabrics are the same for general synthetic fabrics. The process is as follows washing and stentering (Fung & Hardcastle, 2001).

Thumbling can be treated to develop the haptic of fabric just to have a soft touch. The washing must be applied to avoid impurities during the production process. Stentering provides the same tension to have the same properties for all over the fabric. Also, elongation for width, length, and bias direction can be fine-tunned during the stentering process.

Beside standard final treatments, for specific demand of OEMs, different chemical applications can be applied to fabrics. These chemicals can be applied by foulard process or direct coating methods. Additionally, rarely a foam coating application can be used for a specific usage area to have a liquid-proof property. Commonly used chemical treatments are shown in Table 1.

Table 1. Chemical treatments used in automotive fabrics

Chemical property	Application method
Anti-stain	Foulard
Abrasion resistance	Direct coating or Foulard
Flame retardant	Direct coating and/or Foulard
Waterproof	Direct coating and/or Foulard
Anti-bacterial	Foulard

Lamination Techniques for Automotive Upholstery Fabrics

The lamination process is used to produce a composite structure with 2 or more layers that will meet the required technical and aesthetic properties while preserving the properties of the face fabric (Bulut & Sular, 2008). Automotive upholstery fabrics are generally used by laminating with flexible and resilient materials. Upper surface fabrics laminated with high-resilience materials such as foam or nonwoven increase the comfort feeling of vehicle users (Koc et al., 2016). The flame lamination technique and the hot-melt lamination technique are widely used in automotive interior upholstery fabric production (Horrocks & Anand, 2016).

In the flame lamination technique, woven or knitted face fabric, polyurethane lamination foam, and backing fabric are fed into the flame lamination machine. The PU foam surface in the intermediate layer is melted by applying a flame. The melted PU foam acts as an adhesive, and the upper surface fabric, foam, and backing fabric are bonded together to produce a 3-layer composite structure. This lamination method is low cost and fast, and the technical specifications expected from the final product are successfully met (Cakmak et al., 2023). However, during the flame lamination process, toxic gases are released as the lamination foam burns and melts. In order to reduce the environmental impact of the flame lamination method, the use of the hot-melt lamination technique has recently expanded (Horrocks & Anand, 2016).

In the hot-melt lamination technique, hot-melt adhesive is applied between the layers to be laminated, then these materials, which are brought to the form of sandwiches, are fed into electrically heated cylinders at a temperature that will melt the adhesive material in the intermediate layer. Thermoset or thermoplastic adhesives can be used as adhesive material in hot-melt lamination. One-component, moisture-curing/cross-linking reactive polyurethanes (PURs) are generally used as thermoset adhesives. These adhesives do not become plastic again when reheated. The drum-melter is necessary auxiliary equipment in hot-melt lamination machines using thermoset adhesives. CoPA and coPES are generally used as thermoplastic hotmelt adhesive. Thermoplastic adhesives become plastic when heated and solidify when cooled. The process is completely reproducible. In order to work with such adhesives, an extruder is used in hot-melt lamination machines to apply the melted adhesive on the material. Adhesive material selection and hot-melt lamination machine type preference should be made according to the feature expected from the product. It is an environmentally friendly lamination method since no gas emission is released during the hot-melt lamination process (Glawe et al., 2003; Bulut & Sular, 2008).

Test Methods

Seat covers are expected to preserve their properties during the lifetime of the car. When vehicles are offered for sale as second-hand, it is thought that the damaged seats give the impression that the engine is also damaged. Therefore, the test standards applied to upholstery fabrics are quite heavy. Some of the tests applied to car upholstery are common to all textiles, whereas some are unique to technical textiles (Berber, 2014)

Weight: It is the weight of the unit fabric area.

Thickness: It is the distance between the front and back layer of the composite. Thickness test is important because different thicknesses of lamination foam are used in seat fabrics (Berber, 2014).

Breaking Strength: It is the maximum force applied to the fabric sample that results in breaking. It is applied to specimens cut in width, length, bias, and reverse bias direction (Figure 3) (Berber, 2014).

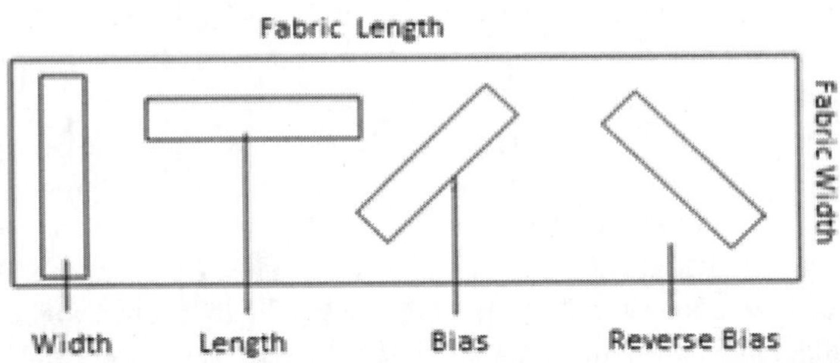

Figure 3. An example of a fabric sample.

Elongation at Break: The elongation at break value is the ratio of the length of the sample at the end of the tensile test to its original length under preload. It is applied to specimens cut in width, length, bias, and reverse bias direction (Berber, 2014).

Seam Strength: It is the strength of the seam against rupture as a result of a force applied vertically to the direction of stitching in sewn fabrics. Sewing should be done with the needle and thread used in reality. It is applied to specimens cut in width, length, bias, and reverse bias direction (Berber, 2014).

Tear Strenght: Under certain conditions, it is the force required to initiate, continue, and extend a tear (Berber, 2014).

Adhesion Strength: It is the resistance of the three-layer laminated structure of upholstery fabrics against separation from each other. This test is carried out to test the force required to separate both the upper surface fabric and the lamination foam and the lamination foam and the backing fabric. Depending on the request of the OEM, the test is performed on samples in the initial state, exposed to a certain humidity and temperature, and treated with cleaning chemicals (Berber, 2014).

Bursting Strength: Under certain conditions, it is the force required to penetrate the fabric by means of the force applied at a perpendicular angle to the fabric surface (Berber, 2014).

Elongation under constant load and permanent elongation: It is carried out to determine the elongation of fabrics under a constant load and the amount of this elongation to be restored in a certain period of time (Berber, 2014).

Abrasion Resistance: Abrasion is the deformation caused by the friction of one surface with another. The abrasion resistance is the resistance of a material to wear. Abrasion resistance testing is one of the most important tests for automobile upholstery, as the upholstery fabric is subjected to the friction of different materials that the occupants wear during the time spent in the vehicle. Abrasive materials are used to simulate the abrasion effect of various materials, and each of these is a different test method. Yarn breakage/pile and loss/color change are observed using abrasives such as fabric/sanding, stone/velcro, etc. The test methods using different abrasive materials are shown below. Each OEM develops its own test methods. Tests can also be carried out with different methods on different test devices using similar abrasive materials (Berber, 2014).

Taber: Abrasion test with cylindrical stone.

Martindale/Cesconi/MIE: Abrasion test with standard fabric and Velcro.

Stoll: Abrasion test with sandpaper.

Mace-Snag Test: This test is carried out with the help of an abrasive mace to determine the degree of snagging of the fabric in the width and length direction. This test simulates the damage to the surface caused by sharp objects (human-animal nails, screwdrivers, etc.). It is aimed to predict the damages that may occur in vehicles that are used in regions where pets are common, especially in Europe and the USA (Berber, 2014).

Trim Test: Depending on the form of the foam under the seat frame, the upholstery fabrics should be trimmable. It is a test to determine whether a wrinkle appearance will occur in upholstery fabrics. The test is based on wrapping the fabric around the hemisphere with different diameters from large to small and compressing them with a ring. The test starts in the hemisphere with the largest diameter. The fabric is expected not to wrinkle around the ring. The test continues by moving to the smaller hemisphere until the fabric is wrinkled (Figure 4). The hemisphere diameter at the point of fracture is the result of the test (Berber, 2014).

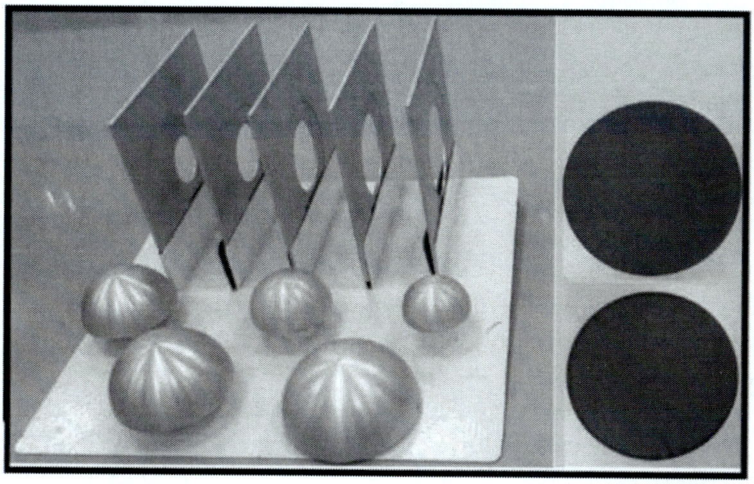

Figure 4. Trim test hemisphere, fit fabric appearance, wrinkled fabric appearance.

Combustion rate: It is the calculation of the combustion rate in mm/min after the test fabric is burned in a special cabinet for a certain period of time. It is one of the tests prioritized by all OEMs. The flame started on the sample in the combustion chamber is required to reach the determined point as slowly as possible (Berber, 2014).

Fogging: It is a test to measure the permanent fog layer caused by volatile compounds on the windshield of the vehicle (Berber, 2014).

Odor Test: It is carried out to determine the intensity and degree of the unpleasantness of the odor released by fabrics in dry and wet conditions when exposed to heat. The odor test can be performed by panelists who have been previously selected and whose odor perception has been tested (Berber, 2014).

Light Fastness: This test is performed to determine the degree of deterioration of the yarn color against the UV effect of light. Evaluation is carried out according to grayscale (Berber, 2014).

Heat Aging Test: Heat aging tests are carried out in order to check the dimensional stabilization of fabrics and to observe the level of deterioration, yellowing, stiffness, or discoloration in the general appearance of the fabric (Oylar, 2021).

Rubbing Resistance Test: It is made to determine the degree of staining of the standard fabric as a result of rubbing the upholstery fabric. The crock meter device is used for the test. The test is carried out under dry conditions and after the standard fabric has been exposed to an acid and alkali solution (Berber, 2014).

Entry-Exit Test: After the automotive interior fabrics are produced, a cut and sew operation is applied to produce the seat cover. One of the most important tests that is performed on a seat trimmed with a seat cover is the entry-exit test. This test method simulates the movements drivers or passengers as they enter and exit the automobile. This simulation is made by a robot donated with equipment that mimics the body of the occupant. Pressure and movement mapping is set up by software defined by the OEM. After the test is performed with a determined cycle, the final aspect is evaluated by authorized employees. No tear on fabric surfaces, no separation on sewing lines, and no design defects are one of the most important criteria while evaluating.

Sustainability Trends and Approaches for Automotive Upholstery Fabrics

Since plastic materials are light, inexpensive, easily processable, and durable materials, their use has increased in the last 100 years. However, these increases in petroleum-based plastic production lead to a waste management crisis. In 2015, it was determined that 302 million tons of plastic accumulation were generated globally and 42 million tons of this amount comes from petroleum-based textiles. Only 15% of textile-derived plastic waste is recycled. The recycling rate has not reached a sufficient level, and this rate needs to be increased. According to the research, it is predicted that the accumulation of plastic waste will reach 850 million tons by 2050. In case of not increasing recycling studies, plastic waste in the oceans and on land will continue to threaten the life of living things on earth by triggering global warming by causing environmental pollution and increasing carbon dioxide emissions (Tripathi et al., 2021; Manshoven et al., 2021).

Automotive textiles are also generally produced based on synthetic fibers. In order to reduce the contribution of automotive fabrics to plastic accumulation, the use of bioplastic, the use of recycled raw materials, the use of natural fibers should be increased and a monomaterial production philosophy should be adopted in order to easily implement the recycling process.

The sustainability goals of the OEMs include being carbon neutral by 2050 within the scope of all GHG emissions. In order to achieve this goal, the use of clean energy should also be increased and a zero-waste philosophy should be adopted.

As the use of recycled fabrics (and the use of any recyclable textile component) push up the final costs of the car, the consumer has to be interested in buying recycled/sustainable, and this is not easy to predict: there is a large interest in the North Americas, but not so much of a concern within Europe (Sinha et al., 2015).

Conclusion

Automotive upholstery fabrics and standard fabrics are not different in terms of production technology, but they are different in terms of technical expectations. To meet these expectations, materials and fabric structures are differentiated. With additional chemicals and processes, their strength, flame retardancy, and resistance to UV lights can be easily improved. However, expectations are changing. In the forthcoming period, the expectation of low weight in electric vehicles, additional functions in connected and autonomous vehicles, extra high strengths in shared vehicles, customization in luxury segment vehicles, and the hygiene expectation arising with the pandemic will lead automotive upholstery manufacturers to innovative applications in the future.

Disclaimer

None.

References

Adanur, S. (1999). *Wellington sears handbook of industrial textiles*. Technomic Publishing Company Inc.

Akgun, M., & Kamacı H. A. (2021). Effect of fabric structural parameters on various comfort properties of automobile seat cover fabrics. *European Mechanical Science*, *5*(3), 105-108. https://doi.org/10.26701/ems.797054.

Akkus, I. (2019). *Investigation of factors determining design and fabric design for automotive upholstery.* [Master dissertation]. Eskisehir Technical University.

Avcu, O., & Unal, P. G. (2018). Mechanical properties of double-layered woven fabrics used in car seat upholstery. *The Journal of The Textile Institute, 109*(11), 1409-1417. https://doi.org/10.1080/00405000.2018.1423901.

Berber, R. O. (2014). *Development of high performance fabrics for automotive upholstery*. [Doctoral dissertation]. Ege University.

Bulut, Y., & Sular, V. (2008). General properties and performance tests of fabrics produced by coating and lamination techniques. *Journal of Textile and Engineer*, *15*, 5-16.

Cakmak, S., Kaplangiray, M. B., Dal, M., Sarısac, S., & Mecit, D. (2023). Finite element analysis of the tensile strength of laminated fabric for automotive seat covers. *Textile Research Journal, 93*(1-2), 409-421. https://doi.org/10.1177/00405175221118838.

Camlıbel, N. O. (2018). *Polyester production, characterization and innovative applications*. IntechOpen. http://dx.doi.org/10.5772/intechopen.74422.

Cuc, S. (2017). Recycled textiles used in automotive interiors. Case study- Ford Motor Company. *Annals of the University of Oradea. Fascicle of Textiles, Leatherwork*, *18*(2). Retrieved May 5, 2023 from: http://textile.webhost.uoradea.ro/Annals/Vol%20XVIII-No%202-2017/Management/Art.nr.%20277-pag.%20171-176.pdf.

Dolle, J. E. (2013). *Supplier management in the automotive industry. Structure and development of supplier relationships of automotive manufacturers*. Springer Gabler.

Erth, H., Gulich, B., & Fuchs, H. (2005). Mobile textiles future trends to 2010. *Man Made Fibers Congress,* Dornbirn.

Fung, W., & Hardcastle, M. (2001). *Textiles in automotive engineering* (1st ed.). Cambridge: Woodhead Publishing Limited.

Glawe, A., Reuscher, R., Koppe, R., & Kolbusch, T. (2003). Hot-melt application for functional compounds on technical textiles. *Journal of Industrial Textiles, 33*(2), 85-92. https://doi.org/10.1177/152808303038587.

Gunay, M., & Sezen, M. (2016, September 16). *Polyester use in automotive textiles, effects of market changes on supplier textile companies*. Retrieved May 5, 2023 from: https://silo.tips/queue/otomotyv-tekstyllerynde-polyester-kullanimi-pazardaky-deyymleryn-tedaryky-teksty?&queue_id=-1&v=1683634357&u=MTk1LjE3NS40NS44Ng==.

Hada, J. S., & Garg, Y. (2015). Seat upholstery/ fabrics used in automobiles. *Man-made Textiles in India, 43*(2), 51-55.

Hoeval, B. T. (1999). New worldwide trends in automotive textiles. *Technical Textiles,* August, *42*(37).

Horrocks, A. R., & Anand, S. C. (2016). *Handbook of technical textiles* (2nd ed.). Woodhead Publishing Limited England.

Jerkovic, I., Pallares, J. M., & Capdevila, X. (2010). Study of the abrasion resistance in the upholstery of automobile seats. *AUTEX Research Journal*, *10*(1), 14-20.

Khalifa, T. F., Gawad, M. A., El-Homossani, M., & Hafez, H. (2016). Effect of some construction elements on the functional properties of car seat fabrics. *International Journal of Science and Research (IJSR), 5*(4), 674-684.

Koc, S. K., Mecit, D., Boyaci, B., Ornek, M., & Hockenberger, A. (2016). Effects of filament cross section on the performance of automotive upholstery fabrics. *Journal of Industrial Textiles, 46*(3), 756–770. https://doi.org/10.1177/1528083715598652.

Manshoven, S., Smeets, A., Arnold, M., & Mortensen, F. L. (2021, January 28). *Plastic in textiles: Potentials for circularity and reduced environmental and climate impacts*. https://www.eionet.europa.eu/etcs/etc-wmge/products/etc-wmge-reports/plastic-in-textiles-potentials-for-circularity-and-reduced-environmental-and-climate-impacts.

Oylar, S. (2021). *Design and development of fully recyclable automotive seat cover*. [Master dissertation]. Bursa Uludag University.

Pamuk, G. (2006). *The properties of fabrics used for automotive seat covers.* [Master dissertation]. Dokuz Eylul University.

Powell, N. B. (2004). Design driven the development of new materials of automotive interiours. *Journal of Textile and Apparel, Technology and Management*, *3*(4), 1-19.

Shishoo, R. (2008). *Textile advances in the automotive industry* (1st ed.). Woodhead Publishing in Textiles.

Sinha, P., Muthu, S. S., Taylor, J., Schulze, R., Beverley, K., Day, C., & Tipi, N. (2015). Systems thinking in designing automotive textiles. *Textiles and Clothing Sustainability, 1*(6), 1-13. https://doi.org/10.1186/s40689-015-0004-z.

Smith, W.C. (2013, September 1). *The advanced textiles knits niche, specialty fabrics review*. https://specialtyfabricsreview.com/2013/09/01/the-advanced-textiles-knits-niche/.

Tripathi, N., Misra, M., & Mohanty, A. K. (2021). *Durable polylactic acid (PLA)-based sustainable engineered blends and biocomposites: Recent developments, challenges, and opportunities*. ACS Engineering Au.

Chapter 6

A Sustainable Approach to Clothtech

Ayşe Şevkan Macit[1,2,*], PhD
[1]Department of Textile Engineering, Uşak University, Uşak, Türkey
[2]Leather, Textile and Ceramics Design Application and Research Center, Uşak University, Uşak, Türkey

Abstract

Technical textiles have an important place in the textile industry. Technical textiles, which affect many aspects of daily life, also include many strong products in terms of both functionality and design. On the other hand, sustainability is extremely important in technical textiles, as is the case in every field of textiles. Today, it is inevitable that a technical textile product that offers advantages in terms of functionality also offers advantages in terms of sustainability. In this chapter, clothtech, which is a segment of technical textiles, has been discussed, and the products in this field have been reported in detail. These products have been examined with a sustainable approach, and in this context, innovations developed in terms of sustainability in these products have been revealed.

Keywords: clothtech, sewing thread, interlining, footwear, sustainability

Introduction

Technical textiles have an important place in the textile industry. Technical textiles, which affect many aspects of daily life, also include many strong products in terms of both functionality and design. Technical textiles are divided into different segments according to the products they contain. Sporttech, buildtech, agrotech, medtech, hometech, protech, mobiltech, indutech, geotech, packtech, oekotech, and clothtech are these segments. Clothtech, one of the mentioned segments, is a technical textile group that includes technical parts of clothing and shoes, and involves interlining, wadding, sewing threads, footwears, and insulation materials (Mecit et al., 2007). Therefore, in terms of the product variety it contains, it is seen to appeal a wide range of uses.

---

* Corresponding Author's Email: ayse.sevkan@usak.edu.tr.

In: Textile Insights and Future Aspects
Editors: Gamze D. Tetik, Gizem Celep and Fulya Yilmaz
ISBN: 979-8-89113-289-4

On the other hand, sustainability is extremely important in technical textiles, as is the case in every field of textiles. Today, it is inevitable that a technical textile product that offers advantages in terms of functionality also offers advantages in terms of sustainability. Considering the efficient use of resources in the globalized world, the use of raw materials and process applications to contribute to sustainability has become a great necessity.

The interlining and wadding are important parts that affect the appearance of the garment. Production of interlining in a way that contributes to sustainability is of great importance, because of its frequent use in apparel. Further, the sewing threads in this group are prominent materials in clothing that are used to transform two-dimensional fabrics into three-dimensional garments. Sustainable sewing thread draws attention in terms of its diversity and wide usage area. Additionally, footwears also have an important place in clothtech. Insulation materials in clothtech are also important in terms of giving insulation properties to technical clothes and shoes. Today, the necessity of developing these products in terms of sustainability is an issue that needs to be taken into consideration.

In this chapter, clothtech, which is a segment of technical textiles, has been discussed, and the products in this field have been reported in detail. These products have been examined with a sustainable approach, and in this context, innovations developed in terms of sustainability in these products have been revealed. Literature studies and good practices highlighting sustainable products for efficient use of resources have been also included in this chapter.

Sewing Thread

Sewing threads are as important as fabric in a garment. Therefore, it can be said that sewing threads is the most important issue among clothtech product groups.

In the apparel industry, sewing thread is an important material used in converting two-dimensional fabric into three-dimensional clothing. The properties of sewing thread, which finds many uses outside the field of clothing, also affect the properties of the garment it is used in. A quality sewing thread should have balanced twist, flawlessness, regular lubrication, optimum elongation, sufficient tensile strength, low rate of wastage, abrasion resistance, and resistance to friction heat (Material Control, 2009; Özkurt Fıçıcıoğlu, 1996). In addition, the sewing thread should have seam strength, high elasticity, shrink resistance, minimum diameter, and color fastness properties that contribute to durability (Hunter & Cawood, 1979).

Sewing threads are produced from various raw materials by different methods. They can be obtained from natural, synthetic, and regenerated fibers (Doğan, 2012). Natural ones have varieties of cotton (soft, glazed, and mercerized), linen, and silk sewing threads. Synthetic ones are generally based on polyester and polyamide. Synthetic sewing threads come in the form of staple fiber, filament fiber, and corespun sewing threads. Regenerated ones are based on rubber or cellulose (Doğan, 2012).

Blended sewing threads are also obtained in different ways. Blended sewing threads can be obtained by spinning fibers together, folding, and twisting threads together, or wrapping natural or synthetic staple fibers around a thread (corespun thread) (Kaya, 2005).

There are different finishing processes applied to sewing threads. Some of these processes are lubrication finish, mercerized finish, glazed finish, water resistance finish, soil resistance finish, and flame resistance finish (Mandal, 2008). Sewing threads can be sized according to their linear density, or they can be sized according to ticket numbers (Mandal, 2008).

In addition to the performance characteristics of the sewing threads, their aesthetic appearance should also be appropriate. There are some factors that determine the aesthetic properties of the sewing threads. These can be given as color and color match, color fastness, brightness, fineness/thickness, suitable for stitch type, and stitch formation feature (Demir, 2017).

In addition, optimization of the fabric and sewing thread is important in order to obtain a smooth seam. The smoothness of the seam is an important parameter in determining the quality of the garment. In this respect, the selection of suitable sewing thread for the fabric is important in this regard. Characteristics of fabric and sewing thread influence the seam quality of the garment (Mandal & Abraham, 2010). Various studies have been conducted on the effects of sewing thread on sewing quality. Fabric and sewing thread compatibility is important in terms of seam appearance, and fabric and sewing thread optimization was examined in a study (Lin, 2004). It has been examined in some studies on the subject effect of sewing thread on seam strength (Akter & Khan, 2015), on seam pucker (Dobilaite & Juciene, 2006), on needle thread tension (Koncer et al., 2013), effect of sewing thread size on garment seam characteristic (Choudhary & Goel, 2013), and effect of organic cotton and polyester sewing thread on seam strength (Erdem et al., 2012). Different effects were observed in these studies. Mandal and Abraham (2010) reported sewing thread properties, such as tensile, friction, dimensional stability, and bending, are related to seam quality properties.

When looking at sewing threads in terms of sustainability, it is obvious that there is an awareness in this regard as in other materials of textiles. The use of organic fiber, recycled fiber, and biodegradable fiber is an approach that contributes to sustainability. The most sustainable thread is one made from natural, recycled, or organic materials. This type of thread is also environmentally friendly as it does not require as much energy as other types of yarn (Wayner Arthur Gallery, 2022). Many thread companies use recycled sewing threads. The use of recycled polyester sewing threads is quite common. In addition, instead of traditional cotton, companies obtain and use organic cotton, which is obtained with less water and pesticide-free agriculture. On the other hand, the use of biodegradable sewing threads contributes to sustainability in terms of being an eco-friendly material. Coats, one of the leading companies in sewing thread, has taken very serious steps in terms of sustainability. Coats is making a commitment to net zero emissions in 2050 (Coats Group plc Sustainability Report, 2022). The company, Coats, realizes an eco-friendly and sustainable process by using Lyocell thread and contributes to sustainability by using 100% recycled polyester (Coats, 2021). The use of Lyocell fiber as a sewing thread from the regeneration of wood fiber makes it a natural and biodegradable choice for sustainable raw material. The use of biodegradable nylon or organic cotton can also be given as similar choices (MIC spa, n.d.). In the sewing thread production, Coats has not only approached sustainability as thread production but also continued its environmental approach in other materials in the production process. According to the Coats sustainability report (2022), Coats used recycled plastic in cones and vicones instead of plastic sleeves in black and dark-colored thread cones and increased recycled plastic use by 36% in 2022 (Coats Group plc Sustainability Report, 2022).

During the sewing process, different wastes such as overlock waste with sewing thread occur (Güngör et al., 2009). Tons of waste are generated in a certain period of time in apparel companies. Overlock waste, sewing thread waste, and recycling are issues that should be considered.

On the other hand, there are two types of recycling processes in the form production waste and post-consumer waste of the garments (Türemen et al., 2019). In the post-consume recycling process of the garment, the sewing threads on the product make this process very expensive and time-consuming. Resortecs sewing thread has been developed for this issue. Resortecs sewing threads can be bio-based or synthetic that melt at different melting points (150 °C, 170 °C, or 190 °C) and they have been developed for the recycling post-consume wastes. Recycling resortecs sewing threads, which is an innovative technology, is very important because it saves time and cost. It is soluble in recycling post-consumer wastes (Recycling, Sorting, Technologies, n.d.). This sewing thread, which allows easy separation from the garment by melting at a certain temperature at the end of the useful life of the product, is used by large companies with well-known brands.

Interlining

Interlining is a kind of accessory material, and it is an important factor that affects the quality and performance of garments and influences consumer purchases (Zhang & Kan, 2018). Interlinings are produced with a suitable handle and weight for the weight of the fabric used. Although interlining is not visible in the products used, it undertakes important tasks such as giving form to the product, supporting the fabric, preventing weft and warp dispersion in small pieces, and maintaining the shape of the garment (Özipek, 2018). A good interlining should give the desired shape to the fabric, make the fabric look hard or soft depending on the purpose, and have similar characteristics as the fabric (Material Control, 2009).

The interlining has some performance problems. These problems are dimensional change during fixation, regional or complete dispersion, hardening of the interlining parts of the garment and decreasing of the surface drape, yellowing of the fabric, and undesirable changes in the interlining after washing (Material Control, 2009). In order to avoid undesirable situations in the use of interlining, it is necessary to mention the points to be considered in the selection of interlining. Interlinings must be suitable for the desired dry cleaning, washing, ironing, and drying values, must be suitable for the effect expected from the garment and usage area, must be suitable for the desired handle, must be suitable for the type, model, and color of the garment, and also the fabric and the interlining must give the same reactions to the external effects they will be exposed to after they are combined (Textile Surfaces, 2011). In particular, the suitability of the fabric and the interlining is an issue that needs attention. Lai (2001) researched this issue and conducted a study to provide the optimal combination of face fabric and interlining fabric.

Interlinings can be examined in two main groups, sew-in interlinings and fusible interlinings. While sew-in interlinings are applied to the fabric by stitching, fusible interlinings are bonded to the fabric with adhesive (Yıldız et al., 2011). The interlinings are in the structure of knitted, woven, or nonwoven (Zhang, 2018). On the other hand, there are three basic characteristics of interlinings, and these are fiber content, weight, and fabrication. These factors contribute to the aesthetics and performance properties of interlinings (Zhang, 2018).

Fusible interlinings are produced by coating various materials with a thermoplastic structure on the substrate chosen as the base fabric. A wide variety of thermoplastic adhesives, including polyester, polyamide, and polyethylene, are generally used as a bonding material for interlinings (Seyidzade, 2022). These materials are petroleum-based synthetic materials. The processes carried out with these materials are not environmentally friendly production. Seyidzade et al. (2023) conducted a study on this issue. The researchers investigated environmentally friendly biodegradable polymers instead of petroleum-based polymers in interlining production, and they determined the optimum interlining coating and application parameters with these materials. In another study that deals with interlinings in terms of sustainability, Telli (2014) investigated the probability of waste yarns in the hair interlining industry and the author intended to make contributions to the environment by bringing innovations for hair interlining production. It has also been reported that waste yarns can be used in the production of hair interlining.

Today, companies that have adopted sustainable interlining production have turned to natural or recycled materials to produce environmentally friendly interlinings.

Footwear

One of the clothtech product groups is footwear which covers the technical materials used in this group. Shoes have an important place in human life. The users of the shoes, whose main purpose is to protect the foot from external conditions, are in expectation of both performance and aesthetics. The shoes, which emerged for the purpose of protection from climate and surface conditions, have passed through many phases until today (Somçağ & Akçakale, 2016).

Shoes both protect the foot from the external environment and complement clothing (Sökmen & Balkanal, 2018). Shoes are important humans' wear. Different materials such as metal, textile, natural leather, artificial leather, and elastomer are used in the production of shoes, and shoes consist of two main parts: the upper part and the sole that comes into contact with the ground (Akçakale, 2016). According to the purpose of use, shoes can be classified as classic shoes, casual shoes, casual sports shoes, active sports shoes, special work shoes, and orthopedic shoes (Yüksel, 2019). Sports shoes take up a lot of space in the clothtech product group. Because there are too many technical parts to sports shoes. Sports shoes are used a lot in daily life apart from sports activities. Sports shoe designs used in daily life are symbols of the social, cultural, economic, and technological changes of the period, as well as reflecting the user identity (Enes, 2011).

Different textile materials used in sports shoes are required to have abrasion-resistant, sweat-absorbent, waterproof, and breathable properties in accordance with the part they form (Akçalı, 2016). The properties expected from the sole are durability, flexibility, water resistance, lightness, slip resistance, and homogeneity (Akçakale, 2016). On the other hand, looking at the applications of leading brands, it can be said that there are many different technologies and materials used in the uppers and soles of shoes. While applying these technologies, it is aimed to achieve performance in a way that creates features such as comfort, ground grip and balance, lightness, ventilation feature, breathing, waterproofing, windproofing, and water resistance.

With the emergence of environmental challenges and concerns, there has been an increase in the development of environmentally friendly footwear products in recent years. While large companies are creating new sustainable design initiatives, medium and small companies have emerged with the goal of producing more environmentally friendly shoes (Jacques et al., 2010). Footwear companies that will contribute to sustainability are turning to natural, organic, biodegradable, or recycled materials. Biodegradable shoes refer to shoes that use biodegradable materials. These materials include natural biodegradable polymers, synthetic biodegradable polymers, or biodegradable blends (Sustainable Materials and Components for Footwear, 2016).

There are studies on sustainable footwear production in the literature. Considering that it is an alternative to chemical processes in shoe production, the use of plasma technology has been studied (Ruzafa-Silvestre et al., 2020; Ruzafa-Silvestre et al., 2022). There is a study on the use of natural and synthetic rubber/waste Ethylene-Vinyl Acetate composites for sustainable production in the footwear industry (Lopes et al., 2015).

Natural rubber is obtained from latex in the rubber tree. These soles have very high abrasion resistance, and good thermal insulation properties, and are flexible. They were gradually replaced by vulcanized rubbers because of the price. Thermal insulation is a measure of the rate at which heat passes through a material or several layers of different materials. High thermal insulation means that heat moves very slowly through the material, and heat accumulates on the side of the heat source. Methods exist to evaluate the thermal insulation of shoes when placed on hot and cold surfaces, respectively (Sustainable Materials and Components for Footwear, 2016). In terms of shoe materials, –upper-lining, and sock sets are of great importance and are responsible for the overall thermal insulation property. The ratio between the thermal insulation of the upper shoe and the bottom of the shoe is another parameter that shows the qualitative criterion of warmth retention (Serweta et al., 2018).

According to the Leather Products and Footwear Industry Sustainability Report (2022), which highlights the vision of the sector in Türkiye, it is emphasized that exporters of the sector must meet their energy needs from renewable sources, increase the use of biodegradable recycled raw materials and biodegradable plant-based packaging, and create supply chains that will allow recycling (Leather Products and Footwear Industry, 2022).

The sustainability steps taken by footwear companies with leading brands in their fields are presented. Nike is a big brand in the footwear industry. The brand has a great deal of work in sustainability. Nike produces footwear designs for men, women, and kids and the company produces shoes marked with "Sustainable Materials" that contain at least 20% recycled materials by weight. Sustainable shoes are generally made from materials such as recycled rubber, recycled plastic, recycled PET, recycled polyester, and recycled plastic bottles. In addition, the company uses sustainably sourced natural materials such as organic cotton, natural rubber, or animal-friendly vegan leather to contribute to sustainability (Sustainable and Eco-Friendly Shoes for Every Occasion, 2022).

Environmentally friendly production principles of Nike; reducing carbon footprint, eliminating waste by creating new products, turning to processes that will reduce water pollution and the use of clean water, and using clean chemicals that have a low impact on the environment (Move to Zero, 2021).

With Nike Air technology, all Nike Air soles produced at Nike's Air Manufacturing Innovation facilities have been made from at least 25% recycled production waste since 2008. In 2020, all Air MI facilities are powered by 100% renewable wind energy. With Nike Flyknit technology, it is a specially developed lightweight fabric with an average of 60% less waste than traditional footwear upper production. Each footwear upper made of Flyknit contains 6-7 plastic bottles (Sustainable Materials + Innovation, n.d.). With Nike Grind technology, 58 thousand tons of materials have been recycled since 1992. Nike Flyleather is a combination of genuine leather and synthetic leather. The material, which resembles genuine leather with its properties, creates much less waste and less environmental impact than genuine leather (Move to Zero, 2021). Nike has also created a sustainable process in packaging. Nike One-Box, the box, arising from the need for a future-oriented design and an environmentally friendly box, is produced from recycled materials between 25% and 50% of the shoe weight (Başpınar, 2022).

With the slogan Move to Zero, Nike's 2025 sustainability goals are as follows: to reduce its greenhouse gas emissions by 0.5 million tons by increasing environmentally preferred materials to 50% of all key materials: polyester, cotton, leather, and rubber. The other is to recycle at least 80% of the waste into Nike products and other products, with 100% diverted from landfill in the extended supply chain. The other is to achieve a 25% reduction in the use of fresh water per kilogram in textile dyeing and finishing (Move to Zero, n.d.).

Considering another important footwear brand, Adidas, in terms of sustainability, the company states that 96% of the polyester they use in their products is recycled (New Personal Best, 2023). Adidas has created a "Three Loop Strategy" that will contribute to sustainability. Adidas partnered with Parley for the Oceans in 2015 and produced more than 30 million pairs of shoes with Parley Ocean Plastic (Loop 1). The idea of sustainability in Loop 2 is that the end of one thing is the beginning of the next and that products can be transformed into new products to extend their lifespan. Loop 3 involves making products with renewable resources and natural processes (The Three Loop Strategy, 2021).

The sustainability goals of Adidas are to replace all virgin polyester with recycled polyester wherever possible by 2024, to reduce its carbon footprint (CO_2e) by an average of 15% per product by 2025 compared to its baseline emissions in 2017, by 2025 achieve climate neutrality (CO_2e) in its retail and operations, reduce combined greenhouse gas emissions from both its own retail and operations and suppliers by 30% compared to 2017 by 2030, and become climate neutral (CO_2e) by 2050 (Our Mission, n.d.).

Another of the footwear brands, Reebok, has developed a product called "NPC UK Cotton+Corn" made of corn and cotton. In shoes with 75% bio-based content, the uppers are made of 100% cotton, the soles are corn-based, and the removable insoles are made from castor bean oil. Furthermore, the packaging is made from recycled materials (Bell, 2018).

An example of sustainable footwear is the Rens company, which mixes waste coffee with recycled plastic to produce shoes. It is stated that materials equivalent to 21 cups of coffee and 6 plastic bottles are used in each shoe. Shoes that prevent bad odors due to waste coffee are also breathable, lightweight, and waterproof shoes (Sezgin, 2021).

In the footwear industry, it is possible to contribute to sustainability in the production process, as well as with the waste of shoes that have completed their useful life. From this point of view, in the sector, research on recycling and recovery is increasing, and application areas are being created when the life of the products ends. It is of great importance that users are informed about environmentally friendly and sustainable production processes of shoes (Imre, 2021). The collection of post-consumer footwear, the separation of materials, and finally the recovery of the value of recyclable materials are among the key factors in achieving sustainable end-of-life product recovery in the footwear industry (Staikos & Rahimifard, 2007).

Wadding

The wadding is the part of the garment that is added internally to the shoulders in order to keep the shoulders upright and appear straight. In recent years, apart from the products such as coats, jackets, and wedding dresses, wadding has also been used for different purposes in many different garments. Wadding models are designed in accordance with the place of use and fashion trends (Özipek, 2018). Sponge, felt, or foam is used as raw material in the wadding. Today, shoulder pads that are cut from durable, washable, and ironable synthetic sponge sheets are common (Material Control, 2009). In other cases, the wadding can be shaped with fiber. Waddings are usually covered with the fabric of the garment in which they are used (Özkurt Fıçıcıoğlu, 1996). It is important to choose the proper size and thickness of the wadding. Otherwise, a taut arm problem will occur (Ünal & Acar, 2015).

The filling part of the wadding and the part covered by the wadding can be made of natural, biodegradable, or recycled material in order to contribute to sustainability.

Insulation Materials

Insulation materials, one of the clothtech product groups, is a group that contains technical products for footwear and clothing. Insulation in footwear has been mentioned in the footwear section. Insulation materials used in clothing are formed with fiber and yarn properties, fabric technologies, or filling structures. When looking at the fiber group, hollow fibers, core yarns, tencel, micromodal, and modified viscose fibers can be mentioned. As fabric technologies, fleece from polyester, wind, and water-resistant soft-shell structures, waterproof and breathable Gore-tex membrane can be mentioned. Apart from these, when it comes to fiberfill, the use of goose down, bead fiber, and micro gel fiber can be mentioned (Uluçınar, 2022).

Fiberfills are used in bedding, clothing, upholstery materials, and some special materials. Depending on the place of use, the properties expected from a good fiberfill are not always the same. The physiological properties of fiberfill that provide comfort in sleeping or wearing are heat insulation and air and moisture permeability. Although natural materials were used as fiberfills in the past, today the use of synthetic fibers, especially polyester fibers, has increased (Keser, 2016). When fiberfill structures are considered in terms of sustainability, Benetton produced a padding jacket by using recycled PET plastic bottles (Padded Jacket with Recycled Wadding, n. d.). There are many companies except Benetton that implement this. Companies are sensitive to sustainability, and they announce this sensitivity. David Jones produced a contemporary jacket with a water-repellent outer and light wadding made with 60% recycled fibers (Water-Repellent Blouson Jacket with Recycled Wadding, n. d.).

The recycled soft-shell jacket offered by Calvin Klein is an example of sustainability from the fabric technologies of the insulation materials (Recycled Soft Shell Jacket, n. d.). Many companies such as Calvin Klein apply sustainable production processes in this way.

Conclusion

Technical textiles have entered human life in many areas. Clothtech is one of the segments of the technical textile. Considering the product groups included in clothtech, it is included in many of the products used in daily life. Therefore, developments related to the subject are also of great importance.

In this chapter, a sustainable approach to clothtech product groups has been included. The issue of sustainability, which is very important globally, has revealed its importance in the field of technical textiles. It is seen that there are sustainable products in many areas of clothtech. It is obvious that these kinds of products will continue to increase, as this issue gains global importance. Looking at the subject headings, it is seen that there are similar approaches in terms of sustainability. On the basis of orientation towards sustainability, the use of natural fiber as a material, the use of biodegradable materials, and the use of recycled materials are quite common. In addition, it is necessary to determine production processes with more environmentally friendly approaches. It is of great importance to evaluate waste and raise awareness of consumers about this issue.

According to the report of the Technical Textile Fair held in Frankfurt in 2015 and attended by many clothtech companies, it can be stated that clothtech companies attach great importance to the concepts of 100% recyclable and optimal carbon footprint (Kavuşturan, 2015). With the awareness created by both producers and consumers about sustainability, it is inevitable that more sustainable processes and products will emerge in the coming years.

Disclaimer

None.

References

Akçakale, N. (2016). The use of rubber for shoe sole. *Technological Applied Sciences, 11*(3), 86-97. https://www.doi.org/10.12739/NWSA.2016.11.3.2A0096.

Akçalı, K. (2016). The investigation of the usage of technical textiles in sports. *International Journal of Sport Culture and Science, 4*(Special Issue 2), 533-546.

Akter, M., & Khan, M. R. (2015). The effect of stitch types and sewing thread types on seam strength for cotton apparel. *International Journal of Scientific & Engineering Research, 6*(7), 198-205.

Başpınar, S. (2022, July 19). Nike, sürdürülebilirlik hedefini kutudan çıkarıyor [Nike gets sustainability goal out of the box]. *MarketingTürkiye*. Retrieved May 5, 2023 from: https://www.marketingturkiye.com.tr/haberler/nike-one-box/.

Bell, J. (2018, August 17). *The Story Behind Reebok's New Sustainable Cotton + Corn Unisex Sneakers*. Footwearnews. Retrieved May 5, 2023 from: https://footwearnews.com/2018/focus/athletic-outdoor/reebok-npc-uk-cotton-corn-unisex-sneakers-1202664597/.

Choudhary, A. K., & Goel, A. (2013). Effect of some fabric and sewing conditions on apparel seam characteristics. *Journal of Textiles, 2013*. http://doi.org/10.1155/2013/157034.

Coats Group plc Sustainability Report 2022. (2022). *Sustainability report 2022*. https://assets.lsegissuerservices.com/original-5a713518-a3db-499a-bb1c-9e21378463cc.pdf.

Coats. (2021). *How is coats pioneering for a sustainable future*?. Retrieved May 5, 2023 from: https://www.coats.com/en/sustainability/case-studies/how-is-coats-pioneering-for-a-sustainable-future---interview-with-rajiv-sharma.

Demir, M. (2017). *An investigation on production possibilities of three-roving sewing threads* [Master dissertation]. Dokuz Eylül University.

Deri ürünleri ve ayakkabi sanayi [Leather products and footwear industry]. (2022, May 5). https://www.iso.org.tr/surdurulebilirlik/docs/print/AYAKKABI_DERI_URUNLERI_printpermit.pdf.

Dobilaite, V., & Juciene, M. (2006). The influence of mechanical properties of sewing threads on seam pucker. *International Journal of Clothing Science and Technology, 18*(5), 335-345. https://doi.org/10.1108/09556220610685276.

Doğan, S. (2012). *Calculating the amount of sewing thread consumption for woven fabrics with different thicknesses* [Master dissertation]. Ege University.

Enes, Ö. (2011). *Design of sneakers through technical and functional change aimed to daily use* [Doctoral dissertation]. Dokuz Eylül University.

Erdem, M., Demirbağ, A., & Özyazgan, V. (2012). Investigation of strength performances of two different sewing threads on denim fabric. *İstanbul Aydın University Journal of Social Sciences*, *4*(14), 23-46.

Güngör, A., Palamutçu, S., & İkiz, Y. (2009). Cotton textiles and the environment: Life cycle assessment of a bathrobe. *Textile & Apparel*, *19*(3), 197-205.

Hunter, L., & Cawood, M. P. (1979). *Textiles: Some technical information and data IV: Sewability, sewing needles, threads and seams.* South African Wool and Textile Research Institute of the CSIR.

Imre, H. M. (2021). Environmental effects of the shoe sector in terms of manufacturing processes, waste and recycling applications. *Aydın Adnan Menderes University Journal of Institute of Social Sciences*, *8*(2), 52-61.

Jacques, J. J., Agogino, A. M., & Guimara˜ es, L. B. (2010, January). Sustainable product development initiatives in the footwear industry based on the cradle to cradle concept. In *International Design Engineering Technical Conferences and Computers and Information in Engineering Conference: Vol. 44144* (pp. 473-481).

Kavuşturan, Y., (2015). Giyim teknik tekstilleri (Clothtech). In Technical Textiles 2015 Frankfurt Report Butekom Technological Evaluation Report, Bursa: Dekupe Yayıncılık, 2015, pp.26-38.

Kaya, G. (2005). *The effects of different sewing thread properties and stitch condensations on seam resistance with different sewing angels* [Master dissertation]. Marmara University.

Keser, F. (2016, November 17). *Filler fibers*. https://en.derstekstil.name.tr/dolgu-lifleri.html.

Koncer, P., Gürarda, A., Kaplangiray, B., & Kanik, M. (2013). The effects of sewing thread properties on the needle thread tension in an industrial sewing machine. *Textile and Apparel*, *24*(1), 118-123.

Lai, S. S. (2001). Optimal combinations of face and fusible interlining fabrics. *International Journal of Clothing Science and Technology*, *13*(5), 322-338. https://doi.org/10.1108/09556220110405073.

Lin, T. H. (2004). Construction of predictive model on fabric and sewing thread optimization. *Journal of Textile Engineering*, *50*(1), 6-11.

Lopes, D., Ferreira, M. J., Russo, R., & Dias, J. M. (2015). Natural and synthetic rubber/waste–Ethylene-Vinyl Acetate composites for sustainable application in the footwear industry. *Journal of Cleaner Production*, *92*, 230-236. https://doi.org/10.1016/j.jclepro.2014.12.063.

Malzeme kontrolü [Material control]. (2009). Retrieved May 5, 2023 from: http://megep.meb.gov.tr/mte_program_modul/moduller_pdf/Malzeme%20Kontrol%C3%BC.pdf.

Mandal, S. (2008). *Studies on seam quality with sewing thread size, stitch density and fabric properties*, [Master dissertation]. The Hong Kong Polytechnic University.

Mandal, S., & Abraham, N. (2010). An overview of sewing threads mechanical properties on seam quality. *Pakistan Textile Journal*, *1*, 40-43.

Mecit, D., Ilgaz, S., Duran, D., Başal, G., Gülümser, T., & Tarakçıoğlu, I. (2007). Technical textiles and applications (Part 2). *Textile and Apparel*, *17*(3), 154-160.

MIC spa, (n.d.). *Eco-friendly yarns sustainable products*. https://www.micspa.com/en/sustainability/244/eco-friendly-yarns.

Move to Zero. (2021). https://www.sportinn.com.tr/niketan-dunya-kaynaklarina-ve-gelecegine-destek-move-to-zero.

Move to Zero. (n.d). Retrieved May 5, 2023 from: https://www.nike.com/sustainability.

New Personal Best. (2023). https://news.adidas.com/planet/new-personal-best--adidas-announces-96--of-polyester-used-in-its-products-is-now-recycled-polyester-/s/b0a60ae7-92db-4300-825a-f033217c021c#:~:text=96%25%20of%20all%20polyester%20used,be%20achieved%20earlier%20than%20expected.

Our Mission. (n.d.). Retrieved May 5, 2023 from: https://www.adidas.com/us/sustainability/our-mission.

Özipek, B. (2018). Türkiye konfeksiyon yan sanayi sektörü: Vizyon 2030 [Turkish apparel sub-industry sector: Vision 2030]. *Sektörel Etütler ve Araştırmalar*, 2018-5.

Özkurt Fıçıcıoğlu, A. (1996). *A research on sub-industry products used in ready-made clothing products.* [Master dissertation]. Gazi University.

Padded Jacket with Recycled Wadding. (n. d.). https://be.benetton.com/padded-jacket-with-recycled-wadding-military-green-2BA2UN028_1Z9.html.

Recycled Soft Shell Jacket. (n. d.). https://ro.calvinklein.com/recycled-soft-shell-jacket-k10k111026prw.

Recycling, Sorting, Technologies, n.d., https://resortecs.com/about-us/.

Ruzafa-Silvestre, C., Carbonell-Blasco, M. P., Pérez-Limiñana, M. A., Arán-Ais, F., & Orgilés-Calpena, E. (2022). Robotised atmospheric plasma treatment to improve the adhesion of vulcanised and thermoplastic rubber materials for a more sustainable footwear. *International Journal of Adhesion and Adhesives*, *117*, 103010. https://doi.org/10.1016/j.ijadhadh.2021.103010.

Ruzafa-Silvestre, C., Carbonell-Blasco, P., Orgiles-Calpena, E. & Aran Ais, F. (2020, October). Low-pressure plasma treatment applied to polymeric materials for a sustainable footwear industry. In *Proceedings of the 8th International Conference on Advanced Materials and Systems, Bucharest, Romania* (pp. 1-3). https://doi.org/10.24264/icams-2020.IV.19.

Serweta, W., Matusiak, M., Olejniczak, Z., Jagiełło, J., & Wójcik, J. (2018). Proposal for the selection of materials for footwear to improve thermal insulation properties based on laboratory research. *Fibres & Textiles in Eastern Europe, 5*(131), 75-80. https://doi.org/10.5604/01.3001.0012.2535.

Seyidzade, L. (2022). *Development of adhesive interlining with biodegradable polymers and investigation of application parameters,* [Master dissertation]. Bursa Uludag University.

Seyidzade, L., Gurarda, A., Kanik, M., & Manasoglu, G. (2023). Development of fully biodegradable fusible interlinings for eco-friendly garments and investigation of their performance on a shirt. *Journal of Applied Polymer Science*, *140*(21), e53888. https://doi.org/10.1002/app.53888.

Sezgin, O. (2021). *Atık kahve çekirdekleri ve plastiklerden, su geçirmez spor ayakkabısı Rens Nomad [Rens Nomad waterproof sneakers made of waste coffee beans and plastics].* https://www.gercekbilim.com/atik-kahve-cekirdekleri-ve-plastiklerden-su-gecirmez-spor-ayakkabisi-rens-nomad/.

Sökmen, S., & Balkanal, Z. (2018). Traditional shoe-making in Bitlis: "Harik" and the last master Haydar Yılmaz. *Trakya University Journal of Social Science*, *20*(1), 383-400.

Somçağ, H., & Akçakale, N. (2016). Shoe making with the method of planking out of wool fiber. *International Journal of Euroasian Researches*, *4*(9), 212-226.

Staikos, T., & Rahimifard, S. (2007). Post-consumer waste management issues in the footwear industry. *Proceedings of the Institution of Mechanical Engineers, Part B: Journal of Engineering Manufacture*, *221*(2), 363-368. https://doi.org/10.1243/09544054JEM732SC.

Sustainable and Eco-Friendly Shoes for every Occasion. (2022). https://www.nike.com/a/sustainable-eco-friendly-shoes.

Sustainable Materials + Innovation. (n.d.). https://www.nike.com/si/sustainability/materials.

Sustainable Materials and Components for Footwear. (2016). https://research.cnr.ncsu.edu/ sustainablebio products/wp-content/uploads/sites/55/2020/08/Sustainable-Materials-and-Components-for-Footwear.pdf.

Tekstil yüzeyleri [Textile surfaces]. (2011). Retrieved May 5, 2023 from: http://megep.meb.gov.tr/ mte_program_modul/moduller_pdf/Tekstil%20Y%C3%BCzeyleri.pdf.

Telli, A. (2014). Utilization of waste yarns in hair interlining industry.*Journal of Natural Fibers*, *11*(4), 352-364. https://doi.org/10.1080/15440478.2014.880102.

The Three Loop Strategy. (2021). https://www.adidas.com/us/blog/643269-the-three-loop-strategy-adidas-plan-to-help-end-plastic-waste.

Türemen, M., Demir, A., & Özdoğan, E. (2019). Recycling and importance for textile industry. *Pamukkale University Journal of Engineering Sciences*, *25*(7), 805-809.

Uluçınar, E. (2022). *Development of military clothing used in cold climate regions using different methods and examination of their performance*, [Master dissertation]. Bursa Uludağ University.

Ünal, Z. B., & Acar, E. (2015). A research regarding the improvement of some operations that cause defects related to the suit production. *Tekstil: časopis za tekstilnu i odjevnu tehnologiju*, *64*(1-2), 37-41.

Water-Repellent Blouson Jacket with Recycled Wadding. (n.d.). https://www.davidjones.com/product/hugo-boss-water-repellent-blouson-jacket-with-recycled-wadding-24855323?nav=883718.

Wayner Arthur Gallery. (2022, November 20). *What thread is biodegradable?*. https://www.waynear thurgallery.com/what-thread-is-biodegradable/.

Yıldız, E. Z., Pamuk, O., & Öndoğan, Z. (2011). A study about the effects of interlinings to sewability properties of the woven fabrics. *Textile and Apparel*, *21*(1), 87-90.

Yüksel, N. (2019). *The use of textiles in heeled women's shoes from the 20th century to the present day,* [Master dissertation]. Marmara University.

Zhang, Q. (2018). *Potential of printable interlining in place of fusible interlining in garment manufacture,* [Doctoral dissertation]. The Hong Kong Polytechnic University.

Zhang, Q., & Kan, C. W. (2018). A review of fusible interlinings usage in garment manufacture. *Polymers*, *10*(11), 1230. http://dx.doi.org/10.3390/polym10111230.

Chapter 7

Waterless Dyeing in the Textile Industry

Necla Yaman Turan*, PhD

Department of Textile Engineering, Uşak University Engineering Faculty, Uşak, Türkey

Abstract

Due to the release of chemicals into the environment, environmental concerns reaching the highest levels, the necessity of reducing waste, and the desire to maximize efficiency bring more environmentally friendly approaches to the fore. In the textile dyeing sector, which causes intense freshwater consumption and wastewater problems, studies on waterless dyeing have increased in recent years. In terms of sustainability, it is important that both carbon dioxide and waste dyestuff can be recycled. After dyeing without the use of water and solvents, the drying step is also eliminated, thus reducing energy consumption. Supercritical carbon dioxide (Sc-CO_2) dyeing has the capacity to be a good alternative to conventional textile dyeing methods. In the case of successful commercialization of Sc-CO_2 dyeing, it will be possible to reduce water use, eliminate discharges from wastewater, and economize in dyeing processes. At the same time, it is foreseen that shortening the processing time, reducing the chemicals used, energy consumption, and air emissions will increase production. In this study, studies on the use of supercritical Sc-CO_2 in the textile industry, which started in the 1980s, and developments in its commercial application were evaluated. During the Sc-CO_2 dyeing process, the solubility and equilibrium distribution of the dyes, mass transfer, and physicochemical events occurring during the coloring process are the studied aspects of this method. Although dyeing with this method has generally been successful in synthetic materials, its use in dyeing with natural fibers has also become a separate focus. The plasticizing effect of CO_2 in glass and semi-crystalline polymers and, therefore, the change in the glassy region also causes changes in the morphology of semi-crystalline polymers such as textile fibers. Among the factors that are effective in dyeing with Sc-CO_2 are the ability to dissolve dyes belonging to the dispersed dye group and the ability of hydrophobic polymers to reduce the glass temperature (Tg). It is promising in terms of natural dyeing that the extraction of pure natural dyestuffs by using Sc-CO_2 in a shorter time and, at the same time, reduces the use of solvents. This study is an evaluation of environmentally friendly dyeings using Sc-CO_2 until 2022.

Keywords: eco-friendly dyeing, supercritic carbon dioxide, waterless dyeing

---

* Corresponding Author's Email: necla.yaman@usak.edu.tr.

In: Textile Insights and Future Aspects
Editors: Gamze D. Tetik, Gizem Celep and Fulya Yilmaz
ISBN: 979-8-89113-289-4

Introduction

Sustainability is an important concept in terms of water. The increase in the world's population, the change, and the increase in needs, on the other hand, climate change is reducing clean water resources. Sustainable designs should be made to reduce water consumption in order to prevent a water crisis that is likely to occur in the near future. However, the obligations in the legislation related to environmental protection, such as increasing water treatment and energy costs, reducing water consumption with new applications, and reusing water, have mandated the development of sustainable designs (Raja et al., 2019).

The textile industry is a sector that is the locomotive of the development of many countries in the world. Although the textile industry basically consists of fibers, yarn production, weaving-knitting, finishing, and ready-made clothing departments, it has become a sector in which more than 60 million people are directly employed, due to the progress in nonwovens, composites, technical textiles, medical textiles, and industrial textile materials. China, the United States of America, India, Pakistan, Brazil, Indonesia, Taiwan, Türkiye, Bangladesh, and South Korea are among the countries that play a major role in the world textile industry. According to studies, almost 2 tons of water are used for jeans production (Naik, 2013).

Reducing the amount of water used in this sector, where water consumption is intense, becomes the main target for studies. Therefore, different approaches have emerged to reduce water consumption (Patel et al., 2014). Notable technologies in this context are focused on Sc-CO_2, plasma, zero liquid discharge systems, and digital printing.

The Supercritical Fluids

The explanation of the phenomenon of supercritical liquids dates back to the 1800s (Berche et al., 2009). Supercritical fluids have the ability to disperse as a gas in a solid under certain temperatures and pressures (due to the lack of surface tension) and dissolve solids as a liquid. The fact that supercritical fluids have properties between gas and liquid seems promising for many applications due to the fact that these properties can be adjusted by pressure and temperature. In a closed container, when the density of gas and liquid reaches an equal value with an increase in pressure and heat, the boundary between the liquid and gas and a supercritical state appears (Fernandez-Cid et al., 2005).

Supercritical liquids show both gas and liquid properties at critical temperatures. For example, at a temperature of 304.13 K and a pressure of 73.773 bar, the density of Sc-CO_2 is the same, and the viscosity of steam is pleasing (Ilshin Autoclave, n.d.)

These properties ensure that the reduction of supercritical CO_2 is high while at the same time ensuring that it has low viscosity, face strain, and high heat communication.

Supercritical liquids have many uses, such as solvent, solvent, drying medium, solute, and fouling agent (Banchero et al., 2019; Promila & Madan, 2018; Soh & Lee, 2019; Esfandiari, 2015; Tabernero et al., 2012; Badens et al., 2018; Jung & Perrut, 2001; Bahrami & Ranjbarian, 2007).

It is possible to use more than one substance as a supercritical liquid. The properties of alternative substances are given in Table 1.

Table 1. Critical properties of various solvents (Ilshin Autoclave, n.d.)

Solvent	Molecular mass (g/mol)	Critical temperature (K)	Critical pressure (atm)	Critical density (g/cm^3)
Carbon dioxide	44.01	304.1	72.8	0.469
Water	18.015	647.096	217.755	0.322
Methane	16.04	190.4	45.4	0.162
Ethane	30.07	305.3	48.1	0.203
Propane	44.09	369.8	41.9	0.217
Ethylene	28.05	282.4	49.7	0.215
Propylene	42.08	364.9	45.4	0.232
Methanol	32.04	512.6	79.8	0.272
Ethanol	46.07	513.9	60.6	0.276
Acetone	58.08	508.1	46.4	0.278
Nitrous oxide	44.013	306.57	72.5	0.452
Comparison of gases, supercritical fluids, and liquids				
		Density (kg/m^3)	Viscosity (μPa.s)	Diffusivity (mm^2/s)
Gases		1	10	1-10
Supercritical fluids		100-1000	50-100	0.01-0.1
Liquids		1000	500-1000	0.001

The most important factor in determining the substance to be used as a supercritical solvent is the polarity of the substance to be used, and the temperature and pressure value required to pass the substance to the critical region plays an important role in the selection of a supercritical fluid (Bach et al., 2002; Phelps et al., 1996).

The most widely used of these supercritical liquids is CO_2 due to its ease of access in its critical conditions, non-toxicity, viscosity similar to gas, density similar to liquids, non-flammable, antimicrobial properties, recyclability, and relatively cheap (Soh & Lee, 2019; Banchero, 2013; 2020; Knez et al., 2014; Promila & Madan, 2018).

Since 1980, intensive studies on the solubility of supercritical liquids have been carried out (Phelps et al., 1996). Since Sc-CO_2 does not have a dissolving effect on polar compounds, its ability to dissolve polar compounds can also be increased by slightly increasing the polarity of Sc-CO_2 (addition of polar compounds such as acetone, and methanol by <5%). In this way, the polarizing ability of Sc-CO_2, which has no dipole moment and can be polarized in a very limited amount, increases (White & Lira, 1992; Yonker et al., 1986). In addition, the solute substance present in the aqueous system or the water content in its structure leads to acid-base interaction, increasing its ability to dissolve (Toews et al., 1995).

The Use of Water in the Textile Industry

The use of supercritical liquids has been the focus of attention, especially on the emergence of environmental concerns. The share of the total water consumption of the textile sector on a sectoral basis is shown to be approximately 2% (Kumar & Agarwal, 2011).

However, in the 2018 data of the World Water Assessment Program, it is stated that the water consumption of the textile industry has reached 10% (World Water Assessment Programme, 2018).

Supercritical liquids are not only a good alternative to water for the dissolution medium of dyestuffs, but also an alternative that provides high efficiency for the extraction of dyestuffs that are born (Borges et al., 2012). It is a good solvent for hydrophobic materials with increased vapor density (Phelps et al., 1996; Zosel, 1976; Laintz et al., 1991; Laintz, 1992; Reetz et al., 1993; Jessop et al., 1994).

Excessive water consumption in textile processes such as washing, bleaching, dyeing, and finishing, and the release of the used water is very important for the sustainability of water (Kant, 2012; Wu et al., 2016; Roy Chodhury, 2014). Water is used as the solvent, chemical, and dyestuff transfer in textile processes, as the medium for washing.

The amount of water used in textile processes is different (Arputharaj et al., 2016; Banchero, 2020; Zheng et al., 2016; Saka, 2019). Although the amount of water used in wool finishing may seem excessive, cotton is the fiber type that consumes the most water for the textile industry due to the worldwide consumption of cotton and the fact that cotton production is an agricultural process.

The processes with the highest water consumption in textile finishing processes are the pre-finishing and dyeing processes (Hussain & Wahab, 2018), the heating of the water to be used requires energy (8-18 MJ per/kg) (Gulzar et al., 2019) which constitute the justified requirements for the use of supercritical liquids.

The factor in the use of supercritical fluids in the textile dyeing industry is not only to reduce water consumption, but it also does not damage fibers, besides, auxiliary substances are not used, shorter dyeing time (due to viscosity decrease and diffusion ability increase due to pressure and heat effect), elimination of drying step, less energy consumption for reasons such as not using flotte, dyeing with classical dyeing depth can be done (Banchero, 2013; 2020; Bach et al., 2002; Kiron, 2012).

The Use of Supercritical Liquids in Textiles

The increase in use of water and wastewater in conventional dyeing of textile materials is increasing rapidly due to consumption and the increase in the world population, and this increase also increases environmental problems. For this reason, the use of supercritical liquids is just the beginning of studies aimed at developing new procedures that cause less water consumption.

Sc-CO_2 Dyeing

Due to changing climatic conditions, it is increasingly difficult for people to find a clean water source. The textile industry, especially the dyeing industry, is a sector with a lot of water consumption. The Sc-CO_2 method is of great importance for the sustainability of water.

Historical Overview of Dyeing with Supercritical Fluid

In 1986 (Sand, 1986) and 1987 (Beres et al., 1987), on the use of Sc-CO_2 in the production of thermoplastic polymers containing ingredients such as fragrance and dyestuffs, two patents have been obtained. Although the patents are primarily on the application of the desired active substance or dyestuff to PVC and rubber with Sc-CO_2, it is stated that it can be used for PET, polyamide, polyacrylonitrile, polyurethane, and polyolefins (Beres et al., 1987). In the first patent, the dyeing of synthetic textile material with supercritical fluid with an added polar modifier such as water, alcohol, and salt was studied (Schollmeyer et al., 1990; Schollmeyer & Knittel, 1993; Knittel & Schollmeyer, 1995b; Knittel et al., 1993b).

In the master's thesis, a laboratory scale dyeing device was prepared. Upon the successful conclusion of PET dyeing, studies have started on the commercial production of the machine (Polulakis et al., 1991; Knittel et al., 1993b; Knittel & Schollmeyer, 1995a). The first dyeing machine using supercritical fluid Jasper GmbH & Co is a 4-coil dyeing machine designed by Velen (Knittel et al., 1993a; Knittel & Schollmeyer, 1995a). A pilot plant for yarn dyeing with a capacity of 30 lt was established in 1995. In addition to dyeing, the removal of excess paints and oils and the recycling of CO_2 have also been provided at the facility (Bach et al., 2002).

Principle of Sc-CO_2 Dyeing and Its Machine

Since CO_2 has values close to normal atmospheric temperature, it is a supercritical fluid that causes little damage to fibers.

The fabric is wrapped in a cylinder with evenly spaced holes and placed in a closed container. Since the movement of fabric and the movement of machine parts are very limited, they can be used even for delicate fabrics (Reji et al., 2019).

The flow characteristics of the liquid in the machine are determined by a main pipe, a series of pipe clusters that are thinner and arranged in an angular/axial direction. On the other hand, the small holes on the cylinder are placed in more than one row. Although smooth distribution is provided by using a large-diameter main pipe, its use remains limited in practice. The amount of supercritical liquid passing through the pipes is determined by the static pressure, which varies depending on the accumulation of liquid in the main pipe, the surface friction of the liquid, and the decrease in liquid momentum (Reji et al., 2019).

The disadvantage associated with Sc-CO_2 dyeing is uneven dyeing in the radial/axial directions. In order to eliminate this limitation, it has been stated that a bracelet assembly on the cylinder surface and side covers will be suitable (Reji et al., 2019; Karst et al., 2004). With the simulation study, a more uniform dyeing was obtained when the direction of the liquid flow was applied from inside to outside and vice versa (Reji et al., 2019; Scharf et al., 2002; Banchero, 2020).

The dyeing liquid dissolved in Sc-CO_2 in a high-pressure environment is pumped into a perforated cylinder (Coelho et al., 2011). Sc-CO_2, which has a high diffusion ability containing pressure-transmitted dye, spreads easily in the fabric when passing through the cylinder holes and allows dyestuff molecules to enter the fiber. With pressure reduction, Sc-CO_2 passes into the vapor phase, loses its solubility, and leaves the dyestuff on it in the fiber, and moves away from the fabric in the gas phase. A strong interaction occurs between the dyestuff trapped in the fiber, especially in the free volume zones, in this way a fast and effective dyeing takes place (Reji et al., 2019). The proper distribution of the dyestuff depends on the smooth flow of Sc-CO_2, the properties of the dyeing liquid, the fabric properties, and the geometric design of the dyeing machine. However, smoothness depends on the diffusion ability, viscosity, and density of Sc-CO_2 rather than the machine design (Reji et al., 2019; Raman & Kim, 2018).

The causes of uneven dyeing can be considered as uneven distribution of supercritical fluid, ambient pressure, and temperature (Banchero, 2013; Long et al., 2011), unresolved dyestuff (Banchero, 2013), incorrect circulation rate (Banchero, 2013; Banchero & Ferri, 2005), too rapid lowering of pressure and rapid removal of CO_2 (Banchero, 2013; Okubayashi et al., 2011). Since the presence of undissolved dyestuff molecules and aggregates in the dyeing flotte causes stains on the fabric, it is necessary to take measures to ensure molecular dissolution of the dyestuff in machine construction (Banchero, 2013). A simulation conducted found that uneven dyeing increases with increasing liquid flow rate, low inlet temperature, and high inlet pressure positively affect the dyeing smoothness due to the flow distribution, dyeing smoothness improves with the bidirectional flow (Reji et al., 2019; Banchero, 2013; Banchero & Ferri, 2005).

Dyeing of Synthetic Fibers

While the first studies on the dyeing of polyester fibers date back to the 1980s (Banchero, 2013; 2020; Bach et al., 2002), the main reason why Sc-CO_2 is widely investigated is that the solubility of water-insoluble disperse dyestuffs in Sc-CO_2 is much higher. At the same time, small CO_2 molecules easily enter the free volume in the amorphous region, increasing the free volume, increasing the mobility of macromolecules (plasticizing), reducing the T_g (20-30ºC), and increasing the diffusion of dyestuff (Banchero, 2013; 2020; Park et al., 2010; Schnitzler & Eggers, 1999; Bach et al., 2001). By reducing the pressure after dyeing, the dyestuff molecule is trapped in the fiber, and dyeing takes place.

Laboratory studies and simulations help to explain the Sc-CO_2 dyeing process (Banchero, 2020; Gao et al., 2015; Reji et al., 2019). The process takes place in the form of condensation of CO_2, storage in a tank and increasing pressure, dissolving the dyestuff in Sc-CO_2, and sending it to the machine after reaching the appropriate pressure. At the end of the process, Sc-CO_2 is completed by recycling the dyestuff and sending it to the separation system to recycle the CO_2. It is stated that up to 95% of the CO_2 (Banchero, 2020; Zheng et al., 2016; Yang et al., 2014; Huang et al., 2017) and 40% of the dyestuff used (Huang et al., 2019) have been recycled within the scope of the studies. In studies carried out at 80-140ºC temperature and 17-29 MPa pressure, both single color and mixture dyes PET dyeing were performed at high K/S values and good fastness (Banchero, 2020; Zheng et al., 2015; 2016; Huang et al., 2013; 2017; 2019; Long et al., 2014; Yang et al., 2014; Gao et al., 2015; Tušek et al., 2000).

Since the investment cost of Sc-CO_2 dyeing is high, the availability of the method depends on the low cost of dyeing. Although it is generally accepted that the operating cost is lower due to the simpler prescription operation, shorter dyeing times, and the elimination of the drying step (Banchero, 2013; 2020; Bach et al., 2002; Kraan, 2005).

The effect of temperature, pressure, dyeing time, and dye/fiber ratio on dyeing depth was studied using CI Disperse Orange dyestuff and it was determined that all factors have a positive effect on color depth, while only temperature has a positive effect on washing fastness, and it should be dyed at a certain critical temperature (Banchero, 2020; Cardozo-Filho et al., 2014; Kong et al., 2019). When dispersed dyestuffs are made in the form of combinations, a lower color efficiency is obtained in combination dyeing. While the dissolution rate of the mixture in dyestuffs is 8-15% lower (Banchero, 2020; Dong et al., 2010; Tamura & Shinoda, 2004), satisfactory fastness values have been achieved (Banchero, 2020; Abate et al., 2019). In a study conducted, it was reported that high color efficiency, high dyestuff consumption, and good fastness values were obtained in Sc-CO_2 dyeing when dyeing with both Sc-CO_2 and classical HT dye with the combined use of turmeric and curcuminoid dyes (Banchero, 2020). Instead of changing the dyestuff ratios in mixed dyes, dyestuffs can be made in different shades by changing the pressure and temperature (Hussian & Wahab, 2018).

The reusability of the recycled solid dyestuffs after Sc-CO_2 dyeing has been studied and a gradual change in the crystal structure of the dyestuff has occurred at temperatures above 120ºC, but no significant changes in color darkness and color fastness have occurred in the second dyeing (Cardozo-Filho et al., 2014; Kong et al., 2019; Zheng et al., 2018).

In the simulation studies conducted, it was determined that there is a positive effect between the deceleration of the supercritical mass flow and homogeneity (Reji et al., 2019). This is also supported by an increase in the Sc-CO2 flow rate and an improvement in the uniformity of yarn dyeing (Banchero & Ferri, 2005).

Removal of dispersed dyestuffs by fiber depending on pressure and temperature (Working conditions: 60-150ºC, 100-350 bar), the amount of dyestuff removed by Sc-CO_2 dyeing is between 3-10% by weight and has a linear relationship with increasing temperature (Banchero, 2013; Bach et al., 2002; Van der Kraan et al., 2007a; Bao & Dai, 2005; Park et al., 2010; Park & Bae, 2002; Ferri et al., 2006). The effect of pressure and temperature on dyeing uptake can be explained by a quasi-dual-mode (Park et al., 2010) and an expanded liquid model (Park & Bae, 2002). A decrease in the dye-fiber affinity occurs in Sc-CO_2 dyeing, which is an exothermic reaction with an increase in pressure (Van der Kraan et al., 2007b; Ferri et al., 2006), as there is an increase in the solubility of the dye and the tendency to switch to flotte, since there is an increase in the dye-fiber affinity (Ferri et al., 2006). The addition of a polar auxiliary solvent causes a decrease in dyestuff-fiber affinity due to an increase in CO_2 sorption and swelling of the polymer (Banchero et al., 2009; 2010; Hirogaki et al., 2005; Ellington et al., 1994).

The conditions for dyeing polyamide, polypropylene (PP), acrylic, and meta-aramid fibers using Sc-CO_2 with new group dyestuffs have been summarized by Banchero (Banchero, 2013). Although studies with polyamide fibers are very limited, in the first studies, a lower K/S value was obtained in polar polyamide dyeing with high crystallinity (Liao et al., 2000a, 2000b; Van der Kraan et al., 2007b; Bach et al., 2002). In order to obtain higher K/S and fastness values in polyamide fibers, it is necessary to dye with dispersed reactive dyestuffs (Park et al., 2010; Elmaaty et al., 2015). During the dyeing of polyamide at a pressure of 100-120ºC and 230-245 bar with vinylsulfone azo-dyestuff, a bond is formed through the terminal amine group of polyamide (Liao, 2005; Bach et al., 2002; Van der Kraan et al, 2007a). Recent studies have shown that polyamide can be dyed with newly synthesized dispersed dyestuffs without the need for the addition of water (Zheng et al., 2015, 2016; Cardozo-Filho et al., 2014; Elmaaty et al., 2015, 2018, 2019).

The dyeing of aromatic polyamide (meta-aramid) fibers is quite difficult due to their high T_g value. Good color efficiency is obtained when is performed disperse dyeing in Sc-CO_2+CYNDYE DNK environment (Zheng et al., 2015; 2017; Kim et al., 2006).

Reverse micellar system due to the plasticizing effect of Sc-CO_2 in dyeing acrylic fibers with basic dyestuffs; darker dyestuffs can be made at lower temperatures and without the addition of auxiliaries (Jun et al., 2004).

The method of dyeing with Sc-CO_2 is considered a suitable dyeing method for PP fibers that cannot be dyed using hydrophobic and classical methods (Miyazaki et al., 2012a; 2012b).

The dyeing depth of the fibers improves as the number of carbon atoms in the alkyl atom in the dyestuff molecule increases, and if it has an excessively long chain, the color yield decreases because it cannot enter the free volume zones (Miyazaki et al., 2012b). Although PP does not have high fastness values with Sc-CO_2 disperse dyestuffs, dyeing with higher K/S values has been performed compared to those performed under HT conditions (Banchero, 2013; Miyazaki et al., 2012a; Elmaaty et al., 2019).

Dyeing of Natural Fibers with Supercritical Fluids

Natural fibers, which are hydrophilic due to their structure, are quite difficult to dye with Sc-CO_2 (Banchero, 2013). Recently, in order to increase the dyeability of natural fibers with Sc-CO_2, a pretreatment with a surfactant such as polyethylene glycol (Bach et al., 2002; Maeda et al., 2002; Gebert et al., 1994; Beltrame et al., 1998), chemical modification with substances containing a hydrophobic group (Bach et al., 2002; Guzel & Akgerman, 2000; Gebert et al., 1994; Ozcan et al., 1998a, 1998b), although many methods such as reverse micellar method (Sawada & Ueda, 2003), use of microemulsions (Liu et al., 2004; Sawada et al., 2002; 2004; 2005), synthesis of new reactive dyestuffs has been tried, the most promising one, for now, is dyeing with newly synthesized reactive dyestuffs (Banchero, 2020). With the use of Sc-CO_2, an increase in the dyeing depth occurs, since the T_g of the cotton fiber will decrease.

The use of water as an auxiliary solvent allows the solubility of the dyestuff in CO_2, the breaking of bonds between the polar groups, and the increase of the free volume (Banchero, 2013).

In addition to the fact that there are very few studies on chemical modification in the literature (Schmidt et al., 2003; Liu et al., 2006; 2008), the fact that it reduces the attitude and physical properties of fibers, chemical use, and excessive energy consumption limits the use of Sc-CO_2 (Liu et al., 2008; Sawada & Ueda, 2003; Fernandez-Cid et al., 2007; Long et al., 2012).

While reverse micelle formation using anionic and cationic surfactants can be used successfully (Sawada et al., 2003; 2004; Jun et al., 2004), it is expensive and its biodegradability is low (Sawada et al., 2005; Sawada & Ueda, 2007), commercial dyes are too large compared to micelles (Sawada & Ueda, 2004; Sawada et al., 2008). The disadvantages of this method are that it is expensive and has low biodegradability (Sawada et al., 2005; Sawada & Ueda, 2007).

While reverse micelle formation using anionic and cationic surfactants can be used successfully (Sawada & Ueda, 2003; 2004; Jun et al., 2004; 2005a; 2005b), it is expensive and its biodegradability is low (Sawada et al., 2005; 2007), commercial dyes are too large compared to micelles (Sawada & Ueda, 2004; Sawada et al., 2008). The disadvantages of this method are that it is expensive and has low biodegradability (Sawada et al., 2005; Sawada & Ueda, 2007).

The working conditions for dyeing cotton and wool from natural fibers with supercritical liquids and the dyestuffs that can be used are summarized by Banchero (Banchero, 2013). When dyeing wool material with acid dyestuffs at a temperature of 45-55ºC and a pressure of 140-160 bar, a higher color yield was obtained than classical dyeing without using any auxiliary substances (without the need to add acid, since CO_2 is partially dissolved in water in the micellar system) (Jun et al., 2005a). Similar results have been observed for silk fibers. While the hydrophilic structure of silk swells easily with water micelles, the swelling of wool, which has a hydrophobic cuticle layer, is much slower. When the amount of water in the micelles is limited, it is necessary to increase the dyeing temperature in order to obtain a sufficient dyeing depth (Sawada et al., 2002; Sawada & Ueda, 2003). When methanol is used as an auxiliary chemical to dye darker cotton materials, a pressure of 300 bar and a temperature of 120°C are required (Fernandez-Cid et al., 2007). Improvement in dyestuff fixation is achieved by adding a small amount of phosphoric acid or acetic acid to the dyeing bath (Fernandez-Cid et al., 2005).

In a study conducted in 1999, wool products were dyed with Sc-CO_2 using mordant dyestuff (Guzel & Akgerman, 1999; 2000). However, in this application, Sc-CO_2 dyeing, which is considered a green method, is limited by the use of pre-mordanting.

Dyestuffs Used in Dyeing with Supercritical Liquids

Since the 1990s, about 90 studies have been conducted on the dissolution of dispersed dyestuffs in Sc-CO_2 (Banchero, 2013). Some of these studies explain only the use of Sc-CO_2, while others have examined the interaction between decolorizer, Sc-CO_2, and auxiliary solvent molecules in the case of polar auxiliary solvent addition (Ellington et al., 1994; Bae et al., 2004; Cabral et al., 2007). It is stated that the solubility of Sc-CO_2 increases due to this interaction (Banchero et al., 2006; 2009; Muthukumaran et al., 1999; Tsai et al., 2007, 2008, 2009). On the other hand, there are many studies that explain the solution of dyestuffs in Sc-CO_2 with Peng-Robinson equality (Gordillo et al., 2003; 2005; Banchero et al., 2006; Tamura & Shinoda, 2004; Shinoda & Tamura, 2003; Santos et al., 2012; Shamsipur et al., 2008), Redlich-Kwong equality (Gordillo et al., 2003; 2005), Soave-Redlich-Kwong equality (Mishima et al., 2002; Coelha & Stateva, 2011; Gordillo et al., 2003; 2005; Coelho et al., 2011; Baek et al., 2004). The solubility of the dyestuff depends on the solvent density and the process temperature determined by the ambient pressure and the auxiliary solvent (Sung & Shim, 1999).

While PET dyes are easily made with Sc-CO_2, there are studies on the synthesis of new-generation dyestuffs to eliminate the difficulty of dyeing polar natural fibers. In the modification of dyestuffs, functional groups that can form chemical bonds with natural fibers are added (Long et al., 2012).

Researchers are focusing on increasing the number of dyestuffs soluble in Sc-CO_2 (Elmaaty et al., 2018), synthesizing them from commercially inexpensive materials (Penthala et al., 2019) or using hypoallergenic molecules (Abate et al., 2019). In addition, it is also aimed at synthesizing special dyes to give antibacterial, fluorescent, or photochromic properties to textiles (Banchero, 2020).

In studies conducted on reactive disperse dyestuffs for hydrophilic fibers, success was achieved with anthraquinone and azo dyestuffs containing vinylsulfone (suitable for wool and silk) and triazine (suitable for cotton) (Banchero, 2013; Elmaaty et al., 2018). Working conditions for vinylsulfone and chlorotriazine dyestuffs are 100-116°C temperature and 225-278 bar pressure (Van der Kraan, 2007b). However, in any case, the necessity of using an auxiliary solvent to be able to dye with Sc-CO_2 restricts the sustainability property of Sc-CO_2. Chlorotriazine and fluorotriazine disperse dyestuffs have been successfully used for cotton dyeing in the last ten years (Fernandez-Cid et al., 2007; 2005; Maede et al., 2002). With azochlorotriazine-type dyestuffs, it has been found that there is no need to use an auxiliary solvent in cotton dyeing, and high K/S and fastness values have been achieved (Gao et al., 2014). In Sc-CO_2 dyes made with different azo structure chlorotriazine type dyestuff, low K/S values but high fastness values could be obtained for both protein and cellulosic fibers in different azo structure chlorotriazine type dyestuff (Banchero, 2020). It has been emphasized that an acetone auxiliary solvent should be used in order to obtain high K/S values (Zhang et al., 2014). It has been stated that anthraquinone-based chlorotriazine dyestuffs give better results in dyeing protein-based fibers (Banchero, 2020). Although vinylsulfone-based dyestuffs can bond with both amino and hydroxyl, there is a need for pre-activation (Banchero, 2020). Due to the fact that Sc-CO_2 does not support the dissolution of inorganic anions, the fixation of the dyestuff does not occur in the celluloses. Using a *perfluoroalkylsulfonylquaternaryammoniumiodide* catalyst and/or carbonate, cotton products can be dyed with high K/S values by using vinisulfone disperse dyestuffs and Sc-CO_2. Cotton products can be dyed with high K/S values using a perfluoroalkylsulfonyl quaternary ammonium iodide catalyst and/or carbonate (Long et al., 2012). In another study conducted with the same sulfone dyestuff, both good color efficiency and high fastness values were obtained in wool by dyeing wool and cotton fibers pre-treated with water, while in the study conducted with cotton, an activation requirement emerged (Banchero, 2020). It has been determined that reactive disperse dyestuffs with the bromoacetyl group provide more effective dyeing on celluloses than dyestuffs with the chloroacetyl group (Gao et al., 2014). The dyeing of cotton, polyamide, wool, and silk fibers under 10-30 MPa pressure at temperatures of 80-140ºC with the tetrabutylammoniumbromide-modified reactive and acid dyestuffs was studied, and dyeing of polyamide, wool, and silk fibers under 100-120ºC temperature and 20 MPa pressure was performed. The necessity of using auxiliary chemicals for higher color darkness has been reported (Maeda et al., 2004).

Conclusion

Environmentally friendly, especially the use of supercritical fluids in basic textile processes, which are expected to cause a significant decrease in the water and carbon footprint of the textile industry, adds a separate value to the industry. There are many studies on the dyeability of textile materials with supercritical fluids. It is seen that this method is especially important for the dyeing of synthetic fibers with hydrophobic characteristics. On the other hand, the use of mordant in the application of a dyestuff with hypoallergenic properties using classical methods also increases environmental concerns. With the use of SCO_2, this restriction is eliminated.

Although there are wet advantages in terms of painting, there are also some disadvantages, such as the fact that the batches to be painted are small and the dyeing uniformity is not fully ensured. For this reason, it is envisaged that the studies to be conducted should be based on simulations that aim both to increase the uniformity of dyeing and to explain the dyeing mechanism.

According to the information obtained, since the chromophore of the dyestuff has an effect on the dyeing efficiency with Sc-CO_2, it will be appropriate to determine the diffusion rates due to the chromophore in the studies to be conducted.

Disclaimer

None.

References

Abate, M. T., Ferri, A., Guan, J., Chen, G., & Nierstrasz, V. (2019). Colouration and bio-activation of polyester fabric with curcumin in supercritical CO_2: Part I-Investigation colouration properties. *Journal of Supercritical Fluids, 152*, 104548. https://doi.org/10.1016/j.supflu.2019.104548.

Arputharaj, A., Raja, A. S. M., & Saxena, S. (2016). Developments in sustainable chemical processing of textiles. In S. Muthu & M. Gardetti (Eds.), *Green fashion, part of the environmental footprints and eco-design of products and processes book series (EFEPP)*. Springer, Singapore. https://doi.org/10.1007/978-981-10-0111-6_9.

Bach, E., Cleve, E., & Schollmeyer, E. (2002). Past, present and future of supercritical fluid dyeing technology - an overview. *Review Progress Coloration, 32,* 88-102. https://doi.org/10.1111/j.1478-4408.2002.tb00253.x.

Bach, E., Cleve, E., Schüttken, J., Schollmeyer, E., & Rucker, J. W. (2001). Correlation of solubility data of azo disperse dyes with the dye uptake of poly(ethyleneterephthalate) fibres in supercritical carbondioxide. *Coloration Technology, 117,* 13-18. http://dx.doi.org/10.1111/j.1478-4408.2001.tb00329.x.

Badens, E., Masmoudi, Y., Mouahid, A., & Crampon, C. (2018). Current situation and perspectives in drug formulation by using supercritical fluid technology. *Journal of Supercritical Fluids, 134,* 274–283. https://doi.org/10.1016/j.supflu.2017.12.038.

Bae, H. K., Jeon, J. H., & Lee, H. (2004). Influence of co-solvent on dye solubility in supercritical carbon dioxide. *Fluid Phase Equilibria, 222-223,* 119-125. https://doi.org/10.1016/j.fluid.2004.06.009.

Baek, J. K., Kim, S., Lee, G. S., & Shim, J. J. (2004). Density correlation of solubility of C.I. disperse orange 30 dye in supercritical carbon dioxide. *Korean Journal of Chemical Engineer, 21,* 230. https://doi.org/10.1007/BF02705403.

Bahrami, M., & Ranjbarian, S. (2007). Production of micro- and nano-composite particles by supercritical carbon dioxide. *Journal of Supercritical Fluids, 40,* 263–283. https://doi.org/10.1016/j.supflu.2006.05.006.

Banchero, M. (2013). Supercritical fluid dyeing of synthetic and natural textiles – a review. *Coloration Technology, 129,* 2-17. https://doi.org/10.1111/cote.12005.

Banchero, M. (2020). Recent advances in supercritical fluid dyeing. *Coloration Technology, 136(4),* 1-19. http://doi.org/10.1111/cote.12469.

Banchero, M., & Ferri, A. (2005). Simulation of aqueous and supercritical fluid dyeing of a spool of yarn. *Journal of Supercritical Fluids, 35,* 157-166. https://doi.org/10.1016/j.supflu.2004.12.009.

Banchero, M., Ferri, A., & Manna, L. (2009). The phase partition of disperse dyes in the dyeing of polyethylene terephthalate with a supercritical CO_2/methanol mixture. *Journal of Supercritical Fluids, 48,* 72-78. https://doi.org/10.1016/j.supflu.2008.09.005.

Banchero, M., Ferri, A., Manna, L., & Sicardi, S. (2006). Solubility of disperse dyes in supercritical carbon dioxide and ethanol. *Fluid Phase Equilibria, 243*(1-2), 107-114. https://doi.org/10.1016/j.fluid.2006.02.010.

Banchero, M., Manna, L., & Ferri, A. (2010). Effect of the addition of a modifier in the supercritical dyeing of polyester. *Coloration Technology, 126*(3), 171–175. https://doi.org/10.1111/J.1478-4408.2010.00243.X.

Banchero, M., Mohamed, S. S. Y., Leone, F., Lopez, F., Ronchetti, S., Manna, L., & Onida, B. (2019). Supercritical solvent impregnation of different drugs in mesoporous nanostructured ZnO. *Pharmaceutics, 11*(7), 340-354. https://doi.org/10.3390/pharmaceutics11070340.

Bao, P., & Dai, J. (2005). Relationships between the solubility of CI disperse red 60 and uptake on PET in supercritical CO_2. *Journal of Chemical & Engineering Data, 50*(3), 838-842. https://doi.org/10.1021/je0496847.

Beltrame, P. L., Castelli, A., Selli, E., Mossa, A., & Bonfatti, A. M. (1998). Dyeing of cotton in spercritical carbon dioxide. *Dyes Pigments, 39,* 335-340. https://doi.org/10.1016/S0143-7208(98)00016-3.

Berche, A., Rado, C., Rapaud, O., Guéneau, C., & Rogez, J. (2009). Thermodynamic study of the U–Si system. *Journal of Nuclear Materials, 389*(1), 101-107. https://doi.org/10.1016/j.jnucmat.2009.01.014.

Beres, A. R., Huvard, G. S., & Korsmeyer, R. W. (1987). EP0222207.

Borges, M. E., Tejera, R. L., Díaz, L., Esparza, P., & Ibáñez, E. (2012). Natural dyes extraction from cochineal (Dactylopius coccus). New extraction methods. *Food Chemistry, 132*(4), 1855-1860. https://doi.org/10.1016/j.foodchem.2011.12.018.

Cabral, V. F., Santos, W. L. F., Muniz, E. C., Rubira, A. F., & Cardozo-Filho, L. (2007). Correlation of dye solubility in supercritical carbon dioxide. *Journal of Supercritical Fluids, 40*(2), 163-169. https://doi.org/10.1016/j.supflu.2006.05.004.

Cardozo-Filho, L., Mazzer, H. R., Santos, J. C., Andreaus, J., Feihrmann, A. C., Beninca, C., Cabral, V. F., & Zanoelo, E. F. (2014). Dyeing of polyethylene terephthalate fibers with a disperse dye in supercritical carbon dioxide. *Textile Research Journal, 84,* 1279-1287. http://dx.doi.org/10.1177/0040517513515317.

Coelho, J. P., Mendonc, A. F., Palavra, A. F., & Stateva, R. P. (2011). On the solubility of three disperse anthraquinone dyes in supercritical carbon dioxide: new experimental data and correlation. *Indian Engineering Chemical Research, 50,* 4618–4624. http://dx.doi.org/10.1177/0040517513515317.

Dong, P., Xu, M., Lu, X., & Lin, C. (2010). Measurement and correlation of solubilities of c.i. disperse red 73, c.i. disperse yellow 119 and their mixture in supercritical carbon dioxide. *Fluid Phase Equilibrum, 297,* 46-51. https://doi.org/10.1016/j.fluid.2010.05.026.

Ellington, J. B., Park, K. M., & Brennecke, J. F. (1994). Effect of local composition enhancements on the esterification of phthalic anhydride with methanol in supercritical carbon dioxide. *Indian Engineer Chemical Research, 33,* 965-974. https://doi.org/10.1021/ie00028a026.

Elmaaty, T. A., Abd El-aziz, E., Ma, J., El-Taweel, F., & Okubayashi, S. (2015). Eco-friendly disperse dyeing and functional finishing of nylon 6 using supercritical carbon dioxide. *Fibers, 3,* 309-322. https://doi.org/10.3390/fib3030309.

Elmaaty, T. A., El-Taweel, F. M., & Elsisi, H. G. (2018). Water-free dyeing of polyester and nylon 6 fabrics with novel 2-oxoacetohydrazonoyl cyanide derivatives under a supercritical carbon dioxide medium. *Fibers and Polymers, 19,* 887–893. https://doi.org/10.1007/s12221-018-7766-2.

Elmaaty, T. A., Sofan, M., Kosbar, T., Elsisi, H., & Negm I. (2019). Green approach to dye PET and nylon 6 fabrics with novel pyrazole disperse dyes under supercritical carbon dioxide and its aqueous analogue. *Fibers and Polymers, 20,* 2510–2521. https://doi.org/10.1007/s12221-019-1208-7.

Esfandiari, N. (2015). Production of micro and nano particles of pharmaceutical by supercritical carbon dioxide. *The Journal of Supercritical Fluids, 100,* 129-141. https://doi.org/10.1016/j.supflu.2014.12.028.

Fernandez-Cid, M. V., Gerstner, K. N., van Spronsen, J., Van der Kraan, M., Veugelers, W. J. T., Woerlee, G. F., & Witkamp, G. J. (2007). Novel process to enhance the dyeability of cotton in supercritical carbon dioxide. *Textile Research Journal, 77*(1), 38-46. https://doi.org/10.1177/0040517507076328.

Fernandez-Cid, M. V., Spronsen, J., Van der Kraan, M., Veugelers, W. J. T., Woerlee, G. F., & Witkamp, G. J. (2005). Excellent dye fixation on cotton dyed in supercritical carbon dioxide using fluorotriazine reactive dyes. *Green Chemistry, 7,* 609-616. https://doi.org/10.1039/B503801D.

Ferri, A., Banchero, M., Manna, L., & Sicardi, S. (2006). Dye uptake and partition ratio of disperse dyes between a PET yarn and supercritical carbon dioxide. *The Journal of Supercritical Fluids, 37*(19), 107-114. https://doi.org/10.1016/j.supflu.2005.07.001.

Gao, D., Yang, D. F., Cui, H. S., Huang, T., & Lin, J. (2014). Synthesis and measurement of solubilities of reactive disperse dyes for dyeing cotton fabrics in supercritical carbon dioxide. *Indian Engineer Chemical Research, 53,* 13862–13870. https://doi.org/10.1021/ie5025497.

Gao, D., Yang, D., Cui, H., Huang, T., & Lin, J. (2015). Supercritical carbon dioxide dyeing for pet and cotton fabric with synthesized dyes by a modified apparatus. *ACS Sustainable Chemical Enginer, 3*(4), 668-674. https://doi.org/10.1021/sc500844d.

Gebert, B., Saus, W., Knittel, D., Buschmann, H. J., & Schollmeyer, E. (1994). Dyeing natural fibers with disperse dyes in supercritical carbon dioxide. *Textile Research Journal, 64,* 371-374. https://doi.org/10.1177/004051759406400701.

Gordillo, M. D., Pereyra, C., & Martínez de la Ossa, E. J. (2003). Measurement and correlation of solubility of Disperse Blue 14 in supercritical carbon dioxide. *The Journal of Supercritical Fluids, 27*(1), 31-37. https://doi.org/10.1016/S0896-8446(02)00215-2.

Gordillo, M. D., Pereyra, C., & Martínez de la Ossa, E. J. (2005). Solubility estimations for Disperse Blue 14 in supercritical carbon dioxide. *Dyes Pigment, 67,* 167–173. https://doi.org/10.1016/j.dyepig.2004.12.013.

Gulzar, T., Farooq, T., Kiran, S., Ahmad, I., & Hameed, A. (2019). Green chemistry in the wet processing of textiles. In Shahid-ul-Islam & B. S. Butola (Eds.), *The impact and prospects of green chemistry for textile technology* (1st ed., pp. 1-20). Woodhead Publishing. https://doi.org/10.1016/B978-0-08-102491-1.00001-0.

Guzel, B., & Akgerman, A. (1999). Solubility of disperse and mordant dyes in supercritical carbon dioxide. *Journal of Chemical Engineering, 44*(1), 83-85. https://doi.org/10.1021/je980157v.

Guzel, B., & Akgerman, A. (2000). Mordant dyeing of wool by supercritical processing. *Journal of Supercritical Fluids, 18,* 247-252. https://doi.org/10.1016/S0896-8446(00)00065-6.

Hirogaki, K., Tabata, I., Hisada, K., & Hori, T. (2005). An investigation of the interaction of supercritical carbon dioxide with poly (ethylene terephthalate) and the effects of some additive modifiers on the interaction. *Journal of Supercritical Fluid, 36,* 166-172. https://doi.org/10.1016/j.supflu.2005.05.004.

Huang, G., Dai, J. J., Dong, F. C., Wang, J. H., & Jia, Y. T. (2013). Compatibility of a disperse dye mixture in supercritical carbon dioxide dyeing. *Coloration Technology, 129,* 305–311. https://doi.org/10.1111/cote.12018.

Huang, T., Cui, H., Yang, D., & Kong, X. (2017). Continuous dyeing processes for zipper tape in supercritical carbon dioxide. *Journal of Cleaner Production, 158,* 95-100. https://doi.org/10.1016/j.jclepro.2017.04.167.

Huang, T., Kong, X., Cui, H., Zhang, T., Li, W., Yu, P., & Lin, J. (2019). Waterless dyeing of zipper tape using pilot-scale horizontal supercritical carbon dioxide equipment and its green and efficient production. *Journal of Cleaner Production, 233,* 1097-1105. https://doi.org/10.1016/j.jclepro.2019.06.189.

Hussain, T., & Wahab, A. (2018). A critical review of the current water conservation practices in textile wet processing. *Journal of Cleaner Production, 198,* 806-819. https://doi.org/10.1016/j.jclepro.2018.07.051.

Ilshin Autoclave. (n.d.). *Supercritical fluid system.* http://sufluxhost.cafe24.com/EN/products/Supercritical_Fluid_System_sum.html.

Jessop, P., Ikariya, T., & Noyori, R. (1994). Homogeneous catalytic hydrogenation of supercritical carbon dioxide. *Nature, 368,* 231–233. https://doi.org/10.1038/368231a0.

Jun, J. H., Sawada, K., & Ueda, M. (2004). Application of perfluoro polyether reverse micelles in supercritical CO_2 to dyeing process. *Dyes and Pigments, 61,* 17-22. https://doi.org/10.1016/j.dyepig.2003.08.003.

Jun, J. H., Sawada, K., Takagi, T., Kim, G. B., Park, C. H., & Ueda, M. (2005a). Effects of pressure and temperature on dyeing acrylic fibres with basic dyes in supercritical carbon dioxide. *Coloration Technology, 121,* 25-28. http://dx.doi.org/10.1111/j.1478-4408.2005.tb00244.x.

Jun, J. H., Ueda, M., Sawada, K., Sugimoto, M., & Urakawa, H. (2005b). Supercritical carbon dioxide containing a cationic perfluoropolyether surfactant for dyeing wool. *Coloration Technology, 121,* 315-319. https://doi.org/10.1111/j.1478-4408.2005.tb00375.x.

Jung, J., & Perrut, M. (2001). Particle design using supercritical fluids: Literature and patent survey. *Journal of Supercritical Fluids, 20,* 179–219. http://dx.doi.org/10.1016/S0896-8446(01)00064-X.

Kant, R. (2012). Textile dyeing industry an environmental hazard. *Natural Science, 4*(1), 22-26. http://dx.doi.org/10.4236/ns.2012.41004.

Karst, D., Rapp, W. A., & Yang, Y. (2004). The modeling of liquor flow in textile beam dyeing. *Journal* of *Textile Institute, 95,* 243–250. http://dx.doi.org/10.1533/joti.2003.0020.

Kim, T., Kim, G., Park, J. Y., Lim, J. S., & Yoo, K. P. (2006). Solubility measurement and dyeing performance evaluation of aramid nomex yarn by dispersed dyes in supercritical carbon dioxide. *Industrial & Engineering Chemistry Research, 45*(10), 3425-3433. https://doi.org/10.1021/ie0507171.

Kiron, M. I. (2012, February 12). *Supercritical fluid dyeing and its advantages*. https://textilelearner.net /supercritical-fluid-dyeing-and-its-advantages/.

Knez, Z., Markocic, E., Leitgeb, M., Primozic, M., Znez Hrncic, M., & Skerget, M. (2014). Industrial applications of supercritical fluids: A review. *Energy, 77,* 235-243. https://doi.org/10.1016/j.energy.2014.07.044.

Knittel, D., & Schollmeyer E. (1995a). Environmentally friendly dyeing of synthetic fibres and textile accessories. *International Journal of Clothing Science and Technology, 7,* 36-45. https://doi.org/10.1108/09556229510080701.

Knittel, D., & Schollmeyer, E. (1995b). *DE4344021. Germany patent.*

Knittel, D., Saus, W., Benken, R., Buschmann, H. J., & Schollmeyer, E. (1993a). *DE4202320. Germany patent.*

Knittel, D., Saus, W., & Schollmeyer, E. (1993b). Application of supercritical carbon dioxide in finishing processes. *Journal of Textile Institute, 84,* 534-552. https://doi.org/10.1080/00405009308658986.

Kong, I. J., Huang, T., Cui, H., Yang, D., & Lin, J. (2019). Multicomponent system of trichromatic disperse dye solubility in supercritical carbon dioxide. *Journal of CO_2 Utilization, 33,* 1-11. https://doi.org/10.1016/j.jcou.2019.04.015.

Kumar, S., & Aggarwal, S. C. (2011). Industrial water demand in India: Challenges and implications for water pricing. In Infrastructure Development Finance Company (Ed.), *India Infrastructure Report* (1st ed., pp. 274-284). Oxford University Press.

Laintz, K. E. (1992). *Separation and spectroscopic studies of metal ions complexed with bis(trifluoroethyl)dithiocarbamate in supercritical carbon dioxide*. [Doctoral dissertation]. University of Idaho.

Laintz, K. E., Wai, C. M., Yonker, C. R., & Smith, R. D. (1991). Solubility of fluorinated metal diethyldithiocarbamates in Supercritical carbon dioxide. *Journal of Supercritical Fluids, 4,* 194-198. https://doi.org/10.1016/0896-8446(91)90008-T.

Liao, S. K. (2005). Dyeing nylon-6,6 with some hydrophobic reactive dyes by supercritical processing. *Journal of Polymer Research, 11,* 285–291. https://doi.org/10.1007/s10965-005-4046-9.

Liao, S. K., Chang, P. S., & Lin, Y. C. (2000a). Analysis on the dyeing of polypropylene fibers in supercritical carbon dioxide. *Journal of Polymer Research, 7, 1*55-159. https://doi.org/10.1007/s10965-006-0115-y.

Liao, S. K., Ho, Y. C., & Chang, P. S. (2000b). Dyeing of nylon 66 with a disperse-reactive dye using supercritical carbon dioxide as the transport medium. *Coloration Technology, 116,* 403-407. https://doi.org/10.1111/j.1478-4408.2000.tb00018.x.

Liu, J., Ikushima, Y., & Shervani, Z. (2004). Investigation on the solubilization of organic dyes and micro-polarity in AOT water-in-CO_2 microemulsions with fluorinated co-surfactant by using UV-Vis spectroscopy. *The Journal of Supercritical Fluids, 32,* 97-103. https://doi.org/10.1016/j.supflu.2004.02.006.

Liu, Z. T., Sun, Z., Liu, Z. W., Lu, J., & Xiong, H. (2008). Benzylated modification and dyeing of ramie fiber in supercritical carbon dioxide. *Journal of Applied Polymer Science, 107,* 1872-1878. https://doi.org/10.1002/app.27220.

Liu, Z. T., Zhang, L., Liu, Z., Gao, Z., Dong, W., Xiong, H., Peng, Y., & Tang, S. (2006). Supercritical CO_2 dyeing of ramie fiber with disperse dye. *Indian Engineer Chemical Research, 45,* 8932–8938. https://doi.org/10.1021/ie061093h.

Long, J. J., Ma, Y. Q., & Zhao, J. P. (2011). Investigations on the level dyeing of fabrics in supercritical carbon dioxide. *Journal of Supercritical Fluids, 57,* 80-86. https://doi.org/10.1016/j.supflu.2011.02.007.

Long, J. J., Xiao, G. D., Xu, H. M., Wang, L., Cui, C. L., Liu, J., Yang, M. Y., Wang, K., Chen, C., Ren, Y. M., Luan, T., & Ding, Z. F. (2012). Dyeing of cotton fabric with a reactive disperse dye in supercritical carbon dioxide. *The Journal of Supercritical Fluids, 69,* 13-20. https://doi.org/10.1016/j.dyepig.2016.12.050.

Long, J. J., Xu, H. M., Cui, C. L., Wei, X. C., Chen, F., & Cheng, A. K. (2014). Anovel plant for fabric rope dyeing in supercritical carbon dioxideand its cleaner production. *Journal of Cleaner Production, 65,* 574-582. https://doi.org/10.1016/j.jclepro.2013.08.008.

Maeda, S., Hongyou, S., Kunitou, K., & Mishima, K. (2002). Dyeing cellulose fibers with reactive disperse dyes in supercritical carbon dioxide. *Textile Research Journal, 72,* 240-244. http://dx.doi.org/10.1177/004051750207200310.

Maeda, S., Kunitou, K., & Mishima, K. (2004). Dyeing cotton and polyamide fibers with modified reactive dyes in supercritical carbon dioxide. *In Asian Pacific Confederation of Chemical Engineering Congress.* 575-575. The Society of Chemical Engineers, Japan.

Mishima, K., Matsuyama, K., Ishikawa, H., Hayashi, K. I., & Maeda, S. (2002). Measurement and correlation of solubilities of azo dyes and anthraquinone in supercritical carbon dioxide. *Fluid Phase Equilibria, 194-197,* 895–904. https://doi.org/10.1016/S0378-3812(01)00720-8.

Miyazaki, K., Tabata, I., & Hori, T. (2012a). Effects of molecular structure on dyeing performance and colour fastness of yellow dyestuffs applied to polypropylene fibres in supercritical carbon dioxide. *Coloration Technology, 128,* 51-59. http://dx.doi.org/10.1111/j.1478-4408.2011.00344.x.

Miyazaki, K., Tabata, I., & Hori, T. (2012b). Relationship between colour fastness and colour strength of polypropylene fabrics dyed in supercritical carbon dioxide: effect of chemical structure in 1,4-bis(alkylamino)anthraquinone dyestuffs on dyeing performance. *Coloration Technology, 128,* 60-67. https://doi.org/10.1111/j.1478-4408.2011.00331.x.

Muthukumaran, P., Gupta, R. B., Sung, H. D., Shim, J. J., & Bae, H. K. (1999). Dye solubility in supercritical carbon dioxide. Effect of hydrogen bonding with cosolvents. *Korean Journal of Chemical Engineer, 16,* 111–117. https://doi.org/10.1007/BF02699013.

Naik, D. J., Desai, H. H., & Desai, T. N. (2013). Characterization and treatment of untreated wastewater generated from dyes and dye intermediates manufacturing industries of Sachin Industrial Area, Gujarat, India. *Journal of Environmental Research and Development, 7,* 1602-1605.

Okubayashi, S., Suzuma, T., Zhao, C., Miyazaki, K., Hirogaki, K., Tabata, I., (2011). Hori, T. Supercritical dyeing of polyester fibers in a mini-plant possessing internal circulator. *Fiber, 67,* 27-33. http://dx.doi.org/10.2115/fiber.67.27.

Ozcan, A. S., Clifford, A. A., Bartle, K. D., & Lewis, D. M. (1998a). Dyeing of cotton fibres with disperse dyes in supercritical carbon dioxide. *Dyes Pigments, 36,* 103-110. https://doi.org/10.1016/S0143-7208(97)00005-3.

Ozcan, A. S., Clifford, A. A., Bartle, K. D., Broadbent, P. J., & Lewis, D. M. (1998b). Dyeing of modified cotton fibres with disperse dyes from supercritical carbon dioxide. *Journal of the Society of Dyers and Colourists, 114,* 169-173. https://doi.org/10.1111/j.1478-4408.1998.tb01977.x.

Park, M. W., & Bae, H. K. (2002). Dye distribution in supercritical dyeing with carbon dioxide. *Journal of Supercritical Fluids, 22,* 65-73. https://doi.org/10.1016/S0896-8446(01)00102-4.

Park, S. C., Tuma, D., Kim, S., Lee, Y. R., & Shim, J. J. (2010). Sorption of C. I. disperse red 60 in polystyrene and PMMA films and polyester and Nylon 6 textiles in the presence of supercritical carbon dioxide. *Korean Journal of Chemical Engineering, 27,* 299-309. https://doi.org/10.1007/s11814-010-0098-6.

Patel, D., Kellici, S., & Saha, B. (2014). Green Process Engineering as the Key to Future Processes. *Processes, 2,* 311-332. http://dx.doi.org/10.3390/pr2010311.

Penthala, R., Kumar, R. S., Heo, G., Kim, H., Lee, I. Y., Ko, E. H., & Son, Y. A. (2019). Synthesis and efficient dyeing of anthraquinone derivatives on polyester fabric with supercritical carbon dioxide. *Dyes and Pigments, 166,* 330-339. https://doi.org/10.1016/j.dyepig.2019.03.027.

Phelps, C. L., Smart, N. G., & Wai, C. M. (1996). Past, present, and possible future applications of supercritical fluid extraction technology. *Journal of Chemical Education, 73*(12), 1163-1168. https://doi.org/10.1021/ed073p1163.

Poulakis, K., Spec, M., Schneider, G. M., Knittel, D., Buschmann, H. J., & Schollmeyer, E. (1991). Färbung von polyestern in überkritischen gasen. *Chemiefasern/Textilindustrie, 41*(93), 142–147.

Promila Madan, V. K. (2018). Supercritical Carbon Dioxide as greener solvent of 21st century. *Asian Journal Chemistry, 30,* 719-723. http://dx.doi.org/10.14233/ajchem.2018.21150A.

Raja, A. S. M., Arputharaj, A., Saxena, S., & Patil, P. G. (2019). Water requirement and sustainability of textile processing industries. In S. S. Muthu (Ed.), *Water in textiles and fashion: Consumption, footprint, and life cycle assessment* (pp. 155-173). Woodhead Publishing. https://doi.org/10.1016/B978-0-08-102633-5.00009-9.

Raman, S. K., & Kim, H. D. (2018). Solutions of supercritical CO_2 flow through a convergent-divergent nozzle with real gas effects. *International Journal Heat Mass Transfer, 116,* 127–135. https://doi.org/10.1016/j.ijheatmasstransfer.2017.09.019.

Reetz, M. T., Konen, W., & Strack, T. (1993). Supercritical carbondioxide as a reaction medium and reaction partner. *Chimia, 47,* 493-493. https://doi.org/10.2533/chimia.1993.493.

Reji, R. V., Raman, S. K., Kim, T. Y., & Kim, H. D. (2019). Computational fluid dynamics simulation of the supercritical carbon dioxide flow in beam dyeing. *Textile Research Journal, 89*(13), 2604-2615. https://doi.org/10.1177/0040517518798647.

Roy Choudhury, A. K. (2014). Environmental impacts of the textile industry and its assessment through life cycle assessment. In S. Muthu (Ed.), *Roadmap to sustainable textiles and clothing. textile science and clothing technology.* Springer, Singapore. https://doi.org/10.1007/978-981-287-110-7_1.

Saka, Z. (2019). *Hydrophilization and bleaching of cotton in a supercritical carbon dioxide fluid medium (anhydrous).* [Master dissertation]. Bursa Uludağ University.

Sand, M. L. (1986). *Method for impregnating a thermoplasticpolymer.* (United states Patent No: 4598006).

Santos, J., Mazzer, H., Machado, G., Andreaus, J., Cabral, V., Zabaloy, M., & Filho, L. (2012). High-pressure phase behaviour of the system (CO_2 + C.I. Disperse Orange 30 dye). *The Journal of Chemical Thermodynamics, 48,* 284–290. https://doi.org/10.1016/j.jct.2011.12.040.

Sawada, K., & Ueda, M. (2003). Adsorption and fixation of a reactive dye on cotton in non-aqueous systems. *Coloration Technology, 119,* 182-186. http://dx.doi.org/10.1111/j.1478-4408.2003.tb00170.x.

Sawada, K., & Ueda, M. (2004). Evaluation of the dyeing mechanism of an acid dye on protein fibers in supercritical CO_2. *Dyes Pigments, 63,* 77-81. https://doi.org/10.1016/j.dyepig.2004.01.008.

Sawada, K., & Ueda, M. (2007). Chemical fixation of disperse dyes on protein fibers. *Dyes and Pigments, 75*(3), 580-584. https://doi.org/10.1016/j.dyepig.2006.07.006.

Sawada, K., Oshima, M., Sugimito, M., & Ueda, M. (2005). Microemulsions insupercritical CO_2 utilizing the polyethyleneglycol dialkylglycerol andtheir use for the solubilization of hydrophiles. *Dyes Pigments, 65,* 67-74. https://doi.org/10.1016/j.dyepig.2004.07.001.

Sawada, K., Oshima, M., Sugimoto, M., Urakawa, H., & Ueda, M. (2008). Complexation of ionic substances with surfactants for solubilisation in supercritical carbon dioxide. *Dyes and Pigments, 76*(1), 1-6. https://doi.org/10.1016/j.dyepig.2006.08.004.

Sawada, K., Takagi, T., Jun, J. H., Ueda, M., & Lewis, D. M. (2002). Dyeing natural fibres in supercritical carbon dioxide using a nonionic surfactant reverse micellar system. *Coloration Technology, 118*(5), 233-237. https://doi.org/10.1111/j.1478-4408.2002.tb00105.x.

Scharf, S., Cleve, E., Bach, E., Schollmeyer, E., & Naderwitz, P. (2002). Three-dimensional flow calculation in a textile dyeing process. *Textile Research Journal, 72*(9), 783-788. https://doi.org/10.1177/004051750207200906.

Schmidt, A., Bach, E., & Schollmeyer, E. (2003). The dyeing of natural fibres with reactive disperse dyes in supercritical carbon dioxide. *Dyes Pigments, 56,* 27–35. https://doi.org/10.1016/S0143-7208(02)00108-0.

Schnitzler, J., & Eggers, R. (1999). Mass transfer in polymers in a supercritical CO_2-atmosphere. *The Journal of Supercritical Fluids, 16*(1), 81-92. https://doi.org/10.1016/S0896-8446(99)00020-0.

Schollmeyer, E, & Knittel, D. (1993). *Germany patent. DE4.200352.*

Schollmeyer, E., Knittel, D., Buschmann, H. J., Schneider, G. M., & Poulakis, K. (1990). *Germany patent. DE3906724.*

Shamsipur, M., Fasihi, J., Khanchi, A., Yamini, Y., Valinejad, A., & Sharghi, H. (2008). Solubilities of some 9,10-anthraquinone derivatives in supercritical carbon dioxide: A cubic equation of state correlation. *Journal of Supercritical Fluids, 47,* 154-160. https://doi.org/10.1016/j.supflu.2008.07.002.

Shinoda, T., & Tamura, K. (2003). Solubilities of C.I. Disperse Orange 25 and C.I. Disperse Blue 354 in supercritical carbon dioxide. *Journal of Chemical Engineer Data, 48*(4), 869–873. https://doi.org/10.1021/je0256131.

Soh, S. H., & Lee, L. Y. (2019). Microencapsulation and nanoencapsulation using supercritical fluid (SCF) techniques. *Pharmaceutics, 11*(1), 21-37. https://doi.org/10.3390/pharmaceutics11010021.

Sung, H. D., & Shim, J. J. (1999). Solubility of C. I. Disperse Red 60 and C. I. Disperse Blue 60 in supercritical carbon dioxide. *Journal of Chemical Engineer Data, 44*(5), 985–989. https://doi.org/10.1021/je990018t.

Tabernero, A., Valle, E. M., & Galán, M. A. (2012). Supercritical fluids for pharmaceutical particle engineering: Methods, basic fundamentals and modelling. *Chemical Engineering Process, 60,* 9–25. https://doi.org/10.1016/j.cep.2012.06.004.

Tamura, K., & Shinoda, T. (2004). Binary and ternary solubilities of disperse dyes and their blend in supercritical carbon dioxide. *Fluid Phase Equilibria, 219*(1), 25-32. https://doi.org/10.1016/j.fluid.2004.01.009.

Toews, K. L., Schroll, R. M., Wai, C. M., & Smart, N. G. (1995). pH defining equilibrium between water and supercritical CO_2: Influence on SFE of organics and metal chelates. *Analitical Chemistry, 67,* 4040-4043. https://doi.org/10.1021/ac00118a002.

Tsai, C. C., Lin, H., & Lee, M. J. (2007). Solubility of disperse yellow 54 in supercritical carbon dioxide with or without cosolvent. *Fluid Phase Equilibria, 260*(2), 287-294. https://doi.org/10.1016/j.fluid.2007.07.070.

Tsai, C. C., Lin, H., & Lee, M. J. (2008). Solubility of C.I. disperse violet 1 in supercritical carbon dioxide with or without cosolvent. *Journal of Chemical Engineering Data, 53*(9), 2163–2169. https://doi.org/10.1021/je8003673.

Tsai, C. C., Lin, H., & Lee, M. J. (2009). Solubility of 1,5-diaminobromo-4,8-dihydroxyanthraquinone in supercritical carbon dioxide with or without cosolvent. *Journal of Chemical & Engineering Data, 54*(5), 1442-1446. https://doi.org/10.1021/je800646h.

Tušek, L., Golob, V., & Knez, Ž. (2000). The effect of pressure and temperature on supercritical co_2 dyeing of pet-dyeing with mixtures of dyes. *International Journal of Polymeric Materials, 47,* 657-665. http://dx.doi.org/10.1080/00914030008031320.

Van der Kraan, M., Fernandez Cid, M. V., Woerlee, G. F., Veugelers, W. J. T., & Witkamp, G. J. (2007a). Equilibrium study on the disperse dyeing of polyester textile in supercritical carbon dioxide. *Textile Research Journal, 77,* 550-558. http://dx.doi.org/10.1177/0040517507077483.

Van der Kraan, M., Fernandez-Cid, M. V., Woerlee, G. F., Veugelers, W. J. T., & Witkamp, G. J. (2007b). Dyeing of natural and synthetic textiles in supercritical carbon dioxide with disperse reactive dyes. *Journal of Supercritical Fluids, 40*(3), 470-476. https://doi.org/10.1016/j.supflu.2006.07.019.

White, G. L., & Lira, C. T. (1992). Multiphase behavior of carbon dioxide with solid aromatics. *Fluid Phase Equilibria, 78,* 269-284. https://doi.org/10.1016/0378-3812(92)87041-K.

World Water Assessment Programme. (2018). New York.

Wu, J., Ma, L., Chen, Y., Cheng, Y., Liu, Y., & Zha, X. (2016). Catalytic ozonation of organic pollutants from bio-treated dyeing and finishing wastewater using recycled waste iron shavings as a catalyst: removal and pathways. *Water Research, 92,* 140–148. https://doi.org/10.1016/j.watres.2016.01.053.

Yang, M., Liu, J., Zhang, Y., & Chen, C. (2014). Rope dyeing of fabric in supercritical carbon dioxide for commercial purposes. *Coloration Technology, 130,* 102-111. https://doi.org/10.1111/cote.12076.

Yonker, C. R., Frye, S. L., Kalkwarf, D. R., & Smith, R. D. (1986). Characterization of supercritical fluid solvents using solvatochromic shifts. *Journal of Physical Chemistry, 90*(13), 3022-3026. https://doi.org/10.1021/j100404a046.

Zhang, H., Yan, H., & Mao, N. (2014). Functional modification with TiO2 nanoparticles and simultaneously dyeing of wool fibers in a one-pot hydrothermal process. *Industrial & Engineering Chemistry Research, 53*(5), 2030-2041. https://doi.org/10.1021/ie402966z.

Zheng, H. D., Zhang, J., Yan, J., & Zheng, L. J. (2017). Investigations on the effect of carriers on meta-aramid fabric dyeing properties in supercritical carbon dioxide. *RSC Advances, 7*(6), 3470-3479. https://doi.org/10.1039/C6RA26479D.

Zheng, H., Zhang, J., Yan, J., & Zheng, L. (2016). An industrial scale multiple supercritical carbon dioxide apparatus and its eco-friendly dyeing production. *Journal of CO_2 Utilization, 16,* 272-281. https://doi.org/10.1016/j.jcou.2016.08.002.

Zheng, H., Zhong, Y., Mao, Z., & Zheng, L. (2018). CO_2 utilization for the waterless dyeing: Characterization and properties of Disperse Red 167 in supercritical fluid. *Journal of CO_2 Utilization, 24,* 266-273. https://doi.org/10.1016/j.jcou.2018.01.014.

Zheng, L. J., Zhang, J., Du, B., & Zhao, Y. B. (2015). Supercritical CO_2 for color graphic dyeing; theoretical insight and experimental verification. *Thermal Science, 19,* 1287-1291. https://doi.org/10.2298/TSCI1504287Z.

Zosel, K. (1976). *Process for the separation of mixtures of substances* (United States Patent No: 3969196).

Chapter 8

New Sustainable Techniques in the Denim Washing Process: A Theoretical Approach

Ayşe Şevkan Macit[1,3,*]
and Elif Kalkanlı Kurt[2,3]
[1]Department of Textile Engineering, Uşak University, Uşak, Türkey
[2]Department of Textile Engineering, Tekirdağ Namık Kemal University, Tekirdağ, Türkey
[3]Leather, Textile and Ceramics Design Application and Research Centre, Uşak University, Uşak, Türkey

Abstract

Increasing the world population requires sustainable ways to use the limited resources of the world. In the denim manufacturing process, the key resources are energy, water, chemical substance, and time. Hence, effective management of water, energy, and other resources are crucial to make the manufacturing process sustainable. Therefore, environmentally friendly chemicals and water, energy, and time-saving methods have been improved over the years. However, today the awareness of and demand for sustainability has been widespread around the world, and this has also forced the firms to find more sustainable solutions in their manufacturing processes. In this chapter, previous and current research on denim washing has been reviewed and its sustainability aspects are highlighted.

Keywords: denim, denim bleaching, stone washing, sustainability

Introduction

Denim manufacturing dates back to the late 1800s. The use of denim has significantly increased due to its conformability and fashionable designs with different visual effects on the fabric after the weaving process. These effects are managed by denim dry fading processes and washing techniques. In the beginning, sandblasting, stone washing and bleaching were used for denim washing as conventional processes (Doshi & Shelke, 2001; Campaign, 2011; Xu et al., 2021). However, these processes have some disadvantages with regard to human health and

* Corresponding Author's Email: ayse.sevkan@usak.edu.tr.

In: Textile Insights and Future Aspects
Editors: Gamze D. Tetik, Gizem Celep and Fulya Yilmaz
ISBN: 979-8-89113-289-4

environmental issues. Therefore, new sustainable techniques have been developed over the years (Batur & Ayranpınar, 2021; Nayak et al., 2022; Maryan & Montazer, 2013).

Accordingly, this chapter aims to focus on new sustainable techniques in denim washing. To do so, it first necessary to provide a general information about the conventional chemicals and methods used for denim washing with their disadvantages in terms of environmental impact and human health. Subsequently, alternative chemicals, enzymes, and methods have been reported to compare conventional chemicals and methods with the current body of the relevant literature. The effects of sustainable methods and chemicals applied in denim processes on various properties of the denim will also be reported. On the other hand, their advantages when considering energy, water, time-saving, environmental impact, and human health are presented as sustainable solutions.

Denim Dry Fading Processes

Sand Blasting

Sandblasting is a dry process that is used in denim. The sandblasting process consists of removing indigo dye from denim fabric mechanically, thereby changing the colour or abrasing the denim (Campaign, 2011). Abrasive materials, generally silica, are blasted through a nozzle at very high speed and pressure onto the denim and the desired abrasion effect is achieved by the duration of exposure (Şahbaz et al., 2007). Thus, the denim looks like pre-worn and used (Choudhury, 2017). The fact that it is water-free and chemical-free are the most important advantages of this method. During the sandblasting process, labour safety is an important issue. Sand as an abrasive material can be in granular, powdered, or other form, and the particles of blasted sand under pressure are released during the process. Therefore, sealed blasting cabinets with ventilation should be used to prevent the health of the labour. In other cases, the labour may experience serious diseases such as silicosis and lung cancer (Akgun et al., 2008; Akgun et al., 2015; Barmania, 2016). Therefore, the sandblasting process is banned in some countries due to its negative impacts on human health. To eliminate the disadvantages of the sandblasting on health and the environment, eco-friendly and sustainable alternatives have been developed over the years. Lasers and ozone techniques are accepted as efficient alternatives to sandblasting (Amutha, 2017).

On the other hand, the sandblasting process may cause a decrease in the strength of denim fabric due to the creation of stress by sand or metal granules to obtain an abraded look (Choudhury, 2017). Thus, Schimper et al. (2011) developed a sustainable process instead of sandblasting called surface activation. Basically, indigo dye is intense on the surface of the yarn and surface activation can achieve a used look by screen printing a concentrated alkali paste (Amutha, 2017; Schimper et al., 2011). The washing process in the surface activation is short and the fabric strength is not lost. Another method that can be used instead of sandblasting is the dry-ice blasting process. CO_2 pellets are used to achieve a fading effect on denim. This process is a non-abrasive process, and hence it does not cause a decrease in the strength of the denim. After blasting with dry ice, the contaminant is only CO_2 gas. Therefore, this method is a water-free and environment-friendly process (Pal et al., 2017).

Whisker

Whisker is a fashioned fading effect created on the hips and thighs of denim. The whiskering effect can be two-dimensional (2D) or three-dimensional (3D) (Kan, 2015). The 2D whisker effects are achieved by hand brushing or mechanical brushing. The design of the whisker effect is engraved on rubber and inserted into the desired area. The garments are then mounted on balloons and the engraved pattern is then scrubbed with sandpaper (Hasan et al., 2021b; Khoa et al., 2021). The 3D whisker affects such as crease or fold are created by spraying resin, catalyst, silicone, and polyurethane in the right proportion (Cheriaa & Baffoun, 2015; Kan, 2015). The steps of the 3D whisker effect comprise spraying with resin, creating the 3D effect, drying, and finally curing (Kan, 2015). These are labour intensive processes, and the experience of the operators is essential. In the manual process, the use of masks is required to keep labour from the inhalation of toxic particles during the process.

Recently, CO_2 laser technology has also been used to achieve a whisker effect on denim. The design is created in a CAD program, and the whisker effect is transferred onto the denim automatically (Khoa et al., 2021). Although both processes successfully create a whiskering effect on denim, the manual whisker process allows a more natural look than the laser process (Kandhavadivu & Parthiban, 2021). On the other hand, nano bubble technology can also be used to create 3D effects by resin on denim. The consumption of water is considerably less than conventional fading methods; thus, it makes a contribution to sustainable denim fashion (Khalil, 2016).

Hand Sand/Scraping

Hand sand/scraping is a dry fading process to obtain a used-look, abrasion on the desired area of denim (Periyasamy & Militky, 2017b). The distressed effect is given by sandpaper, and hence the choice of the right number of papers is a critical parameter for this process. Scrubbing can be easily accomplished on inflated rubber balloons or on flat wooden boards (Arjun et al., 2013). This is a labour intensive process and highly depends on the experience of the labour.

Grinding

Another used look style on denim is given by grinding method (Periyasamy & Militky, 2017b). In this method, a stone grinding machine is mostly used to obtain a grinding effect on denim. Recently, pen-type grinding equipment also prefers to give a worn look to denim (Periyasamy & Militky, 2017b). On the other hand, ripped denim, frequently used for a fashioned effect, can be achieved with sharp-edged tools such as blades or knives.

Laser

The laser process is highly used to give a fading effect to denim. The effect is obtained by oxidising and disintegrating indigo dyestuff according to certain designs with computer connection with laser beams (Batur & Ayranpınar, 2021). The laser process is used to transfer shapes or patterns to the denim fabrics without applying sanding or any other mechanical treatment (Demirel Özkan, 2019). Effects very close to processes such as sandblasting, or sanding can be achieved through this method. In addition, the differences in the fading effects are less in this method compared to other methods. So, the repeatability of the process is very high (Tarhan, 2005). On the other hand, effect quality is relatively more acceptable in the manual system than laser; the laser system produces less hairiness compared to conventional methods (Ahmad et al., 2022).

The fading effect on the laser process depends on the following parameters: laser power, scanning speed, and spot diameter (Tarhan, 2005). There are various studies that examine the effect of variable parameters on the laser process. Dascalu et al. (2000) studied on three different laser wavelengths and determined the best wavelengths and the appropriate laser power density and laser fluency. Dalbaşı et al. (2019) investigated the research on the effect of various laser fading parameters (laser resolution, laser density, and pixel time) on the physical and surface properties of denim fabric. Ortiz-Morales et al. (2003) determined the effectiveness of laser-based technologies, and investigated the properties of denim like colorfastness, strength resistance, tearing strength, and dimensional changes. Özgüney et al. (2009) studied the laser pulse time parameter of the laser process and determined the suitable pulse time in terms of both colour and mechanical properties. The researchers have tried to determine how the laser process affects the fabric properties.

Laser technology has great advantages such as being environmentally friendly, water-free, and reducing the cost of the process (Periyasamy & Periyasami, 2023; Nayak et al., 2022). It can be said that there is water, energy, chemical, and time-saving compared to the equivalent processes (Csanák, 2014).

Denim Washing Processes

Rinse Washing

Rinse washing is a basic wash for denim garments. With this type of wash, starch, dust, and dirt are removed from garments, and the removal of unfixed dyes can be achieved for colour garments. Also, with this washing, giving the garment a used look can be achieved. In addition, due to the sizing process applied to the warp yarns before weaving, the fabric stiffness is high, and the desizing process is also done in rinse washing. Some softening agents can be applied to improve garment handling (Choudhury, 2017).

Back-staining problems may occur during the washing process. Basically, back staining is the dyeing of undyed weft yarns by the dye removed from the fabric. Dispersing agents are used to keep the indigo dyestuff in water. The use of dispersing agents prevents the colour change of materials, such as white labels and pockets on the product (Çalışkan, 2019).

Stone Washing

Denim stone washing has been used over the years. Stone washing is a process that provides a fading effect. It is a water-based process and it is based on washing denim with stones in a machine. Stone washing gives a faded effect or worn look as it abrades the surface of the denim, removing indigo dye (Doshi & Shelke, 2001).

Pumice, which is a volcanic stone, is used in the stone washing method. The main countries that supply pumice stones are the United States, Türkiye, Italy, Iceland, New Zealand, Japan, Indonesia, and the Philippines (Choudhury, 2017). Stone size, stone ratio, liquor ratio, duration of washing, amount of garments, and their grams per square meter are some parameters of the stone washing process, and these parameters affect the degree of fading (Choudhury, 2017).

It is important to choose the type of stone suitable for the product and the desired effect with stone washing. Large and hard stones suitable for heavy-weight fabrics endure more to washing and give a more aged effect on the product after washing. Smaller and softer stones are preferred for lightweight fabrics and delicate products. The choice of the right stone according to the properties of denim is an important issue. Otherwise, it may cause damage to the denim fabric (Çakır, 2010).

The stone washing process has different effects on the fabric properties depending on the process parameters. Various studies on this subject are available in the literature. Feki et al. (2003) studied the effects of stone wash treatment parameters (pumice stone weight, duration of treatment, temperature, etc.) on the mechanical properties of denim. In another study, the authors examined the effect of three kinds of stone washing processes on the objective hand of denim (Feki et al., 2006a). Khan and Mondal (2013) studied the physico-mechanical properties of denim garments finished by the stone-bleach process.

The pumice stone washing process creates a very good worn look effect, but this process also has some disadvantages. The necessity of unloading the stones from machines after the process, and the separation of stones and fabric from each other are labour intensive processes and cause time loss. On the other hand, pumice stones may damage the machine over time. In addition, the environmental impact of this process should not be ignored. This is not an eco-friendly process, and it creates waste in the form of sludge. Also, wastewater treatment is also becoming a necessity.

Regarding the environmental impact of stone washing, some studies have been examined. There is a study on the removal of dye from wastewater with the adsorption by pumice stone used in washing (Körlü et al., 2015). On the other hand, with the idea of a less polluting sustainable washing system, Shibly et al. (2021) observed the wash effects of natural reagents such as lemon, pomelo, guava, ginger, and spondias mombin for the development of eco-friendly denim washing, and they examined the effects of the reagents on different properties of denim fabric.

It is also necessary to mention perlite, which is used in stone washing and is an alternative to pumice stone. With the process of washing denim products with mineral particles called perlite instead of pumice stone, the washed product gives a fading and worn look effect. It is a water-insoluble mineral. It has advantages such as being sterile and light, not giving odour to the environment, not abrading the product applied, not clogging the machine (Karakaya Şen, 2019). Perlite has the same function as stones. Perlite treatment reduces damage caused by pumice stones in washing machines and gives denim better suppleness and a softer finish (Choudhury, 2017).

In search of eco-friendly resources and process solutions, artificial stones have been developed to partially or completely eliminate the disadvantages of pumice stone. Artificial stones give denim a similar faded appearance to pumice stones (Sağırkaya, 2021). These stones have a long lifetime, so they are environmentally friendly compared to pumice stones. The durability of artificial stones is higher and can be used repeatedly between 50 to 300 cycles, depending on the type of stone (Choudhury, 2017). The repeatability of the washing is manageable. The process is economical and ecological, as there is much less stone discharge. There is less damage to the machine and garments (Choudhury, 2017).

In addition to being environmentally friendly, there are studies comparing pumice stone and artificial stone to see the effects of stones on the properties of the fabric and to observe if they provide the expected fading effect (Chattopadhyay & Pachauri, 2012; Sağırkaya, 2021).

Enzyme Washing

Enzyme washing is a laundry process in which enzymes are used, especially for denim and other garments with a worn look (Periyasamy & Militky, 2017a). The enzyme breaks down the surface cellulose fibers of denim and removes them with washing. A certain amount of indigo dye and cellulose fibers are removed from the fabric surface during enzyme washing (Arjun et al., 2013). The enzyme used for dye removal in denim washing is cellulase (Köksal, 2015). Enzyme washing is environmentally friendly because enzymes are of natural origins (Arjun et al., 2013).

This method partially eliminates the disadvantages of stone washing. Using cellulase enzymes eliminates the necessity of frequent rinse washing, prevents the damage of the stones to the machine and the product, and also eliminates the problem of disposal of used stones and wastewater (Doshi & Shelke, 2001).

In order to obtain more vibrant and undamaged denim products, enzyme washing processes are of great importance. Thus, the lifespan of denim products will be longer (Toksöz & Mezarcıöz, 2013). There are two methods for enzyme washing. One of them is bio-polishing and acidic cellulase enzymes are preferred in this process. In the other process, neutral cellulase enzymes are preferred in the enzymatic stone washing process or washing processes carried out only with enzymes without stones applied to denim products (Toksöz & Mezarcıöz, 2013).

While the pumice stones abrade only the surface of denim, enzymes penetrate into the fibers (Tarhan, 2005). One problem with using cellulase enzymes is back staining. Back staining is the precipitation of indigo dye, which is separated from the fabric during washing, on the fabric again. To avoid this problem, the use of dispersing agents in the washing or rinsing steps is a solution (Tarhan, 2005).

To reduce water and energy consumption, there has been carried out research about enzyme washing that can be performed by using amylase, cellulase, and laccase enzymes and their combinations (Maryan & Montazer, 2013). In another study to reduce resource consumption, anhydrous enzymes are used in denim washing (Tufan et al., 2021).

Enzyme and Stone Washing

In the enzymatic stone washing method, enzymes and stones are used together, so the disadvantages of stones are reduced with the use of fewer stones, and the advantages of the cellulase enzyme are added (Tarhan, 2005). This washing method has some advantages such as shorter washing times, high contrast, vibrant and bright colours, surface abrasion, and low machine damage (Tarhan, 2005). Enzymatic stone washing helps to increase the softness and flexibility of denim products (Mondal et al., 2016). This combined process also has environmental effects.

There are many studies investigating stone washing, enzyme washing, or enzymatic stone washing performances. These studies were applied with different parameters and the properties of fabrics such as tensile strength and elongation, bending rigidity, compressibility, tear strength, dimensional stability to washing, colour fastness to washing, colour fastness to perspiration, colour fastness to light, colour fastness to rubbing, changes in fabric weight, shrinkage were compared in the studies (Shahriar, 2020; Hasan et al., 2021a; El-Dessouki, 2015; Montazer & Maryan, 2007; Feki et al., 2006b; Toksöz & Mezarcıöz, 2013; Zerin et al., 2019; Elahi et al., 2019; Mir et al., 2014; Bhattacharjee et al., 2019). Some of these studies reported indicating that enzymatic washing or enzymatic stone washing are environmentally friendly methods and these methods can provide superior properties in comparison.

Bleaching Process

Bleaching is used to obtain the desired shade in denim by decolorizing the indigo dye. The effectiveness of the discoloration depends on the following parameters: bleach liquid/water ratio, duration of treatment, and temperature. In the bleaching process, strong oxidative bleaching agents are preferred to decrease the treatment time to achieve the required colour in denim. In this part, the conventional oxidative bleaching agents used in the denim industry are detailed. Their disadvantages with regard to environmental and health issues and their sustainable alternatives are also determined within the framework of this part.

Sodium Hypochlorite

Sodium Hypochlorite is a strong oxidising agent and is widely used in conventional bleaching processes (Xu et al., 2021). Sodium hypochlorite is mostly preferred in denim bleaching due to its being easy to use and inexpensive (Bechtold et al., 2005). Sodium hypochlorite reacts easily even at low temperatures due to its high oxidation characteristic (Iracheta-Cárdenas et al., 2016). However, the exposure time to sodium hypochlorite is important to control the desired effect on denim. This feature can also cause fiber damage and a reduction in the strength of denim fabric (Periyasamy & Militky, 2017b). During the bleaching with hypochlorite, the presence of chlorine causes yellowing problems in the product. After achieving the desired effect on denim, it should be washed with peroxide to stop the bleaching process (Ahmed et al., 2021).

The bleaching agents release toxic fumes during the process, and it is very harmful to human health. On the other hand, the presence of chlorine also causes environmental pollution. In this line, innovative technologies, new sustainable substances, and processes are required instead of harmful oxidising agents for the denim industry. Bleaching with enzymes and ozone processes are used as sustainable alternative denim bleaching methods instead of sodium hypochlorite bleaching. The details of these sustainable methods are given under the related titles.

Potassium Permanganate

Potassium permanganate (PP) is an inorganic substance and strong oxidising agent (Khalil et al., 2015). Duration of treatment, PP liquid ratio, temperature, and pH are critical parameters to obtain the desired fashion effect on denim (Du et al., 2019; Siddique et al., 2017). PP can be applied to the denim with spray or water in a bath. However, PP is preferred mostly when local bleaching is desired in denim fabric. Therefore, a spray gun or brush is used instead of the bath as in sodium hypochlorite bleaching. The spray gun converts PP chemicals to microparticles that can be easily absorbed by inhalation (Khan et al., 2021). Therefore, ventilation and the use of masks are crucial precautions to protect operator health and/or minimise the harmful effect during the process (Amutha, 2017).

The use of biodegradable chemicals and sustainable methods in the denim industry has increased with the increase of awareness of environmental issues. Today, eco-friendly enzymes known as laccases and ozone are widely used in place of PP (Samanta et al., 2017).

Hydrogen Peroxide

Hydrogen peroxide is a relatively effective oxidising agent in denim. It is preferred when a slight loss of colour is desired in denim. Therefore, the process ends with a fading effect (Choudhury, 2017). Hydrogen peroxide is applied at a high temperature, and hence it causes high energy consumption. The bleaching conditions also lead to an increase in fiber damage (Haq et al., 2015). The toxic fumes formed during the bleaching process are harmful to human health. The sustainable alternatives to hydrogen peroxide can be enzymes and ozone bleaching as in sodium hypochlorite and potassium permanganate.

Enzyme Bleaching as Sustainable Treatment

Laccases, which are used in enzyme bleaching, have become more attractive in the denim industry in recent decades (Rodríguez-Couto, 2012). Laccases are environmentally friendly oxidizer enzymes and catalyse the oxidation of phenolic substances, including indigo. However, laccases can provide a bleaching effect with the contribution of a mediator, which transfers electrons from indigo to laccase (Colomera & Kuilderd, 2015). Laccase enzyme was first commercially launched by a Denmark company in 1996 as a bleaching enzyme acting with a mediator (Rodríguez-Couto, 2012). Then, various laccase enzymes have been released into the denim market over the years (Madhu & Chakraborty, 2017). Although conventional denim bleaching techniques are cheap and efficient, they cause environmental pollution, the consumption of natural resources, and the loss of denim strength (Periyasamy & Militky, 2017b). There have been carried out various research about the effectiveness of laccase enzymes (Araujo et al., 2008; Dwivedi et al., 2011; Iracheta-Cárdenas et al., 2016). The results demonstrate that laccase enzyme bleaching leads to better strength properties in denim due to less fiber damage. The resue and recovery of resources are also efficient in this bleaching method.

Acid Washing (Ice or Snow Washing)

Acid wash, also called ice wash or snow wash, is highly used on denim to obtain significant contrast in colour. It is a chemical washing process and the top layer of colour is stripped during the process, while the bottom layer keeps its colour (Haq & Khan, 2014; Hossain et al., 2020). Thus, it can give a faded effect to the denim. In acid wash on denim, pumice stone is soaked with strong oxidising agents such as potassium permanganate or sodium hypochlorite (Kan, 2015). Then, the acid-soaked pumice stones with denim are put directly into the machine

(Ansari, 2017). Following this, the denim is neutralised to stop oxidation (Arjun et al., 2013). In the acid washing, strong oxidation agents are used, and hence it is a time-efficient process. However, the acid concentration highly affects the mechanical properties of denim fabric. Research shows that this process causes a reduction in the weight, stiffness, and strength of the fabric due to the strong oxidation effect of the agents (Amutha, 2017; Ansari, 2017; Hossain et al., 2020). In addition, the negative impacts of acid washing on human health and the environment are also similar to the bleaching process due to the use of the same agents. Therefore, it should be more careful while using this process.

Plasma

Denim washing methods are water and chemical-intense processes (Cheung et al., 2013). Exposure to chemicals and harsh processes may result in a loss in the strength of denim. In addition to them, excessive time and energy consumption are also other drawbacks of these methods (Kan & Yuen, 2012). Therefore, the denim industry studies for alternative denim fading processes to eliminate these drawbacks. Recently, plasma technology used for colour fading in denim has been developed as a water-free process. Plasma technology is based on surface modification at the nano level through an electrical discharge utilising polymerizing to non-polymerizing gaseous molecules (Cheung et al., 2013; Samanta et al., 2017). The plasma technology can be even used for denim, which is difficult to modify their surface and the treatment duration is much greater than conventional fading processes (Tudoran et al., 2020).

Ozone Technique

To reduce the use of water, various methods and technologies have been developed over the years for circular and sustainable denim fashion (Nayak et al., 2022). Ozone treatment has been developed instead of the conventional bleaching methods that use highly use strong oxidation agents that are harmful to the environment and human health. Ozone treatment is a natural process that provides a bleaching effect due to the mild oxidising agent ozone (O_3) in denim fabric (Ben Hmida & Ladhari, 2016; Marieta et al., 2005). Ozone ensures the bleaching effect on denim by separating the olefin groups of the indigo dye (Körlü, 2018). The ozone technique can be performed in a closed chamber or in a washing machine. Ozone gas treatment in the closed chamber does not require water and harsh chemicals; thus, this process is accepted as eco-friendly when compared to conventional denim bleaching processes that consume large amounts of water and use harsh chemicals (Epelle et al., 2023). In the case of bleaching in washing machines by dissolving ozone in water, water is also not used in large amounts, and the ozonated water can be deionised due to UV radiation (Choudhury, 2017).

Water Jet Fading

The water jet process was developed to create patterns and improve the surface finish, texture, and durability of denim products and other properties. A water jet has water jet nozzles and using the nozzles creates a type of surface finishing on denim products (Periyasamy & Militky,

2017b). One or both sides of the garment are exposed to hydro-jet nozzles in the process. The quantity and manner of fluid impact energy and the type of dye affect the degree of fading, the clarity of the patterns, and the softness of the product (Khalil, 2015; Periyasamy & Periyasami, 2023). Especially good results are obtained with blue indigo-dyed denim products (Khalil, 2015).

Water and chemical consumption are very important issues for sustainability. The water jet fading process does not involve any chemicals, so it is a pollution-free process. A water recycling system can make this a very economical and environmentally friendly way of denim processing (Choudhury, 2017).

Tint Washing

In this process, the garment is lightly coloured to give the final appearance of the denim. This process only gives the impression of a change in the overall colour of the fabric (Choudhury, 2017). This process is done to change the hue, cast, or tone of the indigo dye, and this washing is used to give the denim a worn, vintage look. The tinting of denim garments is usually done after the stone washing process. This process consumes a large amount of water and chemicals. To make this process economical and eco-friendly, the tinting and stone washing effects can be achieved in a single process with less water and chemical consumption by using some novel colour-based enzymes (Choudhury, 2017).

Conclusion

Fashionability in denim, which has been in human life for decades, has always been an important issue. Denim washing processes are carried out in order to catch the fashion in denim, and there have been great developments in this regard over the years. There are many methods that can be applied to denims in washing processes. In general, conventional methods consume too much energy, water, chemicals, and time. It is also known that some conventional methods are harmful to human health. Considering these negative factors, sustainable methods have been developed that can be considered as an alternative to conventional methods.

Sustainability, which covers the effective use of limited resources in an environmentally friendly way, has become an issue that should be considered in all sectors. Today, the necessity to obtain fading effects by using new environmentally friendly processes in the denim washing industry is very important. In these processes, many important issues, such as energy and water consumption, should be considered. These issues in sustainable denim washing can be detailed as follows. Reuse of water, less water consumption, elimination of wastewater, less energy, less chemical, environmentally friendly chemicals, short process, protection of human health, process optimization, quality, productivity, and longevity of washing equipment are the critical issues to consider (Khan & Jintun, 2021). In terms of these issues, laser, water jet, ozone, plasma, enzymatic stone washing, enzymatic bleaching, and nanobubble technologies which are detailed in this chapter, can be considered environmentally friendly methods when compared to the conventional methods. However, laser and ozone technologies have shown the potential to meet the expectations of the denim industry by using less water and chemicals.

Disclaimer

None.

References

Ahmad, S., Ashraf, M., Abid, S., Jabbar, M., Shafiq, F., and Siddique, A. (2022). Recent developments in laser fading of denim: a critical review. *Journal of Natural Fibers*, 19(15), 11621-11631. https://doi.org/10.1080/15440478.2022.2029793.

Ahmed, M., Ahmed, M. P., and Hassan, M. (2021). An explanation of bleach wash on denim cotton fabrics. *Journal of Textile Engineering & Fashion Technology*, 7, 87–90. https://doi.org/10.15406/jteft.2021.07.00272.

Akgun, M., Araz, O., Akkurt, I., Eroglu, A., Alper, F., Saglam, L., Mirici, A., Gorguner, M., and Nemery, B. (2008). An epidemic of silicosis among former denim sandblasters. *European Respiratory Journal*, 32(5), 1295–1303. https://doi.org/10.1183/09031936.00093507.

Akgun, M., Araz, O., Ucar, E. Y., Karaman, A., Alper, F., Gorguner, M., and Kreiss, K. (2015). Silicosis appears inevitable among former denim sandblasters: a 4-year follow-up study. *Chest*, 148(3), 647–654. https://doi.org/10.1378/chest.14-2848.

Amutha, K. (2017). Environmental impacts of denim. In S. S., Muthu (Ed.), *Sustainability in denim* (1st ed., pp. 27–48). Elsevier. https://doi.org/10.1016/B978-0-08-102043-2.00002-2.

Ansari, I. Z. (2017). Impact of stone wash and acid wash on the physical properties of denim. *International Journal of Engineering Research*, 6(12), 499–501. https://doi.org/10.5958/2319-6890.2017.00073.3.

Araujo, R., Casal, M., and Cavaco-Paulo, A. (2008). Application of enzymes for textile fibres processing. *Biocatalysis and Biotransformation*, 26(5), 332–349. https://doi.org/10.1080/10242420802390457.

Arjun, D., Hiranmayee, J., and Farheen, M. N. (2013). Technology of industrial denim washing: Review. *International Journal of Industrial Engineering & Technology*, 3(4), 25–34.

Barmania, S. (2016). Deadly denim: sandblasting-induced silicosis in the jeans industry. *The Lancet Respiratory Medicine*, 4(7), 543. https://doi.org/10.1016/S2213-2600(16)30102-3.

Batur, G., and Kozbekçi Ayranpinar, S. (2021, June 21-22). *Laser technology: ecological solution for sustainable denim production*. Art&Design-2021, International Congress on Art and Design Research and Exhibition book proceedings.

Bechtold, T., Maier, P., and Schrott, W. (2005). Bleaching of indigo-dyed denim fabric by electrochemical formation of hypohalogenites *in situ*. *Coloration Technology*, 121(2), 64–68. https://doi.org/10.1111/j.1478-4408.2005.tb00252.x.

Ben Hmida, S., and Ladhari, N. (2016). Study of parameters affecting dry and wet ozone bleaching of denim fabric. *Ozone: Science & Engineering*, 38(3), 175–180. https://doi.org/10.1080/01919512.2015.1113380.

Bhattacharjee, M., Dhar, A. K., Islam, M. M., and Rashid, M. A. (2019). Development of washing effects on reactive dyed denim fabrics: A value added approach of denim wash. *International Journal of Textile Science*, 8(2), 41-48. https://doi.org/10.5923/j.textile.20190802.03.

Campaign, C. C. (2011). *Fashion Victims-A Report on Sandblasted Denim*. Retrieved May 5, 2023 from: https://archive.cleanclothes.org/resources/national-cccs/fashion-victims-a-report-on-sandblasted-denim/view.

Chattopadhyay, D. P., and Pachauri, R. D. (2012). Preparation of synthetic stones for denim washing and analysis of wash effects. *Textiles and Light Industrial Science and Technology*, 1(2), 24-28.

Cheriaa, R., and Baffoun, A. (2015). Effects of cross linkers combination, for three dimensional effects, on denim garment properties. *Fibers and Polymers*, 16, 1150–1155. https://doi.org/10.1007/s12221-015-1150-2.

Cheung, H. F., Kan, C. W., Yuen, C. W. M., Yip, J., and Law, M. C. (2013). Colour fading of textile fabric by plasma treatment. *Journal of Textiles*, 2013. https://doi.org/10.1155/2013/214706.

Choudhury, A. K. R. (2017). Environmental impacts of denim washing. In S. S., Muthu (Ed.), *Sustainability in Denim* (1st ed., pp. 49–81). Elsevier. https://doi.org/10.1016/B978-0-08-102043-2.00003-4.

Colomera, A., and Kuilderd, H. (2015). Biotechnological washing of denim jeans. In Paul, R. (Ed.), *Denim* (1st ed., pp. 357–403). Elsevier. https://doi.org/10.1016/B978-0-85709-843-6.00012-3.

Csanák, E. (2014, November). *Sustainable concepts and eco-friendly technologies in the denim industry.* International Conference on Design and Light Industry Technologies. Budapest, Hungary. Óbuda University (pp. 1-6).

Çakır, N. (2010). *Effects of finishing and different denim fabrics to fit at jean production*. [Master dissertation]. Pamukkale University.

Çalışkan, E. (2019). *Comparison of ecological and traditional production methods in denim washing*. [Master dissertation]. Pamukkale University.

Dalbaşı, E. S., Özçelik Kayseri, G., and İlleez, A. A. (2019). A research on the effect of various laser fading parameters on physical and surface properties of denim fabric. *Optics & Laser Technology*, 118, 28-36. https://doi.org/10.1016/j.optlastec.2019.04.030.

Dascalu, T., Acosta-Ortiz, S. E., Ortiz-Morales, M., and Compean, I. (2000). Removal of the indigo color by laser beam–denim interaction. *Optics and Lasers in Engineering*, 34(3), 179-189. https://doi.org/10.1016/S0143-8166(00)00087-7.

Demirel Özkan, P. (2019). *Investigation of the effects of fixing materials on fastness and strength of denim fabrics*. [Master dissertation]. University of Kahramanmaraş Sütçü İmam.

Doshi, R., and Shelke, V. (2001). Enzymes in textile industry-An environment-friendly approach. *Indian Journal of Fibre & Textile Research*, 26, 202-205.

Du, W., Zuo, D., Gan, H., and Yi, C. (2019). Comparative study on the effects of laser bleaching and conventional bleaching on the physical properties of indigo kapok/cotton denim fabrics. *Applied Sciences*, 9(21), 4662. https://doi.org/10.3390/app9214662.

Dwivedi, U. N., Singh, P., Pandey, V. P., and Kumar, A. (2011). Structure–function relationship among bacterial, fungal and plant laccases. *Journal of Molecular Catalysis B: Enzymatic*, 68(2), 117–128. https://doi.org/10.1016/j.molcatb.2010.11.002.

Elahi, S., Hosen, M. D., Islam, M., Hasan, Z., Helal, M. M., Al, M. S., and Rakin, S. (2019). Analysis of physical & chemical properties of cotton-jute blended denim after a sustainable (industrial stone enzyme) wash. *Journal of Textile Science and Fashion Technology*, 3(2), 1-8. https://doi.org/10.33552/JTSFT.2019.03.000558.

El-Dessouki, H. A. (2015). Effect of different washing methods on mechanical properties of Egyptian denim fabrics. *International Design Journal*, 5(3), 1099-1107. https://doi.org/10.21608/idj.2015.101517.

Epelle, E. I., Macfarlane, A., Cusack, M., Burns, A., Okolie, J. A., Mackay, W., Rateb, M., and Yaseen, M. (2023). Ozone application in different industries: A review of recent developments. *Chemical Engineering Journal*, 454, 140188. https://doi.org/10.1016/j.cej.2022.140188.

Feki, I., Ghith, A., and Sakli, F. (2003, June). *Effect of stone wash treatment on the denim's mechanical properties*. 4th AUTEX Conference, Roubaix, France (pp. 22-24).

Feki, I., Ghith, A., and Saklı, F. (2006a, November 30-December 2). Effect of three kinds of stone wash processing on objective hand of denim fabric. *2nd International Conference of Applied Research in Textile (CIRAT-2),* Monastir, Tunisia.

Feki, I., Ghith, A., and Sakli, F. (2006b, April). Comparison of mechanical properties of different stone wash treated denim fabrics. *3rd International Conference of Textile Research Division NRC*, Cairo, Egypt (pp. 2-4).

Haq, U. N., and Khan, M. M. R. (2014). Technology of acid wash on woven denim apparel with damp pumice stone. *International Journal of Science, Environment and Technology*, 3(6), 2090-2095.

Haq, U. N., Khan, M. M. R., and Khan, M. M. R. (2015). Investigation of the bulk, surface and transfer properties of chlorine bleached denim apparel at different condition. *European Scientific Journal*, 11(12), 1–15.

Hasan, M. Z., Asif, A. A. H., Razzaque, A., Hasan, M. R., Sur, S., and Faruque, M. O. (2021a). An experimental investigation of different washing processes on various properties of stretch denim fabric. *Journal of Materials Science and Chemical Engineering*, 9(01), 1. https://doi.org/10.4236/msce.2021.91001.

Hasan, M. Z., Rahaman, M. T., Islam, T., and Pranta, A. D. (2021b). An empirical analysis of sustainable denim washing technology in the apparel industries. *International Journal of Industrial and Manufacturing Systems Engineering*, 6(2), 20. https://doi.org/10.11648/j.ijimse.20210602.11.

Hossain, M. J., Hoque, M. S., and Rashid, M. A. (2020). Scopes of acid washing with varying concentrations of phosphoric acid vis-à-vis bleach wash. *Journal of Textile and Apparel, Technology and Management*, 11(2), 1-14.

Iracheta-Cárdenas, M. M., Rocha-Peña, M. A., Galán-Wong, L. J., Arévalo-Niño, K., and Tovar-Herrera, O. E. (2016). A Pycnoporus sanguineus laccase for denim bleaching and its comparison with an enzymatic commercial formulation. *Journal of Environmental Management*, 177, 93–100. https://doi.org/10.1016/j.jenvman.2016.04.008.

Kan, C. W. (2015). Washing techniques for denim jeans. In R. Paul (Ed.), *Denim* (1st ed., pp. 313–356). Elsevier. https://doi.org/10.1016/B978-0-85709-843-6.00011-1.

Kan, C. W., and Yuen, C. W. M. (2012). Effect of atmospheric pressure plasma treatment on the desizing and subsequent colour fading process of cotton denim fabric. *Coloration Technology*, 128(5), 356–363. https://doi.org/10.1111/j.1478-4408.2012.00388.x.

Kandhavadivu, D. P., and Parthiban, D. M. (2021). Innovative and sustainable design development for denim fabric using laser techniques, tie and dye and manual whiskering. *International Journal of Aquatic Science*, 12(2), 4325–4337.

Karakaya Şen, M. (2019). *A study of the significance of the parameters of washing process on the denim surface effect*. [Master dissertation]. Tekirdağ Namık Kemal University.

Khalil, E. (2015). Sustainable and ecological finishing technology for denim jeans. *AASCIT Communication*, 2(5), 159-163.

Khalil, E. (2016). *Nano bubble technology: a new way to sustainable jeans finishing*. 56th Convention of Institution of Engineers, Bangladesh (IEB).

Khalil, E., Rahman, A., and Solaiman, M. (2015). Investigation of the influence of potassium permanganate on denim jeans processing during acid wash. *AASCIT Communication*, 2(6), 271–275.

Khan, M. M. R., and Mondal, M. I. H. (2013). Physico-mechanical properties of finished denim garment by stone-bleach treatment. *Journal of Chemical Engineering*, 28(1), 36-40.

Khan, M., Rahman, K., and Jintun, S. (2021). Sustainability issues of various denim washing methods. *Textile & Leather Review*, 4(2), 96–110. https://www.doi.org/10.31881/TLR.2021.01.

Khoa, T. H. D., Hien, N. T. T., Trung, B. Q., Thach, D. S., and Van Tuong, N. (2021). Application of CO2 laser technology in creating whisker and handsand effects on denim wash. *Journal of Science Technology and Food*, 21(2), 62–72.

Köksal, F. (2015). *The investigation of the causes of color variations in denim laundry (washing) processes*. [Master dissertation]. Tekirdağ Namık Kemal University.

Körlü, A. (2018). Use of ozone in the textile industry. In A., Körlü (Ed.), *Textile Industry and Environment* (1st ed., pp. 1–23). http://dx.doi.org/10.5772/intechopen.81774.

Körlü, A. E., Yapar, S., Perinçek, S., Yılmaz, H., and Bağıran, C. (2015). Dye removal from textile waste water through the adsorption by pumice used in stone washing. *Autex Research Journal*, 15(3), 158-163. https://doi.org/10.1515/aut-2015-0012.

Madhu, A., and Chakraborty, J. N. (2017). Developments in application of enzymes for textile processing. *Journal of Cleaner Production*, 145, 114–133. http://dx.doi.org/10.1016/j.jclepro.2017.01.013.

Marieta, C., Schulz, E., Irusta, L., Gabilondo, N., Tercjak, A., and Mondragon, I. (2005). Evaluation of fiber surface treatment and toughening of thermoset matrix on the interfacial behaviour of carbon fiber-reinforced cyanate matrix composites. *Composites Science and Technology*, 65(14), 2189–2197. https://doi.org/10.1016/j.compscitech.2005.05.008.

Maryan, A. S., and Montazer, M. (2013). A cleaner production of denim garment using one step treatment with amylase/cellulase/laccase. *Journal of Cleaner Production*, 57, 320-326. http://dx.doi.org/10.1016/j.jclepro.2013.05.041.

Mir, S., Hossain, M., Biswas, P., Hossain, A., and Idris, M. A. (2014). Evaluation of mechanical properties of denim garments after enzymatic bio-washing. *World Applied Sciences Journal*, 31(9), 1661-1665.

Mondal, M. I. H., Khan, M. M. R., and Ahmed, M. F. (2016). Physico-mechanical properties of finished denim garment by stone-enzymatic treatment. *Journal of Textile and Apparel, Technology and Management*, 10(1), 1-6.

Montazer, M., and Maryan, A. S. (2007). A comparative study of different stone and bio-stone washing of denim. *9th International Conference of the Asia-Pacific Region for Textile*, Taiwan.

Nayak, R., George, M., Jajpura, L., Khandual, A., and Panwar, T. (2022). Laser and ozone applications for circularity journey in denim manufacturing-A developing country perspective. *Current Opinion in Green and Sustainable Chemistry*, 100680. https://doi.org/10.1016/j.cogsc.2022.100680.

Ortiz-Morales, M., Poterasu, M., Acosta-Ortiz, S. E., Compean, I., and Hernandez-Alvarado, M. R. (2003). A comparison between characteristics of various laser-based denim fading processes. *Optics and Lasers in Engineering*, 39(1), 15-24. https://doi.org/10.1016/S0143-8166(02)00073-8.

Özgüney, A. T., Özçelik, G., and Özkaya, K. (2009). A study on specifying the effect of laser fading process on the colour and mechanical properties of the denim fabrics. *Textile and Apparel*, 19(2), 133-138.

Pal, H., Chatterjee, K. N., and Sharma, D. (2017). Water footprint of denim industry. In Muthu, S. S. (Ed.), *Sustainability in denim* (1st ed., pp. 111–123). Elsevier. http://dx.doi.org/10.1016/B978-0-08-102043-2.00005-8.

Periyasamy, A. P., and Militky, J. (2017a). Denim and consumers' phase of life cycle. In S. S., Muthu (Ed.), *Sustainability in denim* (1st ed., 257-282). Woodhead Publishing. http://doi.org/10.1016/B978-0-08-102043-2.00010-1.

Periyasamy, A. P., and Militky, J. (2017b). Denim processing and health hazards. In S. S., Muthu (Ed.), *Sustainability in denim* (1st ed., pp. 161–196). Elsevier. http://doi.org/10.1016/B978-0-08-102043-2.00007-1.

Periyasamy, A. P., and Periyasami, S. (2023). Critical Review on Sustainability in Denim: A Step toward Sustainable Production and Consumption of Denim. *ACS Omega*, 8(5), 4472-4490. https://doi.org/10.1021/acsomega.2c06374.

Rodríguez-Couto, S. (2012). Laccases for denim bleaching: an eco-friendly alternative. *Sigma*, 1, 10–12.

Sağırkaya, F. S. (2021). *Sustainable practises in denim*. [Master dissertation]. Pamukkale University.

Şahbaz, S., Inönü, H., Öcal, S., Yılmaz, A., Pazarlı, C., and Yeğinsu, A. (2007). Denim sandblasting and silicosis two new subsequent cases in Turkey. *Tüberk Toraks*, 55(1), 87–91.

Samanta, K. K., Basak, S., and Chattopadhyay, S. K. (2017). Environmentally friendly denim processing using water-free technologies. In S. S., Muthu (Ed.), *Sustainability in denim* (1st ed., pp. 319–348). Elsevier. http://doi.org/10.1016/B978-0-08-102043-2.00012-5.

Schimper, C. B., Ibanescu, C., and Bechtold, T. (2011). Surface activation of dyed fabric for cellulase treatment. *Biotechnology Journal*, 6, 1280-1285. https://doi.org/10.1002/biot.201100002.

Shahriar, Asif. (2020). An investigation of eco-friendly enzyme washing process in denim apparel. *SSRG International Journal of Polymer and Textile Engineering (SSRG-IJPTE)*, 7(1), 29-38.

Shibly, M. A. H., Hoque, M. M., and Miah, S. (2021). Development of eco-friendly denim fabric washing by natural resources. *International Journal of Textile Science*, 10(1), 1-6.

Siddique, A., Hussain, T., Ibrahim, W., Raza, Z. A., Abid, S., and Nazir, A. (2017). Response surface optimization in discharge printing of denim using potassium permanganate as oxidative agent. *Clothing and Textiles Research Journal*, 35(3), 204–214. https://doi.org/10.1177/0887302X17701600.

Tarhan, M. (2005). *Effect of worn out methods on denim products performance properties*. [Master dissertation]. Dokuz Eylül University.

Toksöz, M., and Mezarciöz, S. (2013). Application of special washing processes to denim fabrics. *Çukurova University Journal of the Faculty of Engineering and Architecture*, 28(2), 141-148.

Tudoran, C., Roşu, M., and Coroş, M. (2020). A concise overview on plasma treatment for application on textile and leather materials. *Plasma Processes and Polymers*, 17(8), 2000046. https://doi.org/10.1002/ppap.202000046.

Tufan, H. A., Emekdar, E., Şahin, U. K., Şensoy, M., Tunalı, Y., and Güngör, H. (2021). Effects of anhydrous enzymes usage in denim washing on fabric tear strength. *European Journal of Science and Technology*, 32, 1203-1206. https://doi.org/10.31590/ejosat.1055060.

Xu, J., Liu, F., He, Z., Zhang, Z., and Li, S. (2021). Cost optimization of sodium hypochlorite bleaching washing for denim by combining ensemble of surrogates with particle swarm optimization. *Journal of Engineered Fibers and Fabrics*, 16, 15589250211022332. https://doi.org/10.1177/15589250211022331.

Zerin, I., Foisal, A. B. M., Ansari, I. Z., and Hoque, S. M. I. (2019). Effect of enzyme wash with stone on different properties of stretch denim fabric. *SEU Journal of Science and Engineering*, 13(1), 76-83.

Chapter 9

Recent Developments in Flame Retardants for Cotton Fabrics

Seda Yuksekdanaci[1,*], PhD and Gokhan Kok[2], PhD

[1]Department of Occupational Health and Safety, Uşak University Faculty of Health Sciences, Uşak, Türkey
[2]Department of Chemistry, Ege University Faculty of Science, Izmir, Türkey

Abstract

Cotton fabrics are widely used in many fields due to their excellent properties. However, the poor thermal stability and flammability of cellulose fibers limit their use in the production of high-performance fire-protective textiles. Many different inorganic and organic flame retardants have been prepared and applied to provide flame retardancy to cellulose fibers. The history of flame retardancy for cellulosic structures goes back to the use of alum in ancient Egypt and Rome about 400-250 BC. Nowadays, cellulose functionalization methods mainly focus on surface modification, such as coating, layer-by-layer, and grafting. Although the main purpose is to prevent burning, flame retardant cotton products should be comfortable, human movement without restriction, be compact and light, strong effect, longevity on resistant to repeated washing, drying, and dry cleaning. However, nowadays, in addition to these features, non-toxic and eco-friendly are also important requirements. There is a clear emphasis on discontinuing the use of halogenated flame retardants in general, partly under pressure from non-governmental organizations. The key issue for risk managers to be concerned with is whether the benefits to both the environment and human health outweigh the risks. The world of cotton flame retardancy has undergone and is still undergoing many evolutions. These evolutions are largely dependent on different "driving forces." A great deal of effort is continually put into the development of new flame retardants by scientists. Efforts to develop new flame retardants by using alternative methods such as nanoparticle adsorption, layer by layer assembly, and sol-gel processes continue intensively. However, in terms of factors such as eco-friendly and industrial-scale manufacturability, phosphorus-nitrogen systems still remain the most promising, effective and industrializable flame retardants for cotton fabrics. However, efforts to develop more suitable and efficient methods continue. In this chapter, the advantages and disadvantages of these current methods in terms of relevant criteria will be discussed.

---

* Corresponding Author's Email: seda.yuksekdanaci@usak.edu.tr.

In: Textile Insights and Future Aspects
Editors: Gamze D. Tetik, Gizem Celep and Fulya Yilmaz
ISBN: 979-8-89113-289-4

Keywords: cotton fabric, durable, flame retardants

Introduction

The height of cotton plant is 80-120 cm in length, it is a pile-rooted plant with separate cocoon branches, and it has a long flowering and harvest period. Cotton, which can be grown in hot climatic conditions, is a one-year plant that does not have much soil selectivity but needs more water. The structure of cotton fiber consists of four layers called cuticle, primary wall, secondary wall, and lumen. The outermost part of the cotton fiber is the cuticle layer. The cuticle is a waxy layer containing fatty acids, alcohols, glycerides, hydrocarbons, sterols, aldehydes, acyl compounds, cutin, resins, and suberin. The inert structure of wax in the cuticle structure protects the fiber against chemicals and other decomposing substances. The primary wall is a layer of very thin cellulose threads called fibrils just below the cuticle. These walls include less than 30% cellulose, different sugars, non-cellulosic polymers, uronic acid, and different proteins. This layer contains cellulose with a repeating number of monomeric units between 2000 and 6000. Since the primary wall consists of spirals, it provides the cotton fiber to be durable. Below the primary wall is the secondary wall, which forms the body of the fiber. This secondary wall consists of approximately 100% cellulose with a high degree of polymerization (about 14 000). The lumen, which is the fourth layer of the cotton fiber, is a hollow channel that runs along the fiber. This channel is filled with protein, sugar, and mineral solutions during the development process of cotton.

It has been determined that cotton cultivation in India has been made at least 5000 years ago, and its use in fabric weaving coincided with 3000 BC. Cotton, which has been one of the basic goods of the global economy for about 5000 years, is also a renewable raw material source that meets the essential needs of people by being transformed into textiles, clothing, and many other products as the main natural fiber in the world (Küçük, 2015). While the share of synthetic fibers has increased in the global textile fiber market, the share of cotton has started to decrease since the 1960s. Despite this, cotton is still the most used natural textile fiber in the world. The reasons why cotton is preferred so much are that it gives a soft and comfortable feeling to the skin, regulates body temperature well, is cheap, absorbs sweat easily, and cotton clothes can be used comfortably in both tropical and cold climates.

Cotton is an important agricultural crop that is produced for both fiber and oil. Its fibers are the most important input of the textile industry. Its seed is evaluated in the vegetable oil industry, and the remaining pulp after the oil is extracted is used in the feed industry. Every part of the cotton can be used as pulp from its seed, yarn from its fiber, and fuel from its pulp. Furthermore, cotton farming helps create jobs in rural areas while providing an income to local farmers who depend on this crop for their livelihoods. In addition to these economic advantages, cotton production can provide environmental benefits due to its ability to capture carbon dioxide from the atmosphere, thereby helping reduce global warming effects.

The rapid increase in the world population and the increase in the standard of living in industrialized developing societies have increased the consumption and need for cotton. Cotton, which is produced in many countries worldwide, is in high demand in terms of meeting the need of people for clothing, being used as feed in the livestock sector, and as oil in the food sector (Karlı et al., 2018). In recent years, due to environmental concerns caused by petroleum-derived fuels, oil obtained from cotton seeds has been used as a raw material in biodiesel production in increasing amounts. Thus, cotton has also become a part of energy agriculture.

Cotton gains a strategic dimension by making significant contributions to the country's economy through the added value and employment opportunities it creates with this wide area of use. Due to these features, it contributes positively to the development of both agriculture and industry in the regions and countries where it is grown. Considering the world cotton cultivation area and production from 1950 to today, it is seen that although the cultivation areas have increased by approximately 20%, cotton production has increased by approximately four times. In the 2019/2020 cotton season, 26 million tons of cotton were manufactured in an area of 34.2 million hectares worldwide (Özüdoğru, 2021). In the last 70 years, in parallel with the increase in cotton production worldwide, its consumption has increased approximately three times. Approximately 63% of cotton cultivation areas in the world are in Asia, 20% in America, and 14% in Africa. Almost all of the world's cotton production (99.5%) is carried out by the ten largest cotton-producing countries, listed as China, India, USA, Brazil, Australia, Turkiye, Pakistan, Uzbekistan, Argentina, and Greece. World cotton use is also carried out by eight major cotton consumer countries, and these countries are listed as China, India, Pakistan, Turkiye, Vietnam, Bangladesh, Uzbekistan, and Indonesia. While 80% of the world's cotton export is made by six major exporting countries (USA, Brazil, Australia, India, Greece, and West African-CFA countries: Mali, Benin, Burkina Faso), 70% of the world's cotton import is made by five major importing countries (China, Vietnam, Bangladesh, Türkiye, and Pakistan).

Thermal Properties of Cotton Cellulose

The chemical event that occurs as a result of the combination of a flammable substance with oxygen under the influence of heat at a certain rate is called combustion. In order to start the combustion, the combustible material, the combustible substance (oxidizer), and the heat source must be together.

Textile products are one of the most widely used materials in human life. Textiles consist of natural polymers (cellulose and protein), or synthetic polymers (polyesters, polyolefins, and polyamides). Since these polymers are a rich source of carbon-based compounds, they are excellent flammable substances during the combustion process (Horrocks & Price, 2001). Combustion in textile materials occurs in four stages (Weil & Levchik, 2008). These stages are; the heating of textile material, pyrolysis, flaming, combustion, and growth of combustion.

With a given energy supply, the surface temperature of the textile material increases rapidly in relation to the thermal properties of the textile product. After the material has warmed up to its decomposition temperature, the macromolecules that make up the fibers begin to break off. Up to a certain temperature, pyrolysis proceeds at a slow rate. However, after a certain temperature is reached, the pyrolysis rate increases and a flammable gas mixture is formed. At this stage, a spark or flame contact can ignite the textile material (Horrocks & Price, 2001).

The continuation of the combustion mechanism depends on the amount of energy required to increase the heat of the textile material. As long as the increased heat of the material is sufficient to thermally break the fibers and ignite the formed pyrolysis products, the combustion mechanism will proceed. Glassy pyrolysis (Tp), combustion (Tc) temperatures, and limiting oxygen index (LOI) values for some textile fibers are shown in Table 1.

When the textile material is burned, compounds such as CO, CO_2, HCl, HCN, SO_2, NO_2, sulfur, hydrogen, acetaldehyde, and isocyanates can be formed depending on their organic structure.

The flammability of a textile material depends on several variants. The properties of the textile polymer, the structure of the textile, the weight, the additives in the fiber, and the oxygen concentration are some of these factors. Both fiber and fabric properties affect the combustion properties of fabrics. The chemical properties of the fiber, the texture of the fabric, fiber composition, weight, surface shape, and thermal properties affect flammability.

The thermal behavior of natural polymers is studied in three phases. In the first phase, liquid and gaseous products are formed by heating natural polymers. When the natural fibers are heated to 105°C, the moisture in them is removed. The reactions in the first phase are endothermic and occur slowly. When the temperature rises, thermal decomposition begins in natural fiber compounds. When the temperature reaches 150-200°C, gaseous products are formed and intermolecular bonds begin to weaken. Around 200–260°C, exothermic reactions begin and increase in gaseous products is observed. The ignition of some hydrocarbons with a low boiling point occurs in this phase. The second stage occurs between temperatures of 260-450°C. In the second stage, ignition of the substances formed as a result of thermal decomposition is observed. This stage includes the steps in which there is an increase in mass loss, the flame spreads with flammable gases, and the amount of heat increases. The third phase occurs at temperatures above 500°C. The final phase includes the slow combustion of residual waste materials and the formation of ash. Coal formation is high in this phase. The ignition of natural fibers depends on many parameters such as the density of the material, its chemical content, moisture content, thermal conductivity and thermal capacity, and the geometry of the material.

Table 1. Pyrolysis (Tp), combustion (Tc) temperatures, and limiting oxygen index (LOI) values of some textile fibers (Tomasino, 1992)

Fiber	T_p (°C)	T_c (°C)	LOI (%)
Wool	245	600	25
Cotton	350	350	18.4
Viscose rayon	350	420	18.9
Nylon 6	431	450	20-21.5
Nylon 6.6	403	530	20-21.5
Polyester	420-427	480	20-21
Acrylic	290	250	18.2
Polypropylene	469	550	18.6
PTFE	400	560	95
Nomex	410	500	28.5-30
Kevlar	590	550	29
PBI	500	500	40-42

Cotton, which is counted at the beginning of natural fibers, is pure cellulose and is formed by the condensation of glucose units. The amount of polymerized glucose in cotton fiber is around 2000-2200 on average. If the cotton is cross-sectioned and examined microscopically, it is seen that the cotton fibers are tubular. The polymeric chains shown in Figure 1 are arranged sequentially in layers that are helically coiled on top of each other and form the tubular fiber.

Cellulose is hydrophilic, so it can trap more water than its mass in its structure. When it is squeezed, it gives back this water under pressure. The three hydroxyl groups it carries act like acids against strong bases. It decomposes with concentrated and strong acids in hot and cold temperatures. It is completely soluble in concentrated sulfuric acid.

When cellulose is exposed to high temperatures, it goes through a process known as pyrolysis, where it decomposes into volatile gases, tar, and coal. During pyrolysis, cellulose is broken down into smaller molecules, including water, carbon monoxide, carbon dioxide, and other organic compounds. If the temperature is high enough, the cellulose will continue to burn, producing flames and releasing even more heat and light energy. However, in its natural state, cellulose does not burn easily because its ignition temperature is high and it takes a lot of heat to start burning.

Cellulose leads to decomposition reactions at different temperatures. Decompositions at a low temperature are usually oxidative or hydrolytic. When the temperature rises above 200°C, cellulose absorbs heat, and loss of water is observed. When the temperature reached 250°C, pyrolysis began to appear. Cellulose is chemically broken down to form products with a lower molecular weight through pyrolysis. The products to be formed are volatile alcohols, ketones, and aldehydes, less volatile but highly flammable tars, and solid carbonaceous ashes. As the temperature increases further, bond breaks and breaks in the cellulose chain are seen. The next step after this stage is combustion.

Cellulose fibers begin to decompose when their temperature rises to around 300-400°C by being exposed to heat or flame for a certain period of time. In this degradation, straight-chain carbonized cellulose and volatile products begin to be released. When the temperature of the cellulose fibers rises to around 400-600°C, straight-chain organic molecules break down again. At this stage, oxidized coal, carbon dioxide, carbon monoxide, methane, ethylene, and aromatic cellulose are released. If the temperature rises even more and reaches around 600-800°C, the aromatic carbonized cellulose is broken down into acetylene. When the temperature reaches 800-900°C, products such as methane and ethylene are oxidized, releasing carbon monoxide and carbon dioxide (Mathews, 2007).

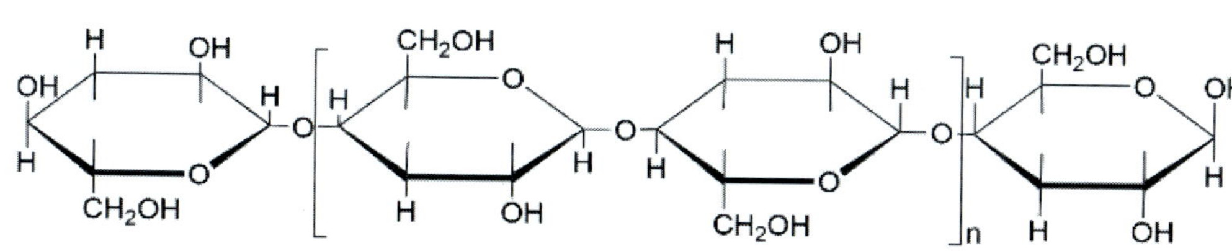

Figure 1. The structure of cellulose.

Flammability Tests

Numerous parameters are considered in the evaluation of the flame retardancy of a textile material. The combustion behavior of textile products is evaluated by considering parameters such as ignition time, self-continuation of combustion, flame spread rate after the product is ignited, internal combustion time, amount of carbonization, and amount of toxic gas released. Therefore, there are different test methods to determine the combustion behavior of textile products.

Limiting Oxygen Index (LOI)

The limiting oxygen index (LOI) is widely used to determine flame retardancy. With the LOI method, the minimum O_2 concentration required in the atmosphere for a textile product to continue to burn is determined. The textile sample is placed vertically in the atmosphere while being tested and then ignited at the top edge with a flame. The oxygen and nitrogen flows are adjusted and measured until the sample is completely burned. After the test, the LOI value is calculated with the formula $\% \text{ LOI} = V_{oxygen} \text{ X } 100/(V_{oxygen} + V_{nitrogen})$. The classification of materials according to the LOI value is shown in Table 2.

Vertical Flammability Test

It is a test method used to determine the combustion behavior of fabrics in the vertical position. The fabrics placed vertically on the device assembly are exposed to the burner flame for 12 seconds. With this test, after the flame source is removed, the burning length, flaming duration, and smoldering time are measured. The results vertical burn-in test are more realistic, especially since the position of the fabric in the garment state is closer to the vertical position compared to all burn angles.

45° Angle Test

In this procedure, the test sample placed at an angle of 45°C to the apparatus is exposed to the standard burner flame for 12 seconds from its lower edge. The evaluation at the end of the test is made by measuring data such as flaming burning time, smoldering time, and time to burn to a certain length after the flame source is removed. This test is mostly applied to woven, knitted, and surface nonwoven fabrics.

Table 2. Classification of flame retardants according to LOI values

LOI value (%)	Class
<24	Combustible, flammable
24-28	Limited power igniter
29-34	Power igniter
>34	Extra power igniter

Cone Calorimeter Test

The cone calorimeter test is based on the fact that the amount of heat released from a burning product is directly related to the amount of oxygen consumed during combustion. With this method, the heat that will emerge depending on the amount of oxygen used up during the burning of the textile material is calculated. Since the amount of heat that will emerge during combustion is effective in the spread of the fire, it is very important to determine the amount of heat released by a material with this method. In addition, the mass loss rate, smoke emission-oxygen consumption-carbon monoxide, and carbon dioxide production amounts of the substance can be measured by this method (Levchik, 2006).

Past, Present, and Future Perspectives of Cotton Flame Retardancy

Cotton, the most widely used textile material, is also highly flammable. Flame retardant finishing of cotton fabrics and clothing and, in many cases, has become necessary today to improve human safety. Most of the studies in the textile flame retardancy industry have concentrated on reducing the flammability of cotton (Horrocks, 1986). This chapter is focused on the only durable methods due to their industrial value.

From the Past to the 1950s

Flame retardant materials first appeared with the use of alum and vinegar in Ancient Egypt in 400 BC and in the Romans in 250 BC. However, its production as an industrial product began in the 17th century with the use of plaster and clay-applied fabrics in theaters in Paris to reduce the risk of fire. Perkin first described a flame retardancy finish for cotton using a mixture of sodium stannate and ammonium sulfate (Esch, 1997).

From the 1950s to the 1980s

During the 1950-1980 period, the fire safety issue and federal regulations in European countries and the United States led to the development of new, and effective flame-retardant chemicals to reduce fire hazards and meet product flammability standards. The first studies on organo-phosphorus flame retardants for cotton fabrics were granted during this period (Tesoro et al., 1968; Aenishanslin et al., 1969). Many flame retardants still used today were developed during this "golden period."

Tetrakis (hydroxymethyl) phosphonium salts (THPX), and Pyrovatex CP (Aenishanslin et al., 1969), developed during this period, continue to be the most popular and widely used flame retardants for cotton fabrics still.

From the 1980s to the 2000s

During this period, much less progress has been made in the development of new flame retardants. However, the mechanisms of action and various environmental and toxic effects of previously developed flame retardants were revealed in these years.

On the other hand, the first legal regulations (1988) were released in order to prevent cigarette-induced fires in furniture fabrics in England. These regulations expanded the use of fabrics coated with flame retardants and led to the development of products such as decabromodiphenyl ether (decaBDE) (Horrocks, 2003; Weil & Levchik, 2008).

From the 2000s to Today

Studies after 2000 focused on the production of less expensive and more eco-friendly new flame retardants (Stowell & Young, 2003; Wu & Young, 2004). Partial progress has been made in the development of halogen-free and formaldehyde-reduced flame retardants. On the other hand, many studies have been carried out to modify the eco-toxic properties of old but gold methods, THPX and Pyrovatex CP (Wu & Young, 2004).

THPX method

THPX has been one of the most widely used durable flame retardants for nearly 50 years. The method is based on tetrakis(hydroxymethyl) phosphonium salts (THPX) (Weil & Levchik, 2008). The reaction between formaldehyde and phosphanide is the fundamental of the THPX method. The chemical reaction of the method is shown in Figure 2.

The proban process is the most widely used process to apply the THPX method. During the proban process, THPX is reacted with urea in a solution whose pH is adjusted to 5-8 before being applied to cotton. The cotton fabric is dried and cured with ammonia gas in a specially designed-ammonia plant. Then, the product is reacted with hydrogen peroxide to produce the final fabric with a NCH_2-P(=O)-linking network. This final cotton fabric has no hydrolyzable functional groups and contains a very stable phosphine oxide skeleton. Therefore, the final fabric can resist 100 washes with alkaline detergents (Kozlowski et al., 2007; Weil & Levchik, 2008).

In the case of using chlorinated THPX, free formaldehyde in the final fabric can produce a volatile carcinogen bis(chloromethyl) in working places (Weil & Levchik, 2008). The most important disadvantage of this method is the need for a special ammonia chamber, which is costly to build. Therefore, the method can only be used by a few major companies that can afford the costs. It is also known that the method gives some side reactions during the application of some dyes. In addition, formaldehyde release occurred during the application. However, this release is lower than that of N-methylol dimethylphosphonopropionamide derivatives.

$$[P(CH_2OH)_4]X + NH_2CONH_2 \longrightarrow (HOCH_2)_2POCH_2NHCONH_2 + HX + HCHO + H_2 + H_2O$$

Figure 2. THPX reaction.

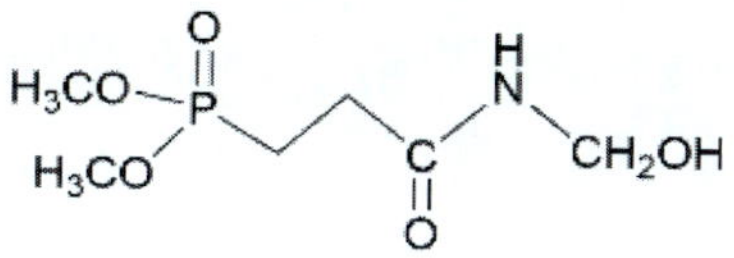

Figure 3. Molecular structure of MDPA.

On the other hand, modified methods for THPX plants that cannot use the gaseous ammonia process has been investigated. The American Cyramid company has developed a modified THPX method using aqueous ammonia (Weil & Levchik, 2008).

Pyrovatex CP and Its Modifications

N-methylol dimethylphosphonopropionamide (MDPA), with its various commercial versions, such as "Pyrovatex CP" and CP New, has been one of the most popular durable flame retardants for cotton since its introduction in 1968 (Tesoro et al., 1968; Aenishanslin et al., 1969). The structure of MDPA is shown in Figure 3.

During the covalent bonding of MDPA's *N*-methylol groups to cellulose, trimethylol melamine is added as a co-reagent to increase bonding yield (Horrocks et al., 2005; Kozlowski et al., 2007).

Since formaldehyde is formed as a by-product of covalent bonding, formaldehyde release continues both during application and throughout the life of the product. The presence of moisture and acidic residues in cotton causes the formation of formaldehyde during storage and service through the hydrolysis reaction. For this reason, it is not used in sensitive products such as children's clothing (US Government of Commerce, 1971).

Horrocks and Hunysman Corp. developed a low formaldehyde grade Pyrovatex CP LF in their research to reduce the formaldehyde release of Pyrovatex CP. Grade Pyrovatex CP LF (Low Formaldehyde) can meet the Oeko-Tex European environmental standard (Weil & Levchik, 2008). Pyrovatex CP LF and Pyrovatex CP New (released later) can covalently bond to cellulose via a phosphoric acid catalyzed reaction to form aminole-bonded structures. The bonding occurs mostly on the primer hydroxyl groups of cellulose because of the sterically hindrance of other positions.

A cotton textile treated with Pyrovatex CP by this method can be durable after 50 or more washes. However, it is more affected by hypochlorite bleaching processes and is not as durable as THPXs.

However, Pyrovatex CP does not have special requirements, such as the patented ammonia gas curing system and a high-cost ammonia plant in the proban method. Instead, it can be easily applied to cotton fabrics with the traditional pad-dry-cure system.

Although there are some commercial and ecological restrictions on the application of these methods, today Pyrovatex and THPX are still widely used. In order for the newly developed alternative methods to replace these two methods, they must meet the following features.

- equal or better convenience of application,
- release zero formaldehyde,
- have a long textile service life on fabric properties (durability, grip, and shrinkage),
- comparable cost and preferably less expensive,
- have equal or less toxicological and environmental effects.

The most important alternative methods developed for these purposes to the present day are surface modification by plasma method, microcapsulation, and nanotechnology applications.

Nanotechnology

Some nano-surface materials have flame-retardant properties. The use of these flame-retardant materials for a combustible product has created a new field for flame-retardant finishing. It is possible to obtain flame retardant products with the special dimensions and structural effects of these nano-surfaced materials (Zhang, 2009). Inorganic-organic hybrid (IOH) nanoparticles have allowed the development of flame-repellent polymers, which have the ability to extinguish when ignited, limit flame formation, and provide low smoke and toxic gas emissions. Most examples of the use of nanoparticles to improve the fire performance of polymer coatings have focused on back coatings for fabrics. Bourbigot et al., (2002; 2005) indicate the addition of nanoclays and poly(silsosesquioxanes) can reduce peak heat releasing in polyurethane-coated cotton.

Microencapsulation

A microcapsule consists of a polymeric wall and a liquid substance trapped within this wall. The capsule wall is an inert substance to the liquid inside. Preparing an emulsion from two immiscible liquids is the first step in order to disperse the core material in the polymer at a micron size. There are various microcapsule methods such as

- coacervation,
- interfacial polymerization
- *in-situ* polymerization
- spray drying
- solidification of liquids
- heat drying of the liquid system
- melt cooling.

according to the preparation procedures (Övez & Yüksel, 2002; Salaün et al., 2011). It is possible to obtain flame-retardant textile products by trapping the flame-retardant material in the capsule and applying this capsule to the textile material.

For example, the combination of polyurethane and phosphate provides a good flame-retardant effect, but the flame-retardant finish with phosphate is not permanent due to the dissolution and migration of phosphate in water. In a study by Girauda et al., diammonium hydrogen phosphate (DAHP) was encapsulated in a polyurethane shell (PUR). The study reported a significant flame retardant effect of the capsules (Girauda et al., 2002).

Plasma Technology

Plasma technology, which is included in surface modification, is a method that is applied to add functionality to textile materials and has many advantages compared to conventional methods. Plasma technology has many advantages over wet processes such as it is a dry process (no water requirement), the amount of chemicals used is very small, short processing time, does not cause industrial and toxic waste, does not damage the mechanical properties of the textile material, and the process is effective only on the fiber surface and energy savings (Schindler & Hauser, 2004; Morent et al., 2008; Tsafack & Grützmacher, 2006; Jahagirdar & Tiwari, 2004).

Plasma application of a polyester fabric with oxygen gas (Ömeroğulları & Kut, 2012), the application of hexamethylenedisiloxane and hexamethyldisilane to cotton and polyamide fabrics by plasma polymerization (Kılıç et al., 2007) can be given as other promising examples. If the plasma treatment is used as a pre-treatment, it significantly reduces the amount of chemicals required in the next process and the amount of pollution (Morent et al., 2008; Tsafack & Grützmacher, 2006; Ömeroğulları & Kut, 2012).

A recent study has resulted in patents showing that glare flame resistance can be increased by surface treatment of both pure cotton, flame retardant cotton, and poly(meta-aramid) fabrics using atmospheric plasma in the presence of clay and a silicon-containing monomer such as hexamethylene disiloxane (HMDSO) (Horrocks et al., 2009; 2011).

Tsafack and Grützmacher (2006) grafted 4 flame retardant acrylate monomers induced by argon plasma to impart flame retardancy to cotton fabrics. They showed that the highest LOI values of 28.5 and 29.5 belonged to diethyl (acryloxyethyl) phosphoramide and acryloxy-1,3 bisdiethylphosphoramide) propane.

Toxic and Environmental Effects of Flame Retardancy Chemicals

The toxic effect of flame retardant compounds and combustible gases should be taken into account, especially if flame retardant finishes are used using halogens and heavy metals. Toxic problems caused by these compounds;

- the ability of halogenated compounds, especially aromatic ones, to form polyhalogenated dioxins and furans,
- mixing of halogenated aromatic flame retardants, into waste water (they usually biodegrade slowly and therefore cause high AOX (adsorbable organic halogen) pollution),
- formaldehyde is released during condensation, and free formaldehyde is found in the finished fabric.

On the other hand, in a study conducted on pig ear skin to model human skin, no phosphonic acid permeation was observed after 24 hours of contact with the Pyrovatex-treated fabric (RCC, 1992). However, it has been reported that phosphonic acids extracted from the fabric can cross the dermal barrier at some concentrations.

In another study, the application of pure THPC to the skin resulted in weight loss and death in rats (Aoyama, 1975). However, there is no data on human trials on this subject.

One of the top issues for flame retardants is the eco-toxicity of chemicals. The addition of chitosan during the phosphorylation of cotton fabric provides an environmentally friendly textile at a reasonable price. In this study, chitosan was added during the phosphorylation process and acted as a nitrogen source, creating a synergistic effect with phosphorus (El-Tahlawy, 2008).

Conclusion

In today's ever-increasing turmoil, fire risks are increasing, and the safety of life and property is decreasing. Therefore, the use of flame-retardant cotton fabrics in daily life is inevitably increasing. Researchers conduct many important studies considering these demands.

It is also being investigated whether the textile material can fulfill several functions together instead of providing a single function. Multifunctional cotton fabrics such as water and oil repellent flame retardant cotton products, the flame-retardant cotton with antistatic properties, and even the flame-retardant cotton with anti-crease properties are in demand today.

In addition to a good flame retardant effect, a flame retardant cotton fabric does not have toxic or allergic effects, is environmentally friendly, is resistant to washing and dry cleaning, does not cause yellowing, does not release toxic gases during pyrolysis, does not deteriorate the technological properties of the fiber and has an affordable price are among the expectations. Research studies should continue taking into consideration all of these expectations.

From this perspective, microencapsulation and nanoparticle applications show promising results as a final product in academic studies. However, they are not competitive with THPX and especially Pyrovatex CP methods in terms of ease of application and cost-effectiveness on an industrial scale.

Plasma technology, on the other hand, emerges as a more promising method. The fact that the patents produced are in demand by many companies shows that the method is promising.

However, despite all the disadvantages of the THPX and Pyrovatex CP methods, the fact that their use is still widespread in the industrial scale shows that more studies are needed in this area.

Disclaimer

None.

References

Aenishanslin, R., Guth, C., Hofmann, P., Maeder, A., & Nachbur, H. (1969). A New chemical approach to durable flame-retardant cotton fabrics. *Textile Research Journal, 39*(4), 375-381. https://doi.org/10.1177/004051756903900413.

Aoyama, M. (1975). Effect of anti-flame treating agents on the skin. *Nagoya Medical Journal, 20*(1), 11–19.

Bourbigot, S., Devaux, E., & Rochery, M. (2002). Polyurethane/clay and polyurethane/POSS nanocomposites as flame retarded coating for polyester and cotton fabrics. *Fire and Materials, 26*(4-5), 149-154. https://doi.org/10.1002/fam.792.

Bourbigot, S., Le Bras, M., Flambard, X., Rochery, M., Devaux, E., & Lichtenhan, J. D. (2005). Polyhedral oligomeric silsesquioxanes: Applications to flame retardant textiles. In M. Le Bras, C. A. Wilkie, S. Bourbigot, S. Duquesne, & C. Jama (Eds.), *Fire retardancy of polymers: New applications of mineral fillers*. London: Royal Society of Chemistry.

El-Tahlawy, K. (2008). Chitosan phosphate: A new way for production of eco-friendly flame-retardant cotton textiles. *Journal of the Textile Institute, 99*(3), 185-191. https://doi.org/10.1080/00405000701584311.

Esch, G. J. van, World Health Organization & International Programme for Chemical Safety. (1997). *Flame retardants: A general introduction*. World Health Organization. Retrieved from: www.inchem.org/documents/ehc/ehc/ehc192.htm.

Girauda, S., Bourbigota, S., & Rocherya, M. (2002). Microencapsulation of phosphate: Application to flame retarded coated cotton. *Polymer Degradation and Stability, 2*(77), 285-297. https://doi.org/10.1016/S0141-3910(02)00063-0.

Horrocks, A. R. (1986). Flame retardant finishing of textiles. *Review of Progress in Coloration and Related Topic, 16*, 62-101. https://doi.org/10.1111/j.1478-4408.1986.tb03745.x.

Horrocks, A. R. (2003). Flame retardant finishes and finishing. In D. Heywood (Ed.), *Textile finishing* (pp. 214-250). Society of Dyers and Colourists.

Horrocks, A. R., & Price, D. (2001). *Fire Retardant Materials*. Cambridge, Woodhead Publishing.

Horrocks, A. R., Kandola, B. K., Davies, P. J., Zhang, S., & Padbury, S, A. (2005). Development in flame retardant textiles. *Polymer Degradation and Stability, 88*(1), 3-12. https://doi.org/10.1016/j.polymdegradstab.2003.10.024.

Horrocks, A. R., Kandola, B. K., Nazaré, S., & Price, D. (2009). *UK Patent Application 0900069.6 Flash fire resistant fabric*. 5 January.

Horrocks, A. R., Nazaré, S., Masood, R., Kandola, B. K., & Price, D. (2011). Surface modification of fabrics for improved flash-fire resistance using atmospheric pressure plasma in the presence of a functionalised clay and polysiloxane. *Polymers for Advanced Technologies, 22*(1), 22-29. https://doi.org/10.1002/pat.1707.

Jahagirdar, J., & Tiwari, L. B. (2004). Study of plasma polymerization of dichloromethane on cotton and polyester fabrics. *Journal of Applied Polymer Science, 94*(5), 2014–2021. https://doi.org/10.1002/app.21128.

Karlı, B., Kadakoğlu, B., & Gül, M. (2018, July 24-26). *Cotton production and foreign trade structure in the world and Turkey*. V. International Multidisciplinary Congree of Eurasia. pp. 129-136. Barcelona, Spain.

Kılıç, B., Akşit, A., & Mutlu, M. (2007). Surface modification and characterization of cotton and poliamid fabrics by plazma polymerization of hexamethyldisilane and hexamethyldisiloxane. *International Journal of Clothing Science and Technology, 21*(2/3), 137-145. https://doi.org/10.1108/09556220910933862.

Kozlowski, R., Wesolek, D., Wladyka-Przybylak, M., Duquesne, S., Vannier, A., Bourbigot, S., & Delobel, R. (2007). Intumescent flame-retardant treatments for flexible barriers. *Multifunctional Barriers for Flexible Structure*, 39-61. https://doi.org/10.1007/978-3-540-71920-5_3.

Küçük, N. (2015). The world of cotton, global actorsa and policies. *ASSAM International Refereed Journal, 2*(4), 60-85.

Mathews, M. C. (2007). *Durable and non-toxic topical flame retardants for cotton and cotton blends*, Master's thesis, North Carolina State University, North Carolina.

Morent, R., Geyter, N. D., Verschuren, J., Clerck, K. D., Kiekens, P., & Leys, C. (2008) Nonthermal plasma treatment of textiles. *Surface & Coatings Technology, 202*(14), 3427–3449. https://doi.org/10.1016/j.surfcoat.2007.12.027.

Ömeroğulları, Z., & Kut, D. (2012). Application of low-frequency oxygen plasma treatment to polyester fabric to reduce the amount of flame retardant agent. *Textile Research Journal, 82*(6), 613-621. https://doi.org/10.1177/0040517511420758.

Övez, B., & Yüksel, M. (2002). Slow release of perfumes from crosslinked microcapsules, *Ecology Environment Journal, 10*(43), 26-29.

Özüdoğru, T. (2021). Cotton production economy in the world and Turkey. *Journal of Textiles and Engineer, 28*(122), 149-161. https://doi.org/10.7216/1300759920212812208.

RCC (Research & Consulting Company AG). (1992). Contact Hypersensitivity to FAT 80'001/I in Albino Guinea Pigs Maximization-Test. RCC Project 316710. Study conducted by RCC Research & Consulting Company AG for Ciba-Geigy AG.

Salaün, F., Lewandowski, M., Vroman, I., Bedek, G., & Bourbigot, S. (2011). Development and characterisation of flame-retardant fibres from isotactic polypropylene melt-compounded with melamine-formaldehyde microcapsules. *Polymer Degradation and Stability, 96*(1), 131-143. https://doi.org/10.1016/j.polymdegradstab.2010.10.009.

Schindler, W. D., & Hauser, P. J. (2004). *Flame-retardant finishes*. Chemical Finishing of Textiles Cambridge, Woodhead Publishing.

Stowell, J. K., & Yang, C. Q. (2003). A durable low-formaldehyde flame retardant finish for cotton fabrics. *AATCC Review, 3*(2), 17-20.

Tesoro, G. C., Sello, S. B., & Williard, J. J. (1968). Flame retardant properties of phosphonate derivatives of cotton cellulose. *Textile Research Journal, 38*, 245–255. https://doi.org/10.1177/004051756803800305.

Tomasino, C. (1992). *Chemistry and Technology of Fabric Preparation and Finishing*. Master's thesis, North Carolina State University, North Carolina.

Tsafack, M. J., & Grützmacher, L. J. (2006). Flame retardancy of cotton textiles by plasma induced graft-polymerization (PIGP). *Surface & Coatings Technology, 201*(6), 2599–2610. https://doi.org/10.1016/j.surfcoat.2006.05.002.

US Government of Commerce. (1971). *Standards for the flammability of children's sleepwear* (DOC FF 3-71). Washington, D.C.

Weil, E. D., & Levchik S. V. (2008). Flame retardants in commercial use or development for textiles. *Journal of Fire Sciences*, *26*(3), 243-281. https://doi.org/10.1177/0734904108089485.

Wu, W., & Yang, C. Q. (2004). Comparison of DMDHEU and melamineeformaldehyde as the binding system for a hydroxy-functional organophosphorus flame retarding agent on cotton. *Journal of Fire Sciences 22*(2), 125-142. https://doi.org/10.1177/0734904104039695.

Yang, C. Q., & Wu, W. (2005). The combination of a hydroxy-functional organophosphorus oligomer and melamineeformaldehyde as a flame retarding system for cotton. *Fire and Materials*, *29*, 109-120. https://doi.org/10.1002/fam.877.

Zhang, H., & Sun, G. (2009). Textile flame retardant review. *Modern Applied Science, 2*(3), 129-133. https://doi.org/10.5539/mas.v3n2p129.

Chapter 10

The Utilization of Textile Materials as Alternative Adsorbents for Wastewater Treatment

Gamze D. Tetik*, **PhD**
Department of Materials Science and Nanotechnology Engineering, Uşak University, Uşak, Türkey

Abstract

As product variety and production diversity increased with industrialization, the aqueous waste used for wet processes can be disposed of in the environment as wastewater in a way that disrupts habitat and harms living things, despite some policies to protect the environment. In this sense, engineering products have a great role in the process of releasing wastewater to the environment through various processes in a way that will keep environmental damage to a minimum. Biological, physical, and chemical treatments are applied at this stage, and well-analyzed engineering products are used in this context. Adsorption, which can be performed in wet and dry environments, is a kind of chemical or physical process utilized for this purpose. The adsorbents used in the treatment of wastewater, in this case wet adsorption, have the ability to adsorb the dissolved components from the medium with their high surface area and high affinity. This chapter presents a current review of textile-based adsorbents, which are advantageous with their porous structures, high surface areas, and generally hydrophilic structures used in wastewater treatment. Moreover, the literature lacks studies that draw on the current framework and focus specifically on textile-based adsorbents. Finally, it is thought that textile-based adsorbents are not received the necessary academic attention, and this chapter is an enlightening work for the studies to be done in this field.

Keywords: adsorption, textile-based adsorbents, wastewater treatment

Introduction

Wastewater is defined as the effluent discharged from houses or industries by domestic uses or industrial processes, respectively. In the textile industry, textile wastewater means an effluent containing organic dyes, phosphates, nitrates, and/ or heavy metals (Andrade & Andrade, 2018). As these substances are hazardous to both the environment and living beings, the

* Corresponding Author's Email: gamze.tetik@usak.edu.tr.

In: Textile Insights and Future Aspects
Editors: Gamze D. Tetik, Gizem Celep and Fulya Yilmaz
ISBN: 979-8-89113-289-4

treatment of textile wastewater is of great importance and is controlled by international/national laws and regulations. It is impossible for industrial wastewater to be released into the environment without any post-processing. The contents of wastewater are brought to the regulated values by chemical, biological, and/ or physical treatment processes. In this context and period when industrialization gradually increased, these purification processes formed the basis for various research in terms of the use of diversified engineering products. Adsorption, which can be chemical or physical, is an indispensable method for removing dissolved impurities from textile wastewater.

Techniques for textile wastewater purification can be divided into three groups: The first group consists of coarse separation techniques. Suspended substances are removed by taking advantage of the density difference. In the second group of treatment, where the biological mechanisms have an important place, different technologies are used and the color, the BOD, and the COD values of the wastewater are reduced. In a tertiary group of treatment, pollutants are removed, and organic color compounds are minimized. Adsorption is found in this group in textile wastewater treatment (Kumar & Saravanan, 2017). In the first place, it is tried to find the most suitable adsorbent in the adsorption process. The selection of optimum conditions that will maximize the "separation factor" in separation by adsorption is made by considering the temperature and composition parameters (Ruthven, 1984).

Adsorption is a phenomenon of accumulation of substances that are desired to be removed in a fluidic medium onto solid adsorbents through physical or chemical interactions (Lameire & De Vriese, 2001). In physical adsorption surface atoms of adsorbents and adsorbate molecules interact with van der Waals and electrostatic attraction forces, while in chemical adsorption chemical bonds are formed between adsorbents and adsorbates (Suzuki, 1990; Deaconu et al., 2016). By this means, many toxic compounds and dyes are removed from the wastewater before the final discharge.

The adsorption process is performed to make unsaturated and unbalanced molecular forces saturated and balanced. A solid adsorbent in wastewater medium balances these forces by trying to attract environmentally unfriendly compounds to its surface (Saaidpour, 2014; Demir & Yalçın, 2014). Silica, clays, granular or powder activated carbon (AC), natural and synthetic bio-adsorbents fulfill this task (Srebrenkoska et al., 2014). Since the literature is deprived of a comprehensive study limited to textile-based adsorbents, this chapter presents studies on textile-based adsorbents and emphasizes the surface properties required for the use of textile materials as adsorbents.

Textile-Based Adsorbents

Although many corporate companies have drawn attention to the field with approaches such as sustainability, environmentalism, protecting nature, minimizing the chemicals they use, benefiting from harmless natural resources, and choosing green methods in production, there are still some unfriendly chemicals discharged as a result of many industrial processes around the world. Thus, nature and living beings are harmed.

Adsorbent types used in adsorption, where industrial wastewater is treated before final disposal, are being developed to remove organic compounds, heavy metal ions, and dyes containing numerous types of these. Dutta et al. (2021) classified the adsorbents for dye removal as activated carbon, metal oxide, carbon, and polymer-based adsorbents. Textile-based adsorbents are found in the polymer-based group.

The adsorption process is influenced by various factors. These factors can be classified as adsorbate, adsorbent, and medium/process parameters. Concentration, polarity, and chemical composition are classified in adsorbate parameters. The initial concentration of dye/hazardous material is important in terms of the amount of compounds to be remediated. Since the saturation of the adsorption sites on adsorbents is determined by the initial concentration of the adsorbates, it directly affects the adsorption capacity. Another important factor for both physical and chemical adsorption is the polarity of the adsorbents. The type and strength of the interaction between the adsorbate molecules and adsorbents depend on their polarity (Suzuki, 1990). The composition and surface properties of the adsorbents are determined by their origins (sources). As the quantity of the adsorbent increases, the adsorption capacity increases due to the increase in the adsorption sites. The surface properties of textile-based adsorbents are given in the following subsection in detail. The nature of the adsorbate and adsorbent are key factors in determining the optimum process parameters. For example, the adsorption of copper onto flax fiber adsorbents at lower and higher pH values showed that at lower pH values adsorption sites bonded to hydrogen (H^+), and at higher pH values adsorption of copper was more efficient (Abbar et al., 2017). A positive temperature change increases the kinetic energy of the adsorbate molecules and the probability of collision and convergence of the adsorbate molecules with the adsorbent surfaces. Thus, it enhances adsorption. The mobility of the medium can be achieved by stirring or applying ultrasonic energy. There are studies that indicate the adsorption development with ultrasonic frequency (Heidarinejad et al., 2018). The crucial and interactive parameters of adsorption are listed in Figure 1.

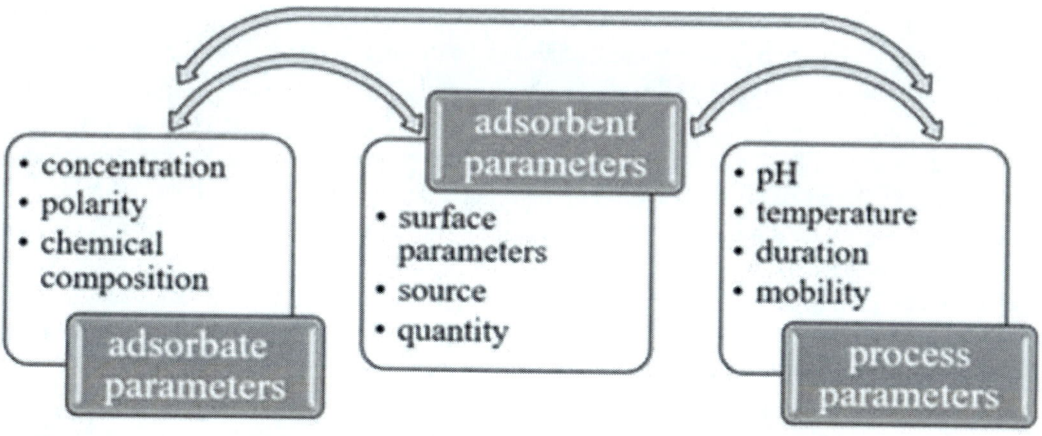

Figure 1. Crucial factors affecting adsorption.

When all of these parameters are considered, it can be concluded that the textile-based porous adsorbents are preferable materials for the removal of various pollutants from wastewater.

Textile-Based Adsorbents in Forms of Fibers, Fabrics, and Felts

The utilization of natural fibers as adsorbents is a new research field in wastewater treatment. The easy availability of natural fibers and their environmental friendliness compared to those of polymeric materials are the advantages. In some cases, several physical and chemical modifications are required in the design of natural fibers as adsorbents (Rahman et al., 2018). In the production of high porosity activated fibers, plant fibers such as jute and coconut can be a good choice; however, depending on the fiber structure, process conditions, and activating agent, fibrous activated carbons of various structures can be obtained (Phan et al., 2006).

Although commercial activated carbon is a favored adsorbent for wastewater treatment due to its performance characteristics, its broad application is limited due to cost considerations (Dias et al., 2007). The use of different waste materials in the preparation of activated carbon is preferred to reduce costs. To date, different studies have emphasized the importance of using waste materials in low-cost adsorbent applications (Chuah et al., 2005; Daouda et al., 2021; Ezeonuegbu et al., 2021; Igwegbe et al., 2021). Activated carbon can be produced from many materials such as coffee wastes (Figueroa Campos et al., 2021), tea wastes (Peng et al., 2013), or municipal wastes (Nagano et al., 2000), as well as from textile wastes (Duman, 2021). Textile wastes containing cotton and polyester have been carbonized, physically, and chemically activated to obtain adsorbents that can be used in oxytetracycline removal. Cotton waste has exhibited ~620 mg/g adsorption capacity (Duman, 2021).

Textile wastes are not only used to synthesize activated carbons but are also utilized for the creation of some types of adsorbents within the framework of environmental approaches. Mendoza-Castillo et al. (2016) synthesized adsorbent structures by pyrolysis for low-cost water defluoridation from denim waste and reported that denim-based adsorbents are effective in the treatment of water and purification. The temperature range of 500–900 °C has been chosen under the N_2 atmosphere. The maximum fluoride adsorption capacity has been specified by the authors as 4.25 mg/g. Mia et al. (2022) reported that the adsorbents prepared by modifying waste silk fabric with tannic acid were effective in the removal of heavy metals from wastewater. In addition to these, there are studies on the production of carbonaceous adsorbents from waste cotton. In these studies, 74 mg/g Alizarin red adsorption capacity (Wanassi et al., 2017) and ~70 mg/g Cr (VI) adsorption capacity (Xu et al., 2020) have been reported.

Hemp is a natural fiber used for metal ion removal from aqueous synthetic solutions (Pejić et al., 2020). The hemp fibers in waste form have been purified by treating them with boiling soap and soda ash, washing and drying processes, and the adsorbents obtained resulted in a cobalt (II) adsorption capacity of ~15.5 mg/g (Tofan et al., 2013). Another study with hemp fibers was conducted in 2014 and focused on copper adsorption in acidic aqueous media. The natural and modified forms of hemp fibers were compared, and the modified ones were found to be effective in copper removal (Balintova et al., 2014). In another study, the influence of the hemp fibers (in different chemical compositions) on the adsorption of heavy metal ions such as Pb (II), Cd (II), and Zn (II) ions was investigated. It has been revealed that when the lignin and hemicelluloses were removed by the chemical treatment, the hemp fibers showed better sorption characteristics. This has been attributed to the increase in roughness of the hemp fiber surfaces that leads to the formation of more capillary gaps (Pejic et al., 2009). There are also studies that hemp is used as an adsorption felt material. According to the study, the parameters for the adsorption of metals played an important role. These outstanding parameters have been determined as the adsorbent dosage (quantity), contact time (duration), and pH for metal sorption in the study. The importance of the optimum parameters of adsorption has been emphasized. Moreover, it has been reported that the affinities of metal ions variated while the Pb (II) ions were more adsorbed compared to others (Loiacono et al., 2018). Furthermore, the hemp fibers have also been investigated as loose fiber and felted forms for removing metallic contents from aqueous solutions containing Al, Co, Cr, Cu, Ni, and Zn. These forms have been composed of raw hemp, hemp washed with water, and hemp washed with sodium bicarbonate. Acidic medium has been found to be effective in sorption. The felt sample washed with sodium bicarbonate has been found to be effective up to 99% (Bugnet et al., 2017).

Agave Americana L. fibers, which are recommended to be used in technical fields and in the production of geotextiles rather than for use in the field of pure textiles (Msahli & Sakli, 2006), are also fibers studied for adsorption in aqueous solutions. This fiber has been tested to determine its effectiveness in the adsorption of a dye (Malachite Green, MG) from aqueous solutions and is suggested as a low-cost adsorbent to remove MG (Altinisik et al., 2015).

The chitosan and kenaf hybrid fiber composites have been analyzed for the removal of heavy metal Nickel (II) ions. According to the results of the study, the chitosan/kenaf fiber composite has a sufficient metal-binding capacity, and thus its adsorption efficiency has been high. The maximum adsorption ratio has been gained at pH 5 (Florence et al., 2015). Kenaf alone is recommended as an effective adsorbent in another study, including batch adsorption tests with copper and nickel ions (Hasfalina et al., 2010).

Jute fiber is the other type of natural fiber that is investigated for determination of its effectiveness in heavy metal adsorption such as Cu (II), Ni (II), and Zn (II). Chemically modified jute fibers by loading with a dye with a specific structure and oxidizing with hydrogen peroxide have presented better metal ion adsorptions. Besides, the study has asserted that when the pH values lowered, the adsorption ratio also decreased (Shukla & Pai, 2005).

In a study examining fibers such as coconut coir, cotton, and kenaf in terms of adsorption of heavy metal ions, it has been revealed that the cell chemistry and structure of the fibers, as well as their lignocellulosic content, have a considerable effect on the adsorption efficiency (Lee & Rowell, 2004).

Electrospun textile nanofiber membranes are the other class of adsorbents used in wastewater treatment. Özturan and Tetik (2022) recently reviewed the electrospun nanofiber adsorbents utilized in methylene blue removal from aqueous solutions. Due to the high specific surface area, high porosity, and functionality of electrospun fiber membranes, they have an important place in wastewater treatment as well as in many other areas. Nano-fiber membranes show superior properties in the treatment of oils, dyes, and heavy metal ions in a water medium (Cui et al., 2020). Carbon nanofibers give good results in Cu (II) removal from wastewater (García-Díaz et al., 2018). Polyethylene terephthalate (PET) electrospun nanofibers chemically modified, two hours stirring at room temperature, with dibenzo crown ether have been studied for the removal of methylene blue (MB) from aqueous solutions. The modified PET nanofiber membrane has resulted in good MB adsorption that is fitting to the Langmuir model (Ahmed et al., 2021).

Considering that seas are water bodies where wastewater is sometimes discharged after the final disposal, it would be appropriate to mention a study carried out for seawater treatment. Five types of polyethylene (PE) fibers with high surface areas have been used in the designs of braided, knitted, and leno woven fabric samples as adsorbents that can be used for uranium removal from seawater. The uranium adsorption capacity has been calculated at 120-185 g/kg under laboratory conditions (Janke et al., 2014).

The studies conducted on textile-based adsorbents, which have advantages in adsorption due to their unique properties such as lignocellulosic contents, compositions, high surface areas, biodegradability, and hydrophilicity, are summarized in Table 1.

Table 1. Studies on textile-based adsorbents in the literature

Textile-based adsorbent type	Treatments	Adsorbate/ Industrial process	Key findings	Refs
Cotton, polyester textile waste	Hydrothermal carbonization, physical and chemical activation	Oxytetracycline	*higher surface area carbon with $ZnCl_2$ *higher number of acidic surface groups with CO_2 *higher C content over 90% with KOH *620 mg/g adsorption capacity	Duman, 2021
Denim waste	Pyrolysis	Defluoridation	*Denim waste as a low-cost and value-added product *4.25 mg/g adsorption capacity	Mendoza-Castillo et al., 2016
Waste silk fabric	Tannic acid modification	Heavy metals	*8.03 mg/g Cd (II) *7.42 mg/g Cu (II), *7.47 mg/g Ni (II) adsorption capacities at pH 9	Mia et al., 2022
Cotton waste	Carbonization and $ZnCl_2$ chemical activation	Alizarin red	*74 mg/g adsorption capacity (at 25 °C, pH 3) *hollow fiber structured AC (Ø 6.5 μm) *292 m^2/g surface area	Wanassi et al., 2017
Cotton waste	Pyrolysis using iron salts	Cr (VI)	*max. 70.4 mg/g adsorption capacity with $FeCl_3$	Xu et al., 2020
Hemp fibers (in waste form)	Purification using soda ash and soap	Co (II)	*low-cost and ecofriendly adsorbent *15.44 mg/g column adsorption capacity 13.58 mg/g batch adsorption capacity	Tofan et al., 2013
Hemp fibers (the waste in short fiber from)	NaCl modification	Pb (II), Cd (II), Zn (II)	*adsorption in single ion solutions is 0.078 mmol/g *NaCl modified hemp fibers give the max. adsorptions as 0.074 mmol/g for Pb (II), 0.035 mmol/g for Cd (II) and 0.035 mmol/g for Zn (II)	Pejic et al., 2009
Hemp fibers (0.063-4.00 mm in length)	unmodified + modified forms (with NaOH and CH3COOH)	Cu (II)	*slight decrease in surface area of modified hemp fiber (from 0.99 m^2/g to 0.93 m^2/g) *8.10 and 8.21 mg/g adsorption capacity for raw and modified fibers	Balintova et al., 2014

Textile-based adsorbent type	Treatments	Adsorbate/ Industrial process	Key findings	Refs
Agave Americana L. fibers (2-10 µm Ø of fibrils)	Washing, combing, drying, grinding treatments	Malachite Green	*chemical composition determination *33 mg/g adsorption capacity at ~45 °C	Altinisik et al., 2015
Chitosan/ kenaf bio-composite	Chitosan solution obtained with 8% acetic acid + kenaf fibers	Ni (II)	*70.55 mg/g adsorption capacity for chitosan pH 5	Florence et al., 2015
Jute	Unmodified, dye-loaded, oxidized forms	Cu (II), Ni (II), Zn (II)	*removal ratios up to 80-84% for Cu and Ni with dye (C.I. Reactive Orange 13) loaded jute fibers *removal ratio up to 66% for Zn with dye (C.I. Reactive Orange 13) loaded jute fibers	Shukla & Pai, 2005
Modified PET nanofibers	dibenzo crown ether modification	Methylene blue	*optimal conditions of adsorption: 30 °C temperature, pH 6, 100 min contact time	Ahmed et al., 2021

Surface Properties of Textile-based Adsorbents

The main reason why the adsorbents are primarily characterized by their surface properties is that adsorption is a surface process and adsorbents must have the optimal surface properties for maximum efficiency (Suzuki, 1990). Since high surface area means the creation of lower-size pores between adsorption surfaces and increases adsorption capacity, several studies focused on modifying textile-based adsorbent surfaces by various agents as mentioned above. In this subsection of the chapter, surface-related properties of textiles are given with measurement methods.

Surface Area

The total area of the surfaces of a material gives the surface area of that material (in m^2), and the ratio of the total surface area to its mass gives the specific surface area (in m^2/g) (Kaewprasit et al., 1998). If this material is a 2D (two-dimensional) non-porous structure, the rectangular prism surface area calculation can be used, considering its thickness. If it is a 1D (one-dimensional) material with a high aspect ratio, the cylinder surface area is calculated. However, things are a little more complicated when fabrics in the 2D material class and fibers in the 1D material class are considered. It is because textile structures contain pores with interconnectivity that change the surface area.

Two critical parameters that directly affect the fabric surface area are the fineness and cross-sectional shapes of the fibers that compose the fabric. If finer fibers are used, the fabric unit weight decreases and the total fiber surface area increases. Increasing the surface area provides an increase in the number of adsorption sites to which the adsorbate will bind and, as a result, an enhancement in the adsorption capacity is encountered (Tascan & Vaughn, 2008). Besides, the surface area of the fabrics is influenced by the tightness and compactness of the fabric.

There are different techniques for surface area measurements of textile materials. The calculated surface area is presented considering the total fiber length in the fabric with software that measures the perimeter of the fibers on scanning electron microscope images of fabrics (Tascan et al., 2011). In addition, MB adsorption is used to determine the surface areas of different cotton fibers. The total amount adsorbed from the first and final concentration the difference of MB in the medium is calculated over the absorbance values recorded by a UV (Ultraviolet)-visible spectrophotometer, and it is stated that the method can be used to determine the surface areas of cotton fibers. In the study, it has been observed that the cotton fiber surface area is highly correlated with its crystallinity (Kaewprasit et al., 1999).

Regarding the experimental measurement methods of surface area, it can be said that the Brunauer-Emmett-Teller (BET) measurement method is the most common method. In this method, measurements are performed by means of a surface area analyzer. Samples are degassed at temperatures that do not damage the material. BET surface area values are calculated through the physically adsorbed -and desorbed- amount of gas (N_2) (which is sufficient for the formation of a gas layer) on the textile sample (Tetik et al., 2022). The larger the surface area, the more interaction sites there will be between textile adsorbents and adsorbate molecules.

Furthermore, when a constant temperature is reached during the execution of the surface area measurement, isotherms are obtained from the relationship between the concentration of the adsorbate on the adsorbent surface and its concentration in the fluidic medium. The

International Union of Pure and Applied Chemistry (IUPAC) classifies these adsorption isotherms into six groups (Abebe et al., 2018). If it is known which of these six classes the adsorption isotherm fits, qualitative information about the interaction between the adsorbate and adsorbent can be obtained and adsorption can be predicted (Demiral & Güngör, 2016; Abin-Bazaine et al., 2022).

Porosity and Pore Size Distribution

There are different techniques for determining porosity and pore size distribution of textile materials. Fabric porosity measurements can be classified as theoretical computational methods that differ according to the type of fabric (woven (Dubrovski, 2000), knitted (Ogulata & Mavruz, 2010; Siddiqui & Sun, 2015), or nonwoven (Epps & Leonas, 2000)), experimental methods, immersion methods (Jakšić & Jakšić, 2007), and methods that calculate (estimate) surface porosity using optical principles or image processing (Ragab et al., 2017; Celep et al., 2022; Angelova, 2012; 2020).

The packing density gives information on fabric porosity and is formulated as $Packing\ density\ of\ a\ fabric\ fabric\ density/density\ of\ fiber\ used\ to\ form\ fabric$. From this point of view, the formulation of porosity is given as $Porosity = open\ pore\ area/total\ area$ or $Porosity\ volume\ of\ voids\ in\ the\ fabric/volume\ of\ fabric$ (Chu & Welch, 1985; Ragab et al., 2017).

Besides the identification of porosity, another crucial parameter of textile materials for adsorption is the pore size distribution (PSD). The mercury porosimeter, which is one of the experimental porosity measurement equipment, gives average pore diameter values in terms of volume (V), area (A), and 4V/A. The variation in these three values is the sign of a wide-ranged PSD (Tetik et al., 2022). Since woven fabrics, knitted fabrics, or nonwovens like felts are not rigid materials comprised of totally uniform units or perfect patterns, the PSD values of textile materials are distributed in a certain range. A wide-range of PSD means the presence of capillary spaces of different diameters in the textile material structure, and it directly affects the entire adsorption process and isotherm type. Therfore, it is essential to know the PSD of the textile adsorbent.

Conclusion

The development of technology and the increasing consumer demands raise the hazards of industrialization by means of discharging wastewater in numerous wet processes in different kinds of industries around the world. This wastewater pollutes water bodies and damages the habitat, and even human life. Therefore, nowadays investigating the possibilities and applications of alternative and effective adsorbents has become a very important issue.

Although the use of environmentally friendly or green materials and methods has come into prominence within the scope of the sustainability approaches, it is essential to develop and research engineering products in order to minimize the damage to nature and living beings. In this sense, textile-based adsorbents stand out with their unique properties such as porosity, high surface area, biodegradability, hydrophilicity, easy availability, and processability. It is seen that textile-based adsorbents, which allow them to obtain high adsorption capacity in many studies, do not receive the attention they deserve, and there are still limited studies carried out in the last two decades. It is expected that this chapter will provide an insight on other studies that will be conducted on this limited topic.

Disclaimer

None.

Acknowledgment

The author thanks Fulya YILMAZ (PhD) and Gizem CELEP (PhD) for their support in the literature survey.

References

Abbar, B., Alem, A., Marcotte, S., Pantet, A., Ahfir, N. D., Bizet, L., & Duriatti, D. (2017). Experimental investigation on removal of heavy metals (Cu2+, Pb2+, and Zn2+) from aqueous solution by flax fibres. *Process Safety and Environmental Protection*, *109*, 639-647. https://doi.org/10.1016/j.psep.2017.05.012.

Abebe, B., Murthy, H. C. A., & Amare, E. (2018). Summary on adsorption and photocatalysis for pollutant remediation: Mini review. *Journal of Encapsulation and Adsorption Sciences, 8*, 225-255. https://doi.org/10.4236/jeas.2018.84012.

Abin-Bazaine, A., Trujillo, A. C., & Olmos, M. (2022). Adsorption isotherms: Enlightenment of the phenomenon of adsorption. In M. Ince & O. K. Ince (Eds.), *Wastewater Treatment* (1st ed., pp. 1-16). Intechopen. https://doi.org/10.5772/intechopen.104260.

Ahmed, H. A., Saleem, P. H., Yasin, S. A., & Saeed, I. A. (2021) The application of modified polyetheleneterphthalate (pet) nanofibers; characterization and isotherm study. *Journal of Physics: Conference Series*, 1853 012006. https://doi.org/10.1088/1742-6596/1853/1/012006.

Altinisik, A., Seki, Y., Ertas, S., Akar, E., Bozacı, E., & Seki, Y. (2015). Evaluating of Agave americana fibers for biosorption of dye from aqueous solution. *Fibers and Polymers*, *16*, 370-377. https://doi.org/10.1007/s12221-015-0370-9.

Andrade, C. J. & Andrade, L. M. (2018). Microalgae for bioremediation of textile wastewater: An overview. *MOJ Food Processing and Technology*, *6*(5), 432-433. https://doi.org/10.15406/mojfpt.2018.06.00200.

Angelova, R. (2012). Determination of the pore size of woven structures through image analysis. *Open Engineering, 2*(1), 129-135. https://doi.org/10.2478/s13531-011-0045-2.

Angelova, R. A. (2020). Determination of porosity of textiles with an uneven pore size. *Comptes rendus de l'Acade'mie bulgare des Sciences*, *73*(1), 96-102. https://doi.org/10.2478/s13531-011-0045-2.

Balintova, M., Holub, M., Stevulova, N., Cigasova, J., & Tesarcikova, M. (2014). Sorption in acidic environment–biosorbents in comparison with commercial adsorbents. *Chemical Engineering, 39*. https://doi.org/10.3303/CET1439105.

Bugnet, J., Morin-Crini, N., Cosentino, C., Chanet, G., Winterton, P., & Crini, G. (2017). Hemp decontamination of poly-metallic aqueous solutions. *Environmental Engineering & Management Journal (EEMJ)*, *16*(3). https://doi.org/10.30638/eemj.2017.054.

Celep, G., Yilmaz, F., & Tetik, G. D. (2022). Evaluation of some physical and tensile properties of commercial surgical masks. *Indian Journal of Fibre & Textile Research*, *47*, 125-130. https://doi.org/10.56042/ijftr.v47i1.64938.

Chu, C. C., & Welch, L. (1985). Characterization of morphologic and mechanical properties of surgical mesh fabrics. *Journal of Biomedical Materials Research*, *19*, 903-916. https://doi.org/10.1002/jbm.820190803.

Chuah, T. G., Jumasiah, A., Azni, I., Katayon, S., & Choong, S. T. (2005). Rice husk as a potentially low-cost biosorbent for heavy metal and dye removal: An overview. *Desalination, 175*(3), 305-316. https://doi.org/10.1016/j.desal.2004.10.014.

Cui, J., Li, F., Wang, Y., Zhang, Q., Ma, W., & Huang, C. (2020). Electrospun nanofiber membranes for wastewater treatment applications. *Separation and Purification Technology, 250*, 117116. https://doi.org/10.1016/j.seppur.2020.117116.

Daouda, M. M. A., Ońesime Akowanou, A. V., Mahunon, S. R., Adjinda, C. K., Aina, M. P., & Drogui, P. (2021). Optimal removal of diclofenac and amoxicillin by activated carbon prepared from coconut shell through response surface methodology. *South African Journal of Chemical Engineering, 38*(1), 78-89. https://doi.org/10.1016/j.sajce.2021.08.004.

Deaconu, M., Senin, R., Stoica, R., Athanasiu, A., Crudu, M., Opriou, L., Ruse, M., & Filipescu, C. (2016). Adsorption decolorization technique of textile/leather – dye containing effluents. *International Journal of Waste Resources, 6*(2), 1000212. https://doi.org/10.4172/2252-5211.1000212.

Demir, E., & Yalçın, H. (2014). Adsorbents: Classification, Properties, Use and Projections. *Turkish Journal of Scientific Reviews*, 7(2), 70-79.

Demiral, H., & Güngör, C. (2016). Adsorption of copper(II) from aqueous solutions on activated carbon prepared from grape bagasse. *Journal of Cleaner Production, 124*, 103-113. https://doi.org/10.1016/j.jclepro.2016.02.084.

Dias, J. M., Alvim-Ferraz, M. C., Almeida, M. F., Rivera-Utrilla, J., & Sánchez-Polo, M. (2007). Waste materials for activated carbon preparation and its use in aqueous-phase treatment: A review. *Journal of Environmental Management, 85*(4), 833-846. https://doi.org/10.1016/j.jenvman.2007.07.031.

Dubrovski, P. D. (2000). Volume porosity of woven fabrics. *Textile Research Journal, 7*(10), 915-919. https://doi.org/10.1177/004051750007001011.

Duman, G. (2021). Preparation of novel porous carbon from hydrothermal pretreated textile wastes: Effects of textile type and activation agent on structural and adsorptive properties. *Journal of Water Process Engineering, 43*, 102286. https://doi.org/10.1016/j.jwpe.2021.102286.

Dutta, S., Gupta, B., Srivastava, S. K., & Gupta, A. K. (2021). Recent advances on the removal of dyes from wastewater using various adsorbents: A critical review. *Materials Advances, 2*(14), 4497-4531. https://doi.org/10.1039/D1MA00354B.

Epps, H. H., & Leonas, K. K. (2000). Pore size and air permeability of four nonwoven fabrics. *International Nonwovens Journal, os-9*(2). https://doi.org/10.1177/1558925000OS-900215.

Ezeonuegbu, B. A., Machido, D. A., Whong, C. M. Z., Japhet, W. S., Alexiou, A., Elazab, S. T., Qusty, N., Yaro, C. A., & Batiha, G. E. S. (2021). Agricultural waste of sugarcane bagasse as efficient adsorbent for lead and nickel removal from untreated wastewater: Biosorption, equilibrium isotherms, kinetics and desorption studies. *Biotechnology Reports, 30*, e00614. https://doi.org/10.1016/j.btre.2021.e00614.

Figueroa Campos, G. A., Perez, J. P. H., Block, I., Sagu, S. T., Saravia Celis, P., Taubert, A., & Rawel, H. M. (2021). Preparation of activated carbons from spent coffee grounds and coffee parchment and assessment of their adsorbent efficiency. *Processes, 9*, 1396. https://doi.org/10.3390/pr9081396.

Florence, J. A. K., Gomathi, T., Thenmozhi, N., & Sudha, P. N. (2015). Adsorption study: Removal of nickel ions using Kenaf fiber/chitosan biosorbent. *Journal of Chemical and Pharmaceutical Research, 7*(5), 410-422.

García-Díaz, I., López, F. A., & Alguacil, F. J. (2018). Carbon nanofibers: A new adsorbent for copper removal from wastewater. *Metals, 8*(11):914. https://doi.org/10.3390/met8110914.

Hasfalina, C. M., Maryam, R. Z., Luqman, C. A., & Rashid, M. (2010). The potential use of kenaf as a bioadsorbent for the removal of copper and nickel from single and binary aqueous solution. *Journal of Natural Fibers, 7*(4), 267-275. https://doi.org/10.1080/15440478.2010.527508.

Heidarinejad, Z., Rahmanian, O., Fazlzadeh, M., & Heidari, M. (2018) Enhancement of methylene blue adsorption onto activated carbon prepared from Date Press Cake by low frequency ultrasound. *Journal of Molecular Liquids, 264*, 591-599. https://doi.org/10.1016/j.molliq.2018.05.100.

Igwegbe, C. A., Ighalo, J. O., Ghosh, S., Ahmadi, S., & Ugonabo, V. I. (2021). Pistachio (Pistacia vera) waste as adsorbent for wastewater treatment: A review. *Biomass Conversion and Biorefinery, 13*, 8793–8811. https://doi.org/10.1007/s13399-021-01739-9.

Jakšić, D., & Jakšić N. (2007). Assessment of porosity of flat textile fabrics. *Textile Research Journal, 77*(2), 105-110. https://doi.org/10.1177/0040517506065892.

Janke, C., Das, S., & Mayes, R. (2014). Preparation of most promising braided and/or textile-based adsorbents for seawater testing. *Milestone letter report*. https://www.osti.gov/servlets/purl/1148694.

Kaewprasit, C., Abidi, N., Gourlot, J. P., & Douillard, J. M. (1999). Specific surface area of some standard cotton fiber and its relation to physical properties. *Proceedings of the 1999 Beltwide Cotton Conference*, Orlando, United States, 718-720.

Kaewprasit, C., Hequet E., Abidi, N., & Gourlot, J. P. (1998). Quality measurement, application of methylene blue adsorption to cotton fiber specific surface area measurement: Part I. Methodology. *The Journal of Cotton Science, 2*, 164-173.

Kumar, P. S., & Saravanan, A. (2017). Sustainable wastewater treatments in textile sector. In S.S. Muthu (Ed.), *The Textile Institute Book Series Sustainable Fibres and Textiles* (pp. 323-346). Woodhead Publishing. https://doi.org/10.1016/B978-0-08-102041-8.00011-1.

Lameire, N. H., & De Vriese, A. S. (2001). Adsorption techniques and the use of sorbents. *Contributions to Nephrololgy, 133*. https://doi.org/10.1159/000060121.

Lee, B. G., & Rowell, R. M. (2004). Removal of heavy metal ions from aqueous solutions using lignocellulosic fibers. *Journal of Natural Fibers, 1*(1), 97-108. https://doi.org/10.1300/j395v01n01_07.

Loiacono, S., Crini, G., Chanet, G., Raschetti, M., Placet, V., & Morin-Crini, N. (2018). Metals in aqueous solutions and real effluents: biosorption behavior of a hemp-based felt. *Journal of Chemical Technology & Biotechnology, 93*(9), 2592-2601. https://doi.org/10.1002/jctb.5612.

Mendoza-Castillo, D. I., Reynel-Avila, H. E., Bonilla-Petriciolet, A., & Silvestre-Albero, J. (2016). Synthesis of denim waste-based adsorbents and their application in water defluoridation. *Journal of Molecular Liquids, 221*, 469-478. https://doi.org/10.1016/j.molliq.2016.06.005.

Mia, M. S., Yao, P., Zhu, X., Zhang, J., Xing, T., & Chen, G. (2022). Removal of heavy metals from aqueous solutions by modified waste silk. *Textile Research Journal, 92*(11-12), 1952-1965. https://doi.org/10.1177/00405175221076026.

Msahli, S., & Sakli, F. (2006). Study of textile potential of fibres extracted from Tunisian Agave Americana L. *Autex Research Journal, 6*(1), 9-13.

Nagano, S., Tamon, H., Adzumi, T., Nakagawa, K., & Suzuki, T. (2000). Activated carbon from municipal waste. *Carbon, 38*(6), 915-920. https://doi.org/10.1016/S0008-6223(99)00208-0.

Ogulata, R. T., & Mavruz, S. (2010). Investigation of porosity and air permeability values of plain knitted fabrics. *Fibres & Textiles in Eastern Europe, 18*(5), 71-75.

Özturan, R., & Tetik, G. D. (2022). Nanofibers: Excellent Adsorbents for the Removal of Methylene Blue. *Usak University Journal of Engineering Sciences, 5*(2), 143-153. https://doi.org/10.47137/uujes.1037357.

Pejić, B., Vukčević, M., & Kostić, M. (2020). Hemp Fibers in Serbia: Cultivation, Processing and Applications. In G. Crini & E. Lichtfous (Eds.), *Sustainable agriculture reviews 42: Hemp production and applications* (pp. 111-146). Springer. https://doi.org/10.1007/978-3-030-41384-2_4.

Pejic, B., Vukcevic, M., Kostic, M., & Skundric, P. (2009). Biosorption of heavy metal ions from aqueous solutions by short hemp fibers: Effect of chemical composition. *Journal of Hazardous Materials, 164*(1), 146-153. https://doi.org/10.1016/j.jhazmat.2008.07.139.

Peng, C., Yan, X., Wang, R., Lang, J., Ou, Y., & Xue, Q. (2013). Promising activated carbons derived from waste tea-leaves and their application in high performance supercapacitors electrodes. *Electrochimica Acta, 87*, 401-408. https://doi.org/10.1016/j.electacta.2012.09.082.

Phan, N. H., Rio, S., Faur, C., Le Coq, L., Le Cloirec, P., & Nguyen, T. H. (2006). Production of fibrous activated carbons from natural cellulose (jute, coconut) fibers for water treatment applications. *Carbon, 44*(12), 2569-2577. https://doi.org/10.1016/j.carbon.2006.05.048.

Ragab, A., Fouda, A., El-Deeb, H., & Abou-Taleb, H. (2017). Determination of pore size, porosity and pore size distribution of woven structures by image analysis techniques. *Journal of Textile Science and Engineering, 7*(5), 1000314. https://doi.org/10.4172/2165-8064.1000314.

Rahman, N. S. A., Yhaya, M. F., Azahari, B., & Ismail, W. R. (2018). Utilisation of natural cellulose fibres in wastewater treatment. *Cellulose, 25*, 4887-4903. https://doi.org/10.1007/s10570-018-1935-8.

Ruthven, D. M. (1984). *Principles of adsorption and adsorption processes*. Wiley Interscience Publication.

Saaidpour, S. (2014). Prediction of the adsorption capability onto activated carbon of liquid aliphatic alcohols using molecular fragments method. *Iranian Journal of Mathematical Chemistry*, *5*(2), 127-142. https://doi.org/10.22052/IJMC.2014.7608.

Shukla, S. R., & Pai, R. S. (2005). Adsorption of Cu (II), Ni (II) and Zn (II) on modified jute fibres. *Bioresource technology*, *96*(13), 1430-1438. https://doi.org/10.1016/j.biortech.2004.12.010.

Siddiqui, M., & Sun, D. (2014). Porosity prediction of plain weft knitted fabrics. *Fibers*, *3*(4), 1–11. https://doi.org/10.3390/fib3010001.

Srebrenkoska, V., Zhezhova, S., Risteski, S., & Golomeova, S. (2014, November 21-22). Methods for waste waters treatment in textile industry. *International Scientific Conference*, Gabrovo, Bulgaria.

Suzuki. M. (1990). *Adsorption engineering*. Kodansha Ltd and Elsevier Science Publishers, Tokyo, Amsterdam.

Tascan, M., & Vaughn, E. A. (2008). Effects of total surface area and fabric density on the acoustical behavior of needlepunched nonwoven fabrics. *Textile Research Journal*, *78*, 289. https://doi.org/10.1177/0040517507084283.

Tascan, M., Vaughn, E. A., Stevens, K. A., & Brown, P. J. (2011). Effects of total surface area and fabric density on the acoustical behavior of traditional thermal-bonded highloft nonwoven fabrics. *Journal of the Textile Institute, 102*(9), 746-751. https://doi.org/10.1080/00405000.2010.515731.

Tetik, G. D., Yilmaz, F., & Celep, G. (2022). Sound absorption characteristics of nanofibre web coated foams. *Indian Journal of Fibre & Textile Research, 47*(3), 334-343. https://doi.org/10.56042/ijftr.v47i3.45651.

Tofan, L., Teodosiu, C., Paduraru, C., & Wenkert, R. (2013). Cobalt (II) removal from aqueous solutions by natural hemp fibers: Batch and fixed-bed column studies. *Applied Surface Science*, *Part A, 285*, 33-39. https://doi.org/10.1016/j.apsusc.2013.06.151.

Wanassi, B., Hariz, I. B., Ghimbeu, C. M., Vaulot, C., Hassen, M. B., & Jeguirim, M. (2017). Carbonaceous adsorbents derived from textile cotton waste for the removal of Alizarin S dye from aqueous effluent: kinetic and equilibrium studies. *Environmental Science and Pollution Research*, *24*, 10041–10055. https://doi.org/10.1007/s11356-017-8410-1.

Xu, Z., Gu, S., Sun, Z., Zhang, D., Zhou, Y., Gao, Y., Qi, R., & Chen, W. (2020). Synthesis of char-based adsorbents from cotton textile waste assisted by iron salts at low pyrolysis temperature for Cr(VI) removal. *Environmental Science and Pollution Research, 27*, 11012–11025. https://doi.org/10.1007/s11356-019-07588-4.

Chapter 11

A Review on the Investigation of the Performance Properties of Recycled Textile Surfaces

Eda Acar*, PhD
and Zümrüt Bahadır Ünal, PhD
Department of Textile Engineering, Ege University, Izmir, Türkey

Abstract

Sustainability is the actions taken to protect natural resources and transfer them to future generations and prevent waste by considering environmental, economic, and social responsibilities together. This approach has led to the development of waste management and recycling concepts. In this way, the consumption of natural resources is reduced and it benefits the environment and the economy. Practices that promote waste management and recycling should continue and sustainability principles should be adopted. The recycling of textile waste helps to conserve natural resources and reduce the amount of waste, as in other wastes. The reuse of recycled textile materials saves energy, water, and resources, as well as reduces the carbon footprint. The usage capacity of natural fibers, especially cotton, is quite high all over the world. Water and chemical material consumption are very high in cotton production. This situation reduces the scarce water resources and the chemicals used cause environmental pollution, and thus global warming. Therefore, recycling of textile waste is an important step for a sustainable future. Today, the textile industry focused on the development of innovative technologies for environmentally friendly and sustainable production. The performance properties of fabrics produced with the use of recycled textiles also gain importance. Recycled textiles, especially cotton and polyester-containing fabric scraps, or second-hand materials, are obtained by various processes. Intensive research and studies are carried out to recycle the wastes formed from these raw materials, as well as other textile raw materials. Textile surfaces obtained by recycling can vary in terms of many properties such as strength, flexibility, softness, surface unevenness, thermal comfort, and fastness, depending on their raw materials, blend ratios, knitting structures, and processes. The performance properties expected from the obtained fabrics vary according to the use area of that fabric. Improving the performance properties of recycled textile materials in accordance with their usage areas is important both in terms of increasing sustainability and the use of fabrics produced with recycled materials in the apparel industry. Within the scope of this study, it is aimed to investigate the studies

* Corresponding Author's Email: eda.acar@ege.edu.tr.

In: Textile Insights and Future Aspects
Editors: Gamze D. Tetik, Gizem Celep and Fulya Yilmaz
ISBN: 979-8-89113-289-4

conducted to investigate the performance properties of the surfaces obtained by recycling and to evaluate the results obtained.

Keywords: fabric performance, recycled fabric, sustainability

Introduction

Sustainability is an approach that aims at the efficient use of resources and the protection of the environment in order not to deplete natural resources and leave a healthy world for future generations. Recycling is a process that promotes the efficient use of resources through the reuse and recycling of waste materials. Sustainability and recycling are related concepts, and recycling is considered a part of sustainability.

Recycled textiles are obtained by removing the used textile products produced during the production of apparel from being waste and reusing them through various processes. This process aims to create a sustainable production cycle by preventing the depletion of the raw materials used in textile production. These products go through the process of collecting old clothes and other textile products, sorting, cleaning, separating their fibers, producing new yarns, fabrics, and finally transforming them into new products. In addition, the production of recycled textiles requires less energy, water, and chemicals than traditional textile production. Therefore, recycled textiles provide advantages both environmentally and economically. Almost 75% of textile waste is disposed of in landfills, around 25% is recycled or reused, and less than 1% of all textiles are recycled for garment production (Juanga-Labayen et al., 2022). In this regard, advancing reuse and recycling technologies for textile waste in diverting waste from landfill is quite essential.

Good performance characteristics of fabrics obtained from recycled textiles are an important issue in terms of sustainability. The long and effective use of a resource by converting it is of great importance for the efficient use of existing resources. The performance characteristics of recycled textiles depend on the quality of their raw materials, the type of recycling process, and the subsequent chemical treatment of the fabric. The performance properties expected from the surfaces obtained by the recycling process, on the other hand, vary depending on the properties expected from the garment in which the fabric will be used. In order to determine these expected performance properties, various tests such as strength, thermal comfort, and fastness are performed on recycled fabrics. All these studies contribute to the use of recycled fabrics in the production of high-value-added products by determining and increasing the performance properties.

With the increase in the trend for sustainable fashion, much research has been done and continues to be done on the performance properties of fabrics obtained from recycled textiles. These studies aim and contribute to the evaluation and increase of the use of recycled textiles in the apparel industry by increasing the performance properties of the recycled textile surface. In this study, the recycling methods of textile surfaces are explained. Then, studies that evaluated the performance properties of recycled textile surfaces in recent years were examined. Based on the results of the research, the physical and thermal performance properties of the fabric surfaces obtained by recycling were evaluated.

Recycling Methods for Textile Waste

The recycling methods of textile waste are classified in different ways in the literature. These classification methods can be as given in Table 1 below (Bursa Province Recycling Facility Pre-Feasibility Report, 2021).

Table 1. Classification of recycling in textiles (Sandin & Peters, 2018)

Classification according to the technology/method applied in recycling	*Mechanical*
	Chemical
	Thermal
Classification according to the degree of decomposition of the obtained material	Fiber recycling
	Polymer/oligomer recycling
	Monomer recycling
Classification according to the value of the material obtained	Downcycling
	Upcycling
Classification according to the use area of the obtained material	Open loop recycling
	Closed loop recycling

Mechanical Method for Textile Recycling

Mechanical recycling is the most widely known method. The methods applied in the recycling process basically include the following stages. The fibers obtained in this method can be used in the production of yarn, woven fabric, knitted fabric, and nonwovens.

1) Separation of material and colors
2) Pulling or shredding
3) Mixing
4) Carding
5) Spinning
6) Weaving

In the mechanical recycling process, homogeneous structures are generally preferred in terms of fiber type and color. Thus, the need for repainting is eliminated. Metal accessories such as buttons and zippers, which may cause problems for the recycling machine, are removed from textile products. The fabric is then passed through roller presses filled with sharp scissors to divide it into thin cotton pieces. The recovered fibers are shorter than the original fibers due to the deterioration of the fiber structure as a result of the pulling and shredding processes (Prospering in the Circular Economy, 2020; Damayanti et al., 2021). Therefore, they can be mixed with other fibers to restore the properties of the recycled yarn. Alignment is applied to the fibers by applying a carding process to produce a cotton sliver called a sliver. Woven and knitted fabrics are produced with the yarns obtained.

One of the disadvantages of the mechanical method is the difficulty in separating the materials. Fabrics are generally produced in certain mixtures, and various chemical processes are applied. Their separation into components is technically complex and time-consuming, as well as energy-intensive and costly. Also, the results are lower in quality compared to the original product. For example, cotton can be mechanically recycled into new yarns, but this results in a shorter fiber length and reduced yarn quality. Therefore, recycled cotton fibers are often required to be mixed with other fibers to produce better-quality apparel products.

Chemical Method for Textile Recycling

Chemical recycling refers to the process of depolymerization of polymers as a result of chemical treatment of synthetic-based wastes or to the process of dissolving and returning them to raw materials or intermediate products (Altun, 2016). As a result of the chemical method, polyester is obtained from synthetic textile wastes, and cotton or viscose is obtained from organic textile wastes (Bursa Province Recycling Facility Pre-Feasibility Report, 2021). The fact that polyester does not biodegrade in nature and is found as a large amount of waste constitutes the raw material of the needed polyester fiber and provides a great advantage in terms of environmental and economic aspects.

Thermal Method for Textile Recycling

In the thermomechanical method, the waste is melted and granulated. These granules can then be used in the production of plastics and textile fibers (Altun, 2016).

Upcycling and Downcycling Methods for Textile Recycling

Another classification method proposed for recycling in textiles is a classification based on the value of the recycled material compared to the original material. Downcycling is a type of recycling in which the original product is broken into pieces and reformed; however, the product loses its original quality in the process and is often made into a lower-quality product (Bursaligil, 2019). From this point of view, if the value/quality of the product obtained after recycling is lower than the original product, the process is called downcycling. Today, recycling activities carried out in the textile sector are mostly in the downcycling class due to the wear and tear of textile products as a result of use. Upholstery fabrics, insulation materials, filling materials, agriculture and gardening products, socks, and blankets obtained as a result of recycling can be given as an example of these activities.

On the contrary, upcycling aims to keep the value and quality of the product obtained from the reuse or use of a used product or processed waste high through design (Niinimaki, 2013). If the quality of the product obtained by this method is higher than that of the original product, it is called upcycling.

On the other hand, the production of cotton and cotton yarn from waste textile products, especially due to polymer/oligomer and monomer recycling, is considered upcycling in the textile industry. Although upcycling is the recommended method in terms of the high added value of the product, downcycling, which ensures the effective use of textile waste, is also very important within the framework of the circular economy approach (Bursa Province Recycling Facility Pre-Feasibility Report, 2021).

Open- and Close-Loop Recycling in Textiles

Another classification that stands out in textile recycling is made according to whether the material obtained after recycling in the process is in the form of a closed or open-loop, according to the area of use. According to this approach, if recycled materials (for example, textile or textile product waste) are reused for use in the same sector, closed-loop recycling is in question, and if it provides input to different sectors as raw materials or intermediate products, open-loop recycling is in question.

Although recycling processes for textile wastes have environmental benefits, they also cause some problems. In the recycling process, environmental problems such as the need for a large amount of energy in mechanical, chemical, and biological processes, the emissions to air, water, and soil in all these processes, and the additional raw materials needed during the process occur (Eser et al., 2016). Therefore, all of these factors should be considered together when evaluating the benefits of recycling processes.

Previous Studies Evaluating the Performance of Recycled Textile Surfaces

Along with the frequency of clothing purchasing behaviors, the rate of disposal of waste clothing also increases. Therefore, many researchers conduct research to evaluate waste textile recycling and waste clothing life cycle. Xie et al. (2021) conducted a systematic review study on the recycling and reuse of waste clothing. They summarized the existing waste-clothing collection and recycling methods and evaluated their advantages and disadvantages.

The mechanical and physical properties of recycled textile surfaces have been the subject of numerous studies. Cotton fiber is one of the most widely used textile materials after polyester, and waste cotton has a significant proportion of 24% of total textile waste. Since recycled cotton fiber has some performance problems due to its structure, it can be mixed with other natural and synthetic fibers. Lu et al. (2023) reviewed the properties of waste cotton structure and value-added products produced from waste cotton, such as electronic devices, adsorptive materials, and biofuels using mechanical, chemical, and biological recycling methods. Besides, their advantages and disadvantages were clearly stated. In another study, considering the disadvantages of recycled cotton, the effects of cotton-containing textile wastes on recycled fiber properties after shredding were investigated. For this purpose, pre-consumer cotton textile wastes are classified according to fabric tightness and previous finishing treatments. The waste ratio and length of recycled fibers, spinnability of recycled cotton fibers, and also characteristics of manufactured yarns were tested. According to the test results, it was found that the fabric structure had no effect on the short fiber ratio and the average fiber length and yarn strength were higher in recycled fabrics with a single jersey structure. It has been stated that better quality recycled cotton fibers can be obtained by selecting loosely knit greige cotton fabrics shredded in 3 passages with a large feed size (Utebay et al., 2019). Similarly, it tried to emphasize the potential of recycled cotton fibers in the clothing industry by comparing some basic mechanical properties of recycled cotton fabrics with those obtained from carded cotton. For this purpose, in order to make a comparison, the recycled Ne 18/4 cotton yarn and Ne 18/4 carded cotton yarn were provided, and woven fabrics with plain and 2/2 twill structures were manufactured. Tensile behavior, handle properties, abrasion, and pilling performance were tested. According to the results, the tensile strength values of recycled cotton fabrics can be developed with increasing yarn density. The recycled cotton fabrics had better abrasion resistance, lower pilling performance degree because of the hairy yarn structure, and lower bending rigidity which is an advantage in terms of drapability and clothing comfort. As a result, according to all test results, clues were presented that woven fabrics obtained from recycled cotton could be an alternative in the clothing industry (Pamuk & Yıldız, 2022).

The optimum recycled content ratio was tried to be determined on the basis of the relevant fiber type, and the effect of the increase and decrease of the recycled fiber content in the fabric structure on the test results was evaluated in some of the studies. The material, which was waste during the yarn production stages, was recycled by a mechanical method and mixed with the original cotton fiber in different proportions, to obtain yarns with recycled content. Single jersey knit fabrics were produced with these yarns and the effect of the recycling rate on the selected properties was evaluated. Burst strength, air permeability, and fabric hardness tests of fabrics were carried out. It has been determined that as the recycling rate increases, the bursting strength decreases, the air permeability increases, and the fabric hardness decreases in recycled cotton-containing knitted fabric (Doba Kadem & Sevgi, 2022). Demir (2022) the characteristics of the blended yarns manufactured using viscose fibers and recycled polyethylene terephthalate fibers (r-PET) obtained were examined. For this purpose, different blended yarn ratios (33-67%, 50-50%, 67%-33% r-PET/viscose fibers, and 100% r-PET) were obtained and the physical, mechanical, and structural features are tested. According to the results obtained, it was stated that the blending ratio has a significant effect on the yarn properties. Higher breaking strength values were obtained in yarn structures with high R-PET content. In addition, yarns with a higher R-PET fiber content have better evenness and hairiness values and compact structure.

The burst strength, bursting distension, air permeability, and wale-wise wicking rate properties of single jersey knitted fabrics containing r-PET and v-PET blend CO ring spun yarns in different ratios (70/30%, 50/50%, and 30/70%) were determined. In order to determine the optimum R-PET and V-PET blend ratio and yarn manufacturing technology, numerical optimization analysis was performed using predetermined parameters. According to the numerical optimization desirability function result, a knitted fabric made from a V-PET type compact with a 58.62% blend ratio was found to be the most desirable. R-PET fiber is an alternative to virgin fibers and the use of R-PET fiber will become more common in the textile industry (Sarıoğlu, 2019). In order to determine the effect of the blend ratio of waste fiber on the yarn properties in different yarn counts such as Ne 10, Ne 20, and Ne 30. 8 different cotton yarns were produced containing waste of cotton ring spun yarn in different yarn counts and different blend ratios in rotor spinning. Yarn properties such as unevenness, thin/thick places, neps, hairiness, breaking strength, and breaking elongation were tested. After statistical evaluation, the waste did not have a significant effect on yarn properties negatively and Ne 10 was determined as the most suitable yarn count (Uyanik et al., 2022). In this study, the bursting strength properties of knitted fabrics, which have blend ratios (100% PET, 100% R-PET, 65–35% R-PET-CV, 50–50% R-PET-CV, 35–65% RPET-CV, 100% CV), were studied in different yarn numbers (Ne 10, Ne 20, Ne 30, and Ne 40). According to the results of the study, yarn count and tenacity were found to be the most effective factors in bursting strength. The R-PET fiber has bursting strength values that are close to those of virgin PET fiber for the knitted fabrics produced by coarse yarns. It is emphasized that fine yarns like Ne 40 are not proper for knitted fabrics (Uyanık, 2021).

In a study by Khan et al. (2022), various properties of the yarns and fabrics produced were compared with 100% cotton-based yarns and fabrics, virgin polyester-cotton blend (50:50) based yarns and fabrics, and with a blend of recycled polyester with virgin polyester and cotton (50:50) based yarns and fabrics. Siro spun yarns of recycled polyester fiber can be used as an alternative to virgin polyester in blends with cotton, as they exhibit comparable tensile properties, good yarn evenness, and structure uniformity. Recycled polyester blended fabrics also exhibited equally good tensile properties, tear strength, and bursting strength, which are the result of good tensile and uniform structural properties. Recycled polyester cotton-based fabrics also exhibited better moisture management and drape properties compared to virgin polyester-based fabrics. The air permeability and resistance to pilling of the recycled polyester cotton fabric were higher due to a slight change in morphology after the recycling process. It is concluded that recycled polyester fibers can be used as an alternative to virgin polyester fibers in siro-spun medium-count yarns and consequent fabrics.

The use of recycled products is of great interest in the automotive industry, and it is difficult to use recycled fibers instead of virgin fibers in high-tech applications. High performance, such as high abrasion resistance, light fastness, and a long lifetime, is expected from automotive seat fabrics. In a study conducted in light of this information, the possibilities of using recycled yarns in automotive seat fabrics were investigated. Two different types of fabrics with different yarn linear densities and constructions were produced from recycled and virgin polyester filament fibers and their properties were compared. The tensile strength, tear strength, and seam strength of recycled fabrics are lower than those of virgin fabrics. As a result of the study, it was stated that although there were some differences between the yarn and fabric test results, it was concluded that the values were within the acceptable range for customer criteria (Mecit et al., 2015).

Saini et al. (2020) included different types of fiber in their research. Mechanical properties of yarns and fabrics produced from virgin and recycled wool, polyester, and acrylic fibers have been compared in terms of tensile strength and yarn twist per inch for yarn and tensile strength, elongation, and bending length for fabric. Yarns produced from recycled fibers exhibited better properties in terms of strength as well as twists per inch. The results of this study showed that yarns and fabrics containing recycled fibers have little difference from control products and can also be used for apparel where strength is required. Chauhan et al. (2019) produced needle-punched nonwovens using virgin PET and rPET and investigated the breaking strength characteristics of these structures. As a result, the breaking strength of the nonwoven surface obtained from the rPET fibers was found to be 8-10% lower than the original.

Sewing performance increases the quality of the garment and extends its lifetime, and is another important issue to be evaluated in fabrics. Bestem and Baykal (2022) focused on minimizing the seam slippage value in chenille upholstery with recycled and virgin polyester yarns content in their study. As a material, 32 chenille yarns were manufactured in two different pile lengths by using r-PET and virgin polyester yarns in a binder and/or pile structure. The investigation of the effects of r-PET use on seam slippage in chenille upholstery fabrics constitutes the innovative aspect of the study. According to the results, the usage of recycled yarns (r-PET) in the binder and/or pile yarns of chenille in the fabrics did not affect the warp seam slippage values. Besides, the results of the independent sample t-test revealed that the usage of r-PET or virgin PES yarn in the binder and pile yarns did not have a significant effect on seam slippage of the warp direction.

Comfort is considered one of the important features required in the clothing industry and is one of the subjects studied by many researchers around the world (Tama et al., 2020; Isler et al., 2021). Performance expectations for comfort are increasing day by day as the production and use of garments made from recycled fabric become widespread. There are studies to investigate the comfort properties of recycled fabric surfaces. Gun and Kuyucak (2022) produced knitted fabric from recycled rotor yarn containing waste fiber and investigated the thermo-physiological comfort performance properties of these fabrics taking into consideration of the incorporated covered and PBT elastic yarns. They evaluated the effects of coated PBT in comparison with their equivalent original acrylic counterparts. Acrylic fiber has the wool-like touch and thermal comfort properties. Added value was provided by including coated polyester/lycra and PBT elastic yarns in recycled fabrics. When the thermo-physiological properties of recycled and untreated acrylic fabrics were compared, it was seen that the thermal conductivity and thermal absorbency were lower and the thermal resistance was higher in recycled acrylic fabrics. It has been stated that the values of the thermal physiological properties of recycled acrylic fabrics are suitable for use in cold weather conditions.

Necef et al. (2022) obtained r-Co-contained fabrics and tested them in their research in order to investigate the usability of fabrics in the apparel industry. Ne 30/1 yarns were produced with the blend ratio of 45% r-Co / 55% PES and also Ne 30/1 50% PES/50% Co yarns were preferred for comparison with the recycled yarns. Using the same manufacturing conditions, single jersey and rib fabric structures were obtained. The thermal manikin, Alambeta, and air permeability instruments were used to determine the thermal behavior of the structures. According to the results of the study, it was observed that although 45% r-Co / 55% PES yarn involves more PES, r-Co leads to a drop in the tensile strength value. In addition, while the air permeability values of fabrics containing recycled fiber decreased, an increase was observed in the thermal resistance values. It has been explained that the reason for the decrease in air permeability and the increase in thermal resistance is due to the higher hairiness of recycled yarns and, therefore, a higher fabric thickness. As a result, it was also determined that the thermal manikin tests on clothes are more effective than the thermal resistance tests on fabrics in order to evaluate the thermal insulation of a garment under real-use conditions.

In another study, the thermal comfort properties of socks samples produced from recycled polyester and virgin cotton fibers and their blends (100% recycled polyester, 70% 30% recycled polyester/virgin cotton, 50% 50% recycled polyester/virgin cotton 30% 70% recycled polyester/virgin cotton 100% virgin cotton) were investigated. Statistical evaluations were conducted that focused on the effect of blend ratios on the related properties of socks. According to experimental and statistical results, it is stated that the blend ratio has a significant effect on all the comfort properties of the socks. Recycled polyester fiber socks have higher thermal resistance and lower thermal conductivity values than virgin cotton socks. As a result of comparison blend ratios, 70% 30% recycled polyester/virgin cotton fiber blended socks show better results and it is concluded that this blend ratio can be used for active sportswear clothing. According to the thermal comfort results obtained, a higher percentage of recycled polyester fibers in the blends is preferable for cold environmental conditions (Vasanth Kumar & Raja, 2021).

The comfort properties of recycle fiber-containing textile surfaces have generally been investigated for woven and knitted surfaces. There are a limited number of studies evaluating the thermal comfort performance of nonwovens, as they are used in product groups and applications where different properties are expected. Wazna et al. (2019) obtained nonwoven surfaces with textile wastes in their research. The thermophysical properties of nonwoven fabrics obtained by using acrylic and wool fibers were investigated. As a result of the study, it was observed that recycled nonwoven surfaces showed close properties to the original structures.

There are some performance declines over time due to the repeated laundering process throughout the lifetime of the garments. There are a limited number of studies that evaluate the surface and dimensional properties of recycled fabrics after laundering. Özkan and Gündoğdu (2021) investigated the release of fibers from recycled (R-PET) and virgin polyester (PES) knitted fabrics during domestic washing in terms of polymer type, filament form, and fabric structure. In order to achieve an objective evaluation of filament type, the filament linear density and the parameters for fabric manufacturing were kept as constant as possible. The knitted fabrics were washed three times on the James H. Heal Gyrowash washing machine under the same washing conditions according to the TS EN ISO 105-C06 standard. It was determined that the R-PET knitted fabrics released approximately 2.3 times more fiber than the virgin PES fabrics. The amount of released fibers decreased significantly with an increase in the number of wash cycles for both R-PET and virgin PES samples. This study showed that knitted fabrics made from R-PET released more fibers with a shorter length compared to virgin PES. In another study, fabrics in three different types of knitted structures (single jersey-rib-interlock) were manufactured by blending recycled cotton fibers with both recycled and virgin polyester fibers. The mechanical properties of the yarns and the performance properties of the produced fabrics were investigated and compared before and after repetitive laundering and drying processes. It was found that yarn type and fabric stiffness are not significantly related after the laundering cycle. It was observed that fabrics made of fully recycled cotton-PES yarns show more percentages of dimensional change than fabrics made of semi-recycled and virgin yarns. Fabrics containing fully recycled cotton-pes yarns gave a better performance in terms of fabric weight loss than semi-recycled and virgin yarns. The performance indicators determined in this study showed that the use of recycled cotton fibers in the production of knitted fabrics will not negatively affect the performance properties (Yılmaz & Özgen Keleş, 2023).

Conclusion

The textile sector is one of the sectors that appeals to both traditional and technical fields with a wide range of products. The textile industry includes different production methods such as knitting, woven, and nonwoven fabrics, and wastes are generated at each step of these processes. It is a very important issue that wastes generated during production processes and after the consumption of textile products are evaluated and input into another production process. In this way, planning should be carried out in a way that the system will not generate waste or that the wastes will be included in the cycle and evaluated appropriately.

The mechanical and thermal properties of recycled fabrics may vary depending on the type of raw material used, the recycling method, and the production process. When studies are examined, it has been observed that recycled fabrics generally have mechanical properties similar to quality fabrics, especially in terms of strength. However, it is not possible in current technology for recycled fabrics to be as homogeneous as fabrics made from original raw materials. Since the recycled fibers used in yarn and fabric production also shorten, the pilling problem is the most important problem of knitted fabrics obtained by recycling. The thermal properties of recycled fabrics may also be lower than the thermal properties of the original fabrics. Since these fabrics have low thermal properties, they can sometimes have insufficient protection in cold climate conditions. For this reason, the disadvantages of recycled fabrics can be eliminated by using additional thermal layers or insulation materials to increase thermal performance, especially in clothing designed for outdoor activities.

As a result, although the pilling problem, the lack of color homogeneity, and the lack of thermal properties in recycled fabrics may seem like disadvantages, they are features that can be ignored beside the advantages of recycled fabrics that offer a sustainable fashion option. As stated in many studies in the literature study, the performance values of the recycled fabric structures have reached acceptable levels. Considering the strong properties they have, recycled fabric structures can be recommended to be used in the appropriate area. Studies can be carried out to improve properties such as surface problems and thermal values. Despite all these measures, fabrics that do not have suitable properties for garment production can find use in other production areas as bags, shoes, carpets, insulation materials, etc.

Disclaimer

None.

References

Bestem, E. S. İ., & Baykal, P. D. (2022). Optimization of the yarn slippage at the seam in chenille upholstery fabrics including recycled and virgin polyester yarns. *Textile and Apparel, 32*(1), 1-8. https://doi.org/10.32710/tekstilvekonfeksiyon.928820.

Bursa Eskişehir Bilecik Developlemt Agency. (2021). Bursa province recycling facility pre-feasibility report. https://bebka.org.tr/wp-content/uploads/2021/10/bursa-ili-tekstil-geri-donusum-tesisi-on-fizibilite-raporu2021.pdf.

Bursaligil, G. (2019). Examination of zero waste fashion design methods and practices. *International Journal of Interdisciplinary and Intercultural Art, 4*(7), 81-100. https://doi.org/10.29228/ijiia.7.75.

Chauhan, V. K., Singh, J. P., & Debnath, S. (2020). Tensile behavior of virgin and recycled polyester nonwoven filter fabrics. *Journal of Industrial Textiles, 50*(4), 483-511. https://doi.org/10.1177/1528083719833976.

Damayanti, D., Wulandari, L. A., Bagaskoro, A., Rianjanu, A., & Wu, H. S. (2021). Possibility routes for textile recycling technology. *Polymers, 13*(21), 3834. https://doi.org/10.3390/polym13213834.

Demir, M. A. (2022). Comparative study on mechanic and structural properties of recycled pet (r-pet) and viscose blended yarns. *Journal of Textiles and Engineer, 29*(127), 133-139. https://doi.org/10.7216/1300759920222912703.

Doba Kadem, F., & Sevgi, R. (2022). Determining the effect on performance characteristics of recycled cotton fiber ratio in single jersey knitted fabrics. *Cukurova University Journal of the Faculty of Engineering, 37*(3), 609-616.

Eser, B., Çelik, P., Çay, A., & Akgümüş, D. (2016). Sustainability and recycling opportunities in the textile and apparel sector. *Journal of Textiles and Engineer, 23*(101), 43-60. https://doi.org/10.7216/1300759920162310105.

Gun, A. D., & Kuyucak, C. N. (2022). Thermo-physiological comfort performance of recycled plain knitted fabrics produced from acrylic waste fiber with the effects of incorporated covered and pbt elastic yarns. *Fibers and Polymers, 23*(9), 2762-2771. https://doi.org/10.1007/s12221-022-4231-z.

Isler, M., Tama Birkocak, D., & Abreu, M. J. (2021). The influence of starch desizing on thermal properties of traditional fabrics in anatolia. In M. Á. Gardetti & S. S. Muthu (Eds.), *Handloom sustainability and culture: Entrepreneurship, culture and luxury* (pp. 197-219). Springer, Singapore. https://doi.org/10.1007/978-981-16-5967-6_9.

Juanga-Labayen, J. P., Labayen, I. V., & Yuan, Q. (2022). A review on textile recycling practices and challenges. *Textiles, 2*(1), 174-188. https://doi.org/10.3390/textiles2010010.

Khan, A., Iftikhar, K., Saleem, D., & Akhtar, S. (2022). Recycled polyester and cotton blended siro spun yarns and fabrics: a sustainable and eco-friendly approach. *Journal of Natural Fibers, 19*(16), 13457-13474. https://doi.org/10.1080/15440478.2022.2095551.

Lu, L., Fan, W., Meng, X., Xue, L., Ge, S., Wang, C., Foong, S. Y., Tan, C. S. Y., Sonne, C., Aghbashlo, M., Tabatabaei, M., & Lam, S. S. (2023). Current recycling strategies and high value utilization of waste cotton. *Science of the Total Environment, 856*, 158798. https://doi.org/10.1016/j.scitotenv.2022.158798.

Mecit, D., Berber, R. O., & Boyaci, B. (2015). A study on the application of recycled fabrics as automotive seat upholstery. *Industria Textila, 66*(5), 278-282.

Necef, Ö. K., Tama Birkocak, D., Kılıç, A., Boz, S., Abreu, M. J., & Öndoğan, Z. (2022). Investigating the effect of recycled cotton included fabrics on the thermal behaviour by using a female thermal manikin. *Autex Research Journal*. https://doi.org/10.2478/aut-2022-0020.

Niinimäki, K. (2013). *Sustainable fashion: New approaches*. Aalto University.

Özkan, I., & Gundogdu, S. (2021). Investigation on the microfiber release under controlled washings from the knitted fabrics produced by recycled and virgin polyester yarns. *Journal of the Textile Institute 112*(2), 264-272. https://doi.org/10.1080/00405000.2020.1741760.

Pamuk, G., & Yildiz, E. Z. (2022). Physical properties of plain and twill woven fabrics produced from carded and recycled cotton yarn. *Pamukkale University Journal of Engineering Sciences, 28*(7), 1030-1035.

Saini, N., Yadav, S., & Rose, N. M. (2020). A comparative study of physical properties of yarns and fabrics produced from fresh and recycled fibres. *Current Science, 118*(5), 791-795.

Sandin, G., & Peter, G.M. (2018). Environmental impact of textile reuse and recycling - a review. *Journal of Cleaner Production, 184*, 353-365. https://doi.org/10.1016/j.jclepro.2018.02.266.

Sarıoğlu, E. (2019). An investigation on performance optimization of r-pet/cotton and v-pet/cotton knitted fabric. *International Journal of Clothing Science and Technology, 31*(3), 439-452. https://doi.org/10.1108/IJCST-08-2018-0108.

Tama, D., Isler, M., & Abreu, M. J. (2020). Evaluating the thermal comfort properties of Rize's traditional hemp fabric (feretiko) using a thermal manikin. *Materials Today: Proceedings, 31*, 197-200. https://doi.org/10.1016/j.matpr.2019.10.063.

Uşak Chamber of Commerce and Industry Report. (2016). *Recycling, environmental and economic effects of textile production and usage wastes*. https://usaktso.org/dosya/Kurumsal/Trk_Teks_Ger_Don.pdf.

Utebay, B., Çelik, P., & Çay, A. (2019). Effects of cotton textile waste properties on recycled fibre quality. *Journal of Cleaner Production, 222*, 29-35. https://doi.org/10.1016/j.jclepro.2019.03.033.

Uyanık, S. (2021). The bursting strength properties of knitted fabrics containing recycled polyester fiber. *The Journal of the Textile Institute, 112*(12), 1998-2003. https://doi.org/10.1080/00405000.2020.1862490.

Uyanik, S., Parlakyiğit, P., & Ovali, S. (2022). Closed-loop recycled yarn production from yarn wastes and investigation of their physical properties within the scope of sustainability. *Cukurova University Journal of the Faculty of Engineering, 37*(4), 1087-1102.

Vasanth Kumar, D., & Raja, D. (2021). Study of thermal comfort properties on socks made from recycled polyester/virgin cotton and its blends. *Fibers and Polymers, 22*, 841-846. https://doi.org/10.1007/s12221-021-0471-6.

Wazna, M. E., Gounni, A., Bouari, A. E., Alami, M. E., & Cherkaoui, O. (2019). Development, characterization and thermal performance of insulating nonwoven fabrics made from textile waste. *Journal of Industrial Textiles, 48*(7), 1167-1183. https://doi.org/10.1177/1528083718757526.

Xie, X., Hong, Y., Zeng, X., Dai, X., & Wagner, M. (2021). A systematic literature review for the recycling and reuse of wasted clothing. *Sustainability, 13*(24), 13732. https://doi.org/10.3390/su132413732.

Yılmaz, E., & Özgen Keleş, B. (2023). Effects of recycled fiber usage and laundering processes on the performance properties of knitted fabrics. *Fibers and Polymers*, *24*(4), 1503-1516. https://doi.org/10.1007/s12221-023-00128-5.

Chapter 12

Core-Spun Yarns for Denim Fabric Production

Tuba Bedez Üte*, **PhD**
and Pınar Çelik, PhD

Department of Textile Engineering, Faculty of Engineering, Ege University, Izmir, Türkey

Abstract

Denim garments, which were in demand for durable and long-lasting fabrics, maintained their position among fashion trends over the years as a symbol of comfort, style, and sometimes freedom. The yarns used in denim fabrics also differ with the diversification of the performance and appearance characteristics expected from garments over the years. The fiber type, yarn structure, and fabric production techniques change, but the preference for denim clothes by people of all ages and from all walks of life never changes. In this study, the commonly preferred yarn types in denim fabrics and their production methods will be reviewed. The types of fiber used in denim fabrics differ depending on the end use purpose, climatic conditions, and priority requirements. For example, while technical yarns with high abrasion resistance are preferred for motorcycle drivers, regenerated cellulose fibers are used to provide comfort if they are to be used in summer. Besides, the yarn properties to be used in denim fabrics also differ depending on the performance features expected by the users, productivity, and cost analyses of the manufacturers. According to the prioritized performance feature, rotor, ring, slub, single-core, or double-core yarns are used as both weft and warp yarns.

Keywords: core-spun yarns, dual-core spun yarns, double-core spun yarns, denim, elasticity, bagging

Introduction

The aesthetic and performance properties expected from garments are affected by their components, fiber, yarn, and fabric properties. In this context, the insufficiency in the component properties leads to these expectations needing to be fully met. Although aesthetic

* Corresponding Author's Email: tuba.bedez@ege.edu.tr.

In: Textile Insights and Future Aspects
Editors: Gamze D. Tetik, Gizem Celep and Fulya Yilmaz
ISBN: 979-8-89113-289-4

expectations differ from person to person, performance expectations such as comfort in clothing, elasticity, and strength are essential for everyone.

A wide range of fibers, including cotton, polyester, elastane, linen, and regenerated cellulosic fibers, and their blends can be used for weft yarns (Ertaş et al., 2016). Generally, cotton and regenerated cellulosic fibers are preferred for warp yarns, as they are more suitable for the warp dyeing process. Twill 3/1, twill 2/1, and satin are common denim structures, and the weight of the fabrics is generally between 4-16 Oz/yd^2.

The fiber and yarn properties used in denim fabrics differ depending on the performance features expected by end users and their productivity. The types of fiber used in denim fabrics differ depending on the end use purpose, climatic conditions, and priority requirements. In order to use different yarn structures to increase the use of denim garments in winter, Babaarslan and Telli (2013) compared the properties of denim fabrics produced with ring slub warp, carded, and chenille weft yarns. It has been observed that the chenille yarns show less abrasion than expected in the denim fabric structure with high fabric density. Fashion designers also use different materials in denim fabrics. Park and Kim (2018) used Hanji paper yarn as weft yarn in denim fabrics. For this purpose, they used rope-dyed and cheese-dyed cotton warp yarns. They combined the paper yarn plied with the core spun yarn (sheath: cotton, core: elastane) and single (twisted paper tape yarn) weft yarns or used one Hanji paper yarn/one core-spun yarn as a weft yarn for denim fabric production. They have obtained lighter denim fabrics with higher absorption velocity and properties similar to existing denim fabrics in terms of dimensional stability, stretchability, elongation, recovery rate, and drying velocity fabrics. Hanji denim fabrics were recommended for summer wear.

Some researchers have worked on knitted structures as an alternative to denim fabrics. Değirmenci and Çelik (2016) produced denim-looking, diagonal fleeced knitted fabrics with cellulosic fibers and examined the relationship between extension and bursting strength. Jamshaid et al. (2020) compared the functional properties of woven and knitted (cross-terry structure of 2×1) denim fabrics produced with yarns at different linear densities. Knitted denim fabrics had better comfort properties but worse pilling resistance.

This study investigated the effects of different yarn properties and fabric production parameters, especially core spun yarns used in denim fabrics. Primarily, the concepts of elasticity, permanent elongation, and bagging, which are the main targets of using core-spun yarns, will be defined, and the needs of the core yarns will be met and studies in this field will be examined. Afterward, studies on the post-process of denim fabrics woven by core yarns will be reviewed. Some studies on the subject in the denim sector in the field of sustainability, which is the agenda of our age, will be included.

Elongation, Elastic Recovery, and Growth

Elasticity is the recovery property and the ability of a material to stretch by applying a force and then return to its original shape and size when the force is removed. Fabric growth is defined as the length difference of a fabric with the application of tension and after relaxation (El-Ghezal et al., 2009).

Bagging is a residual deformation that generally occurs in the elbow, knee, hip, etc. parts of garments because of force exertion from body parts to cloth (Şengöz, 2004). Bagging results from the lack of dimensional stability or recovery when repeated or prolonged pressure is exerted on a fabric (Hasani et al., 2012). It is observed that the elasticity, growth, and recovery properties of denim fabrics are tested according to the ASTM D3107 standard test method (Table 1). Since bagging is a 3D deformation and the test method used tests one-way deformation, many researchers have worked on new test methods. Sülar (2011; 2016) presented a testing instrument composed of an artificial arm with an elbow joint to simulate fabric bagging under dynamic conditions. Yeung et al. (2002) used image processing for evaluating and predicting fabric bagging. Zhang et al. (1999a; 1999b) investigated objective and subjective evaluation methods for fabric bagging. They have conducted many studies on bagging. In one of their research projects (Zhang et al., 2000a), during the bagging process in woven fabrics, experiments were carried out on the viscoelastic behavior of the fibers, and bagging was simulated with a mathematical model. The results revealed the importance of the viscoelastic behavior of the fibers in determining the rheological behavior of the fabrics during bagging. Zhang et al. (2000b) further investigated the relative contribution of elasticity and viscoelasticity of fibers and inter-fiber friction to the bagging behavior of wool-woven fabrics and performed mathematical modeling. The experimental results showed that the modeling is suitable for predicting the bagging behavior of wool-woven fabrics. Uçar et al. (2002) reported that the KEF_FB system can be used in evaluating knitted fabric bagging. They showed that by measuring the shear and bending parameters of the fabrics, it is possible to determine the properties of the bagging. They found a relationship between the residual bagging height that is measured with the fabric bagging test and the mechanical parameters that are obtained from the KES_FB system. Abghari et al. (2004) developed a test device having a rectangular clamp with four jaws that simulate the actual conditions of bagging formation in different clothing positions and a method for analyzing the bagging behavior of woven fabrics.

Table 1. Definition of fabric elasticity, recovery, and growth under a specified tension. (ASTM D 3107, 2019)

Fabric Stretch (elasticity) (%)	A certain specified parameter is applied to the fabric specimen and the percentage of the longitudinal dimensional change under tension.
Fabric Growth (%)	The percentage of the longitudinal dimensional change with the tension but after relaxation
Fabric Recovery (%)	The percentage of the fabric growth withdrawal concerning the fabric growth determined after tension was gained at the different relaxation periods.

Baghaei et al. (2010) produced plain woven fabrics using different types of core-spun weft yarns. The difference in yarns is due to the difference in the elastane draw ratio (2.9- 3.1- 3.6- 3.9) and the twist coefficient (α_e 3.4 and 3.9). They have designed a test device for cyclic loading of tensile fatigue. They have concluded that the increase in draw ratio and decrease in twist coefficient of core-spun yarns result in higher elastic recovery. Gazzah et al. (2015) predicted the bagging recovery velocity of bagged denim fabric samples as a function of frictional parameters such as yarn-to-yarn friction as well as metal-to-yarn friction with the help of KES-FB4. They have produced weft yarns with different fibers (cotton, polyester, and Tencel™) within and without elastane filament. They have found a high correlation between inputs and kinetic velocity and, non-elastic fabrics have lower velocity recovery than elastic fabrics.

When studies in this context were examined, the fact that many researchers focused on elasticity and bagging features reveals the importance of the subject for users. What is expected from the garments is that while the elasticity increases, the bagging properties decrease and, at this point, different materials and spinning methods are used.

Core-Spun versus Dual-Core Yarns

As mentioned above, elasticity and recovery are the most important comfort features for users who expect skin-like comfort from clothing. To meet these expectations in denim garments, Özdil (2008) suggested the use of elastic structures. Core-spun yarns are useful because of some problems caused by naked elastane after use. In the production of core yarns, a filament core is usually covered by staple sheath fibers, benefiting from the properties of both components. In brief, at least two components are needed to produce core spun yarns (Invista, 2006). Although the use of elastic yarns in fabric structures has positive contributions, there are also some difficulties when used as warp yarns in woven fabrics. For this reason, it is aimed to provide the required elasticity by using elastic components in weft yarns in denim fabrics.

The core yarns are classified into different aspects. They are classified as soft- and hard-core (rigid) yarns according to whether they are elastic or not; according to the number of core materials, they are classified as single- and double-core yarns. If the core filament is not made of a completely rigid material but provides some elongation, albeit limited, it is defined as semi-rigid core yarn. Dual-core (double-core, duo-core) yarns are made of three components; an elastic polyurethane filament and a multifilament are used in the core, and a staple fiber sheath covers both. Here, it is necessary to mention which features of single-core yarns are insufficient and why double-core yarns are needed. The fact that the produced hard-core yarns are insufficient in terms of elasticity and soft-core yarns in terms of permanent elongation (recovery) has led to the need for double-core yarns. This new generation of yarns is also advantageous regarding the spinning and winding process (Tekstil & Teknik, 2017). Parallel to the developments in core yarn production techniques, there are also developments in the fibers used. Bi-component polyester fibers and polybutylene terephthalate (PBT) have inspired interest as an alternative to elastane. These new-generation fibers are advantageous regarding chemical resistance, strength, recovery, dimensional stability, and elasticity properties (Çataloğlu, 2007). Different raw materials are preferred as core and sheath materials depending on the end use.

The mentioned materials are well-known core materials. To produce soft-core yarns that have high elasticity, elastane (EA) filament is used. Polyester filament (PES) can improve tenacity with rigid core yarns. Polybutylene terephthalate (PBT) is a textured polyester filament yarn with properties of stretch and elasticity. Polytrimethylene terephthalate (PTT) is an aromatic polyester with better recovery bending, which is better than that of nylon 6, PET, PBT, and PP, except nylon 66 (Chen & Tang, 2004). T400® is a trademark of Invista, an elastic elastomultiester polyester fiber that can be used instead of textured yarn for stretch textiles. Regarding sheath fibers, according to the end use, cotton, regenerated cellulosic fibers such as viscose (CV), modal, lyocell, or staple polyester fibers can be used.

In the production of dual-core, ring-spun yarns, elastic and non-elastic filaments are fed from different creels. The ring-spinning machine is equipped with an extra creel above or behind the roving creel (Figure 1a). The tension rollers feed the elastic filament, and the roller speed can be adjusted, depending on the desired elasticity (Figure 1b). Different speeds of the rollers adjust the elastane and second filament draw ratio (Figures 1b and 1c).

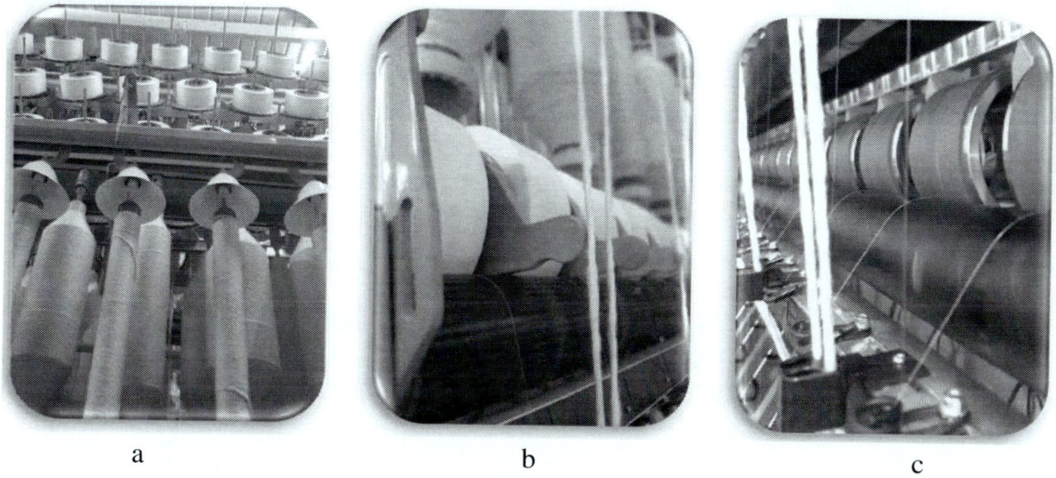

Figure 1. The feeding systems of core-spun yarn spinning: a. the extra creel for dual-core spinning, b. the elastane filament feeding roller, c. the drafting roller for second core in dual-core yarns.

As elastane is fed with the v-grooved roller to the nip point of the delivery roller, the second core can be fed simultaneously to the same point, in the center of the spinning triangle (Figure 2). However, this is not the only option to produce double-core yarns. Before being fed into the ring-spinning machine, two different cores can also be brought together by methods such as intermingling and fed into the system as a single core.

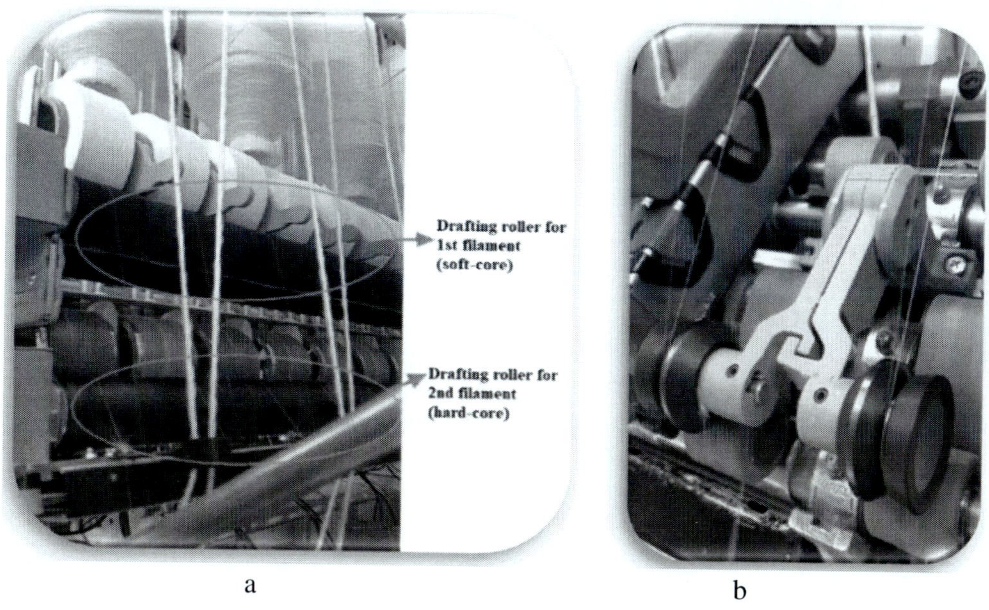

Figure 2. The system for dual-core yarn production; a. drafting rollers b. V-grooved roller.

The properties core-spun and elastic yarn have attracted the attention of many researchers (Babaarslan, 2001; Su et al., 2004; Su & Yang, 2004; Çelik et al. 2009; Ertekin & Kirtay, 2015). Bansal et al. (2019) examined the tensile properties of core spun yarns produced with different yarn production parameters (yarn count, twist multiplier, elastane ratio, etc.). Coarser core-spun yarn had a lower tenacity compared to fiber yarns. The yarn tenacity increased with increasing twist multiplier and decreasing stretch percentage. The finer core-spun yarn had higher strength and elasticity due to its better recovery behavior than the coarser yarns. Sarıoğlu and Babaarslan (2017a) investigated the effect of filament fineness and yarn count on the fatigue behavior of rigid core-spun yarn produced with PET-textured filament yarn (PET DTY).

In a study by Telli et al. (2017), used cotton-covered core-spun (core material: inox, copper, and tungsten wires) and double core yarns (elastane with inox, copper, and tungsten) for electromagnetic shielding effectiveness.

In a study by Vuruşkan et al. (2011), produced cotton-covered soft-core yarns in different yarn counts (Ne 16, Ne 24, Ne 32), twist coefficient (α_e 3,5, α_e 4, α_e 4,5), elastane thickness (44 dtex and 78 dtex) and elastane draft ratio (3, 3.5, 4). They examined their effect on the strength and elongation values. They have reported that the tensile strength increases with increasing in elastane draft ratio and decreasing in filament thickness, whereas yarn elongation is higher with a thicker filament and higher elastane draft ratio.

In a study by Sarıoğlu et al. (2018), used medium, fine, and micro false-twist textured polyester filament yarns with different filament numbers (36, 96, 144, 196, 333) and fineness (3.05, 1.15, 0.76, 0.57, 0.33), at the same linear density (110 dtex) to produce cotton-covered hard core-spun yarn and compared their influence on yarn twist liveliness. They have found that the residual torque of the yarns decreases statistically significantly with the decrease in filament fineness.

Jaouachi et al. (2007, 2010a, 2010b) studied spliced yarn used in denim fabrics. They investigated the effects of yarn linear density, and the duration of air and water joining on the breaking strength of wet spliced yarns and found that yarn linear density and the length of the splice are the most effective parameters. They also evaluated wet pneumatically spliced elastic core-spun denim yarns with fuzzy theory (Jaouachi et al., 2010a). According to their model, core-spun yarn linear density, splice length, and duration of water joining are the most influential parameters, respectively, on splice performance. Doniyorova and Sadikova (2022) explained the production of denim fabrics using warp yarn spun from Indigo dyed fiber.

The effects of core-spun yarn properties on the knitted and woven fabric properties were also investigated by many researchers. Akankwasa et al. (2014) investigated the bursting strength, elongation at break, tensile strength, air permeability, and pilling properties of plain knitted fabrics produced with cotton/T400® and 100% cotton ring spun yarns in the same linear density. Fabrics knitted from 100% cotton fabrics were better in pilling resistance and tensile strength, but worse in breaking elongation, bursting strength, and air permeability, compared to the fabrics knitted with cotton/T400® yarns.

Çelik and Kaynak (2017) examined the effects of the elastane draw ratio in the production of core-spun yarn and pick density on the air permeability of denim woven fabrics. Ertaş et al. (2016) focused on the effect of the weft density of denim fabrics that have different amounts of dual-core yarn quantity and elastane ratio per unit area. They have investigated the effect on the weight, width, cost per unit, resistance, and color properties of the fabric.

Babaarslan et al. (2007) compared some fabric properties, such as the abrasion resistance, strength, extension, and elongation properties of 2/1 twill fabrics produced with PES/CV and

PES/CV/EA yarns. In fabric production, weft yarns were replaced with soft core-spun yarns. They have found that with the help of elastane, the strength, elongation, and elasticity values can be increased, but there was no difference in the abrasion resistance.

Babaarslan et al. (2015) produced denim fabrics using multifilament polyester weft yarns that have different filament numbers. They have found significant differences in fabric stiffness, permanent elongation, and air permeability. Air permeability and stiffness decrease with a decrease in filament fineness with microstructure.

Kaynak et al. (2017) produced denim fabrics using 100% cotton combed ring, compact, sirospun, open-end (OE) rotor, and vortex weft yarns, and investigated fabric performance characteristics. Ring, sirospun, and compact yarns were recommended for better yarn tenacity, fabric breaking, and tear strength; however, sirospun and ring-spun yarns were recommended for better handling. Fabrics with ring-spun yarns exhibit the lowest stretch properties and the worst abrasion resistance among all fabrics, and compact yarns have come to the forefront with their performance properties.

Hailemariam and Muhammed (2022) produced denim fabrics with ring- and rotor-spun cotton yarns and compared the mechanical properties of the fabrics. They have produced fabrics from ring-spun warp yarn and rotor-spun weft yarn, from rotor-spun warp yarn and ring-spun weft yarn, from rotor-spun yarns in both warp and weft directions, and ring-spun yarns in both warp and weft directions. Denim fabrics produced with ring-spun yarns were advantageous in tensile strength and tear strength, whereas denim fabrics made from rotor-spun yarns were better in air permeability, abrasion resistance, and pilling resistance.

Kara and Akgun (2018) investigated the effects of weft yarn fiber content (different blend ratios of cotton, Coolmax, Cordura, thermocool, and elastane) and fabric constructional parameters (weave structure, thickness, weight, bulk density, and cover factor) on the moisture management properties of denim fabrics. It is found that with the increase in the cotton fiber ratio in the weft yarn, the overall moisture management capability (OMMC) and the accumulative one-way transport capability (OWTC) values of denim fabrics also increased. They found no significant relationship between moisture transmission properties and cover factors.

Mezarcıöz et al. (2020) investigated the effects of weft yarn linear density, elastane draft ratio, weft density, elastane fineness, weft yarn production methods (cotton-elastane core-spun, cotton-polyester sirospun) on the elasticity and growth properties of denim fabrics. The use of elastane was the most effective parameter; for better growth performance, elastane thickness should be higher. For greater fabric growth and elasticity, the draft ratio should be increased. The weft density had a contrary effect on the elasticity.

Mangat et al. (2012) the effect of weft yarn materials and washing treatments on the moisture management properties of denim fabrics. They reported that denim fabric moisture management properties showed diversity depending on containing hydrophilic and/or hydrophobic weft yarn materials. They suggested that the use of hydrophilic materials and quick-dry finishes could improve the moisture management properties of denim fabrics.

Mahmud et al. (2022) investigated the effect of a change in sheath cotton count on the performance of denim fabric. They produced five different yarn counts with 40 denier elastane filament yarn in the core. When the cotton sheath count was increased, the tensile and tear strength, elongation, and recovery of the samples increased in the weft direction. In addition, they reported that the pilling resistance property of fabrics became better with the increase of sheath count in core spun yarn.

The effects of yarn parameters together with weaving parameters such as weft density and weave construction on fabric properties were investigated in many studies.

Şekerden and Çelik (2010) produced woven fabrics with different weft yarns, weft densities, and weave structures (different types of plain and twill). They have concluded that weft density significantly affected weft-wise shrinkage, fabric weight, and breaking strength in the weft direction. Weft yarn count was effective on fabric weight and breaking strength. The weave patterns were effective in fabric thickness and weft-wise elasticity.

The effect of weave design and weft density on the bagging properties of cotton woven fabrics was examined by Doustar et al. (2010). They have produced woven fabrics in plain, twill 3/1, twill 2/2, and hopsack structures and used the FAST (Fabric Assurance by Simple Testing) system to measure the mechanical properties of the fabric. The mechanical properties measured with the SiroFAST tester can be listed as compression, fabric thickness, bending length, bending rigidity (in warp and weft directions), shear rigidity and formability, and extensibility (De Boos & Tester, 1994; Minazio, 1995). Their results showed that the shear rigidity and formability decreased, whereas some important bagging parameters, such as bagging fatigue, resistance, hysteresis, and residual hysteresis, increased with the decrease in weft density. They have concluded that shear rigidity and formability mainly affect fabrics' bagging deformation.

Akter et al. (2021) produced denim fabrics using different core-spun weft yarns that had different blend ratios of sheath material and elastane percentages. They found that with the increase of polyester content in the blended weft yarn, weft-wise tensile and tear strength of the fabrics were increased with the increase of elastane percentage in the weft yarn, but the weight of the fabric decreased. Fabrics performed better stretchability with the increase of elastane percentage in the weft yarn, but the growth changed by reversal.

Bedez Üte and Kadoğlu (2019) analyzed the effects of elastane linear density (44, 78, and 117 dtex) and elastane brand on the mechanical properties of double-core (EME, PES, and EA combinations) and single-core (hard- and soft-core) yarns spun with different core materials and their results showed that the type of the core materials and the thickness of the materials have significant effects on yarn properties. Afterward, they investigated the effects of weft yarn type (double-core, soft-core and hard-core) and core material (EME, PES, and EA as single or together) of weft yarns on the elasticity and recovery properties of denim fabrics. They concluded that elastic fabrics with improved body movement comfort, lower growth, and shrinkage values can be obtained by using double-core weft yarns.

Bedez Üte and Kadoğlu (2018) considered the influence of the composition and density of weft yarns (core spun and dual-core spun) on the mechanical properties of denim fabrics. Elastane (EA), polybutylene terephthalate (PBT), and elastomultiester (EME) containing cotton-covered elastic weft yarns were used in fabric production. Core-spun yarns were produced using one filament (PBT, EME, or EA), but dual-core yarns were spun into two filaments (PES+EA, PBT+EA, or EME+EA) in the core. Dual-core yarns can be produced in two ways: previously combined or simultaneously combined just before spinning a triangle. Elastane and the inelastic polyester filament were previously combined with an air-jet texturing process (PES/EA IMG (intermingled)). The weaving process was performed with different weft density values. According to the test results; as the weft density increased, the stiffness of the fabric increased, the elongation of the break, the elasticity, the shrinkage of the fabric, and the growth values decreased. Fabrics woven with elastane-containing weft yarns had higher shrinkage and growth values. Regarding the growth, in more detail, core-spun weft yarns ensured higher growth values compared to dual-core weft yarns. However, the elasticity and elongation of the fabrics are higher, but the lower tensile strength was lower when produced with dual-core weft yarns.

Bedez Üte (2019) used elastane, PBT, and elastomultiester as core materials in the production of elastic weft yarns for denim fabrics. The effect of core-spun and dual-core weft yarn content and design in corporation with weft density on the mechanical properties of the denim fabrics was investigated. Dual-core weft yarn was produced with PES+EA, PBT+EA EME+EA core materials. PES and elastane filaments were previously combined with air intermingling, whereas PBT+EA and EME+EA filaments were simultaneously fed the nip point of the drafting system on a ring-spinning machine. It was found that with the increase in weft density, the breaking elongation, elasticity, and growth values of the denim fabrics decreased. However, stiffness increased, and fabric stiffness is lower for fabrics produced with a core-spun yarn compared to a dual-core yarn. The elongation and elasticity values of the fabrics with dual-core spun weft yarns were higher, but the growth values were lower than the values of the fabrics with core-spun yarns.

Seyrek Kurban and Babaarslan (2019) conducted a literature review on fibers, types, and performance characteristics of super stretch denim fabrics. Afterward, they used different cores (elastane, PBT, T400 (PET/PTT)) and sheath fibers (cotton, modal, EME (elasto multitester) to produce multi-component weft yarns for denim fabrics (Seyrek Kurban & Babaarslan, 2022). It was found that fabric containing PBT in dual-core yarns has lower shrinkage values and fabrics produced with weft yarns EL+T400® in the core and EME/CO in the sheath part showed better stretch properties. The T400®+elastane core material provides better stretch properties compared to the PBT+elastane core material. Regarding sheath fibers, weft yarns covered with EME/cotton fibers were advantageous with the lowest growth values and the highest elongation and elasticity values.

Bansal et al. (2020) produced denim fabrics with broken twill, using elastane containing core spun weft yarns and investigated the effect of yarn count and elasticity percentage of core spun weft yarns on tensile and recovery behavior of denim fabrics. They tested the fabrics using the load cycling method and used image analysis to measure the elastic performance coefficient (EPC). When a deforming load is applied to a textile material and withdrawn, the ability to recover from deformation shortly after removal is defined as IER (Immediate Elastic Recover), whereas the recovery with time is defined as DER (Delayed Elastic Recovery). They have found that the tensile strength of denim fabrics (both in warp and weft directions), IER, DER, and EPC decrease, but PS (permanent set) increases with increasing the linear density and stretch percentage of the fabrics. Their further study (2019) investigated the effects of yarn production parameters (yarn count and twist multiplier), stretch, and extension percentage on the mechanical properties of the yarn. It was found that yarn tenacity is lower for coarser core spun yarns having the same elastane fineness and twist multiplier. The IER and DER decreased with a decrease in yarn count and increased with a decrease in stretch percentage. The yarn tenacity decreased with an increase in the stretch percentage and a decrease in the twist multiplier.

Chen and Tang (2004) performed a mechanical tensile load on PTT, PET, and PBT and compared the changes in instantaneous elastic recovery. They have presented that the PTT filament provides softer touch, due to its lower initial modulus and tenacity but higher breaking elongation and instantaneous elastic recovery, compared to the PBT and PET filament.

Babaarslan et al. (2021) compared the performance properties of denim fabrics produced with core, and dual-core spun weft yarns. They have used PET/PTT (T400®), PTT (Solotex®), PBT, and elastane (Lycra®) as core materials in yarn production. In their research, dual core-spun yarns showed higher elongation, but lower strength values compared to core-spun yarns. The unevenness and hairiness values of elastane-containing core-spun yarns were higher than those of PTT/PBT dual-core-spun yarns. For improved strength, dimensional stability, and stiffness of the fabrics, PBT filament was suggested, whereas for improved elasticity, elastane filament was suggested as the core material in core-spun weft yarns. It was stated that if fabric softness and flexibility are a priority for users and low shrinkage and growth are required, PBT can be preferred with its cost advantage. In their further study, they focused on the effect of elastane fineness on the properties of yarn and denim fabrics (Babaarslan et al., 2023). For higher values of yarn elongation, fabric stiffness, elasticity, and shrinkage, coarser elastane was recommended in dual-core weft yarns. Regarding the type of core component in dual-core yarns; PET/PTT results with more hairiness, the PTT filament showed a higher weight change after washing, higher stiffness, tearing, and tensile strength, whereas PET showed lower stiffness.

Erbil et al. (2022) also focused on the effect of weft yarn structure on the performance properties of denim fabrics. For this purpose, they have produced core spun yarns (core material: EA), and dual-core yarn (core materials: EA and PBT), in which the sheath materials are cotton. They have used rigid, core-spun, and dual-core spun weft yarns for denim fabric production. They have concluded that elastic yarns are effective in the strength properties (tensile and tearing) of fabrics by increasing the fabric weight, decreasing the staple fiber ratio in the yarn, and decreasing the packing density of the yarn. Therefore, they caused lower air permeability and lower seam slippage percentages. Fabrics produced with core-spun weft yarns showed the highest stretch and growth values after washing processes.

The effects of filament fineness and the elastane draft ratio of dual-core weft yarns on the properties of denim fabrics were the scopes of another research (Babaarslan et al., 2018). They have used cotton, core-spun (sheath material: cotton, core material PET, with different filament fineness), and dual-core spun yarns (sheath material: cotton, core material PET, with different filament fineness, and elastane with different draw ratio) as weft yarns. Core spun yarn was recommended for denim fabrics for better mechanical (higher tenacity, elongation, and elastic recovery) and comfort (higher wicking and absorption rating) performance. The elongation and elastic recovery of the fabric increased with the elastane draw ratio.

Luo et al. (2011) examined the effects of bi-component PTT/PET contents, weft densities, and yarn twists on the strain and resilience properties of plain, twill, and satin fabrics. Longer yarn floats resulted in higher elasticity in plain and twill fabrics compared to plain fabrics. Increasing the PTT/PET content in weft yarns up to 66.7% and decreasing weft density can lead to higher fabric elasticity. It has been observed that when PTT/PET filaments are used in the fabric, 6% of their original elastic potential is used.

The use of yarns with high elasticity in garments is quite common. Although polyurethane (PU)/polyamide (PA) core-spun yarns provide elasticity in such garments, there are some problems, such as dimensional stability and aging of elastane. Zhao et al. (2013) used poly trimethylene terephthalate (PTT)/polyester (PET) bi-component filament yarns as an alternative for seamless garments and evaluated the mechanical and physical properties of fabrics. They have found better results in dimensional stability, elastic recovery, and wrinkle resistance with the PTT/PET bi-component filament than the PU/PA core-spun yarn.

Yıldırım et al. (2022) experimented with different production parameters (twist, wool, and elastane draft ratio) for the production of dual-core yarns. The core materials were elastane with wool and the sheath material was cotton in dual-core yarns. A study was conducted to select the most suitable production parameters for the denim industry using Multi-Objective Optimization on the Basis of the Ratio Analysis Method (MOORA) method. They recommended these methods for other decision-making problems with any number of different preferences and yarn quality values.

Türksoy and Üstündağ (2015) compared and evaluated the physical properties and performances of hybrid yarns produced for denim fabrics. They used four-methods to produce hybrid yarn doubling with single yarn, doubling with double yarn, covering with twisted double yarn, and covering with non-twisted double yarn, using commingling elastic filaments. They reported that the maximum stretching value of the fabrics woven with the hybrid yarns produced by covering with non-twisted double yarn and classical twisting with single yarn methods was the highest. In addition, the lowest permanent stretching value was found for denim fabrics woven with hybrid yarn produced by classical twisting with single yarns. Also, denim fabrics woven with covering with non-twisted double yarn have the highest shrinkage values.

Türksoy et al. (2022) studied the comfort properties of denim fabrics containing dual-core weft yarn. The wool yarn and elastane yarn were used as core materials and cotton fiber was used as sheath material to improve the thermal insulation properties of the denim fabric. In addition, denim fabrics were produced with elastic core and dual-core spun yarns. They found that fabrics containing dual-core wool-elastane yarns had lower thickness values and lower water permeability values. However, they had higher elasticity and air-permeability values than the other fabrics. The effects of production parameters (twist level, wool, and elastane draw ratio) on dual-core yarn properties, designed for use in denim fabrics, were investigated (Türksoy & Yıldırım, 2018). They have reported that with the increase in twist level; yarn unevenness and hairiness decreased, and yarn tenacity increased. Increasing the wool draft decreased the yarn hairiness and increased the elongation values, whereas increasing the elastane draft ratio increased both the yarn elongation and the tenacity.

Sarıoğlu and Babaarslan (2017a, 2017b) compared the characteristics of denim fabrics woven with different core-spun weft yarns. They analyzed the effect of filament fineness and yarn core/sheath ratio on denim fabrics. The core material was false-twist textured microfilaments (PET) with different filament fineness (3.05 dtex, 1.15 dtex, 0.76 dtex, 0.57 dtex, and 0.33 dtex), and the sheath material was cotton. Fabrics with core-spun weft yarns had higher elongation values for all yarn counts and filament fineness. It was found that for finer yarns, the core materials, whereas coarser yarns sheath material, was more effective on static tearing resistance. Their further study focused on the porosity and air permeability of denim fabrics produced with the mentioned core-spun weft yarns (Sarıoğlu & Babaarslan, 2019). It was found that the total porosity and air permeability of denim fabrics increase with the increase of yarn count (tex) and filament fineness (dtex), and a high correlation was found between the dependent and independent variables.

Sinha et al. (2017) investigated the tensile and elastic recovery behavior of cotton/elastane core-spun yarn in the manufacturing stretch of denim fabrics. They produced cotton/elastane core-spun yarns with different combinations of linear density, twist multiplier, and stretch ratio. They found better elastic performance when using the lower linear density, lower stretch ratio, and low twist multiplier to produce cotton/elastane core spun yarn.

Rahim et al. (2023) produced dual-core (sheath: cotton, core: EA and T400) and single-core spun yarns (sheath: cotton, core: EA or T400). They used different draft ratios for both T400 (1.01, 1.07, 1.14, 1.2, and 1.3) and elastane (3.0, 3.4, 3.8, and 4.2) filaments. For dual-core yarns, an optimum draft ratio was determined for higher strength and elongation, lower evenness and hairiness, and lower growth (bagging) which is 3.8 for elastane, and 1.2 for T400. With the repetitive dynamic loading test, it is found that there is low growth and high recoverability. Compared to single-core yarns, dual-core yarns display better yarn properties in terms of tenacity, elongation, hairiness, evenness, plastic deformation, and stress decay.

In some studies, the effect of the process parameters after yarn and fabric production was examined. Eryuruk (2019) focused on the effects of fabric content (with or without elastane) and finishing processes (bleaching, rigid, resin, and softener treatments) on the mechanical and thermal comfort characteristics of denim fabrics. Elastane-containing fabrics had better handle and drape properties, but lower tear strength, shear stiffness, and bending rigidity. The compression resilience and thermal resistance were higher, but the surface roughness, thermal conductivity, thermal absorptivity, and thermal diffusivity water vapor permeability values of elastane-containing fabrics were lower than those of the fabrics without elastane. Bleaching and softening treatments improved shear, bending, stiffness, and drape properties, thermal conductivity, thermal absorptivity, and thermal diffusion values. He et al. (2018) also focused on finishing treatments by investigating the effect of ozone on denim fabrics. They compared the effect of the yarn material (lyocell, cotton, and polyethylene terephthalate (PET)) on the color fading process. According to their results, PET fabrics were more convenient for this process compared to cellulosic fibers. Kan and Yuen (2009) produced denim fabrics with elastic and non-elastic weft yarns and tested their performance under the effect of repeated home laundering processes. The elastane ratio in the elastic weft yarns was different. In the first five washing cycles, the dimensional stability of elastic denim fabrics after washing showed a large shrinkage due to relaxation shrinkage, which is also effective on air permeability, mass change, and thickness of fabric, stretchability after tension. After simulated home washing practices, the percentage of elastane in the fabric was also effective in the washing performance. Midha et al. (2017) studied the effect of enzyme washing and repeated washing on comfort properties and moisture management, and the drying rate of denim fabrics containing cotton, polyester, and cotton-elastane core-spun weft yarns. The permeability aspects of the denim fabric were better when using cotton weft yarn than the other fabrics. On the other hand, when using core-spun stretch weft yarn, the moisture management capabilities of the denim fabric were found to be better.

From a different point of view, Padhye and Nayak (2010) examined the sewing performance of denim fabrics containing elastic-core spun yarns. They have compared the effects of two commonly used sewing threads (spun and core-spun polyester) and different elastane percentages (0%, 1%, 3%, and 5%), fabric weight, and silicone finish on seam efficiency, seam pucker, seam slippage, and needle cutting index. Seam efficiency was influenced by increasing elastane percentage, whereas seam pucker and needle cutting index were more affected by fabric weight than elastane percentage. The silicone finish exhibited negative effects on sewing performance. Using coarser spun polyester or finer core-spun polyester sewing threads is recommended for stitching denim fabrics.

Recycling of Denim Fabrics and the Use of Recycled Yarn in Denim

People of all ages consume denim products, and the production of denim fabrics grows very fast yearly. At the same time, awareness of sustainability and recycling is increasing in this sector daily; thus, the use of recycled materials in the production of denim fabric has also been a widespread practice. Many companies have added products containing recycled fibers to their product range. In addition, there are studies and eco-friendly practices during denim production, such as the use of enzymes as an alternative to conventional chemicals for the dyeing and finishing process and the recovery of water (Radhakrishnan, 2017). However, there are problems worldwide regarding the collection and recycling of used denim, and efforts continue to establish a closed loop in denim production.

Around 2900 liters of water and many chemicals and energy are consumed in the production of denim jeans (Uncu Akı et al., 2020; Levi Strauss Co., 2015; Fibre2Fashion, 2014). According to the research result of Levi's 501 life cycle impact factor, 33.4 kg of CO_2 emission contribution to climate change, 48.9 g of PO_4 emission contribution to eutrophication, 12 m^2/year contribution to land occupation, and 3781 liters of water consumption were found (Levi Strauss Co., 2015; Radhakrishnan, 2017).

Zhao et al. (2021) investigated the environmental consequences of denim products. Cotton fiber and denim fabric production were reported to be the most contributors to carbon emissions and water consumption. According to their results, when the proportion of polyester fiber increases in denim products, the carbon footprint increases, but the water footprint decreases.

Fidan et al. (2021) aimed to search the input of mechanically recycled cotton fiber and combined heat and power (CHP) plant versus virgin cotton fiber and grid energy regarding the environmental impacts, cost, and quality of denim fabric products in their study. They used eight scenarios and environmental impacts were analyzed using the life cycle assessment approach (LCA). They reported that the highest environmental impact improvements were obtained with 100% recycled cotton and the use of CHP plants (scenario 8). They suggested using mechanically recycled cotton and a CHP plant for the circular economy for denim production according to their study's results.

Studies evaluate pre-consumer waste generated during the denim fabric and apparel production process. A mechanical process recycles these wastes, especially coarse open-end rotor yarns containing recycled cotton, which are used as weft yarn in the production of denim fabric. However, these mechanically recycled fibers have a considerably short fiber content. It can be spun using carrier fibers. On the other hand, post-consumer products are more challenging to utilize. It often turns into low-value products, such as thermal and sound insulation and pressure distribution, or goes to a solid-waste landfill (Radhakrishnan, 2017).

Some researchers investigated the utilization facilities of waste denim fabric for composite materials (Meng et al., 2020; Sezgin et al., 2021). Contin et al. (2022) investigated the facilities and material characteristics of mechanically recycled pre-consumer denim fabric waste. Mecarciöz et al. (2022) evaluated the effect of recycled cotton with virgin cotton, organic cotton, and better cotton fibers on denim fabric performance. They reported that the fabric contains recycled cotton and that the other fabrics have acceptable technical performance.

Telli and Babaarslan (2016) investigated the manufacturing of commercial denim fabric with 100% recycled raw materials. The waste of the spinning mill (dyed (indigo) and undyed) was used as recycled cotton. Since the fiber lengths are very short in recycled cotton, virgin cotton, Tencel fiber, and recycled PET fibers from bottles were used as carrier fibers. They produced Ne 8 open-end yarns in five different blends. Although the physical properties of the yarns using recycled raw material are worse than of those without recycled material, a very adverse effect has not been observed on the mechanical and physical properties of the fabric. It only harmed the appearance of the fabric.

In another study by Telli and Babaarslan (2017), they produced denim fabrics using different proportions of recycled yarn waste cotton fibers, virgin cotton fibers, and recycled PET fibers from bottles. They applied the singeing, washing, drying, and sanforizing processes to gray fabrics. Enzyme and stone washing were also processed to finish the fabrics. They evaluated the washing performance of denim fabrics related to the recycled fibers content. They found that recycled cotton content has an adverse effect on the tenacity values of fabric; on the other hand, recycled PET content has a positive effect on the tenacity and elongation values of the fabrics, and it has a negative effect on fabric handles. Nevertheless, they reported that the fabric containing recycled PET was more durable than the fabrics containing cotton and recycled cotton in the washing processes.

Kanan et al. (2022) used two different yarn wastes in denim yarn production. One was white yarn waste coming from the winding and the other was blue-dyed yarn waste from the dyeing. The yarn wastes blended two different virgin cotton fibers. They reported that no deterioration in yarn quality results was observed when up to 60% waste fiber was used.

The use of elastane in denim fabrics is also a compelling factor for pre-consumer and post-consumer waste recycling. Some researchers investigated the recycling facilities of textiles containing elastane. Wang et al. (2022) studied the recycling of cotton textiles containing elastane. They converted cotton–elastane fabric waste into regenerated films using different solvent systems. They suggested that the resultant film could be used in some applications, for example packaging materials. Phan et al. (2023) investigated the selective chemical dissolution of elastane from blended textiles for better-quality recycling.

Some studies on sustainability, recycling, and life cycle in the denim fabric and garments sector were mentioned above. As can be seen, this subject attracts much attention, and possibilities for establishing a closed-loop production system are being investigated. Since fiber properties deteriorate with mechanical recycling, studies on chemical recycling possibilities in fabric recycling have recently attracted attention. New designs with a sustainable fashion approach will assist in the collection, classification, and recycling of post-consumer waste.

Conclusion

Since the beginning of use in the 1800s, comfort and flexibility have come to the fore in denim products, employing a significant place in our lives and being indispensable. Due to the work of fashion designers and engineers, significant developments have been achieved in denim fabrics, and more satisfying designs have been made regarding flexibility, durability, and aesthetics. Although the production process has some harmful effects on the environment, it is foreseen that it will contribute to sustainability in terms of prolonging the life of the products and being a suitable sector for evaluating recycled fibers. The developments in fiber and yarn production technologies make it easier to provide the performance characteristics expected from the products. The use of dual-core yarns in denim fabric manufacturing has become one of the critical milestones in the history of denim technology.

Disclaimer

None.

References

Abghari, R., Shaikhzadeh Najar, S., Haghpanahi, M., & Latifi, M. (2004). Contributions of in-plane fabric tensile properties in woven fabric bagging behaviour using a new developed test method. *International Journal of Clothing Science and Technology*, *16*(5), 418-433. https://doi.org/10.1108/09556220410554615.

Akankwasa, N. T., Jun, W., Yuze, Z., & Mushtaq, M. (2014). Properties of cotton/T-400 and 100% cotton plain knitted fabric made from ring spun yarn. *Material Wissenschaft und Werkstofftechnik*, *45*(11), 1039–1044. https://doi.org/10.1002/mawe.201400246.

Akter, N., Repon, M. R., Mikučionienė, D., Jalil, M. A., Islam, T., & Karim, M. R. (2021). Fabrication and characterization of stretchable denim fabric using core spun yarn. *Heliyon*, *7*(12), e08532. https://doi.org/10.1016/j.heliyon.2021.e08532.

ASTM. (2019). *Standard test methods for stretch properties of fabrics woven from stretch yarns* (D3107-07).

Babaarslan, O. (2001). Method of producing a polyester/viscose core-spun yarn containing spandex using a modified ring spinning frame. *Textile Research Journal*, *71*, 367. https://doi.org/10.1177/004051750107100415.

Babaarslan, O., & Telli, A. (2013). A study on usage of chenille yarn in denim fabric production. *Journal of Textile and Engineer*, *20*(92), 1-10. https://doi.org/10.7216/130075992013209201.

Babaarslan, O., Balcı, H., & Güler, Ö. (2007). Effect of elastane on the properties of Pes/Vis blend woven fabrics. *Textile and Apparel*, *2007*(2), 110–114.

Babaarslan, O., Sarıoğlu, E., Çelik, H. İ., & Avcı, M. A. (2018). Denim fabrics woven with dual core-spun yarns. In M. K. Singh (Ed.), *Engineered fabrics* (pp. 19-39). Intech Open. https://doi.org/10.5772/intechopen.82717.

Babaarslan, O., Shahid, M. A., & Doğan, F. B. (2021). Comparative analysis of cotton covered elastomeric hybrid yarns and denim fabric properties. *Journal of Engineered Fibers and Fabrics*, *16*. https://doi.org/10.1177/15589250211059188.

Babaarslan, O., Shahid, M. A., & Doğan, F. B. (2023). Design of hybrid yarn with the combination of fiber and filaments and its effect on the denim fabric performance. *Fibres & Textiles in Eastern Europe*, *31*(1), 25-37. https://doi.org/10.2478/ftee-2023-0004.

Babaarslan, O., Telli, A., & Karaduman, S. (2015). Effect of microfilament yarns on the performance properties of denim fabrics. *Journal of Textile and Engineer*, *22*(99), 7–14. https://doi.org/10.7216/130075992015229902.

Baghaei, B., Shanbeh, M., & Ghareaghaji, A. A. (2010). Effect of tensile fatigue cyclic loads on bagging deformation of elastic woven fabrics. *Indian Journal of Fibre & Textile Research*, *35*, 298-302.

Bansal, P., Maity, S., & Sinha, S. K. (2019). Effects of process parameters on tensile and recovery behavior of ring-spun cotton/lycra denim yarn. *Journal of the Institution of Engineers (India): Series E*, *100*, 37-45. https://doi.org/10.1007/s40034-019-00133-5.

Bansal, P., Maity, S., & Sinha, S. K. (2020). Elastic recovery and performance of denim fabric prepared by cotton/lycra core spun yarns. *Journal of Natural Fibers*, *17*(8), 1184-1198. https://doi.org/10.1080/15440478.2018.1558151.

Bedez Üte, T. (2019). Analysis of mechanical and dimensional properties of the denim fabrics produced with double-core and core-spun weft yarns with different weft densities. *The Journal of the Textile Institute, 110*(2), 179-185. https://doi.org/10.1080/00405000.2018.1470451.

Bedez Üte, T., & Kadoğlu, H. (2018). The effects of core material parameters on the mechanical properties of double core and single core spun yarns. *IOP Conference Series: Materials Science and Engineering, 459*(1), 012032. https://doi.org/10.1088/1757-899X/459/1/012032.

Bedez Üte, T., & Kadoğlu, H. (2019). The effect of single and dual-core yarns produced with different core materials on the elasticity and recovery properties of denim fabrics. *Textile & Leather Review, 2*(4), 174-182. https://doi.org/10.31881/TLR.2019.10.

Chen, K., & Tang, X. (2004). Instantaneous elastic recovery of poly (trimethylene terephthalate) filament. *Journal of Applied Polymer Science, 91*(3), 1967-1975. https://doi.org/10.1002/app.13424.

Contin, B., Kohan, L., Oliveira Duarte, L., Fernandes, P. R. B., Ruchel-Soares, R., Siqueira, M. U., & Baruque-Ramos, J. (2022). Brazilian cotton jeans recycling: Characterization of shredded pre-consumer waste. *Materials Circular Economy, 4*, 19. http://doi.org/10.1007/s42824-022-00060-8.

Çataloğlu, A. (2007). *A study on elasticity and growth properties of elastane composed denim fabrics.* [Master Thesis]. Ege University.

Çelik, H. İ., & Kaynak H. K. (2017). An investigation on effect of elastane draw ratio on air permeability of denim fabrics. *IOP Conference Series: Materials Science and Engineering, 254*(8), https://doi.org/10.1088/1757-899X/254/8/082007.

Çelik, P., Bedez Üte, T., Özden, D., Çömlekçi, H., & Akkale, E. C. (2009). The effect of core/sheath ratio and twist amount on yarn charachteristics of filament core yarns, *Electronic Journal of Textile Technologies, 3*(2), 29-37.

De Boos, A. G., & Tester, D. H. (1994). FAST (Fabric Assurance by Simple Testing) SiroFAST – A System for Fabric Objective Measurement and its Application in Fabric and Garment Manufacture. *CSIRO Textile and Fibre Technology*, Report No. WT92.02, ISBN 0643060251.

Değirmenci, Z., & Çelik, N. (2016). Relation between extension and bursting strength properties of the denim viewed knitted fabrics produced by cellulosic fibers. *Fibres & Textiles in Eastern Europe, 1* (115), 101-106. https://doi.org/10.5604/12303666.1172065.

Doniyorova, M. A., & qizi Sadikova, G. Q. (2022). Use of dyed fiber yarn for warp yarn of denim fabrics. *Texas Journal of Engineering and Technology, 15*, 10-13.

Doustar, K., Najar, S., & Maroufi, M. (2010). The effect of fabric design and weft density on bagging behavior of cotton woven fabrics. *The Journal of the Textile Institute, 101*(2), 135–142. https://doi.org/10.1080/00405000802309584.

El-Ghezal, S., Babay, A., Dhouib, S., & Cheikhrouhou, M. (2009). Study of the impact of elastane's ratio and finishing process on the mechanical properties of stretch denim. *The Journal of the Textile Institute, 100*(3), 245–253. https://doi.org/10.1080/00405000701757925.

Erbil, Y., Babaarslan, O., Islam, M. R., & Sirlibaş, S. (2022). Performance of core & dual-core cotton yarn structures on denim fabrics. *Journal of Natural Fibers, 19*(14), 9500-9513. https://doi.org/10.1080/15440478.2021.1982837.

Eryuruk, S. H. (2019). The effects of elastane and finishing processes on the performance properties of denim fabrics. *International Journal of Clothing Science and Technology, 31*(2), 243-258. https://doi.org/10.1108/IJCST-01-2018-0009.

Ertaş, O. G., Ünal, B. Z., & Çelik, N. (2016). Analyzing the effect of the elastane-containing dual-core weft yarn density on the denim fabric performance properties. *The Journal of the Textile Institute, 107*(1), 116–126. https://doi.org/10.1080/00405000.2015.1016319.

Ertekin, M., & Kırtay, H. E. (2015). Tensile properties of some technical core spun yarns developed for protective textiles. *Textile and Apparel, 25*(2), 104–110.

Fibre2Fashion. (2014). *Denim companies answer to the call for sustainable jeans*, Retrieved May 3, 2023 from: https://www.fibre2fashion.com/industry-article/7245/denim-companies-answer-to-the-call-for-sustainable-jeans.

Fidan, F. S., Aydogan, E. K., & Uza, N. (2021). An integrated life cycle assessment approach for denim fabric production using recycled cotton fibers and combined heat and power plant. *Journal of Cleaner Production, 287*, 125439, https://doi.org/10.1016/j.jclepro.2020.125439.

Gazzah, M., Jaouachi, B., & Sakli, F. (2015). Recovery velocity evaluation of bagged denim fabrics as function of frictional parameters. *International Journal of Clothing Science and Technology, 27*(2), 302-314. https://doi.org/10.1108/IJCST-11-2013-0128.

Hailemariam, A. F., & Muhammed, N. (2022). Effect of ring and rotor spun yarns on denim fabric properties. *Research Journal of Textile and Apparel*. https://doi.org/10.1108/RJTA-11-2021-0137.

Hasani, H., Zadeh, S. H., & Behtaj, S. (2012). Bagging behavior of different fabric structures knitted from blended yarns using image processing. *Journal of Engineered Fibers and Fabrics, 7*(3), 8-15. https://doi.org/10.1177/155892501200700302.

He, Z., Li, M., Zuo, D., & Yi, C. (2018). The effect of denim color fading ozonation on yarns. *Ozone: Science & Engineering, 40*(5), 377-384. https://doi.org/10.1080/01919512.2018.1435259.

Invista (2006). Producing Core-Spun Yarns Containing Lycra® Elastane Fiber.

Jamshaid, H., Asfand, N., Rajput, A. W., Basra, S. A., Zahid, B., & Ali, A. (2020). Comparison of functional properties of woven and knitted denim fabrics. *Industria Textila, 71*(1), 3-7.

Jaouachi, B., Hassen, M. B., & Sakli, F. (2007). Strength of wet spliced denim yarns after sizing using a central composite design. *AUTEX Research Journal, 7*(3), 159-165.

Jaouachi, B., Hassen, M. B., Sahnoun, M., & Sakli, F. (2010a). Evaluation of wet pneumatically spliced elastic denim yarns with fuzzy theory. *The Journal of the Textile Institute, 101*(2), 111-119. https://doi.org/10.1080/00405000802283672.

Jaouachi, B., Louati, H., & Hellali, H. (2010b). Predicting residual bagging bend height of knitted fabric using fuzzy modelling and neural networks. *AUTEX Research Journal, 10*(4), 110-115.

Kan, C. W., & Yuen, C. W. M. (2009). Evaluation of the performance of stretch denim fabric under the effect of repeated home laundering processes. *International Journal of Fashion Design, Technology and Education, 2*(2-3), 71-79. https://doi.org/10.1080/17543260903302329.

Kanan, M., Wannassi, B., Barham, A. S., Ben Hassen, M., & Assaf, R. (2022). The quality of blended cotton and denim waste fibres: The effect of blend ratio and waste category. *Fibers, 10*(9), 76. https://doi.org/10.3390/fib10090076.

Kara, G., & Akgun, M. (2018). Effect of weft yarn fiber contents on the moisture management performance of denim fabrics woven with different constructional parameters. *Textile and Apparel, 28*(2), 151-161.

Kaynak, H. K., Babaarslan, O., Avci, M. E., & Doğan, F. B. (2017). Effects of spinning technology on denim fabric performance. *Industria Textila, 68*(3), 197-203.

Levi Strauss & Co. (2015). *The life cycle of a jean - Understanding the environmental impact of a pair of Levi's 501 ® jeans*. San Francisco. Retrieved May 3, 2023 from: https://www.levistrauss.com/wp-content/uploads/2015/03/Full-LCA-Results-Deck-FINAL.pdf.

Luo, J., Wang, F. M., Li, D., & Xu, B. (2011). Elasticity of woven fabrics made of polytri-methylene terephthalate/polyethylene terephthalate bicomponent filaments. *Textile Research Journal, 81*(8), 865-870. https://doi.org/10.1177/00405175103917.

Mahmud, Md. F., Masudur Rahman, A. N. M., Sarker, A. A. H., Akter, S., Islam, Md. R. (2022). Evaluation of the impact of cotton sheath linear density of core spun yarn on the performance of denim fabrics. *Hellion, 8*(8). https://doi.org/10.1016/j.heliyon.2022.e09998.

Mangat, M. M., Hussain, T., & Bajzik, V. (2012). Impact of different weft materials and washing treatments on moisture management characteristics of denim. *Journal of Engineered Fibers and Fabrics*, 7(1), 38-48. https://doi.org/10.1177/155892501200700104.

Meng, X., Fan, W., Ma, Y., Wei, T., Dou, H., Yang, X., Tian, H., Yu, Y., Zhang, T., & Gao, L. (2020). Recycling of denim fabric wastes into high-performance composites using the needle-punching nonwoven fabrication route. *Textile Research Journal, 90*(5–6) 695–709. https://doi.org/10.1177/0040517519870317.

Mezarcıöz, S., Oğulata, R. T., & Nergis, A. (2020). Investigation of elasticity and growth properties of denim fabrics woven with core and siro spun yarn. *Textile and Apparel, 30*(4), 312-320.

Midha, V., Kumar, S. S., & Kumar, M. N. (2017). Investigation on permeability and moisture management properties of different denim fabrics after repeated laundering, *The Journal of The Textile Institute, 108*(1), 71-77. https://doi.org/10.1080/00405000.2016.1153873.

Minazio, P. G. (1995). FAST – fabric assurance by simple testing. *International Journal of Clothing Science and Technology, 7*(2/3), 43–48. https://doi.org/10.1108/09556229510087146.

Özdil, N. (2008). Stretch and bagging properties of denim fabrics containing different rates of elastane. *Fibres and Textiles in Eastern Europe*, *161*(66), 63–67.

Padhye, R., & Nayak, R. (2010). Sewing performance of stretch denim. *Journal of Textile and Apparel, Technology and Management*, *6*(3), 1-9.

Park, T. Y., & Kim, M. O. (2018). Manufacture and physical properties of the denim fabrics using Hanji paper yarn as weft yarn. *Fashion and Textiles*, *5*, 1-13. https://doi.org/10.1186/s40691-018-0140-6.

Phan, K., Ügdüler, S., Harinck, L., Denolf, R. Roosen, M., O'Rourke, G., Vos, D. D., Speybroeck, V. V., De Clerck, K., & De Meester, S. (2023). Analysing the potential of the selective dissolution of elastane from mixed fiber textile waste. *Resources, Conservation & Recycling*, *191*, 106903. https://doi.org/10.1016/j.resconrec.2023.106903.

Radhakrishnan, S., (2017). Denim recycling. In S. S. Muthu (Ed.), *Textiles and clothing sustainability, Recycled and upcycled textiles and fashion* (pp.79-125). Springer Singapore. https://doi.org/10.1007/978-981-10-2146-6.

Rahim, M. A., Rahman, S., & Uddin, A. J. (2023). Low-bagging (growth) stretch denim yarn production by spinning optimization of cotton-wrapped dual-core elastane and T400 multifilament. *Heliyon*, *9*(3). https://doi.org/10.1016/j.heliyon.2023.e13639.

Sarıoğlu, E., & Babaarslan, O. (2017a). Fatigue behavior of core-spun yarns containing filament by means of cyclic dynamic loading. *IOP Conference Series: Materials Science and Engineering*, *254*(8), https://doi.org/10.1088/1757-899X/254/8/082012.

Sarıoğlu, E., & Babaarslan, O. (2017b). A comparative strength analysis of denim fabrics made from core-spun yarns containing textured microfilaments. *Journal of Engineered Fibers and Fabrics*, *12*(1), https://doi.org/10.1177/155892501701200103.

Sarıoğlu, E., & Babaarslan, O. (2019). Porosity and air permeability relationship of denim fabrics produced using core-spun yarns with different filament finenesses for filling. *Journal of Engineered Fibers and Fabrics*, *14*. https://doi.org/10.1177/1558925019837810.

Sarıoğlu, E., Babaarslan, O., & Avcı, M. E. (2018). Effect of filament fineness on composite yarn residual torque. *Autex Research Journal*, *18*(1), 7–12. https://doi.org/10.1515/aut-2016-0036.

Seyrek Kurban, N., & Babaarslan, O. (2019). Literature review on the properties of super stretch denim fabrics. *Journal of Textile and Engineer*, *26*(113), 104-115. https://doi.org/10.7216/1300759920192611312.

Seyrek Kurban, N., & Babaarslan, O. (2022). The effect of fiber types used as core/wrapper fibers in multi-component yarns on the stretch properties of denim fabrics. *Journal of Natural Fibers*, *19*(15), 11332-11348. https://doi.org/10.1080/15440478.2022.2025973.

Sezgin, H., Kucukali-Ozturk, M., Berkalp, O. B., & Yalcin-Enis, I. (2021). Design of composite insulation panels containing 100% recycled cotton fibers and polyethylene/polypropylene packaging wastes. *Journal of Cleaner Production*, *304*, 127132. https://doi.org/10.1016/j.jclepro.2021.127132.

Sinha, S. K., Bansal, P., & Maity, S. (2017). Tensile and elastic performance of cotton/lycra core spun denim yarn. *Journal of The Institution of Engineers (India): Series E*, *98*(1), 71 – 78. https://doi.org/10.1007/s40034-017-0095-y.

Su, C. I., & Yang, H. Y. (2004). Structure and Elasticity of Fine Elastomeric Yarns. *Textile Research Journal*, *74*(12), 1041–1044. https://doi.org/10.1177/004051750407401202.

Su, C. I., Maa, M. C., & Yang, H. Y. (2004). Structure and performance of elastic core-spun yarn. *Textile Research Journal*, *74*(7), 607–610. https://doi.org/10.1177/004051750407400709.

Sülar, V. (2011). A new testing instrument with artificial arm to simulate fabric bagging by human elbow. *Industria Textila*, *62*(5), 259-264.

Sülar, V. (2016). Prediction of fabric bagging occurred by an artificial arm under dynamic conditions. *Journal of Fashion, Technology and Textile Engineering*, *0*(2). https://doi.org/10.4172/2329-9568.S2-003.

Şekerden, F., & Çelik, N. (2010). Weft elastane weaving and fabric characteristics. *Textile and Apparel*, *2*, 120–129.

Şengöz, N. G. (2004). Bagging in textiles. *Textile Progress*, *36*(1), 1-64. https://doi.org/10.1533/jotp.36.1.1.59475.

Tekstil & Teknik. (2017, July 25). *Fashionable dual core yarns: Perfectly spliced and wound*, Retrieved May 3, 2023 from: https://www.tekstilteknik.com.tr/fashionable-dual-core-yarnsperfectly-spliced-wound/.

Telli, A., & Babaarslan, O. (2016). Commercialized denim fabric production with post-industrial and post-consumer wastes. *Textile and Apparel, 26*(2), 213-216.

Telli, A., & Babaarslan, O. (2017). The effect of recycled fibers on the washing performance of denim fabrics, *The Journal of The Textile Institute, 108*(5), 812-820. https://doi.org/10.1080/00405000.2016.1192236.

Telli, A., Daşan, Y., Babaarslan, O., & Karaduman, S. (2017). Usage of core and dual-core yarns containing tungsten for electromagnetic shielding. *Advance Research in Textile Engineering, 2*(1), 1013.

Türksoy, H. G., & Üstündağ, S. (2015). Elastic hybrid yarns for denim fabrics. *Industria Textila, 66*(5), 306-313.

Türksoy, H. G., & Yıldırım, N. (2018). Effect of process variables on the properties of dual-core yarns containing wool/elastane. *Industria Textila, 69*(5), 352-356.

Türksoy, H. G., Yıldırım, N., & Ertek Avcı, M. (2022). Comparative evaluation of some comfort properties of denim fabrics including dual-core yarns containing wool and elastane yarn, *The Journal of The Textile Institute, 113*(10), 2051-2058, https://doi.org/10.1080/00405000.2021.1964195.

Uçar, N., Realff, M. L., Radhakrishnaiah, P., & Ucar, M. (2002). Objective and subjective analysis of knitted fabric bagging. *Textile Research Journal, 72*(11), 977-982. https://doi.org/10.1177/004051750207201108.

Uncu Akı, S., Candan, C., Nergis, B., & Önder, N. S. (2020). Understanding denim recycling: A quantitative study with lifecycle assessment methodology. In A. Korlu (Ed.), *Waste in textile and leather sectors* (pp. 1-26). Intech Open. https://doi.org/10.5772/intechopen.92793.

Vuruşkan, D., Ilhan, I., & Babaarslan, O., (2011). Effect of production parameters on strength and elongation of the selected yarns containing elastane. *Textile and Apparel, 21*(1), 22–29.

Wang, L., Huang, S., & Wang, Y. (2022). Recycling of waste cotton textile containing elastane fibers through dissolution and regeneration. *Membranes, 12*, 355. https://doi.org/10.3390/membranes12040355.

Yeung, K. W., Li, Y., Zhang, X., & Yao, M. (2002). Evaluating and predicting fabric bagging with image processing. *Textile Research Journal, 72*(8), 693-700. https://doi.org/10.1177/004051750207200808.

Yıldırım, N., Akgül, E., & Türksoy, H. G. (2022) Selection of dual-core yarn production parameters for denim fabric by using MULTIMOORA method, *The Journal of The Textile Institute, 113*(6), 1039-1047. https://doi.org/10.1080/00405000.2021.1914411.

Zhang, X., Li, Y., & Yeung, K. W. (2000a). Viscoelastic behavior of fibers during woven fabric bagging. *Textile Research Journal, 70*(9), 751-757. https://doi.org/10.1177/004051750007000901.

Zhang, X., Li, Y., Yeung, K. W., & Yao, M. (1999a). Fabric bagging: part I: subjective perception and psychophysical mechanism. *Textile Research Journal, 69*(7), 511-518. https://doi.org/10.1177/004051759906900708.

Zhang, X., Li, Y., Yeung, K. W., & Yao, M. (1999b). Fabric bagging: Part II: Objective evaluation and physical mechanism. *Textile Research Journal, 69*(8), 598-606. https://doi.org/10.1177/004051759906900809.

Zhang, X., Li, Y., Yeung, K. W., & Yao, M. (2000b) Relative contributions of elasticity and viscoelasticity of fibres and inter-fibre friction in bagging of woven wool fabrics. *The Journal of The Textile Institute, 91*(4), 577-589. https://doi.org/10.1080/00405000008659129.

Zhao, L., Hu, H., Shen, J., & Rong, H. (2013). The use of a polytrimethylene terephthalate/polyester bi-component filament for the development of seamless garments. *Textile Research Journal, 83*(12), 1283-1296. https://doi.org/10.1177/0040517512471749.

Zhao, M., Zhou, Y., Meng, J. Zheng, H., Yanpeng, C., Shan, Y., Guan, D., & Yang, Z. (2021). Virtual carbon and water flows embodied in global fashion trade - a case study of denim products. *Journal of Cleaner Production, 303,* 127080. https://doi.org/10.1016/j.jclepro.2021.127080.

Chapter 13

Fancy Yarns: Alternatives to Comfort-Enhanced Fabrics

Zehra Evrim Kanat*, **PhD**
Department of Textile and Fashion Design, Canakkale Onsekiz Mart University, Canakkale, Türkey

Abstract

Fancy yarns, which contain various effects in terms of color and/or structure, attract the attention of textile and fashion designers with their aesthetic appearance. Designers use these surface effects, created by some planned irregularities in the yarn, as a design element. Moreover, the use of fancy yarn also enables the economical production of original designs. With the developments in recent years, fancy yarn types have become very diverse, and it is possible to talk about the limitless effects that can be obtained by using both the type and parameter of fancy yarns as variables. In addition to the design differences fancy yarns create in the appearance of the fabric, they also change the surface characteristics of the fabric, sensory and thermophysiological comfort properties. Research has been carried out to examine the effects of raw materials used in yarn production, fancy yarn production parameters, and fabric production parameters on the dimensional, physical, appearance, and comfort properties of the fabric. In this section, fancy yarn structures are defined and classified, and studies on the comfort properties of fabrics produced using fancy yarn are mentioned.

Keywords: fancy yarn, thermophysiological comfort, slub yarn, chenille yarn, boucle yarn

Introduction

Fancy yarns introduced to the market in the 19th century are expressed by local irregularities in their structure and have attracted great interest in recent years. A fancy yarn is defined by the Textile Institute (Textile Terms and Definitions) as "a yarn that differs from the normal construction of single and folded yarns by way of deliberately produced irregularities in its construction. These irregularities relate to an increased input of one or more of its components, or to the inclusion of periodic effects, such as knops, loop, curls, and slubs." (Qotb et al., 2022).

* Corresponding Author's Email: zevrimk@gmail.com.

In: Textile Insights and Future Aspects
Editors: Gamze D. Tetik, Gizem Celep and Fulya Yilmaz
ISBN: 979-8-89113-289-4

With these deliberate, decorative, constantly repeated, programmed color and/or form irregularities in the structure of fancy yarns, it is aimed to create a difference in the aesthetic appearance of the yarn rather than its performance characteristics. Fancy yarns are very diverse in terms of their structural properties, fiber components, and production methods, and their structures have become increasingly complex in recent years (Ulku et al., 2003; Ragaišienė & Kumpikaitė, 2010; Bedeloglu, 2014; Qotb et al., 2022). These color and texture effects have allowed us to make a difference in the aesthetic appearance of fabrics and clothes in the field of fashion and design (Bedeloglu, 2014).

Although fancy yarns do not constitute a large volume in the total textile production, they are of commercial importance for clothing and decorative products due to the optical and/or structural decorative effects they provide. The use of these yarns adds attractiveness to the product and increases its added value (Petrulyte & Petrulis, 2014).

In addition, these yarns also change the end-use characteristics of the product. With the use of fancy yarn, the surface properties of the fabrics can be changed, and their softness and permeability properties can be improved. The usage possibilities of fancy yarns are quite wide. Their aesthetic and decorative features make them preferred in both woven and knitted fabrics. Although hand knitting was thought to be the biggest market for fancy yarns in the beginning, today their use has become widespread in areas such as outerwear, upholstery, knitwear, and trimming (Petrulytė, 2008; Ragaišienė & Kumpikaitė, 2010; Alshukur et al., 2020a). It also has some industrial applications in making floor and wall coverings, covers for car seats and airplane seats, etc., (Ragaišienė & Kumpikaitė, 2010; Alshukur et al., 2020a).

The interest in fancy yarn production and applications of these yarns, which were introduced commercially in the 19th century, is increasing. Slub, curly, and knotted fancy yarns are widely used on the market, while new yarns continue to be developed (Petrulytė, 2008; Fouda et al., 2022). According to a market analysis published by Stratview Research in 2020, the Fancy Yarn Market is predicted to grow from $4.5 billion in 2020 to $6.3 billion in 2026, with a CAGR of 5.8% (Fancy Yarn Market Report, 2023).

Structure and Types of Fancy Yarns

Components of fancy yarns, which can have very different structures and include diameter, bulk, and/or color, irregularities, etc., can be staples or filaments made of natural or artificial fibers (Petrulytė, 2004; Kumpikaitė et al., 2010a; 2010b). The spinning and twisting processes are very important in order to obtain yarns with various effects (Petrulytė, 2008). Most of these yarns with different properties are produced on modified or specially developed spinning and twisting machines. Some fancy yarns are produced from "fancy fiber" or "fancy sliver" on conventional machines, while others are made from filament yarns, while others are produced on knitting machines specially designed for fancy yarn production (Petrulytė, 2004).

In general, fancy yarns consist of "core" (ground) yarn, "effect material" and a "binder" that holds the structure together. The "effect" material is combined with the "core" (ground) to create effects such as loops, piles, waves, knots, snarls, etc. The effect created with these two components is fixed with a "binder" (Figure 1) (Petrulytė, 2004; Qotb et al., 2022). But recently, yarns that use non-woven material as an effect component as well as knitted and braided yarns are available on the market. This diversity in the production process, effect, and material characteristics of fancy yarns and in the number of combinations has revealed the necessity of systematically classifying yarns (Petrulytė, 2004).

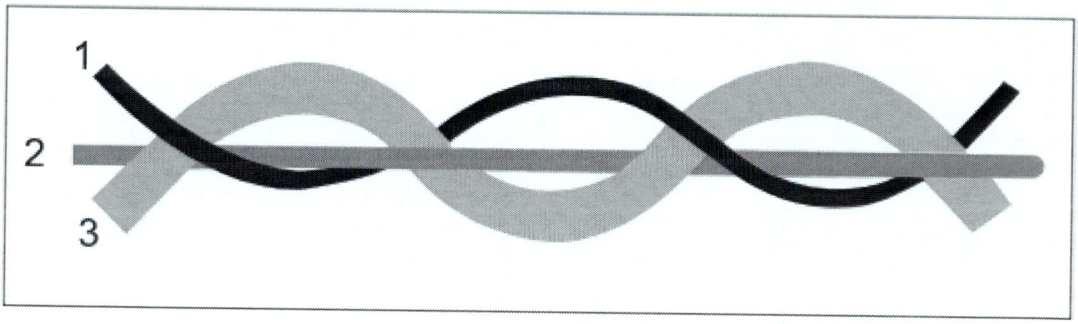

Figure 1. Structure of fancy yarn: 1. binding yarn, 2. core or ground yarn, 3. effect material.

Criteria used for the systematic classification of fancy yarn types:

- production methods,
- use area of fancy yarns and raw material of components,
- number and type of components,
- effect type,
- material for effect component (Petrulytė, 2004).

When classified in terms of production methods of fancy yarns, they can be divided into two classes: direct and indirect (Figure 2). In the direct system, modified or specially developed spinning and twisting machines are used for yarn production, or yarns are produced by weaving, weft, and warp knitting on special equipment. Whereas in the indirect systems, yarns are produced by adapting conventional processing techniques (spinning, carding, drawing, dyeing, raising, etc.) (Petrulytė, 2004; Tvarijonavičienė et al., 2005; Čiukas et al., 2006).

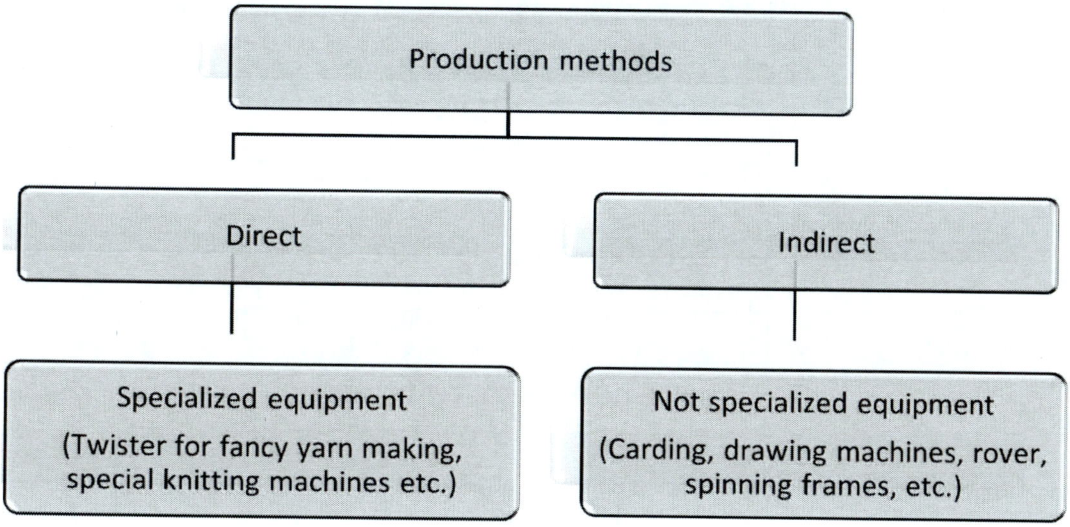

Figure 2. Classification according to the production method of fancy yarn.

If fancy yarns are classified according to their usage areas, there are two groups produced from woven, knitted fabrics or non-woven materials: clothing and decorative fabrics. Fancy yarns can be produced from many different raw materials according to the usage area and desired effect characteristic, and classification could be made according to the raw material (Petrulytė, 2004).

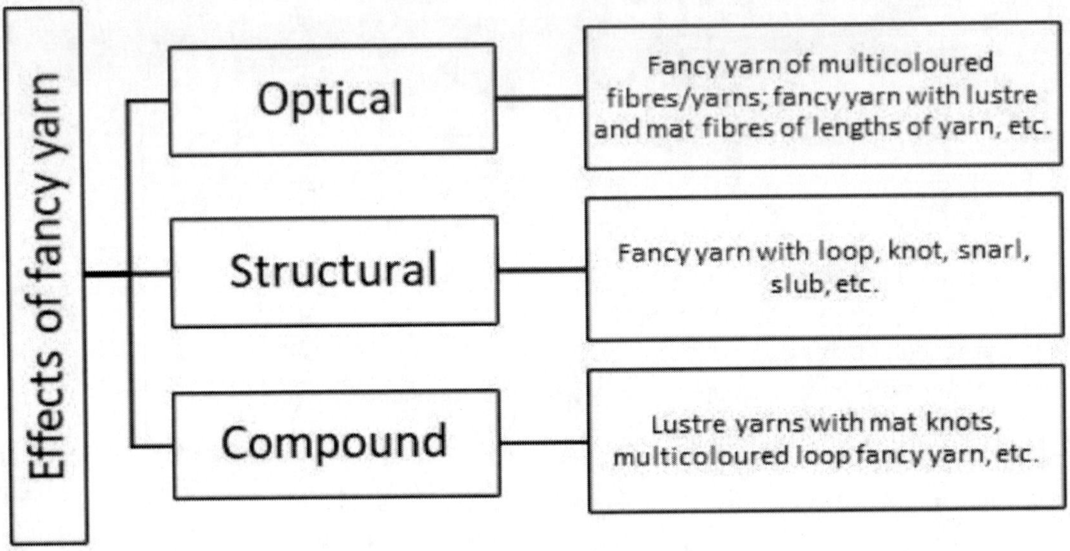

Figure 3. Classification of fancy yarns according to the effect type.

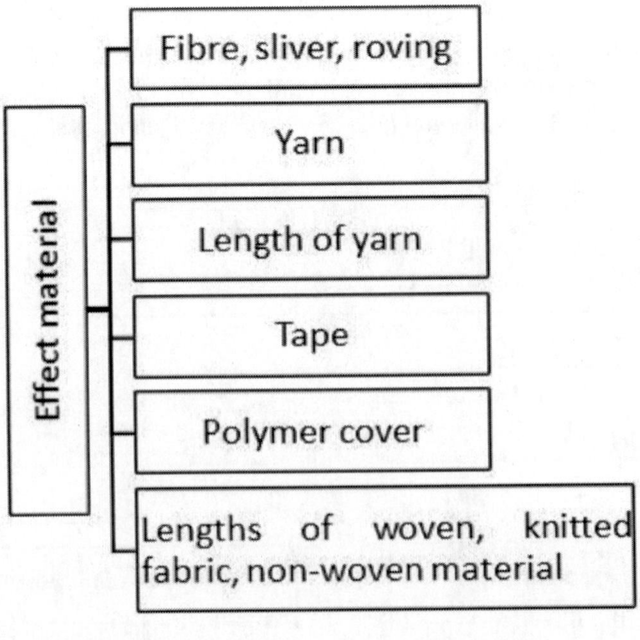

Figure 4. Classification of fancy yarns according to the effect of material.

The number and type of fancy yarn components that can completely change the structure and appearance of the yarn is also a parameter used for classification. The components can be tapes, rovings, non-woven materials, etc., as well as various yarns from staple or filament fibers (Petrulytė, 2004).

The mentioned classification methods may not always be suitable. The appearance of the yarns produced by different methods may be the same, or the effects of the yarn obtained with the same number of components and raw materials may be completely different from each other. For this reason, the classification made according to the effect type and the material used for the effect is considered the most consistent classification for fancy yarns. The classification according to the type of effect is shown in Figure 3 and the classification according to the effect material is shown in Figure 4. The classification made according to these two criteria includes both traditional and new structure fancy yarns, taking into account the yarn character (Petrulytė, 2004).

The classification of fancy yarns facilitates the evaluation of the yarn and fabrics obtained from these yarns. In addition, a systematic approach is important in order to design fancy yarn with the desired properties.

As mentioned, there are many different kinds of fancy yarns in the market and the structural properties of these yarns can be quite different from each other. The types of fancy yarns whose comfort properties are discussed in this chapter are briefly explained below.

Slub Yarns

The slub yarn contains periodically or randomly thick places whose thickness and/or length can be changed. The slub part is coarser than the normal yarn. In addition to the slub thickness, which can be three or four times the basic yarn, the slub length, and frequency are also used as parameters in the yarn design (Gong & Wright, 2002; Thanabal et al., 2021).

Knop Yarns

The knop yarn has bundles of yarn that are wrapped at regular or irregular intervals along its length. Although it has a similar effect to slub yarn, here, unlike slub yarn, thick places consist of yarn (Gong & Wright, 2002; Qotb et al., 2022).

Loop Yarn

The loop yarn consists of a core and an effect yarn. The loop effect is achieved by overfeeding the effect yarn by 200% or more to create circular protrusions wrapped around the core yarn (Gong & Wright, 2002; Tarafder, 2014).

Spiral Yarn

In the spiral yarn, two yarns spiral smoothly around each other. In this yarn, which has a plied yarn structure, the lengths of the two yarns are different from each other (Gong & Wright, 2002; Tarafder, 2014).

Ribbon Yarn

Ribbon-type fancy yarns are a type of fancy yarn in knitted structure and can be produced with braiding weft and warp knitting techniques. Stapel or filament yarns can be used for the production of these yarns (Tvarijonavičienė, 2005; Turay et al., 2009; Bedeloglu, 2014).

Chenille Yarn

The chenille yarn consists of two or more highly twisted fine strong yarns, i.e., short lengths of spun yarn or filament called pile, held together by lock yarns. (Gong & Wright, 2002; Ulku et al., 2003; Tarafder, 2014; Fouda et al., 2022).

Boucle Yarn

The boucle yarn consists of an effect yarn (or roving) and a twisted core, which is combined to obtain wavy protrusions on the surface. The effect is created by different distributions of the effect component (Gong & Wright, 2002; Uygun Nergis & Candan, 2006).

Macaroni Yarns

The effect material in Macaroni yarns, which consists of effect material, main ground yarn, and binding yarn components, is roving. This material improves the aesthetic properties of the yarn but imparts a lower strength. This part of the yarn on the fabric surface provides a different appearance (Ceven & Gunaydin, 2018).

Comfort Properties of Fancy Yarns

Comfort is the physiological, psychological, and physical harmony between people and their environment (Slater, 1985). In order to ensure comfort, the clothing must maintain the body thermophysiological balance. This is defined as thermophysiological comfort, and the most important components of thermophysiological comfort can be considered as thermal characteristics and permeability properties. In addition, it is stated that physical properties such as perceived rigidity, hardness, and softness, which occur in contact with the garment with the body, are effective in the comfort evaluation. Neurological perceptions that occur as a result of the contact of the textile material with the skin are expressed as tactile comfort (Hes et al., 2001; Li, 2001).

These intentional, decorative, constantly repeated, programmed color and/or form irregularities, which create the aesthetic and decorative appearance of fancy yarns that attract attention in textile and fashion design, also change the performance properties of the fabrics produced from these yarns. These irregularities in the structure of the yarns will affect the permeability properties of the fabrics and, definitely, their handle properties. The comfort properties of clothes and some textile structures such as upholstery fabrics are very important to the end user. By changing the structural properties of fancy yarns, it is possible to improve both the aesthetic appearance and the comfort properties of fabrics produced with these yarns. As mentioned in the previous sections, there are many types of fancy yarns with different structures and appearances, and it is also possible to obtain yarns with the same appearance and different structures. For this reason, the comfort properties of fabrics obtained from fancy yarns are as complex as they are interesting.

The thick places in the structure of the slub yarns change the thickness and pore structure of the fabrics obtained from these yarns. Therefore, there are changes in the permeability properties of the fabric according to the slub characteristics. In a study evaluating the air permeability properties of woven fabrics produced with slub yarn; the effects of yarn count and slub parameters such as distance, amplitude, length, and thickness were investigated. While the effect of all parameters on air permeability was found to be significant, it was stated that the effect of slub thickness was more important than other parameters. The air permeability of the fabrics decreases as the slub thickness and length increase and increases with the increase in the yarn count. When the size of the pores decreased, the friction surface increased. The higher slub thickness reduced the air permeability of the fabric as it significantly reduced the pore sizes. The slub distance was found to be the least significant parameter in the air permeability value compared to other parameters (Özgen & Altas, 2017).

Some comfort properties of single jersey fabrics produced by varying the thickness and length of the slub were investigated. As in woven fabric, as the thickness and length of the slub increase, the air permeability decreases. The researchers reported that the reason for this is related to the decrease and restriction of the free air flow through the pore size of the fabric structure, as in the previous study. Furthermore, fabrics with thicker and longer slubs have a higher rate of absorption and evaporation of moisture (Thanabal et al., 2021).

Fouda (2016) examined the effect of the slub yarn ratio used in a single jersey fabric. Fabric thickness was determined with a camera. The images were evaluated, and the results show that there was a significant change in the fabric thickness between the places from the ground and the slub yarns. The effect of the slub yarn ratio on the thermal conductivity and the thermal absorptivity values of the fabric was found to be statistically insignificant. However, the slub yarn ratio and the ground yarn count have a significant effect on thermal resistance. Besides, the effect of the slub yarn ratio on relative water vapor permeability is statistically insignificant, and its effect on air permeability is significant. As a result of another study examining the use of elastane in slub yarns, it was seen that the use of elastane caused a lower air and water vapor permeability, as well as higher thermal conductivity and thermal absorption values. It was found that the effects of slub length and slub thickness are significant on fabric spirality, thermal conductivity, thermal absorption, and relative water vapor permeability, while slub length has no significant effect on air permeability properties (Ertekin et al., 2022). All slub parameters should be examined in studies to improve the comfort properties of fabric with slub yarns, and it should be taken into account that changes in these parameters will also change the aesthetic properties of the fabric to be obtained.

One of the parameters to be considered when comparing the comfort properties of fancy yarns is the fabric structure. In the study, it was examined the effect of raw material (Cotton, Polyester, blended) and the weaving structure (Matt 2/2, Crepe, 1/3 Twill) on the properties of woven fabrics with slub and knop yarns. The thickness and air permeability of the fabrics with knop yarns were higher. Moreover, it was concluded that the weaving structure and raw material directly affect the air permeability of the fabric and there is a direct relationship between the weaving structure and the fabric thickness (Qotb, 2022).

Fancy yarns with different effects may behave differently from each other depending on the raw material and weaving type. Moreover, the properties of the fabrics produced from those yarns may show different changes after use. In this respect, there are studies comparing different types of fancy yarns. In the study in which the abrasion resistance values of woven fabrics produced using fancy yarns with slub, looped, and spiral structures and air permeability after abrasion were compared, different woven fabric structures (twill and satin) and weft yarn raw material were also examined as parameters. While the air permeability of sateen fabrics is highest in fabrics with loop yarn, fabrics with slub yarns have the lowest air permeability. This may be due to the regularity of the sizes and distributions of the fancy yarn effects. It has been determined that the air permeability changes in a different way during the abrasion process, that is, as the number of abrasion cycles increases, the air permeability decreases in the first place and the air permeability increases after a certain point. The fancy yarn effect and the regularity of their distribution are decisive for air permeability. Raw material and fabric weave also affected the air permeability of the fabrics. The air permeability of fabrics with synthetic yarns is higher in both weaving structures; however, synthetic yarn sateen fabrics appear to be less resistant to abrasion. The air permeability of sateen fabrics with the yarns of both raw materials is higher because the weave is less rigid than the twill weave; yarns can slip through the fabric (Kumpikaitė et al., 2010a). In another study, which investigated the effect of abrasion on change of air permeability and mass loss in fabrics woven using fancy yarns with slub, loop, and spiral structures, it was concluded that fancy yarn structure has an effect on the abrasion resistance of fabrics. Although the type of fancy yarn did not have a significant effect on the air permeability value in this study, it did affect the course of the change. With abrasion, the air permeability of the slub fabric changes the most and that of the spiral structure fabric the least. Although the fancy yarn structure does not affect the fabric mass loss, it changes the course of the mass change. The mass of spiral-structured fabrics changes during abrasion. The regular nature of the fancy yarn affects this phenomenon (Ragaišienė & Kumpikaitė, 2010). As can be seen, different types of fancy yarns give different responses depending on the raw material and weaving structure. For this reason, raw material and fabric structure should also be evaluated in research.

Knitted yarns can be produced with different needle numbers and obtained in different thicknesses. In a study examining the thermal properties and air permeability properties of knitted fabrics produced with knitted fancy yarns, it was determined that the effect of the number of machine needles and the material on the thermal properties of the fabrics was statistically significant. The thermal conductivity and thermal absorbency values of fabrics knitted from fancy yarns produced with high needle counts were found to be higher than those knitted with low needle counts. In the study made with knitted yarns, it was determined that the effect of machine drafting speed on thermal conductivity and thermal resistance values was not statistically significant. As the speed of the drafting roller increases, the thermal absorption value generally decreases, while the air permeability values increase due to the finer yarn structure obtained. When evaluated in terms of material, fabrics produced from 100% acrylic first knitted fancy yarns show high thermal conductivity and low resistance values compared to fabrics produced from 50%-50% wool-acrylic yarns (Turay et al., 2009). In another study in which the effect of heat setting on the properties of fabrics obtained from ribbon-type fancy yarns with different needle counts was examined, ribbon-type fancy yarns produced with 3 and

4 needles were used. The air permeability and thermal properties of the fabric were affected by the number of needles, and this effect was found to be significant. The air permeability of the fabrics made with 3-needle yarns and the thermal conductivity values of the fabrics made from 4-needle fancy yarns are higher due to the increasing number of needles creating thicker yarn and denser fabric structure. Due to the higher thickness, the thermal resistance values were also higher in fabrics made from fancy 4-needle yarns. Statistically, the effect of the number of needles on the thermal absorbance value was not significant. However, the thermal absorption values of fabrics made of fancy 3-needle yarns were higher. The heat setting process has affected the properties of fabrics made with ribbon-type fancy yarns, and the effect is statistically significant. Thanks to heat treatment, the important properties of yarns (hairiness and stretch) and fabrics (abrasion resistance, air permeability, and thermal absorbency) are improved. It is stated that fabrics feel warmer at the first touch, especially after the heat setting process (Bedeloglu, 2014). While investigating the comfort properties, the processes after fabric production are also important, and these processes can be used to improve the comfort properties.

Chenille yarn is widely used in the manufacture of upholstery, outerwear, or decorative fabrics. While the pile heights of chenille yarns have an important effect on the aesthetic appearance of the fabrics, the changes they create in the fabric structure and their effect on the comfort properties have also been examined in various studies. In a study, 40 tex chenille yarns with 2 different twists (700 and 850 turns/m) and three different pile heights (1, 1.2, 1.5 mm) were produced to examine the geometric, physical, and thermo-physiological properties of chenille fabrics. Single jersey fabrics with a loop length of 1, 1.2, and 1.4 cm were knitted from these yarns. While the porosity values of all samples increased as the loop length increased, the porosity decreased, due to the decrease in yarn density and fabric weight, as the twist value increased. Higher loop length leads to higher thermal resistance and air permeability and lower abrasion resistance, water vapor resistance, and thermal conductivity. In addition, high-twist chenille yarns have higher thermal resistance, thermal conductivity, water vapor resistance and abrasion resistance, and lower air permeability. Moreover, when the pile height of the chenille yarn is raised; thermal resistance, water vapor resistance, and thermal conductivity raise, air permeability, and abrasion resistance reduce. The study suggested a fabric sample suitable for cold climates; loop length of 1.4 cm, yarn twist of 850 tpm, and pile height of 1 mm (Fouda et al., 2022).

The bending stiffness is a feature used to give an idea when evaluating the fabric handle. In a study, the stiffness value of the fabric, which is one of the most important features that determine the handle characteristics of the garment, was evaluated using the bending rigidity measurement. For this, seven different types of pile yarn and two different types of core yarn materials were tested on chenille yarns. As a result of the study, the effect of the core yarn was found to be insignificant. It is stated that the pile yarn material is the most important factor in the bending rigidity value in both the wale and the course directions in chenille yarns (Kavusturan et al., 2010).

Low bending stiffness values of fabrics and garments are associated with high sensory comfort, and there are several studies investigating the bending stiffness of fabrics made from bouclé yarns. In a study the relationship between the boucle yarn structure and the bending stiffness values. The number of the boucle, the size of the boucle, the circularity ratio of the boucle, and the shape factor of the yarn are the investigated parameters of the yarn structure. The results showed that increasing the value of the stiffness of the effect yarn increased the size of the boucle but decreased both the number of boucle and the shape factor of the yarn (Alshukur et al., 2020a). The structure of multi-thread fancy yarns was mainly affected by the twist stiffness of the effect yarn and, to a lesser extent, the twist stiffness of the core yarn. It was found that the stiffer the effect yarns of the boucle yarn, the larger the boucle forms, and the relationship was linear. Also, the bending stiffness of the effect yarns increased as the number of the boucle decreased. As a result, the stiffer the effect yarns of the boucle yarns, the less the number of boucle and semi-boucle forms. The structure of multi-thread fancy yarns is affected by the variation in the bending stiffness of the effect yarn and the core yarn. It was found that the higher the variation in the bending stiffness of the effect yarns, the higher the variation in the boucle form size. This relationship was also linear (Alshukur et al., 2020b).

It is mentioned that different materials can be used instead of yarn as the effect material. A study was conducted to examine the effect of material type and yarn count on the air permeability of knitted fabrics produced from macaroni yarns. In the study, fabrics produced with 100% cotton, 100% polyester macaroni yarns, and blended fabrics produced from macaroni yarns using acrylic roving as effective material, polyester as binding yarn, and ground yarn were investigated. As a result of this study, the highest air permeability was found in 100% polyester fabrics. The largest difference in air permeability was determined between polyester and cotton fabrics. Air permeability decreases as macaroni yarns become thicker, regardless of the material. This effect is seen both before and after washing (Ceven & Gunaydin, 2018).

There are various parameters that affect the yarn properties, and thus the comfort properties in fabrics produced from fancy yarns. These parameters, which need to be investigated, vary according to the type of fancy yarn. The effect material is also a parameter that needs to be examined. For this reason, the comfort properties of fabrics produced with fancy yarns are quite complex. There are not enough studies in this area.

Conclusion

In this section, fancy yarn structures are defined and classified, and studies on the comfort properties of fabrics produced using fancy yarn are mentioned.

The developments in the field of fancy yarns attract the attention of textile and fashion designers. Estimates show that the fancy yarn market is a very open market for growth.

The previously limited application areas of fancy yarns have expanded a lot. They are widely used in many areas, such as clothing, upholstery, and vehicle textiles. In addition to the performance features of fabrics produced from fancy yarns, comfort has also gained importance with expanding usage areas.

Fancy yarn types are increasing. With different production parameters and the variety of materials used, it is possible to obtain many yarns with very different structures and different properties. It is possible to improve fabric comfort properties with planned irregularities in fancy yarn structures.

For this purpose, the effects of yarn parameters on thermophysiological and sensory comfort according to yarn type should be investigated on a larger scale. Raw material, effect material, and fabric structure should also be considered in research. Previous studies indicate that different types of fancy yarns behave differently.

It is important to determine the comfort demands of the fabric according to the area of use, to choose the type of fancy yarn to be produced for this, and to determine the production parameters of this yarn and the raw material and structure of the materials to be used.

Disclaimer

None.

References

Alshukur, M., Fotheringham, A., & Gong, H. (2020a). The influence of component stiffness on the structure of multithread fancy bouclé yarn. *Journal of Industrial Textiles, 49*(7), 889–905. https://doi.org/10.1177/1528083718801365.

Alshukur, M., Fotheringham, A., & Gong, H. R. (2020b). Relationship between the interaction of bending stiffness of component yarns and the structure of fancy bouclé and semi-bouclé yarns. *Fibers and Polymers, 21*(2), 437-446. https://doi.org/10.1007/s12221-020-8156-0.

Bedeloglu, A. (2014). The influence of heat-setting process on physical properties of ribbon-type fancy yarns and fabrics produced from them. *Industria Textila, 65*(1), 3-9.

Ceven, E. K., & Gunaydin, G. K. (2018). Investigation of selected physical properties of knitted fabrics produced from macaroni yarns. *Fibres & Textiles in Eastern Europe, 26* (4:130), 59-66. https://doi.org/10.5604/01.3001.0012.1314.

Čiukas, R., Tvarijonavičienė, B., & Mikučionienė, D. (2006). Estimating the linear density of fancy ribbon-type yarns and the structure indices of fabrics knitted from them. *Fibres & Textiles in Eastern Europe, 14* (4: 58), 41- 43.

Ertekin, G., Ertekin, M., & Marmaralı, A. (2022). Visual perception and performance properties of fabrics knitted with elastic core cotton slub yarns. *Journal of Natural Fibers, 19*(3), 810-822. https://doi.org/10.1080/15440478.2021.1944428.

Fancy Yarn Market Report. (2023, May 06). *Stratview research*. Retrieved May 06, 2023 from https://www.stratviewresearch.com/2008/Fancy-Yarn-Market.html#market-insights.

Fouda, A. (2016). Effect of slub yarn ratio on single jersey knitted fabric properties. *Journal of Textile Science and Engineering, 6,* 273. https://doi.org/10.4172/2165-8064.1000273.

Fouda, A., Těšinová, P., Khalil, A., & Eldeeb, M. (2022). Thermo-physiological properties of polyester chenille single Jersey knitted fabrics. *Alexandria Engineering Journal, 61,* 7029–7036. https://doi.org/10.1016/j.aej.2021.12.041.

Gong, R. H., & Wright, R. M. (2002). *Fancy yarns: Their manufacture and application*, Woodhead Publishing Limited, Cambridge, UK.

Hes, L., Offerman, P., & Dvorakova, I. (2001, June 26- 29). *The effect of underwear on thermal contact feeling caused by dressing up and wearing of garments* [Conference session]. Tecnitex 2001 Autex Conference, Povoa de Varzim, Portugal, 236-245.

Kavuşturan, Y., Çeven, E. K., & Özdemir, Ö. (2010). Effect of chenille yarns produced with selected comfort fibres on the abrasion and bending properties of knitted fabrics. *Fibres & Textiles in Eastern Europe, 18*(1: 78), 48-53.

Kumpikaitė, E., Ragaišienė, A., & Barburski, M. (2010a). Comparable analysis of the end-use properties of woven fabrics with fancy yarns, Part I: Abrasion resistance and air permeability. *Fibres & Textiles in Eastern Europe*, *18*(3: 80), 56-59.

Kumpikaitė, E., Ragaišienė, A., & Barburski, M. (2010b). Comparable analysis of the end-use properties of woven fabrics with fancy yarns, Part II: Abrasion resistance and mass. *Fibres & Textiles in Eastern Europe* *18*(4 :81), 43-45.

Li, Y. (2001). The science of clothing comfort. *Textile Progress*, *31*(1-2), 1-135.

Özgen, B., & Altaş, S. (2017). Evaluation of air permeability of fabrics woven with slub yarns. *Textile and Apparel, 27*(2), 126-130.

Petrulytė, S. (2004). Fancy yarns: Efforts to methodise, problems, and new suggestions. *Materials Science (Medžiagotyra), 10*(1), 85- 88.

Petrulytė, S. (2008). Influence of technological parameters on the periodical effects of fancy yarns. *Fibres & Textiles in Eastern Europe, 16*(3: 68), 25- 29.

Petrulyte, S., & Petrulis, D. (2014). Influence of twisting on linen fancy yarn structure. *Journal of Natural Fibers, 11*(1), 74-86. https://doi.org/10.1080/15440478.2013.842512.

Qotb, M. M. M. M., Ali, O. E. E., Gawad, M. A., & İbrahim, A. E. S. (2022). The effect of different structural formations of innovative yarns on fabric comfort properties. *International Design Journal, 12*(2), 333-341. https://doi.org/10.12816/IDJ.2022.222688.

Ragaišienė, A., & Kumpikaitė, E. (2010). Investigation of end-use properties of woven fabrics with synthetics fancy yarns. *Materials Science (Medžiagotyra), 16*(4), 337- 340.

Slater, K. (1985). *Human Comfort*. Thomas Sprinfield, USA.

Tarafder, N. (2014). Fancy yarns - application and production methods. *Man-made Textiles in India, 42*(9), 334- 339.

Thanabal, V., Gopalakrishnan, M., & Vijayasekar, R. (2021). Study on the quality characteristics of slub yarn knitted fabrics. *Indian Journal of Fibre and Textile Engineering (IJFTE), 1*(1), 10- 14.

Turay, A., Süpüren, G., Özdil, N., & Özçelik, G. (2009). The effects of the production conditions of ribbon typed fancy yarns on the thermophysiological properties. *Textile and Apparel, 4*, 280-285.

Tvarijonavičienė, B., Mikučionienė, D., & Čiukas, R. (2005). Influence of knitting process conditions and washing on tensile characteristics of knitted ribbon yarns. *Fibres & Textiles in Eastern Europe, 13*(4: 52), 74- 77.

Ulku, S., Ortlek, H. G., & Omeroğlu, S. (2003). The effect of chenille yarn properties on the abrasion resistance of upholstery fabrics. *Fibres & Textiles in Eastern Europe, 11*(3:42), 38-41.

Uygun Nergis, B., Candan, C. (2006). Performance of bouclé yarns in various knitted fabric structure. *Textile Research Journal, 76*(1), 49–56. https://doi.org/10.1177/0040517506059210.

Chapter 14

Composites Reinforced with Spacer-Knitted Fabrics

Mustafa Ertekin*, PhD
and Gözde Ertekin, PhD
Department of Textile Engineering, Ege University, Izmir, Türkey

Abstract

Knitted fabrics are increasingly being used in composite applications due to advances in fiber technology, including high-performance fibers such as carbon, glass, and aramid, as well as the development of three-dimensional (3D) fiber-preforming machines. Knitted preforms have gained significant attention in various industries, including aerospace, medical, automotive, and civil engineering. Spacer-knitted fabrics are the most commonly used knitted textile preform, particularly for lightweight composites. Spacer-knitted fabric composites have better drapability and impact protection compared to other textile-based composites. In this chapter, the recent progress of the spacer-knitted fabrics in composites including composite manufacturing techniques, application possibilities, and some other performance properties compared to other composite materials will be investigated.

Keywords: mechanical properties, spacer-knitted fabrics, textile reinforced composites

Introduction

Primary textile manufacturing methods for fabricating fibrous structures for composites include weaving, knitting, braiding, and nonwoven production techniques (Fangueiro, 2011). Knitting is a widely used technique accounting for more than 30% of global fabric production and can be classified into two categories: weft knitting and warp knitting. Weft knitting produces loops in a horizontal direction, while warp knitting creates loops in a vertical direction (Figure 1). Weft knitting is more elastic, more porous, and offers greater design flexibility than warp knitting (Boisse, 2015; Lemair & Pearsall, 2007; Parker et al., 2006; Sek et al., 2000; Verpoest et al., 1997; Burgess, 1994).

Weft-knitted (Chen & Zhuang, 2020; Delavari & Semnani, 2021; Ding et al., 2021; Dong et al., 2022; Ghatrenabi et al., 2021; Hajjari et al., 2021; Liu et al., 2022c; Ravandi et al., 2022;

* Corresponding Author's Email: mustafa.ertekin@ege.edu.tr.

In: Textile Insights and Future Aspects
Editors: Gamze D. Tetik, Gizem Celep and Fulya Yilmaz
ISBN: 979-8-89113-289-4

Temesgen & Sahu, 2022; Tian et al., 2021; Yan et al., 2022; Zeynivand et al., 2021) and warp-knitted fabric (Chen et al., 2022a, 2022b; Fang et al., 2022; Li et al., 2022; Liu et al., 2022b; Rameshkumar et al., 2022) reinforcements have gained significant attention in the composites industry in the last years due to their distinctive properties compared to other reinforcement fabric structures such as woven and braided fabrics. The interlocking loops in knitted fabrics have the ability to glide over each other and provide high deformability to the fabrics. The deformability of knitted fabric reinforcements allows them to form complex preform shapes for liquid molding, which is useful for producing composite components. Advanced knitting machines enable the production of the near-net-shaped fabrics, reducing material waste and production time. Using near-net shaped preforms in combination with conventional liquid molding techniques has the potential to mass-produce composite materials and reduce production costs, which is particularly important as composite applications shift from high-cost, high-performance aerospace products to low-cost, mass-producible products in general engineering.

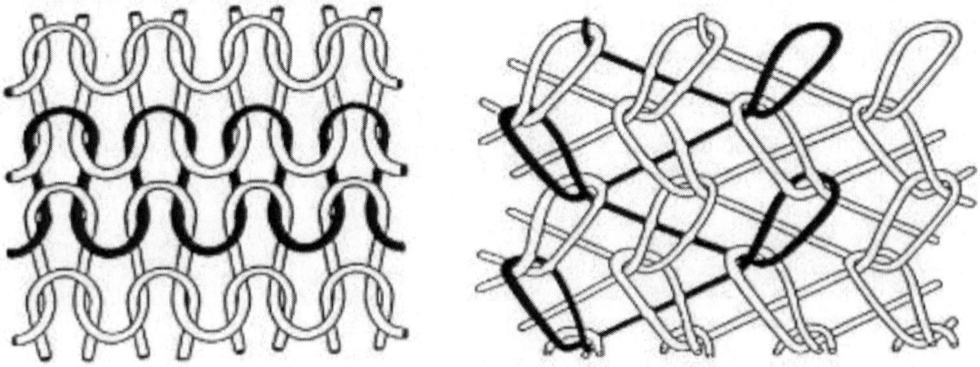
(a) (b)

Figure 1. Schematic diagrams of (a) weft-knitted fabrics and (b) warp-knitted fabrics.

Although the garment industry uses various types of knitted fabrics for fashion-related purposes, a restricted range of knit structures are currently under examination for engineering applications in composites. Table 1 provides a summary of the typical warp- and weft-knitted fabrics that have been studied for engineering applications (Boisse, 2015; Miravete, 1999).

Table 1. Classification of typical warp- and weft-knitted fabrics used in engineering applications (Miravete, 1999)

Type	Fabric classification	Weft-knitted fabric structure	Warp-knitted fabric structure
I	2-D fabric	Plain, rib, Milano rib, inlaid fabrics	Dembigh, Atlas
II	2-D fabric – 3-D shape	Plain, rib	Dembigh, Atlas
III	3-D solid fabric	Plain and rib fabrics with inlay fiber yarns	Multiaxial warp-knitted fabrics or non-crimp fabrics
IV	3-D hollow fabric (spacer fabric)	Single jersey face structure	Single Dembigh face structure

Various textile architectures present opportunities for designing composite properties. Although the two dimensional (2D) biaxial fabric was initially used for composite reinforcement, multilayered 2D fabric structures lack through-the-thickness fibers, resulting in inferior interlaminar properties and damage tolerance. In contrast, 3D textile fabrics that incorporate through-the-thickness fibers exhibit superior interlaminar strength and damage tolerance, which makes them an appealing choice for producing structural components capable of withstand multidirectional mechanical and thermal stresses. 3D textile-reinforced composites offer several benefits, such as toughness, damage tolerance, structural integrity, ease of handling the reinforcing material, and suitability for net-shaped manufacturing, making them a suitable choice for composite reinforcements (Kumar & Thakur, 2017). Knitted textiles are a viable option for specific functions because of their controlled anisotropy, the processability of high-performance fibers, and the ability to be formed into complex 3D shapes with ease. The development of 3D-configured products, such as spacer structures, represents a new technical method for producing high-performance materials, particularly as composite reinforcements (Maity et al., 2021).

This chapter reviews the recent progress in the spacer-knitted fabrics in composites including composite manufacturing techniques, application possibilities, and some other performance properties compared to other composite materials.

Spacer-Knitted Fabric

Spacer-knitted fabrics are intricate 3D constructions composed of two separate fabric layers that are vertically connected with spacer yarn or spacer layers, maintaining a hollow space between adjacent connecting yarns or layers. They are generally classified according to their knitted structure, which can either be weft or warp-knitted (Man, 2016; Triantafillou, 2016) (Figure 2). Weft-knitted spacer fabrics can be produced on either flat or circular knitting machines, whereas warp-knitted spacer fabrics can be only produced on the double-needle-bar raschel machine.

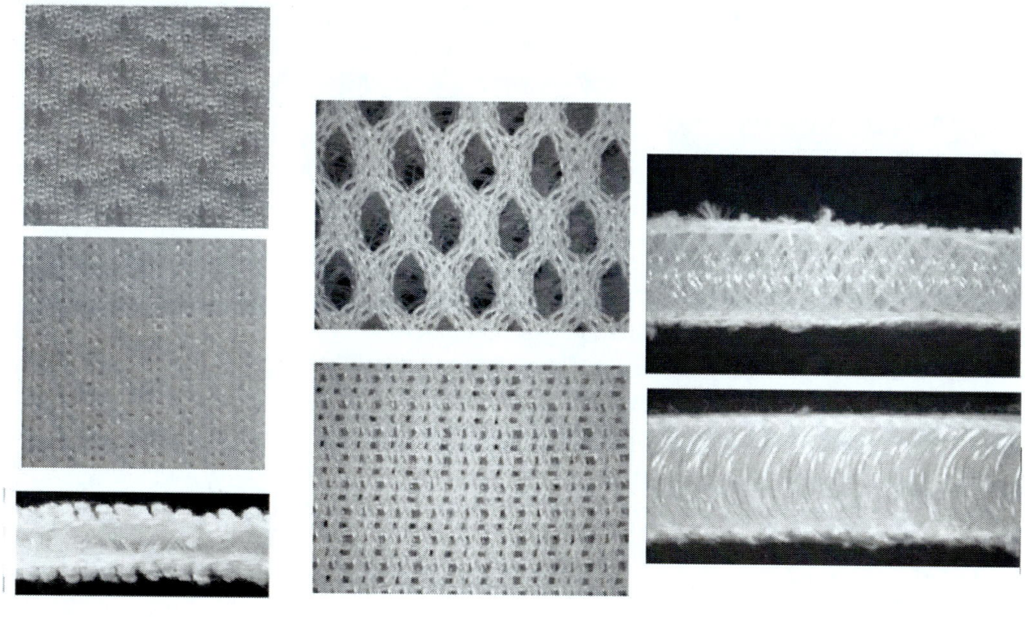

(a) (b)

Figure 2. Surface and cross-sectional views of (a) weft-knitted and (b) warp-knitted spacer fabrics (Ertekin, 2014; 2017).

The properties of spacer fabrics are influenced by various factors, including fabric thickness, the number of inlay yarns per width, spacer yarns per unit area, the properties of inlay and spacer yarns, and the inclination angle of spacer yarns. Spacer fabrics can be manufactured in different structures such as open, closed, or mesh-like. The thickness of the fabric can be modified by changing the length of the spacer yarns between the two surfaces. The outer layers of the spacer fabrics, where the highest tensile and compressive forces occur, have high-strength inlay yarns. This makes spacer fabrics suitable for concrete reinforcement, and also spacer fabrics with metal wire as inlay yarns can also be used for structural reinforcement in hybrid metal or plastic products (Triantafillou, 2016).

Composites

A composite refers to a material formed by combining two or more constituent materials, each possessing distinct physical and chemical properties, to produce a new material with superior properties. The composite comprises a reinforcement and a matrix material, which synergistically function to produce a material with enhanced strength, reduced weight, and improved durability, surpassing the properties of each individual material. The reinforcement material, in the form of fibers, particles, or flakes, generally imparts strength and stiffness, while the matrix material acts as a binding agent, confining the reinforcement in place and shielding it from external factors (Kumar & Thakur, 2017; Rana & Fangueiro, 2016).

Composite materials are generally used when conventional materials, such as metals, ceramics, and polymers, cannot satisfy the specific requirements of a particular application. One of the primary benefits of composites is that their properties can be customized by adjusting the type, proportions, orientation, and process parameters of the constituent materials. Additionally, composites have a high strength-to-weight ratio, making them an excellent option for applications in the automotive and aerospace industries. They also possess desirable properties, such as high fatigue resistance, toughness, thermal conductivity, and corrosion resistance. The primary drawback of composites is their high processing costs, which limit their widespread adoption (Kumar & Thakur, 2017).

Types of Composites

Composite materials have numerous advantages compared to conventional materials. Most composites are lightweight, yet possess significantly high mechanical properties. As a result, they exhibit high specific strength, stiffness, and modulus values. Additionally, they provide excellent chemical and weather resistance. Composites are highly versatile in that there are many reinforcement and matrix materials available, allowing the desired properties to be obtained from the final composite. By aligning the reinforcement fibers in a particular direction, composites can achieve directional properties, which are highly beneficial for structures that require different properties in different directions. Various manufacturing techniques exist to produce polymer composites, making it simple to create composites with complex shapes.

Various types of composites are available, each with its unique combination of constituents, properties, and applications. The most common types of composites based on the type of matrix materials used are polymer matrix composites (PMC), metal matrix composites (MMC), and ceramic matrix composites (CMC). PMC consists of a polymer matrix reinforced with fibers, particles, or other materials, and is widely used in the aerospace, automotive, and sports industries due to its excellent strength-to-weight ratio. MMC is composed of a metal matrix reinforced with ceramic, carbon, or other materials, and is used in the aerospace, defense, and electronic industries due to their high strength, stiffness, and thermal conductivity. CMC is composed of a ceramic matrix reinforced with fibers and is used in high-temperature applications such as gas turbines, thermal insulation, and nuclear reactors due to their excellent thermal and mechanical properties. Additionally, hybrid composites, made from a combination of two or more types of composites, have emerged as promising areas of research for developing new materials with enhanced properties and functionalities (Rana & Fangueiro, 2016; Kumar & Thakur, 2017).

Composites are categorized as either particulate composites or fiber-reinforced composites, based on the type of reinforcement used. Particulate composites consist of a matrix material reinforced with small particles, such as ceramic or metallic particles. On the other hand, fiber-reinforced composites consist of a matrix material reinforced with fibers. Fiber-reinforced composites can be classified into two categories: short fiber-reinforced composites and continuous fiber-reinforced composites, with short fibers either aligned or randomly oriented and continuous fibers being unidirectional, bidirectional, or multidirectional. Continuous fiber-reinforced composites generally offer better performance capabilities than particulate composites. The majority of currently manufactured composites are reinforced with fibers, as they possess several desirable properties compared to particulates. Fibers are particularly favorable as reinforcements due to their high strength, high aspect ratio (i.e., length-to-diameter ratio), and high flexibility.

Structural composites can be classified into two types: laminates and sandwich composites. Laminates are composites made up of multiple layers, in which the orientation of the fibers may be the same or different in each layer. Sandwich composites, on the other hand, comprise composite skin layers with a lightweight foam or honeycomb structure core, which may vary depending on the application. Sandwich composites demonstrate high stiffness and low weight. Moreover, composites can be categorized based on the size of their reinforcing material, which has an impact on the composite's mechanical, thermal, and electrical characteristics:

- Macrocomposites: They are composed of reinforcement materials that are larger than 1 mm in size. Examples include fiber-reinforced composites, laminates, and particulate-reinforced composites.
- Microcomposites: They are composed of reinforcement materials that are between 1 micrometer (μm) and 1 mm in size. Examples include whisker-reinforced composites and short-fiber-reinforced composites.
- Nanocomposites: They are composed of reinforcement materials that are less than 100 nanometers (nm) in size. Examples include carbon nanotube-reinforced composites and graphene-reinforced composites (Rana & Fangueiro, 2016).

Manufacturing Techniques of Composites

Various manufacturing techniques exist to produce composites, some of which are suitable for large-scale production. However, certain techniques may not be capable of producing intricate shapes. When selecting a manufacturing technique, factors such as application, cost, and desired properties must be taken into consideration. Processing MMCs and CMCs typically requires higher temperatures compared to those of PMCs, which means that the methods used for PMCs may not be appropriate for MMCs and CMCs.

Brief explanations of commonly used manufacturing methods for these three types of composites are provided below:

Manufacturing of PMCs

In general, the manufacturing techniques of PMCs can be broadly categorized into two main types: those based on thermoset resins and those based on thermoplastics. Thermoset resins are materials that undergo a chemical reaction, known as curing, when heated, resulting in a 3D network structure that cannot be melted or reshaped after curing. Thermoplastics, on the other hand, are materials that can be melted and reshaped multiple times when heated and solidified when cooled. In the context of composites, thermoset resins are often used as the matrix material for high-performance applications due to their superior mechanical and thermal properties, while thermoplastics are preferred due to their easy processability and recyclability. Thermoset resins are typically manufactured using methods such as hand lay-up, resin transfer molding, resin infusion, autoclave processing, compression molding, filament winding, and pultrusion. On the other hand, thermoplastics are commonly processed using methods such as injection molding, reinforced reaction injection molding, and thermoforming. The following is a brief explanation of these manufacturing methods:

Hand Lay-Up

Hand lay-up is a widely adopted method for manufacturing composites due to its simplicity, low cost, and ability to produce complex shapes with ease. The process involves creating a mold, which can be made from a variety of materials including wood, plastics, and fiber-reinforced plastics, according to the final product shape. The mold can be single- or multi-segmented depending on the complexity of the product. The curing of the mold can be done either at room temperature or by heating to accelerate the curing process.

Resin Transfer Molding (RTM)

RTM is a process that injects a liquid thermoset resin into a closed mold under low pressure and temperature, resulting in a high-quality surface finish and accurate dimensional properties of the composite. It is a semi-automated manufacturing process that offers advantages over hand lay-up, such as improved consistency in product quality and surface finish. Molds made from fiber-reinforced plastics or metals are used, and reinforcement material is stacked over the female mold, which is then closed with the male mold. A resin mixture with a curing agent is mixed and injected into the mold cavity, with vents provided at the farther ends to allow excess resin to flow out. After filling the mold, the resin is left to cure, and the product is removed once the mold is opened.

Vacuum-assisted resin transfer molding (VARTM) is a variant of RTM that uses vacuum pressure to remove trapped air and assist in resin flow (Figure 3). Resin infusion or vacuum bag molding is a process similar to RTM, but instead the resin is introduced into the mold cavity via vacuum pressure. The upper mold may be either flexible or rigid. In order to avoid the resin from entering the vacuum pump after filling the mold, a resin trap must be provided in the vacuum line. This process is commonly used to produce very large products, such as boat hulls.

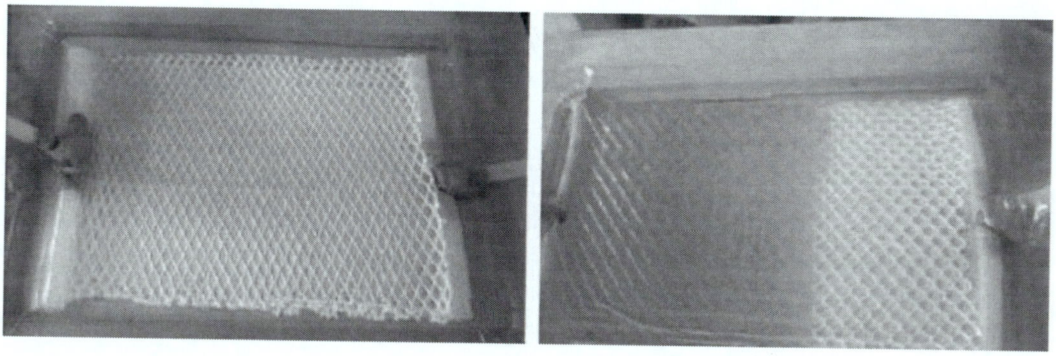

Figure 3. Resin transfer during the production of warp-knitted spacer composite.

Autoclave Process

Autoclave processing is a method used for curing composites made with prepregs and involves the use of a high-temperature pressure vessel. The composite product is placed over a mold and covered with a vacuum bag to remove any trapped air. The autoclave is then slowly heated and pressurized to facilitate the curing and consolidation of the composite product. The addition of bleeder fabric layers can help accurately control the resin content of the product. This method can produce high-quality composite products and can also be used to consolidate products made by the hand lay-up process.

Compression Molding

In compression molding, molding compounds (thermoset resin, fibers, and fillers) are used with hot curing catalysts added to prevent premature curing. To avoid curing, the compounds are stored at a cool temperature. The compounds can be prepared in sheet form for large products or in bulk form for small products. To manufacture a product using compression molding, the molding compound is placed onto a heated compression molding die. The pressure is then applied by moving the top or bottom plate to fill the die cavity. After being cured for a few minutes, the plates are moved apart, and the product is removed from the die. The high pressure and temperature involved necessitate the use of metallic dies, with the resulting products having a good surface finish. Additionally, thermoplastic composite sheets can also be used to produce thermoplastic composites through compression molding.

Filament Winding

Filament winding is a manufacturing process used to create hollow cylindrical structures such as pipes, spheres, tanks, and pressure vessels. The process involves winding continuous fibers in the form of rovings or tows that are typically impregnated with a resin, which may be either thermosetting or thermoplastic onto a mandrel in a specific pattern. The winding pattern, the fiber and resin materials used, and also the angle of winding, which is determined by the rotational speed of the mandrel, can be varied to achieve desired mechanical and chemical properties. After the desired number of layers is applied, the part is cured or solidified by heat, chemical reaction, or radiation. Once cured, the mandrel is removed from the finished product, leaving a hollow structure with a consistent thickness and high strength-to-weight ratio. Filament winding is commonly used in aerospace, automotive, and defense industries due to its ability to produce lightweight, strong, and corrosion-resistant components.

Pultrusion

In pultrusion, fibers that have been impregnated with resin are pulled through a heated die that consolidates and cures the resin. The resulting composite product is then cut to the desired length using a saw. To withstand the high temperature and abrasion, hard-surfaced metallic dies are employed in this process. However, problems such as fiber breakage and improper wetting may arise. The pultrusion process is used to fabricate long composite products with a consistent cross-section. It enables high fiber content to be achieved, resulting in the production of long tubes and rods to construct ladders, walkways, and side grills due to their excellent corrosion resistance.

Thermoforming

The process of producing thermoplastic composite products involves heating the composite sheet above the glass transition temperature of the matrix and then transferring it to a die in a press. The sheet is deformed within the die and shaped to create the final product. This process is commonly used to create packaging, disposable cups, trays, lids, and other consumer products.

Manufacturing of MMCs

The production of MMCs involves the integration of a reinforcement phase, such as ceramic particles, whiskers, or fibers, into a metallic matrix to enhance its mechanical, physical, and thermal characteristics. It depends on the specific properties required for the final composite material, as well as the type and size of the reinforcement material used. A general overview of manufacturing techniques is presented below:

Powder Metallurgy

In this process, the reinforcement material and the metallic matrix material are blended in powder form, compacted, and then sintered to generate the composite material.

In-Situ Processing

In this process, the metal matrix is melted, and the reinforcement material is added to the melt, followed by solidification of the material.

Liquid Infiltration

This process involves pre-placing the reinforcement material in a mold and then infiltrating the metallic matrix material into the mold under pressure. The metal fills the mold, surrounds it, and bonds to the reinforcement material.

Spray Deposition

This process comprises spraying the reinforcement material onto the surface of the molten metallic matrix material. The reinforcement material is melted into the metal surface to generate a composite material.

Casting

In this process, the molten metal is poured into a mold containing the reinforcement material. The metal solidifies, encapsulates, and bonds with the reinforcement material to create the composite material.

Manufacturing of CMCs

The production of CMCs based on the type and size of the reinforcing material used involves the integration of a reinforcing phase, such as ceramic fibers, whiskers, or particles, into a ceramic matrix to improve its mechanical, physical, and thermal properties. The following presents an overview of the various manufacturing processes utilized for CMCs:

Chemical Vapor Infiltration

This process entails passing a gaseous reactant over a preform of ceramic fibers contained in a sealed chamber. The reactant decomposes, depositing a layer of ceramic material on the fibers. This sequence is repeated several times to form the desired thickness of the matrix.

Liquid Phase Infiltration

In this method, a ceramic fiber preform is placed in a mold, and a liquid slurry of ceramic matrix material is poured into it. The mold is then heated to remove the solvent and sinter the ceramic matrix material, which fills the gaps between the fibers.

Melt Infiltration

This process involves melting the ceramic matrix material and infiltrating it into a ceramic fiber preform. The matrix material flows into the spaces between the fibers and fills the voids. The melt is then cooled and solidified to form a composite material.

Hot Pressing

A ceramic fiber preform is placed in a mold, surrounded by ceramic matrix material in powdered form. The mold is then heated and compressed under high pressure, causing the matrix material to melt and flow into the preform, filling the voids between the fibers. The composite material is then cooled and solidified.

Spark Plasma Sintering

In this technique, a ceramic fiber preform is encircled by ceramic matrix material in powdered form. The powder is then compacted and exposed to a high electric current, causing rapid heating and sintering of the powder. The matrix material melts and flows into the preform, creating a composite material (Rana & Fangueiro, 2016; Kumar & Thakur, 2017).

Recent Research on Composites Reinforced with Weft-Knitted Spacer Fabrics

Various studies were conducted on the mechanical properties such as compression, bending, flexural resistance, and low-velocity impact resistance of the composites reinforced with weft-knitted spacer fabrics produced by different materials and structural parameters.

The effect of the cross-sectional shape of the weft-knitted spacer composite on the bending, compression, and flexural resistance (Azadian et al., 2018; Hassanzadeh et al., 2016; Turki et al., 2018) was investigated in some studies. The flexural behavior of weft-knitted spacer composites produced using E-glass fibers and unsaturated polyester resin via the vacuum-assisted resin transfer molding process was evaluated by Azadian et al. (2018). The bending behavior of vacuum-assisted resin transfer molded composites with unsaturated polyester resin was compared to that of sandwich composites with a polyurethane foam core. The effect of cross-sectional shape on bending performance was investigated, and it was found that the V-shape cross-section had the highest bending stiffness. Hassanzadeh et al. (2016) investigated the bending and compression characteristics of the composite panels reinforced with e-glass spacer-knitted fabrics and they stated that, due to the the larger load distribution area, composites with higher thickness exhibit higher resistance and regardless of their geometrical cross-section, the samples provide higher compression resistance than the typical 3D woven and warp-knitted spacer composites. Turki et al. (2018) evaluated the effect of fabric-cross-sectional shape (U- and V-shaped) on the compression and flexural resistance of the composite structures produced using cotton yarn and unsaturated polyester resin. The U-shaped spacer fabric-reinforced composites exhibited superior compression resistance, whereas the V-shaped fabric-reinforced composites displayed better flexural properties.

The resistance against internal and external static pressures of tubular-shaped weft-knitted spacer composites using untwisted glass and HT-nylon yarns and impregnated with epoxy resin via vacuum infusion molding process were investigated by Omrani et al. (2018). According to the findings, in comparison with the plain knitted fabric-reinforced composite, tubular composite samples with non-knitting yarns and weft-knitted spacer fabric-reinforced composites with 5 and 10 connecting layers have stronger resistance to internal and external hydrostatic pressures.

In another study, it was found that, due to their superior mechanical properties, flat-knitted spacer fabrics by producing hybrid yarns containing glass filament and polypropylene filaments have considerable potential to function as reinforcement materials in composite structures (Abounaim et al., 2009).

Hassanzadeh et al. (2019) examined the flatwise and edgewise compression resistance of glass/epoxy composite panels reinforced with different core thicknesses and structural geometries. An increase in the thickness led to a decrease in the compressive strength of the flatwise loaded samples. Variations in structural geometry have a greater effect on the compression resistance of the panels under edgewise compression than variations in thickness. The composite panels showed satisfactory mechanical performance and desirable appearance, suggesting their potential to serve as an alternative to conventional materials.

During the design process of structural components, computer-based modeling, simulation, and virtual testing are often used. Although physical testing is still reliable for verification purposes, virtual testing has advantages, as it significantly reduces labor, cost, and time during "what if" type experiments. Furthermore, numerical simulation can aid in understanding material behavior, including the in-situ performance of components, which may be difficult to test physically (Jawaid et al., 2018). Hasanalizade and Dabiryan (2022b) developed an analytical model to predict indentations on composite materials reinforced with weft-knitted spacer fabric due to an impactor. They found that increasing the number of Z-fibers per unit area reduced indentations in both the X and Y directions, while the higher elastic modulus of Z-fibers led to fewer indentations. Decreasing the angle coefficient of Z-fibers improved their buckling resistance. The model can predict indentations at different points of sandwich-structured composites based on their coordinates. In another study, they developed a method involving the use of various indices such as the relative displacement, energy per pile, damaged volume, damaged pile, and damaged area indexes for characterizing the low-velocity impact behavior of sandwich-structured composites that are reinforced with 1x1 rib, 3x3 rib, and 5x5 rib gaiting weft-knitted spacer fabric (Hasanalizade & Dabiryan, 2022a). Moreover, Delavari and Dabiryan (2021) proposed a mathematical model to determine the compressional behavior of composites reinforced with 1x1 rib gaiting weft-knitted spacer fabric. The effect of fabric geometry on the low-velocity impact behavior of composites reinforced with weft-knitted spacer fabrics was investigated by Dabiryan et al. (2019) using the energy-balance method. A theoretical model was developed to predict the energy dissipation during impact in composites reinforced with fabrics, taking into account the structural parameters of the fabrics. In order to evaluate the proposed model, the low-velocity impact behavior of the composites reinforced with weft-knitted spacer fabrics with two types of spacer yarn orientation was tested by using a drop hammer testing machine. The suggested model's ability to forecast the impact behavior of weft-knitted spacer fabric composites is confirmed by a comparison of theoretical and experimental results.

Recent Research on Composites Reinforced with Warp-Knitted Spacer Fabrics

Several researchers have investigated the mechanical properties (compression, tensile, bending, impact resistance, etc.) of the composites produced from warp-knitted spacer fabrics by varying structural parameters. The structural parameters of the fabric had a significant effect on the compression and impact behaviors of composites produced from warp-knitted spacer fabrics. By using spacer fabrics with a closer surface layer structure, coarser spacer yarn, and a larger spacer yarn inclination angle, warp-knitted spacer composites perform better compression resistance. Due to their enhanced capacity to absorb energy, composites reinforced with warp-knitted spacer fabrics composite having closer surface layer structures, finer spacer yarn, the larger inclination angle of spacer yarn, and higher thickness providing better impact resistance

(Chen et al., 2015). Another researcher stated that the tensile properties of warp-knitted spacer fabrics were not significantly affected by the distribution pattern of the spacer monofilaments, while the compression performance was correlated with the spacer monofilament ratio (Liu et al., 2022a). Hybrid layered structures combined with a natural fabric-reinforced composite plates and warp-knitted spacer fabric were evaluated for acoustic applications. The composite plates were produced using a vacuum infusion technique, with jute and linen woven fabrics used as reinforcing materials and impregnated with epoxy resin. According to the findings, using a single or double layer of spacer fabric combined with a natural fabric-reinforced composite led to sound absorption coefficients that were better than those of non-hybrid layered structures. The top-performing sample consisted of double layers of spacer fabric combined with a jute fabric-reinforced composite plate. When compared to glass fabric-reinforced composites, integrating natural fabric-reinforced composites with warp-knitted spacer fabric resulted in better sound absorption performance, making it a potential solution for interior noise control in vehicles and buildings (Okur & Yaradanakul, 2022).

Flexible composites reinforced with spacer-knitted fabrics have gained increasing acceptance within the last few years (Figure 4). In a study, the researchers have filled room temperature vulcanized silicone rubber with warp-knitted spacer fabrics with filling rates of 30%, 50%, and 70%, and a warp-knitted spacer fabric without silicone rubber. Increasing the filling rate of silicone rubber improves the compressive and impact properties of flexible composite materials reinforced with warp-knitted spacer fabrics. The impact properties of flexible cushioning composites are influenced by various structural parameters, with fabric thickness having the most significant impact, followed by spacer inclination angle, spacer density, and mesh number in descending order of importance (Jiang et al., 2022).

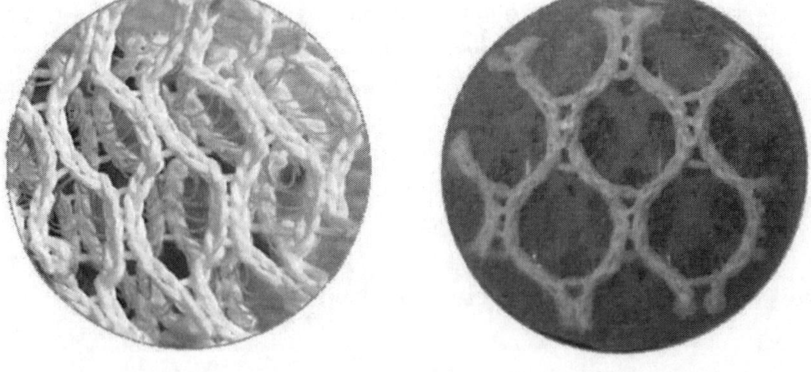

(a) (b)

Figure 4. Warp-knitted spacer fabric (a) untreated and (b) reinforced with silicone rubber.

The damage characteristics and energy absorption capabilities of lattice warp-knitted spacer flexible inflatable composites subjected to low-velocity impact were studied by Yang et al. (2023). According to the findings, these composites are able to achieve an energy absorption capability of up to 90% and the energy absorption rate of the composites decreases with increasing in thickness.

In a study, the flexural properties of polyurethane composites composed of polyurethane foam core, warp-knitted spacer fabrics, and polyurethane resin were analyzed in relation to the surface structure of the spacer fabrics, microsphere types, and contents. The study revealed that the composites exhibited exceptional flexural properties and that modifying the surface structure of the fabric and the type of microspheres could lead to a significant improvement in flexural performance (Lou et al., 2022).

Chen used experimental and theoretical methods (using finite element analysis and ANSYS software) to examine the compression behavior and energy-absorption effectiveness of polyurethane (PU)-matrix composite specimens by impregnating warp-knitted spacer textiles with flexible PU foam. Good consistency was observed between the experimental and FE simulation strain-stress curves. Moreover, these findings suggest that fabric structure parameters play a crucial role in determining the compression behaviors and energy-absorption efficiency of PU-matrix composites (Chen, 2021a). He also predicted the compression performance of the PU-based warp-knitted spacer fabric composite by using the compression meso-mechanics theoretical model based on the Winkler elastic foundation beam theory. He proposed a regular pattern for the compression properties of this composite from a meso-mechanics perspective (Chen, 2021b).

Another example of flexible composites reinforced with warp-knitted spacer fabrics is the fabric impregnated with shear-thickening fluid that improves the low-velocity impact and compressive characteristics of the composites. Shear thickening fluid-filled warp-knitted spacer fabric composites may serve as damping or energy-absorptive materials for human body protection (Gu et al., 2023; Xu et al., 2022).

Conclusion

Recent advances in fiber technology, such as high-performance fibers such as carbon/graphite, glass, and aramid, as well as three-dimensional fiber preforming machines, have resulted in knitted fabrics finding new applications in the composite industry. Knitted preforms have gained significant attention in various industries, including aerospace, medical, automotive, and civil engineering. Spacer-knitted fabrics are the most commonly used knitted textile preform, particularly for lightweight composites. Spacer-knitted fabric composites have better drapability and impact protection compared to other textile-based composites. Future research will focus on developing new fiber and matrix systems, fiber configurations, and fiber processing methods while also advancing state-of-the-art technologies. More research needs to be conducted on the engineering design and production economies of 3D spacer textile composites in order to develop a production method that is both affordable and sustainable.

Disclaimer

None.

References

Abounaim, M., Hoffmann, G., Diestel, O., & Cherif, C. (2009). Development of flat knitted spacer fabrics for composites using hybrid yarns and investigation of two-dimensional mechanical properties. *Textile Research Journal, 79*(7), 596-610. https://doi.org/10.1177/0040517508101462.

Azadian, M., Hasani, H., & Shokrieh, M. M. (2018). Flexural behavior of composites reinforced with innovative 3D integrated weft-knitted spacer fabrics. *Journal of Industrial Textiles, 48*(1), 58-76. https://doi.org/10.1177/1528083717721923.

Boisse, P. (2015). *Advances in composites manufacturing and process design*. Woodhead Publishing.

Burgess, G. (1994). Generation of cushion curves from one shock pulse. *Packaging Technology and Science, 7*(4), 169-173. https://doi.org/10.1002/pts.2770070403.

Chen, J. C., & Zhuang, Y. F. (2020). Tension and compression of sandwich composites with weft-knit fabric cores. *Modern Physics Letters B, 34*(7-9). https://doi.org/10.1142/S0217984920400047.

Chen, J., Gao, X., Zhao, K., & Wu, W. (2022a). Research on influence of matrix component on the mechanical behavior of multiaxial warp-knitted composites. *Fibers and Polymers, 23*(11), 3126-3137. https://doi.org/10.1007/s12221-022-0298-9.

Chen, J., Xia, Y., Zhao, B., Chen, W., Wang, M., & Luo, F. (2022b). Nonlinear characteristics and a new fractional constitutive model for warp-knitted NCF composites under normal loading conditions. *Polymer Testing, 107*, 107464. https://doi.org/10.1016/j.polymertesting.2021.107464.

Chen, S. (2021a). Low-velocity impact properties of polyurethane composites reinforced with warp-knit spacer fabric. *AATCC Journal of Research, 8*(2), 38-47. https://doi.org/10.14504/ajr.8.2.5.

Chen, S. (2021b). The compression meso-mechanics theoretical model and experimental verification of polyure-thane-based warp-knit spacer fabric composites. *AATCC Journal of Research, 8*(5), 30-38. https://doi.org/10.14504/ajr.8.5.4.

Chen, S., Long, H. R., Liu, Y. H., & Hu, F. C. (2015). Mechanical properties of 3D-structure composites based on warp-knitted spacer fabrics. *Autex Research Journal, 15*(2), 127-137. https://doi.org/10.2478/aut-2014-0045.

Dabiryan, H., Hasanalizade, F., & Sadighi, M. (2019). Low-velocity impact behavior of composites reinforced with weft-knitted spacer glass fabrics. *Journal of Industrial Textiles, 49*(4), 465-483. https://doi.org/10.1177/1528083718787533.

Delavari, K., & Dabiryan, H. (2021). Mathematical and numerical simulation of geometry and mechanical behavior of sandwich composites reinforced with 1×1-Rib-Gaiting weft-knitted spacer fabric; compressional behavior. *Composite Structures, 268*, 113952. https://doi.org/10.1016/j.compstruct.2021.113952.

Delavari, K., & Semnani, D. (2021). Effects of processing variables on the mechanical behavior of thermoplastic polypropylene/glass composite reinforced with weft-knitted fabric with various layups. *Fibers and Polymers, 22*(7), 2006-2017. https://doi.org/10.1007/s12221-021-0008-z.

Ding, S., Zou, B., Wang, P., Huang, C., Liu, J., & Li, L. (2021). Geometric modeling and recycling of 3D printed fiber reinforced thermoplastic composite plain weft knitted structures. *Composites Part A: Applied Science and Manufacturing, 149*, 106528. https://doi.org/10.1016/j.compositesa.2021.106528.

Dong, T., Zhang, Y., Yu, M., Zhang, X., & Ma, P. (2022). Surface performance of weft-knitted fabric composites reinforced with polytetrafluoroethylene fibers. *Journal of Applied Polymer Science, 139*(13), 51864. https://doi.org/10.1002/app.51864.

Ertekin, G. (2014). *A research about performance and comfort properties of mechanical impact protection textile products* [Ph.D. thesis]. Ege University.

Ertekin, G. (2017). Analysis of thermo-physiological comfort and moisture management properties of flat knitted spacer fabrics. *Textile and Apparel, 27*(3), 241-250.

Fang, C., Chen, Y., Jia, L., & Yan, R. (2022). Cold nitrogen plasma treated glass fiber warp knitted structural composites: mechanical properties and response surface analysis. *Journal of Industrial Textiles, 52*, 15280837221121966. https://doi.org/10.1177/15280837221121966.

Fangueiro, R. (2011). *Fibrous and composite materials for civil engineering applications*. Elsevier.

Ghatrenabi, M. H., Dabiryan, H., & Nosraty, H. (2021). Modeling the geometry of weft-knitted integrated preforms as reinforcement of composite joints. *Fibers and Polymers, 22*(9), 2572-2580. https://doi.org/10.1007/s12221-021-1180-x.

Gu, L., Du, Z., Liu, G., & Ma, M. (2023). Mechanical properties of shear thickening fluid filled warp-knitted spacer fabrics. *Journal of Industrial Textiles, 53*, 15280837231159939. https://doi.org/10.1177/15280837231159939.

Hajjari, M., Nedoushan, R. J., Dastan, T., Sheikhzadeh, M., & Yu, W. R. (2021). Lightweight weft-knitted tubular lattice composite for energy absorption applications: An experimental and numerical study. *International Journal of Solids and Structures, 213*, 77-92. https://doi.org/10.1016/j.ijsolstr.2020.12.017.

Hasanalizadeh, F., & Dabiryan, H. (2022a). A new approach to characterize the low-velocity impact behavior of sandwich-structured composite reinforced with weft-knitted spacer fabric. *Journal of Textiles and Polymers, 10*(3), 35-41.

Hasanalizade, F., & Dabiryan, H. (2022b). Theoretical modeling of low-velocity impact behavior of sandwich-structured composites reinforced with weft-knitted spacer fabric. *Journal of Industrial Textiles, 52*, 15280837221113915. https://doi.org/10.1177/15280837221113915.

Hassanzadeh, S., Hasani, H., & Zarrebini, M. (2016). Thermoset composites reinforced by innovative 3D spacer weft-knitted fabrics with different cross-section profiles: Materials and manufacturing process. *Composites Part A: Applied Science and Manufacturing, 91*, 65-76. https://doi.org/10.1016/j.compositesa.2016.09.017.

Hassanzadeh, S., Hasani, H., & Zarrebini, M. (2019). Compression load-carrying capacity of 3D-integrated weft-knitted spacer composites. *Journal of Sandwich Structures & Materials, 21*(4), 1379-1405. https://doi.org/10.1177/1099636217716575.

Jawaid, M., Thariq, M., & Saba, N. (2018). *Mechanical and physical testing of biocomposites, fibre-reinforced composites and hybrid composites.* Woodhead Publishing.

Jiang, S., Ding, M., Chen, W., & Wang, J. (2022). Preparation and properties of flexible cushioning composites based on silicone rubber and warp—knitted spacer fabric. *Silicon, 15*, 1-16. https://doi.org/10.1007/s12633-022-02267-5.

Kumar, B., & Thakur, S. (2017). *Textiles for advanced applications.* Intech Open.

Lemair, M., & Pearsall, D. J. (2007). Evaluation of impact attenuation of facial protectors in ice hockey helmets. *Sports Engineering, 10*, 65-74. https://doi.org/10.1007/BF02844204.

Li, J., Shao, H., Shao, G., Su, C., Yu, Q., Huang, Y., Chen, N., & Jiang, J. (2022). Flexible warp-knitted metal mesh-based composites: an effective emi shielding material with efficient joule heating. *ACS Applied Polymer Materials, 4*(10), 7025-7041. https://doi.org/10.1021/acsapm.2c00955.

Liu, L., Lin, H., & Ma, P. (2022a). Influences of spacer monofilament distributions on mechanical properties of warp-knitted spacer fabric/resin composites. *Textile Research Journal, 92*(13-14), 2501-2512. https://doi.org/10.1177/00405175221086041.

Liu, Q., Lanlanwang, Luo, M., Wu, Q., Kang, Y., & Ma, P. (2022b). Stab resistance of flexible composite reinforced with warp-knitted fabric like scale structure at quasi-static loading. *Journal of Industrial Textiles, 51*(5_suppl), 7983S-7998S. https://doi.org/10.1177/15280837211056987.

Liu, Q., Sun, Y., Zhao, J., & Ma, P. (2022c). Failure mechanism of weft-knitted insertion fabric/Surlyn resin flexible composite for stab resistance. *Textile Research Journal, 93*(11-12), 2473-2489. https://doi.org/10.1177/00405175221143544.

Lou, G., Chen, S., & Wu, W. (2022). Study of the flexural properties of polyurethane-foam-core composites reinforced with warp-knitted spacer fabric. *Industria Textila, 73*(4), 347-353.

Maity, S., Rana, S., Pandit, P., & Singha, K. (2021). *Advanced Knitting Technology*. Woodhead Publishing.

Man, Y. W. (2016). *Advances in women's intimate apparel technology*. Woodhead Publishing.

Miravete, A. (1999). *3-D textile reinforcements in composite materials.* Woodhead Publishing.

Okur, N., & Yaradanakul, M. C. (2022). Development of hybrid layered structures based on natural fabric reinforced composites and warp knitted spacer fabric for acoustic applications. *Journal of Industrial Textiles, 51*(2_suppl), 2216S-2245S. https://doi.org/10.1177/1528083721994677.

Omrani, E., Hasani, H., & Esmaeili, F. (2018). Mechanical performance of tubular composites reinforced by innovative 3D integrated knitted spacer fabrics. *Journal of Applied Polymer Science, 135*(14), 46074. https://doi.org/10.1002/app.46074.

Parker, M. J., Gillespie, W. J., & Gillespie, L. D. (2006). Effectiveness of hip protectors for preventing hip fractures in elderly people: systematic review. *British Medical Journal, 332*(7541), 571-574. https://doi.org/10.1136/bmj.38753.375324.7C.

Rameshkumar, M., Malathy, R., Chandiran, P., Paramasivam, S., Chung, I. M., Kim, S. H., & Prabakaran, M. (2022). Study on flexural behaviour of ferrocement composites reinforced with polypropylene warp knitted fabric. *Polymers, 14*(19), 4093. https://doi.org/10.3390/polym14194093.

Rana, S., & Fangueiro, R. (2016). *Fibrous and textile materials for composite applications*. Singapore: Springer.

Ravandi, M., Moradi, A., Ahlquist, S., & Banu, M. (2022). Numerical simulation of the mechanical behavior of a weft-knitted carbon fiber composite under tensile loading. *Polymers, 14*(3), 451. https://doi.org/10.3390/polym14030451.

Sek, M. A., Minett, M., Rouillard, V., & Bruscella, B. (2000). A new method for the determination of cushion curves. *Packaging Technology and Science: An International Journal, 13*(6), 249-255. https://doi.org/10.1002/pts.517.

Temesgen, A. G., & Sahu, O. (2022). Comparative study on the mechanical properties of weft knitted and warp fabric reinforced composites. *Journal of Modern Materials,* 9(1), 21-25. https://doi.org/10.21467/jmm.9.1.21-25.

Tian, Q., Qin, Z., Shi, B., Jia, L., Lu, S., & Yan, R. (2021). Low-velocity impact resistance and acoustic emission evaluation on mechanical failure of carbon fiber weft-knitting-reinforced composites. *Polymers for Advanced Technologies, 32(*8), 3123-3136. https://doi.org/10.1002/pat.5325.

Triantafillou, T. (2016). *Textile fibre composites in civil engineering*. Woodhead Publishing.

Turki, S., Ben Abdallah, S., & Ben Abdessalem, S. (2018). Development of composite materials reinforced with flat-knitted spacer fabrics. *The Journal of the Textile Institute, 109*(10), 1315-1321. https://doi.org/10.1080/00405000.2018.1423897.

Verpoest, I., Gommers, B., Huysmans, G., Ivens, J., Luo, Y., Pandita, S., & Philips, D. (1997, July). The potential of knitted fabrics as a reinforcement for composites. In *Proceedings of the 11th International Conference on Composite Materials (ICCM-11), Gold Coast, 1*, 108-133.

Xu, W., Yan, B., Hu, D., & Ma, P. (2022). Preparation of auxetic warp-knitted spacer fabric impregnated with shear thickening fluid for low-velocity impact resistance. *Journal of Industrial Textiles*, *51*(4_suppl), 7188S-7204S. https://doi.org/10.1177/1528083720927013.

Yan, R., Zhang, Q., Shi, B., Liu, S., Qin, Z., & Jia, L. (2022). Investigation on low-velocity impact and interfacial bonding properties of weft-knitted UHMWPE reinforced composites. *Journal of Industrial Textiles, 51*(3_suppl), 5370S-5388S. https://doi.org/10.1177/1528083720931474.

Yang, T., Xu, J., Lu, Y., Liu, Y., & Ma, P. (2023). Damage characterizations of lattice warp-knitted spacer flexible inflatable composites subjected to low-velocity impact. *Thin-Walled Structures, 182*, 110298. https://doi.org/10.1016/j.tws.2022.110298.

Zeynivand Mojarrad, M., Dabiryan, H., Rezadoust, A. M., & Nosraty, H. (2021). Numerical and experimental study of internal pressure behaviour of integrated and non-integrated weft-knitted reinforced composite elbows. *Journal of Composite Materials, 55*(20), 2765-2775. https://doi.org/10.1177/0021998321990727.

Chapter 15

Textile Preforms for the Design of Advanced Composites

Elif Kalkanlı Kurt[1,2,*], PhD
[1]Department of Textile Engineering, Tekirdag Namık Kemal University, Tekirdag, Türkey
[2]Leather, Textile and Ceramics Design Application and Research Centre, Usak University, Usak, Türkey

Abstract

The use of textile preforms in composite materials has increased over the years. The production of complex shape preforms with lightweight and cost advantages through the automated textile process has brought textiles to the forefront in advanced applications. Today, textile preforms have been used in a wide variety of applications from medical to aerospace. Weaving, knitting, braiding, and stitching technologies have been utilized to create two-dimensional (2D) and three-dimensional (3D) textile preforms from high-performance fibers and/or natural fibers for advanced composite materials. The textile-reinforced composites have different mechanical properties depending on their manufacturing techniques, fiber topology, and design pattern. For instance, 2D textile preforms in which yarn is aligned only in the plane (x- and y- direction) of the laminate have low interlaminar fracture resistance through the thickness (z-direction). To improve the mechanical performance of the 2D fiber-reinforced composites, 3D textile architectures have been introduced which have a binder yarn through their thickness. 3D manufacturing techniques enable to creation of multiaxial complex near-net-shaped textile preforms with cost efficiency. This chapter outlines current and previous research on textile preforms manufacturing techniques: weaving, braiding, knitting, and stitching, and highlights their future aspects.

Keywords*:* woven, braiding, knitting, stitching

Introduction

Textile-reinforced composite structures have attracted considerable attention from aerospace, automotive, marine, and wind turbine industries through their superior mechanical performance (Aiman et al., 2016; Khan, 2019; Xie et al., 2022). Textile composites exhibit high strength and

* Corresponding Author's Email: ekalkanli@nku.edu.tr.

In: Textile Insights and Future Aspects
Editors: Gamze D. Tetik, Gizem Celep and Fulya Yilmaz
ISBN: 979-8-89113-289-4

stiffness at low weight, high impact resistance, high corrosion resistance, and excellent fatigue performance (Priyanka et al., 2017; Zhao et al., 2022). Textile-reinforced composites can meet the demands of advanced applications with different textile preform manufacturing processes. Their near-net complex shape preforms through weaving, knitting, braiding, and stitching technologies can be produced with standard textile machinery systems. Because the preform architecture considerably affects the performance, mechanical properties, and the cost of the composites (Wambua & Anandjiwala, 2011). For instance, the triaxial braiding process may attract interest in complex structural shapes, whereas the stitching process can be used to improve the damage tolerance of the composites (Poe Jr et al., 1999).

Textile structural composites are produced by the impregnation of reinforcing fibers, in weaving, braiding, knitting, or stitching form, with thermoset or thermoplastic matrix materials (Priyanka et al., 2017). Fibers are load-bearing parts in a composite system (Sikkema et al., 2003). Therefore, high-performance fibers such as carbon, glass, and aramid are preferred in textile-reinforced composites. In addition, fiber orientation, number of fabric layers, stacking sequence of the layers, type of textile fabric preform, and the pattern of the textile preforms need to be considered to obtain optimum performance of these structures (Bakar et al., 2013; Priyanka et al., 2017).

In this chapter, textile preforms fabrication technologies will be detailed from the past to the present. Accordingly, 2D textile preforms will be explained considering their advantages and limitations for advanced applications. Then, 3D textile preforms will be discussed with their considerable mechanical performance.

Weaving

The use of woven textile preforms is a well-established technique to produce lightweight and high-performance structures for aerospace, wind turbine, civil engineering, sports equipment, etc. Weaving fabrics are categorized as two-dimensional (2D) and three-dimensional (3D). In 2D weaving, there are two sets of yarns known as warp and weft. The principle of 2D weaving is to interlacing warp and weft yarns at 90° angles. Plain, twill, and five-harness satin are some of the fundamental weaving patterns of the 2D woven textile preforms (Aiman et al., 2016). If a weaving pattern is explained with an example, a plain weave is formed by alternating the warp yarn under or over the weft yarn at each cross-over point on a weaving loom (Gowayed, 2013). On the other hand, triaxial weaving consisting of three sets of yarn offers attractive properties (Tyler, 2011; Zhao et al., 2004). The three sets of yarn interlace with each other at 60° angles in the plane direction (Tyler, 2011). These triaxial fabrics are well known for being lightweight, damage-resistant, mechanically quasi-isotropic, and resistant to shear forces (Tyler, 2011; Wang et al., 2018). The triaxial weaving fabrics have a high degree of porosity due to the alignment of the yarn. This can be a desired feature for some advanced applications such as wing skins of unmanned aerial vehicles, which need to resist impact damage from air loads (Wang et al., 2018).

Although 2D textile preforms exhibit good mechanical performance, the existence of crimp during yarn interlacing and lack of yarn through their thickness are the weakness of these structures. To reduce the effect of these drawbacks, 3D woven fabric, which reinforces z-yarn (binder) through the thickness of the fabric, has been introduced by using a modified conventional weaving loom (Aiman et al., 2016; Bilisik, 2012; Mouritz et al., 1999). Although specialized weaving machines have been developed to produce 3D woven structures, Dobby and Jacquard weaving machines are still the better option due to the cost of the newly developed specialized machines (Wambua & Anandjiwala, 2011). 3D woven preforms are classified based on the angle of the binder yarns and the penetration level of the binder yarns. Accordingly, 3D woven fabric types can be classified as orthogonal, angle interlock, through-thickness interlock, and layer-to-layer interlock (Dau et al., 2021). 3D orthogonal weaving fabrics are produced by placing of three sets of yarn in three mutually orthogonal directions (the angle of the binder yarn is 90°) (Jiao et al., 2020; Tan et al., 2000). In angle interlock weave, the binder yarn angle is less than 90° (Dau et al., 2021). On the other hand, in layer-to-layer interlock, the binder yarns combine only a few layers together, but if the binder yarns go through the whole preform thickness, it is called a through-thickness interlock (Dau et al., 2021; Jiao et al., 2020). The presence of the binder yarns through the thickness of the 3D woven preforms considerably improves the fracture toughness and damage resistance of the textile-reinforced composites compared to 2D textile composites (Isart et al., 2015; Li et al., 2021; Pankow et al., 2011).

Braiding

Braiding is one of the textiles preforms that is highly used in composites. Seamless tubes, circular/round structures, and flat ribbons can be produced by braiding technology. The principle of braiding is that three or more strands intertwine diagonally in a predetermined pattern (Potluri & Nawaz, 2011). Braiding technology is generally preferred in industrial textiles when needed for high shear and torsional strength and stiffness (Ayranci & Carey, 2008). Braided composites also exhibit high damage tolerance and dimensional stability with a multidirectional complex near-net-shaped (Ayranci & Carey, 2008; Bilisik, 2013). The braided textile preforms can be categorized as two-dimensional (2D) and three-dimensional (3D) (Carey, 2017).

Two-dimensional (2D) braids are formed by the intertwining of two or three braider yarns (Naik, 1996). The 2D biaxial braids consist of two yarn sets that are interwoven at a braiding angle (+ϴ, - ϴ) with a 1x1 or 2x2 pattern on a cylindrical mandrel (Naik, 1996; Ogin, 2000). In triaxial braided fabric, a third yarn (axial yarn) paralleled to the mandrel axis is placed in addition to two yarn sets (Birkefeld et al., 2012). Hercules, diamond, and regular patterns are the mainly used braid architectures in braiding technologies (Ayranci & Carey, 2008). The number of spindles, the braiding structure, the cylindrical mandrel radius, and the yarn diameter are essential parameters in the mechanical performance of the braided fabric (Li et al., 2017). 2D braiding fabrics are lack of binder fibers through the thickness (Ayranci & Carey, 2008). In addition, the crimp of yarn due intrinsically to the braiding also causes low impact resistance in 2D braided composites (Ayranci & Carey, 2008; Birkefeld et al., 2012). The design capabilities of the braided fabric highly depend on the width of the braiding machine. Ishmael et al. (2017) defined that the increase in the widths of the braided fabric causes an increase in manufacturing costs.

In recent years, 3D braided structures have received enormous attention for use in various applications from aviation to civil engineering (Alpyildiz, 2012; Rawal et al., 2013). To produce 3D braided textile preforms, two-step, and four-step cartesian braiding machines have been developed (Li et al., 2022). However, the direction of the bobbin movement on the machine bed is restricted on these braiding machines (Emonts et al., 2021). Therefore, rotary braiding machines have been introduced to produce complex-shaped 3D braid preforms (Emonts et al., 2021; Li et al., 2022). 3D braided textile preforms provide high delamination resistance, impact damage tolerance, and outstanding crash-worthiness properties (Bilisik, 2013; Mouritz et al., 1999).

Knitting

Knitting is the second most frequently used textile fabrication method after weaving. Knitting fabric is simply formed by interlocking loops of yarn (Ray, 2012). Loops of yarn, which are also called stitches, are intermeshed using a needle to create a new loop and draw it through the existing loop (Leong et al., 2000). The knitted fabric can be in a line or tube form. In knit technology, yarns running in the width of the fabric are known as wales, while yarns in the length of the fabric are called courses. Course spacing, wale spacing, stitch length, fabric thickness, the number of yarns, and fiber density highly affect the porosity of the knitted fabric (Dias & Delkumburewatte, 2008). Knitting fabrics can be 2-dimensional (2D) and 3-dimensional (3D).

In 2D knitting technology, the knitting process can be categorized as weft knitting and warp knitting according to the direction of the loop formation. In weft-knitting, the loops are created in the course direction of the knitted fabric (Ramakrishna, 1997). The weft-knitted fabric structures are plain, purl, rib, and interlock (Ghosh et al., 2022). In warp-knit fabric, loops are formed in the direction parallel to the fabric selvage (Leong et al., 2000). In this process, the yarns are fed from the warp beam (Elmogahzy, 2019). Flat-bed and circular machines are used in the manufacturing of weft-knitting fabric and for warp-knitted fabrics, tricot, and raschel machines are mostly used (Leong et al., 2000).

Warp-knitted fabrics are more stable when compared to the weft knitting structures, whereas weft-knitting structures are easier to manufacture (Pamuk & Çeken, 2008; Ramakrishna, 1997). Weft and warp-knitting processes make it possible to create complex-shaped structures (Karaduman et al., 2017). However, the extensibility of the knitted fabrics may lead to high energy absorption capability, dimensional stability, and improvements in the strength of the knitted fabric are also required for advanced composites (Fong et al., 2021; Padaki et al., 2006). Therefore, modifications and innovations have been implemented to the knitting technology over the years to meet these demands. One of them, the in-laid process, which is the insertion of straight and uncrimped fibers into either warp or weft knitted structures, improves the mechanical properties and dimensional stability of the knitted composites (Leong et al., 2000; Padaki et al., 2006). The straight yarns are laid during the knitting process with a conventional knitting machine system. Another modified knitting structure is split-film-warp-knitting, in which thermoplastic film is co-knitted with fibers and thus highly oriented structures can be achieved (Stumpf et al., 1998). The main advantage of the split-film-warp knitted fabric is the production of composite structures at high speed and high volume (Leong et al., 2000).

Non-crimp fabrics have attracted much attention in advanced applications. Multi-axial multilayer warp knit (non-crimp) 3D fabric is manufactured by combining layers of straight fiber tows (yarns) oriented at predetermined angles with the knitting loops through the thickness (Wan et al., 2015). Thus, multiaxial warp knit fabric provides superior in-plane mechanical performance and impact damage tolerance in the out-of-plane direction (Kang & Kim, 2000; Li et al., 2020).

Sandwich structures consisting of a core and two skins (top and bottom) are highly preferred in advanced applications due to their high mechanical strength and lightweight. However, delamination between the skin and the core is one of the main concerns (Mostafa et al., 2013). To overcome this issue, different methods have been developed, one of which is 3D knitting. The principle of 3D knitted sandwich structures is to attach the top and bottom skins with pile yarns (Phillips et al., 1997; Velosa et al., 2012). The advantage that makes 3D knitting technology distinctive among other manufacturing techniques of the sandwich panels is that it enables one to produce double-curved panels easily (Phillips et al., 1997).

Stitching

Stitching technology is presented for advanced applications to eliminate the interlaminar and delamination weakness and to improve the mechanical properties through the thickness of the conventional two-dimensional (2D) laminates (Mouritz et al., 1999; Song et al., 2022a; 2022b, Tan et al., 2008). The principle of stitching for advanced composites is simply the insertion of stitching yarn through the thickness of the textile preforms to combine layers of fabric. In this process, stitching density, stitching type, stitching pattern, stitching thread material, and its diameter highly affect the mechanical properties of the stitched composites (Gnaba et al., 2019; Song et al., 2022a).

In stitching, mostly high-performance threads such as carbon, glass, or aramid are preferred, similar to other textile preforms (Gnaba et al., 2019; Mouritz et al., 1997; Song et al., 2022b). In recent years, there have been also used different high strength threads; poly (p-phenylene benzo-isoxazole (PBO)), polyester, and silk to improve the interlaminar strength of the composites (Karahan et al., 2010; Mouritz & Jain, 1999; Song et al., 2022a; Sun et al., 2022).

Another toughening method through the thickness of the fabric is Z-pinning which is based on the insertion of thin metal or composite rods similar to the stitching technology (Koh et al., 2012; Mouritz, 2020). The studies demonstrate that, depending on the end-use applications, each high-strength stitching thread has improved the delamination and interlaminar resistance of the stitched composites. The stitch thread diameter has a significant effect on the size of the delamination at impact damage, but the onset of the delamination is not relevant to the stitch tread diameter (Tan et al., 2013).

The stitch pattern, which is another parameter, also affects the mechanical performance of the stitched composites. Stitching patterns include vertical, tilt 600, horizontal, tilt 300, tilt 300/300, tilt 450/900, box, zigzag, cross, tilt 600/600 (Song et al., 2022a; Yaakob et al., 2019). In turn, the appropriate stitching pattern improves the fracture initiation of the stitched composites. However, the direction and the angle of the stitching have important effects on the load-carrying capacities of the stitched composites. For example, stitching symmetry provides a uniform bearing load (Ma et al., 2022). On the other hand, stitching patterns in which threads are placed in the line of the loading direction, i.e., box and vertical, increase the tensile strength while horizontal stitching patterns cannot fully carry out tensile load (Song et al., 2022a).

Among the stitching types, lock stitch has initially been used. However, stress concentration and the formation of resin-reach areas due to the principle of the lock stitch can cause in-plane damage (Song et al., 2022a). Therefore, a modified lock stitch has been developed. In this stitch, there is no interlock coil, and hence the concentrated stress is reduced, resulting in less in-plane fiber distortion (Karaduman et al., 2017). In addition to them, chain stitch is also commonly used in stitched composites (Dransfield et al., 1994; Mouritz et al., 1997). Over the years, various new stitching types have been implemented in the stitching technology, which provide advantages for complex composite structures. Some of them are as follow: single-sided-stitch, blind stitch, tufting stitch, two-needles-two-threads, and two-needles-one-thread (Ogale & Mitschang, 2004; Song et al., 2022a; Zhao et al., 2009).

Another important parameter in the stitched fabric is the stitch density, which is defined as the number of stitches per area. The studies demonstrate that higher stitch density leads to a lower rate of delamination growth (Tan et al., 2010; 2013).

In the beginning, stitching technology had been performed by hand or without specialized industrial sewing machines (Song et al., 2022a). Today, robotic control and CNC control technologies have also been implemented to obtain near-net-shaped on the stitched composites (Zhao et al., 2009). The stitching technology can be implemented in any reinforcement fabric. However, stitching of thick and large preforms is a problem with current machinery systems (Bannister, 2001).

Several studies have investigated the effect of stitching on the mechanical properties of composites. Damage tolerance of the stitched composites has been improved and the delamination area has decreased compared to 2D composite structures (Drake et al., 2021; Xuan et al., 2019; Zhao et al., 2009). The delamination performance of the 2D composites is highly dependent on interlaminar strength. Strengthening of the fabrics through thickness stitching enables to improvement in the interlaminar strength of the panels. Furthermore, stitching also affects the crack propagation of the composites (Jain et al., 1998). The stitching may delay the final fracture of the composites, as the stitched panel may continue to carry out the load even after initial damage (Dransfield et al., 1994). Composite panels with stitching threads present higher local buckling strength (Parlapalli et al., 2007).

Besides, all those advantages of the stitching technology, the penetration of the needle causes localized fiber damage on the textile preform (Dransfield et al., 1994). The needle insertion can also lead to misalignment of in-plane fibers; therefore, the mechanical performance of the composites is reduced in this direction. Tufting, which is a one-sided stitching process, inserts the thread with a lower tension when compared to the stitching process. Hence, it may decrease the adverse effect of the stitching on the in-plane properties of the textile preforms (Martins et al., 2018). Resin-rich areas due to the creation of the hole in the reinforcement fabric can occur due to the insertion of the needle and they make the panel susceptible to damage (Yudhanto et al., 2013).

Conclusion

This chapter has shown an extensive amount of research that has been performed on textile preforms, stitching, knitting, weaving, and braiding. 2D and 3D fabrics have been described in detail with their advantages and limitations. New developed techniques provide solutions to the limitations of conventional textile preforming techniques. The stitching process improves the delamination resistance of the structures by joining the layers of the composites. However, it may cause fiber damage in the plane direction and this issue can be minimized by the tufting method. 3D knitted fabric can be used for complex shaped components. Multiaxial warp knit techniques have also considerable potential to produce sandwich structures and 3D knitted fabric. On the other hand, the modified knitted structures in-laid and split warp knitting fabrics provide dimensionally stable structures with cost efficiency. 3D braided fabrics exhibit outstanding crash-worthiness properties and high delamination resistance.

The developments and modifications in the manufacturing techniques of textile preforms will gradually increase the use of these structures in advanced applications such as aerospace, marine, wind turbine, and civil engineering. However, there is still a need to increase the research carried out on the newly developed textile manufacturing techniques to obtain more information about their contribution to the mechanical performances of the textile-reinforced composites.

Disclaimer

None.

References

Aiman, D. P. C., Yahya, M. F., & Salleh, J. (2016). Impact properties of 2D and 3D woven composites: A review. *AIP Conference Proceedings*, *1774*(1), 020002. https://doi.org/10.1063/1.4965050.

Alpyildiz, T. (2012). 3D geometrical modelling of tubular braids. *Textile Research Journal*, *82*(5), 443–453. https://doi.org/10.1177/0040517511427969.

Ayranci, C., & Carey, J. (2008). 2D braided composites: A review for stiffness critical applications. *Composite Structures*, *85*(1), 43–58. https://doi.org/10.1016/j.compstruct.2007.10.004.

Bakar, I. A. A., Kramer, O., Bordas, S., & Rabczuk, T. (2013). Optimization of elastic properties and weaving patterns of woven composites. *Composite Structures*, *100*, 575–591. https://doi.org/10.1016/j.compstruct.2012.12.043.

Bannister, M. (2001). Challenges for composites into the next millennium—a reinforcement perspective. *Composites Part A: Applied Science and Manufacturing*, *32*(7), 901–910. https://doi.org/10.1016/S1359-835X(01)00008-2.

Bilisik, K. (2012). Multiaxis three-dimensional weaving for composites: a review. *Textile Research Journal*, *82*(7), 725–743. https://doi.org/10.1177/0040517511435013.

Bilisik, K. (2013). Three-dimensional braiding for composites: A review. *Textile Research Journal*, *83*(13), 1414–1436. https://doi.org/10.1177/0040517512450766.

Birkefeld, K., Röder, M., von Reden, T., Bulat, M., & Drechsler, K. (2012). Characterization of biaxial and triaxial braids: Fiber architecture and mechanical properties. *Applied Composite Materials*, *19*, 259–273. https://doi.org/10.1007/s10443-011-9190-2.

Carey, J. P. (2017). Introduction to braided composites. In *Handbook of Advances in Braided Composite Materials* (pp. 1–21). Elsevier. https://doi.org/10.1016/B978-0-08-100369-5.00001-5.

Dau, F., Dano, M.-L., Vérone, B., Girardot, J., Aboura, Z., & Morvan, J.-M. (2021). In-plane and out-of-plane characterization of a 3D angle interlock textile composite. *Composites Part A: Applied Science and Manufacturing*, *149*, 106581. https://doi.org/10.1016/j.compositesa.2021.106581.

Dias, T., & Delkumburewatte, G. B. (2008). Changing porosity of knitted structures by changing tightness. *Fibers and Polymers*, *9*, 76–79. https://doi.org/10.1007/s12221-008-0012-6.

Drake, D. A., Sullivan, R. W., Clay, S. B., & DuBien, J. L. (2021). Influence of stitching on the fracture of stitched sandwich composites. *Composites Part A: Applied Science and Manufacturing*, *145*, 106383. https://doi.org/10.1016/j.compositesa.2021.106383.

Dransfield, K., Baillie, C., & Mai, Y.-W. (1994). Improving the delamination resistance of CFRP by stitching—a review. *Composites Science and Technology*, *50*(3), 305–317. https://doi.org/10.1016/0266-3538(94)90019-1.

Elmogahzy, Y. E. (2019). *Engineering textiles: Integrating the design and manufacture of textile products*. Woodhead Publishing.

Emonts, C., Grigat, N., Merkord, F., Vollbrecht, B., Idrissi, A., Sackmann, J., & Gries, T. (2021). Innovation in 3D braiding technology and its applications. *Textiles*, *1*(2), 185–205. https://doi.org/10.3390/textiles1020009.

Fong, K. W., Li, S. Q., & Liu, R. (2021). A new geometric model of laid-in weft-knitted fabrics. *Textile Research Journal*, *91*(17–18), 2087–2105. https://doi.org/10.1177/0040517521992765.

Ghosh, S., Chauhan, R., & Roy, S. (2022). Mechanics of loop formation in plain weft knitting machinery. In *Advanced Knitting Technology* (pp. 67–93). Elsevier. https://doi.org/10.1016/B978-0-323-85534-1.00005-2.

Gnaba, I., Legrand, X., Wang, P., & Soulat, D. (2019). Through-the-thickness reinforcement for composite structures: A review. *Journal of Industrial Textiles*, *49*(1), 71–96. https://doi.org/10.1177/1528083718772299.

Gowayed, Y. (2013). Types of fiber and fiber arrangement in fiber-reinforced polymer (FRP) composites. In N. Uddin (Ed.), *Developments in fiber-reinforced polymer (FRP) composites for civil engineering* (pp. 3–17). Elsevier. https://doi.org/10.1533/9780857098955.1.3.

Isart, N., Mayugo, J. A., Blanco, N., Ripoll, L., Solà, A., & Soler, M. (2015). Geometric model for 3D through-thickness orthogonal interlock composites. *Composite Structures*, *119*, 787–798. https://doi.org/10.1016/j.compstruct.2014.09.044.

Ishmael, N., Fernando, A., Andrew, S., & Waterton Taylor, L. (2017). Textile technologies for the manufacture of three-dimensional textile preforms. *Research Journal of Textile and Apparel*, *21*(4), 342–362. https://doi.org/10.1108/RJTA-06-2017-0034.

Jain, L. K., Dransfield, K. A., & Mai, Y.-W. (1998). On the effects of stitching in CFRPs—II. Mode II delamination toughness. *Composites Science and Technology*, *58*(6), 829–837. https://doi.org/10.1016/S0266-3538(97)00186-3.

Jiao, W., Chen, L., Xie, J., Yang, Z., Fang, J., & Chen, L. (2020). Effect of weaving structures on the geometry variations and mechanical properties of 3D LTL woven composites. *Composite Structures*, *252*, 112756. https://doi.org/10.1016/j.compstruct.2020.112756.

Kang, T. J., & Kim, C. (2000). Energy-absorption mechanisms in Kevlar multiaxial warp-knit fabric composites under impact loading. *Composites Science and Technology*, *60*(5), 773–784. https://doi.org/10.1016/S0266-3538(99)00185-2.

Karaduman, N. S., Karaduman, Y., Ozdemir, H., & Ozdemir, G. (2017). Textile reinforced structural composites for advanced applications. *Textiles for Advanced Applications*, *87*. https://doi.org/10.5772/intechopen.68245.

Karahan, M., Ulcay, Y., Eren, R., Karahan, N., & Kaynak, G. (2010). Investigation into the tensile properties of stitched and unstitched woven aramid/vinyl ester composites. *Textile Research Journal*, *80*(10), 880–891. https://doi.org/10.1177/0040517509346441.

Khan, R. (2019). Fiber bridging in composite laminates: A literature review. *Composite Structures*, *229*, 111418. https://doi.org/10.1016/j.compstruct.2019.111418.

Koh, T. M., Feih, S., & Mouritz, A. P. (2012). Strengthening mechanics of thin and thick composite T-joints reinforced with z-pins. *Composites Part A: Applied Science and Manufacturing*, *43*(8), 1308–1317. https://doi.org/10.1016/j.compositesa.2012.03.023.

Leong, K. H., Ramakrishna, S., Huang, Z. M., & Bibo, G. A. (2000). The potential of knitting for engineering composites—a review. *Composites Part A: Applied Science and Manufacturing*, *31*(3), 197–220. https://doi.org/10.1016/S1359-835X(99)00067-6.

Li, C., Wang, X., Zhao, F., Wang, F., Ting, Z., & Wang, L. (2017). Modeling of braiding parameter impact on pore size and porosity in a tubular braiding fabric. *E-Polymers*, *17*(3), 221–226. https://doi.org/10.1515/epoly-2016-0060.

Li, D., Yang, Y., Wang, Z., & Jiang, L. (2020). Experimental investigation on mechanical response and failure analysis of 3D multi-axial warp knitted hybrid composites. *Composite Structures*, *246*, 112340. https://doi.org/10.1016/j.compstruct.2020.112340.

Li, W., Qiao, Y., Fenner, J., Warren, K., Salviato, M., Bažant, Z. P., & Cusatis, G. (2021). Elastic and fracture behavior of three-dimensional ply-to-ply angle interlock woven composites: Through-thickness, size effect, and multiaxial tests. *Composites Part C: Open Access*, *4*, 100098. https://doi.org/10.1016/j.jcomc.2020.100098.

Li, X., He, X., Liang, J., Song, Y., Zhang, L., Wang, B., Ma, J., & Kong, G. (2022). Research Status of 3D Braiding Technology. *Applied Composite Materials*, 1–11. https://doi.org/10.1007/s10443-021-09963-2.

Ma, P.-S., Ye, J.-Y., Tian, K., Chen, X.-H., & Zhang, L.-W. (2022). Fracture phase field modeling of 3D stitched composite with optimized suture design. *Computer Methods in Applied Mechanics and Engineering*, *392*, 114650. https://doi.org/10.1016/j.cma.2022.114650.

Martins, A. T., Aboura, Z., Harizi, W., Laksimi, A., & Khellil, K. (2018). Analysis of the impact and compression after impact behavior of tufted laminated composites. *Composite Structures*, *184*, 352–361. https://doi.org/10.1016/j.compstruct.2017.09.096.

Mostafa, A., Shankar, K., & Morozov, E. V. (2013). Insight into the shear behaviour of composite sandwich panels with foam core. *Materials & Design*, *50*, 92–101. https://doi.org/10.1016/j.matdes.2013.03.016.

Mouritz, A. P. (2020). Review of z-pinned laminates and sandwich composites. *Composites Part A: Applied Science and Manufacturing*, *139*, 106128. https://doi.org/10.1016/j.compositesa.2020.106128.

Mouritz, A. P., & Jain, L. K. (1999). Further validation of the Jain and Mai models for interlaminar fracture of stitched composites. *Composites Science and Technology*, *59*(11), 1653–1662. https://doi.org/10.1016/S0266-3538(99)00027-5.

Mouritz, A. P., Bannister, M. K., Falzon, P. J., & Leong, K. H. (1999). Review of applications for advanced three-dimensional fibre textile composites. *Composites Part A: Applied Science and Manufacturing*, *30*(12), 1445–1461. https://doi.org/10.1016/S1359-835X(99)00034-2.

Mouritz, A. P., Leong, K. H., & Herszberg, I. (1997). A review of the effect of stitching on the in-plane mechanical properties of fibre-reinforced polymer composites. *Composites Part A: Applied Science and Manufacturing*, *28*(12), 979–991. https://doi.org/10.1016/S1359-835X(97)00057-2.

Naik, R. A. (1996). Analysis of woven and braided fabric-reinforced composites. *ASTM Special Technical Publication*, *1274*, 239–263.

Ogale, A., & Mitschang, P. (2004). Tailoring of textile preforms for fibre-reinforced polymer composites. *Journal of Industrial Textiles*, *34*(2), 77–96. https://doi.org/10.1177/1528083704046949.

Ogin, S. L. (2000). Textile-reinforced composite materials. In A. R. Horrocks & S. C. Anand (Eds.), *Handbook of technical textiles* (pp. 264-279). Woodhead Publishing.

Padaki, N. V, Alagirusamy, R., & Sugun, B. S. (2006). Knitted preforms for composite applications. *Journal of Industrial Textiles*, *35*(4), 295–321. https://doi.org/10.1177/1528083706060784.

Pamuk, G., & Çeken, F. (2008). Manufacturing of weft-knitted fabric reinforced composite materials: a review. *Materials and Manufacturing Processes*, *23*(7), 635–640. https://doi.org/10.1080/10426910802316484.

Pankow, M., Salvi, A., Waas, A. M., Yen, C. F., & Ghiorse, S. (2011). Resistance to delamination of 3D woven textile composites evaluated using End Notch Flexure (ENF) tests: Experimental results. *Composites Part A: Applied Science and Manufacturing*, *42*(10), 1463–1476. https://doi.org/10.1016/j.compositesa.2011.06.013.

Parlapalli, M. R., Soh, K. C., Shu, D. W., & Ma, G. (2007). Experimental investigation of delamination buckling of stitched composite laminates. *Composites Part A: Applied Science and Manufacturing*, *38*(9), 2024–2033. https://doi.org/10.1016/j.compositesa.2007.05.001.

Phillips, D., Verpoest, I., & Van Raemdonck, J. (1997). Optimising the mechanical properties of 3D-knitted sandwich structures. *Proceedings of the 11th International Conference on Composite Materials*, *14*(18), V211–V218p.

Poe Jr, C. C., Dexter, H. B., & Raju, I. S. (1999). Review of the NASA textile composites research. *Journal of Aircraft*, *36*(5), 876–884. https://doi.org/10.2514/2.2521.

Potluri, P., & Nawaz, S. (2011). Developments in braided fabrics. In R. H. Gong (Ed.), *Specialist Yarn and Fabric Structures* (pp. 333-353). Woodhead Publishing Limited. https://doi.org/10.1016/B978-1-84569-757-0.50014-3.

Priyanka, P., Dixit, A., & Mali, H. S. (2017). High-Strength Hybrid Textile Composites with Carbon, Kevlar, and E-Glass Fibers for Impact-Resistant Structures. A Review. *Mechanics of Composite Materials*, *53*, 685–704. https://doi.org/10.1007/s11029-017-9696-2.

Ramakrishna, S. (1997). Characterization and modeling of the tensile properties of plain weft-knit fabric-reinforced composites. *Composites Science and Technology*, *57*(1), 1–22. https://doi.org/10.1016/S0266-3538(96)00098-X.

Rawal, A., Saraswat, H., & Kumar, R. (2013). Tensile response of tubular braids with an elastic core. *Composites Part A: Applied Science and Manufacturing*, *47*, 150–155. https://doi.org/10.1016/j.compositesa.2012.12.006.

Ray, S. C. (2012). *Fundamentals and advances in knitting technology*. CRC Press.

Sikkema, D. J., Northolt, M. G., & Pourdeyhimi, B. (2003). Assessment of new high-performance fibers for advanced applications. *MRS Bulletin*, *28*(8), 579–584. https://doi.org/10.1557/mrs2003.169.

Song, C., Fan, W., Liu, T., Wang, S., Song, W., & Gao, X. (2022a). A review on three-dimensional stitched composites and their research perspectives. *Composites Part A: Applied Science and Manufacturing*, *153*, 106730. https://doi.org/10.1016/j.compositesa.2021.106730.

Song, W., Liu, T., Song, C., Wang, S., Jing, Y., Ma, J., Li, G., Zhou, R., Dou, H., & Fan, W. (2022b). Effects of stitching yarn types on flexural fatigue properties of 3D stitched carbon fiber composites. *Polymer Testing*, *115*, 107744. https://doi.org/10.1016/j.polymertesting.2022.107744.

Stumpf, H., Mäder, E., Baeten, S., Pisanikovski, T., Zäh, W., Eng, K., Andersson, C.-H., Verpoest, I., & Schulte, K. (1998). New thermoplastic composite preforms based on split-film warp-knitting. *Composites Part A: Applied Science and Manufacturing*, *29*(12), 1511–1523. https://doi.org/10.1016/S1359-835X(98)00086-4.

Sun, Y., Fan, W., Song, C., Gao, X., Liu, T., Song, W., Wang, S., Zhou, R., Li, G., & Li, S. (2022). Effects of stitch yarns on interlaminar shear behavior of three-dimensional stitched carbon fiber epoxy composites at room temperature and high temperature. *Advanced Composites and Hybrid Materials*, *5*(3), 1951–1965. https://doi.org/10.1007/s42114-022-00526-y.

Tan, K. T., Watanabe, N., & Iwahori, Y. (2010). Effect of stitch density and stitch thread thickness on low-velocity impact damage of stitched composites. *Composites Part A: Applied Science and Manufacturing*, *41*(12), 1857–1868. https://doi.org/10.1016/j.compositesa.2010.09.007.

Tan, K. T., Yoshimura, A., Watanabe, N., Iwahori, Y., & Ishikawa, T. (2013). Effect of stitch density and stitch thread thickness on damage progression and failure characteristics of stitched composites under out-of-plane loading. *Composites Science and Technology*, *74*, 194–204. https://doi.org/10.1016/j.compscitech.2012.11.001.

Tan, P., Tong, L., Steven, G. P., & Ishikawa, T. (2000). Behavior of 3D orthogonal woven CFRP composites. Part I. Experimental investigation. *Composites Part A: Applied Science and Manufacturing*, *31*(3), 259–271. https://doi.org/10.1016/S1359-835X(99)00070-6.

Tan, Y., Wu, G., Suh, S. S., Yang, J. M., & Hahn, H. T. (2008). Damage tolerance and durability of selectively stitched stiffened composite structures. *International Journal of Fatigue*, *30*(3), 483–492. https://doi.org/10.1016/j.ijfatigue.2007.04.008.

Tyler, T. (2011). Developments in triaxial woven fabrics. In R. H. Gong (Ed.), *Specialist yarn and fabric structures: Developments and applications* (pp. 141-163). Woodhead Publishing.

Velosa, J. C., Rana, S., Fangueiro, R., Van Hattum, F. W. J., Soutinho, F., & Marques, S. (2012). Mechanical behavior of novel sandwich composite panels based on 3D-knitted spacer fabrics. *Journal of Reinforced Plastics and Composites*, *31*(2), 95–105. https://doi.org/10.1177/0731684411431354.

Wambua, P. M. and, & Anandjiwala, R. (2011). A review of preforms for the composites industry. *Journal of Industrial Textiles*, *40*(4), 310–333. https://doi.org/10.1177/1528083709092014.

Wan, Y., Zhang, F., Gu, B., Sun, B., & Wang, Y. (2015). Predicting dynamic in-plane compressive properties of multi-axial multi-layer warp-knitted composites with a meso-model. *Composites Part B: Engineering*, *77*, 278–290. https://doi.org/10.1016/j.compositesb.2015.03.060.

Wang, Z., Bai, J., Sobey, A., Xiong, J., & Shenoi, A. (2018). Optimal design of triaxial weave fabric composites under tension. *Composite Structures*, *201*, 616–624. https://doi.org/10.1016/j.compstruct.2018.06.090.

Xie, J., Guo, Z., Shao, M., Zhu, W., Jiao, W., Yang, Z., & Chen, L. (2022). Mechanics of textiles used as composite preforms: A review. *Composite Structures*, 116401. https://doi.org/10.1016/j.compstruct.2022.116401.

Xuan, J., Li, D., & Jiang, L. (2019). Fabrication, properties and failure of 3D stitched carbon/epoxy composites with no stitching fibers damage. *Composite Structures*, *220*, 602–607. https://doi.org/10.1016/j.compstruct.2019.03.080.

Yaakob, M. Y., Husin, M. A., Abdullah, A., Mohamed, K. A., Khim, A. S., Fang, M. L. C., & Sihombing, H. (2019). Effect of stitching patterns on tensile strength of kenaf woven fabric composites. *International Journal of Integrated Engineering*, *11*(6), 70–79. https://doi.org/10.30880/ijie.2019.11.06.008.

Yudhanto, A., Watanabe, N., Iwahori, Y., & Hoshi, H. (2013). Compression properties and damage mechanisms of stitched carbon/epoxy composites. *Composites Science and Technology*, *86*, 52–60. https://doi.org/10.1016/j.compscitech.2013.07.002.

Zhao, N., Rödel, H., Herzberg, C., Gao, S.-L., & Krzywinski, S. (2009). Stitched glass/PP composite. Part I: Tensile and impact properties. *Composites Part A: Applied Science and Manufacturing*, *40*(5), 635–643. https://doi.org/10.1016/j.compositesa.2009.02.019.

Zhao, Q., Hoa, S. V, & Ouellette, P. (2004). Progressive failure of triaxial woven fabric (TWF) composites with open holes. *Composite Structures*, *65*(3–4), 419–431. https://doi.org/10.1016/j.compstruct.2003.12.004.

Zhao, Z., Li, B., & Ma, P. (2022). Advances in mechanical properties of flexible textile composites. *Composite Structures*, 116350. https://doi.org/10.1016/j.compstruct.2022.116350.

Chapter 16

Textile Reinforcements and Innovative Approaches in Composite Material Manufacturing

Önder Yeşil*, PhD
Department of Mechanical Engineering, Uşak University, Uşak, Türkey

Abstract

Composite materials have been widely used by designers and engineers as a solution to obstacles, limitations, and problems that are encountered in engineering and design applications. Conventional isotropic engineering materials have the same mechanical and physical properties and uniformity in each direction and in points within a volume. On the contrary, composite materials can be configured to provide desired mechanical and physical properties at each point in the same direction. The main cause that provides the desired mechanical and physical properties is reinforcing the composite materials with the appropriate textile reinforcements. For this reason, textile reinforcements used in the manufacturing of composite materials are of such significant importance in determining the mechanical and physical characteristics of composite materials. This situation necessitated the production of suitable textile reinforcements in order to manufacture composite materials with the desired properties and caused the categorization of textile reinforcements. In this chapter, the types of textile reinforcement that are used in composite materials and manufacturing methods are categorized and explained their benefits in properties of composite materials.

Keywords: types of textile reinforcement, composite materials, manufacturing methods, physical properties

Introduction

A composite material is a type of material that is obtained by combining at least two or more different materials that have one as a matrix component and the other as reinforcement, at a macroscopic scale without forming alloys, and exhibit the superior properties of its

* Corresponding Author's Email: onder.yesil@usak.edu.tr.

In: Textile Insights and Future Aspects
Editors: Gamze D. Tetik, Gizem Celep and Fulya Yilmaz
ISBN: 979-8-89113-289-4

components. The first use of composite materials in human history dates back to 1500 BC. Since then, it has taken place in human life at an increasing rate. Composite materials have been found in a wide application area in plumbing, home and business equipment, construction industry, electricity distribution, renewable energy sources, automotive industry, aviation and space, marine, defense, industry, sports and entertainment, and agriculture sectors.

The development of composite materials has been followed by the development of new suitable production methods. Tough reinforcements and durable matrix components that are suitable for the field of use and need have taken their place in the industry as superior materials used in many engineering applications (Nagavally, 2017; Starr, 2000; Bhagwan et al., 2015; Fitzer & Manocha, 1998; Balasubramanian, 2014; Mazumdar, 2002; Vinson & Sierakowski, 2002; Yeşil, 2019).

Composite materials had a wide variety of investigation and research fields in science. Composite materials can generally be classified or named by their matrix types, reinforcement types, and their geometry or structure. The classification of composite materials by reinforcement types is given in Figure 1. Composite materials categorized by matrix types are polymer matrix composites (PMC), ceramic matrix composites (CMC), and metallic matrix composites (MMC) (Callister & Rethwisch, 2013). Textile reinforcements are widely used in polymer matrix composite (PMC) materials. Therefore, this chapter is mainly based on PMC materials and textile reinforcements used in the manufacturing of PMC materials.

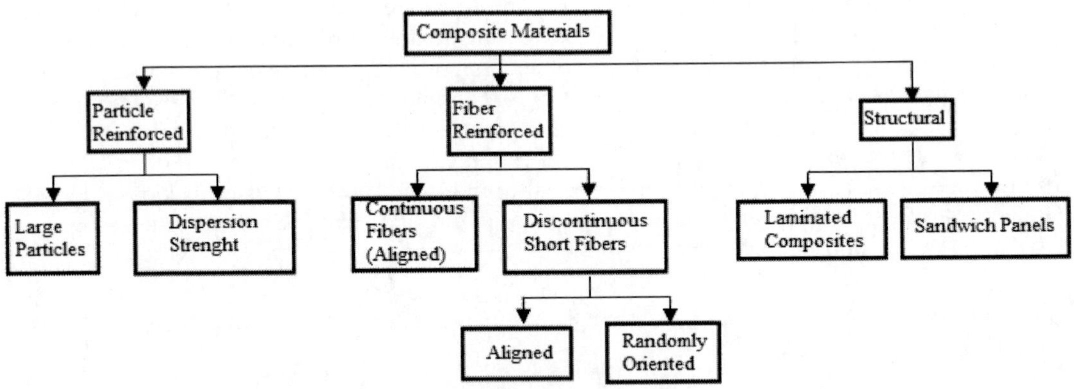

Figure 1. Classification of composite materials by reinforcement types.

PMC Materials Manufacturing Methods

As an engineer or designer, one should take into consideration the selection of the appropriate reinforcement element and matrix to ensure to provide the composite material with the desired properties that are suitable for its task. Polymer matrix composite material manufacturing methods vary depending on the desired properties of the composite material. The selection of suitable reinforcement elements and matrix types in the manufacturing process also has a significant effect on the properties of the desired composite material. Nowadays, composite material production methods have evolved depending on cost effectiveness, desired performance, and properties.

Hand Lay-Up Method

The hand lay-up method is simple, cost effective, and the oldest manufacturing technique for PMC composites. The hand lay-up process largely requires labor and materials costs for fabrication. A mold generally used to form the product and the equipment is cheaper than the other PMC manufacturing methods. Mold can be manufactured from plastics, wood, or reinforced plastics to reduce manufacturing costs. Any complex form can be manufactured by using one-piece or multiple-piece molds which have the final shape of the product. The surface quality and the dimensions of the final product are largely dependent on the mold and the workmanship, making the quality and properties of the products poorer and more variable than other manufacturing methods. Firstly, a mold release agent thin layer is applied to the surface of the mold to easily release the cured product. This process also has a significant effect on the surface quality of the final product (Balasubramanian, 2014; Mazumdar, 2002).

In this method, glass fiber woven fabrics, rovings, or strand mats are used as reinforcements, and isophthalic polyester resin and its curing agents are used as matrix due to their low cost. The surface veil can be preferable to improve the appearance of the product. Laminated composites are also easily manufactured by this method. However, the quality of the product will be poorer than other that of molding methods. The gel coat layer is applied over the release agent film with a thickness of less than 1 mm. The reinforcement element (woven, roving, chopped strand mat, etc.) is placed on the gel-coat layer, and resin charge is applied to the reinforcement sequentially with a cylindrical paintbrush. There are different types of rollers to wet the reinforcements and remove air bubbles from the product between the reinforcement layers. Reinforcement layers are applied sequentially until the desired product thickness is provided. Low production speed and excessive labor requirements are some of the certain disadvantages of this method. In addition, only one surface of the product that contacts the mold has a good and smooth appearance.

Spray-Up Method

Although the spray-up method is very similar to the hand lay-up method, it is an improved PMC production method that reduces labor and increases production speed. The spray-up method needs a spray gun with a cuter mechanism, and compressor, unlike the hand lay-up method. Therefore, the equipment requirements and initial investment costs are higher than those of the hand lay-up method. The mechanical properties of the product are poorer than those of the hand lay-up products. Especially, glass fiber rovings are used as reinforcement elements but also strand mats, woven fabrics, and surface veils can be used outer surface of the products. In general, it is suitable for the manufacturing of products with a high material thickness and resin-rich structure. The fiber volume ratio of spray-up products is lower than that of the hand lay-up products, which significantly affects the mechanical properties. The product thickness variation is not much in flat regions, but it can be observed that a significant increase in radius and sharp corner regions can be observed due to the spraying angle (Cripps et al., 2000).

In general, the spraying process starts after applying an optional gel-coat layer onto mold surface. Simultaneously, the chopped rovings in the spray gun are sprayed with the resin from the nozzles to the mold surface. The spraying process ends when the desired product thickness is reached. A roller is used to remove bubbles and wet fibers, which ensures the quality of the product. Products produced by the spray-up method are classified as randomly oriented short fiber reinforced composite materials. Therefore, the mechanical properties are varied in directions according to the orientation of the fiber.

Resin Transfer Molding (RTM) Method

The RTM method is a solution to produce composite parts with good quality, low-dimensional tolerances, high strength, and desired mechanical properties. The RTM method is a preferred method when the production quantity is medium or high and products with higher quality and mechanical properties are desired compared to the hand lay-up and spray-up method. The equipment requirement and initial investment cost of the RTM method are higher than those of the two mentioned methods. A resin injection system consisting of two metering pumps, a mixing head, and multi-segment heatable metal molds are needed in this method. Reinforcements that are used in this method are mats, woven fabrics, stitched materials, and preforms.

It is a semiautomatic process and needs resin with low viscosity, and resin curing agent. Reinforcements are placed in the mold cavity as a hand lay-up method. Then, the heated mold is closed and the injection process begins under a certain pressure value. After the injection process, a certain amount of time is required for the product to cure. Finally, the cured product is removed from the mold. Since the method involves pressure, wetting of the reinforcement elements, and bubble, gap problems in the product are minimized. In time, the RTM method was developed one step further using a vacuum instead of pressurized resin to maximize wetting performance and minimize trapped air. It is called the vacuum-assisted resin transfer method (VARTM) (Park & Seo, 2011).

Vacuum-Assisted Resin Transfer Molding Method

VARTM method is similar to the RTM method, but the resin is transferred to the mold cavity by vacuum rather than pressure. The autoclave molding method is also similar to this method to some extent. The bagging stage is a crucial and time-consuming stage of this method. Reinforcements that are used in the RTM method can be preferable. Large parts such as boat hulls and wind turbine blades can be produced with this method (Balasubramanian, 2014; Park & Seo, 2011).

Vacuum Impregnation Method

Autoclave molding, vacuum-assisted compression molding, and VARTM methods are all developed from the vacuum impregnation method. The fiber wetting process is performed slowly under a vacuum environment. Thus, there is no air entrapment, resin rich region, and fiber wetting problem. The vacuum environment also helps to ensure uniform part thickness. Vacuum-assisted manufacturing methods can be classified as Vacuum Impregnation Method, Vacuum injection molding method, and Vacuum bag molding (Balasubramanian, 2014). Low labor costs, high surface quality, good dimensional accuracy, cheaper mold cost, and production of large parts capability are some of the advantages of this method. Expensive resin systems and time-consuming mold preparation stages are needed, which are disadvantages of this manufacturing method. Mats, woven fabrics, stitched materials, prepregs, and performs can be used as reinforcement in this method.

There are some conditions to select a suitable resin according to the manufacturing methods. Curing conditions, viscosity, gelation time, and wettability are the important parameters of resins that must be suitable for the manufacturing method. On the other hand, mechanical properties, chemical resistance, service temperature of the part, and electrical resistance properties are the parameters that are desired from the product, which also depend on the resin used in the manufacturing. Polyester, epoxy, vinyl ester, and phenolic resins can be used for this method.

Autoclave Molding Method

Autoclave molding is a method with a heated and controlled atmosphere. Composite parts that are used in aerospace applications are produced under high temperatures and pressure by this method. These conditions provide low porosity, better consolidation, and mechanical properties for products. Prepreg reinforcements are generally preferred for this manufacturing method. The vacuum bagging stage is a time-consuming process and the production speed is low, which are disadvantages of this method. In addition, high equipment cost is a condition that should not be ignored (Balasubramanian, 2014).

Compression Molding Method (SMC-BMC)

Generally, two pieces of matched metal molds are used in this method to manufacture composite materials. Compression molding is also called SMC or BMC depending on the thickness of the manufactured part. SMC stands for sheet molding compound and BMC bulk molding compound. Production speed is very high, and this method can be preferred when it comes to the production of medium-sized parts in large quantities.

Molding compounds are a mixture of short fibers such as chopped fibers, resin, fillers, color agents and etc. A certain amount of molding compound is placed on the die. Molding compounds also can be in the form of sheets or bulk in this method. The molding compound spreads as waffle dough under a pressurized environment and fills the mold cavity. The heated die plates are closed and the curing stage of the product begins. The curing stage takes a few minutes at approximately 120 °C (Park & Lee, 2012). Products produced by this method have high surface quality and low dimensional tolerances with good mechanical properties due to low porosity and thermal curing.

Filament Winding Method

Although rovings are generally used as reinforcement, prepreg tapes, even mats, and woven fabrics can be preferred in the filament winding method. The fiber volume ratio can be obtainable up to 70% or even more (Mazumdar, 2002). Filament winding is a widely used method for the manufacturing of products with axial symmetry such as spheres, cylinders, conical, etc. This method has a production principle similar to thread cutting on a lathe. Rovings that are impregnated in a resin bath are wound on a rotating mandrel at a certain winding angle. Mechanical properties can be configured by using a hoop, helical, and polar winding types. Since the method uses continuous fiber (rovings) as reinforcement and provides to manufacture products with a high fiber volume ratio, better mechanical properties can be achieved. Tanks, silos, pressurized vessels and tubes, and, especially, pipes are the main products of the filament winding method.

Generally, polyester, vinyl ester, and epoxy thermoset resins are used in this method. Resin costs are polyester, vinyl ester, and epoxy from cheap to expensive, respectively. Therefore, isophthalic polyester resin is mostly used in general chemical plant applications. Vinyl ester resin is preferred in chemical plant applications, where chemical resistance and heat resistance have great importance. Epoxy resin is costly, but its applications are aircraft, space, military, and marine industries, where cost is not at the forefront.

The advantages of the filament winding method are

- A semiautomated process, manufacturing can be done with low labor.
- A high fiber volume ratio can be obtainable up to 70%.
- Products with better mechanical properties that are in the accordance of specific loading conditions can be configured by suitable winding type and angle.
- Material wastage is less than in other methods and huge axisymmetric products can be manufactured with low initial investment costs.

The disadvantages of the filament winding method are

- Parts produced by the filament winding method have lower interlaminar shear strength and compressive strength values than other manufacturing methods. Using CSM or woven fabrics can improve strength values.
- The outer surface quality of the product is poorer than that of molding method products (Balasubramanian, 2014).

Pultrusion Method

The pultrusion method was invented by Sir Brandt Goldsworthy, and it had been first developed as a composite material production method based on a pulling profile in a vertical position. Reinforcements that are used in the pultrusion method have been developed over time and high-strength fibers, such as carbon and aramid, have begun to be used in the pultrusion method instead of glass fibers. In the 1980s, two new resin systems (phenolic and modified acrylic) were developed. Polyester, vinyl ester, and epoxy resins which are widely used today were derived from phenolic and modified acrylic resin systems (Starr, 2000). The pultrusion method is classified into two groups thermoplastic pultrusion and thermoset pultrusion according to the type of resin (matrix) used in the production of composite material (Baran, 2015). In the thermoplastic the pultrusion line; there is a preheating chamber, a heated die, and a cooling die that will give the composite material its final shape. Thermoset composite materials constitute

75% of plastic-reinforced composite materials. The curing and shaping process is performed with the use of a heated die in the thermoset the pultrusion method (Meyer, 1985).

The pultrusion method provides to manufacture of composite materials with a high fiber volume ratio and constant cross-section. It is the most suitable method for manufacturing of composite profiles such as tubes, beams, rods, and angles. The pultrusion method involves rovings, mats, woven fabrics, and/or surface veils as reinforcements. Products with a high fiber volume ratio of 70% or even higher can be manufactured by this method. The production rate was approximately 0.6 – 0.9 m/min and, nowadays, it can be achieved up to 4.5 m/min (Balasubramanian, 2014). The production rate depends mainly on the resin curing time and amount of resin which is relevant to material thickness. Pultrusion method in the production line respectively; arranging the reinforcing elements, wetting the reinforcing elements in the prepared resin charge, guised the wetted reinforcing elements to have the desired shape and configuration, curing the oriented composite material components by passing them through a die that has the geometry of the composite profile desired to be produced, preheated to a certain temperature, the processes of pulling the cured and finalized profile out of the die with the help of different types of pullers, and cutting the profile that reaches the desired length are carried out. The production process in the pultrusion line includes the following steps: placement of the rovings onto the guide, immersing the rovings to wet into the prepared resin charge, forming the wetted rovings with mats by guides to provide the desired shape and configuration, curing wetted reinforcements by passing them through a die that is preheated to a certain temperature and has the geometry of the product, pulling the profile, which is mostly cured and formed, with the use of different types of pullers, and finally, cutting the profile that reaches the desired length, respectively (Yeşil, 2019).

The basic elements of a pultrusion line are a roving creel, resin bath, preforming guides, heatable metal die, pullers, and cut-off saw. Figure 2 shows a basic pultrusion line. Figure 3 shows the pultruded box beam and its reinforcements. Unidirectional fiber-reinforced composites that consist mainly of rovings can be manufactured by this method. The mechanical properties of this product are superior in the fiber direction than in the transverse direction. Mats, woven fabrics, and stitched textiles are used to improve the mechanical properties of the product in the transverse direction. Furthermore, the surface veil is preferred to provide a good appearance, chemical resistance, environmental resistance, etc., to the product. Generally, E glass rovings as reinforcement and isophthalic polyester resin as matrix elements are preferred because of their low costs. S-glass, aramid, carbon fibers, and their hybrids are used in production, where their superior properties are needed. Dies are usually manufactured from alloy steel, tool steel, or stainless steel with chrome-plated. It is desirable that the dies are resistant to abrasion wear and have a low surface roughness. For this reason, die surfaces are polished and chrome plated to provide better abrasion wear resistance and low surface roughness, which affects the quality of the product and die life. In addition, a post-cure station can be added after the die in order to increase the mechanical properties of the product by providing more curing time and suitable conditions.

Generally, additives are added to the resin in order to provide solutions to the difficulties that may arise during the production stages. These are classified as reaction accelerators, die-release agents, and thermal shrinkage inhibitors (Işık, 2008). Polymerization agents (reaction accelerators) are used to initiate and accelerate the polymerization reaction of the resin. Organic peroxides are used in polyester and vinylester resins. In order to accelerate the curing process, these resin additives are used in certain proportions as reaction accelerators, as well as the heat energy provided by the hot pultrusion die.

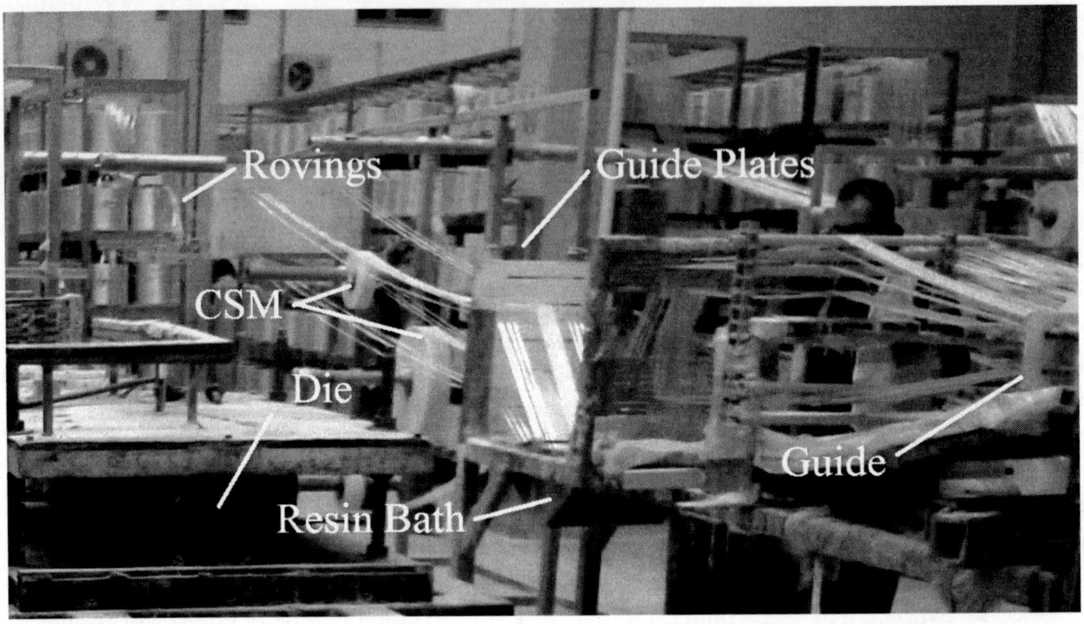

Figure 2. Pultrusion line.

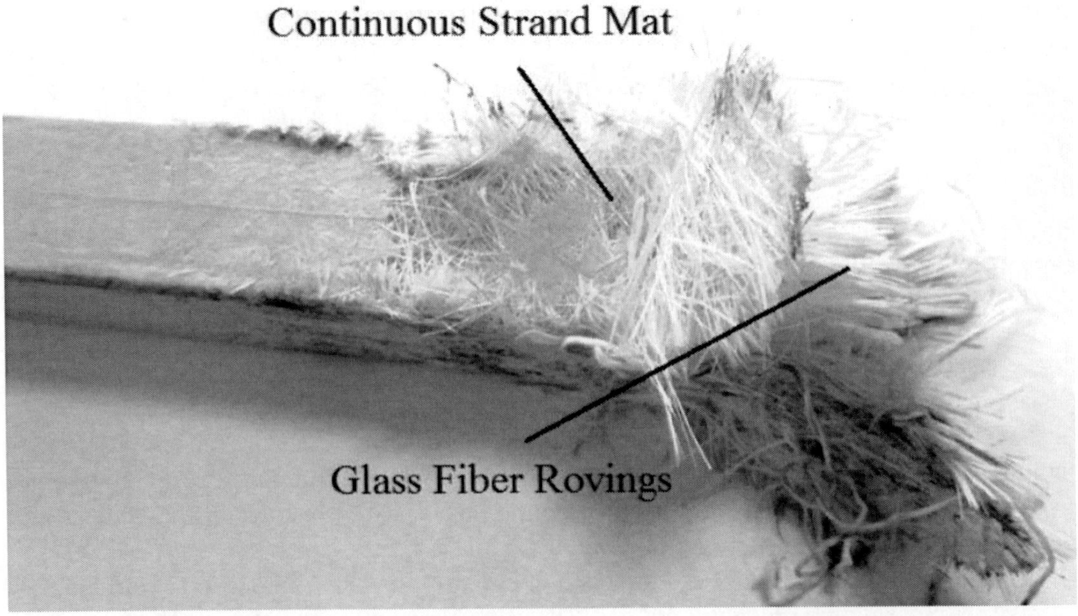

Figure 3. Pultruded box beam and its reinforcements

Acid anhydrides and amines are used as curing agents in epoxy resins (Bai, 2013). Insufficient wetting of fibers, fiber rupture due to the tension, and poor mechanical properties in the transverse direction compared to the fiber direction are disadvantages of this method.

Thermoplastic Composite Manufacturing Methods

Thermoplastic matrix composite manufacturing methods are preferred less than the thermoset matrix composite manufacturing methods mentioned above due to resin-based process difficulties. Thermosets and thermoplastics are both classified in polymers. Polymers consist of mer molecules, which are chained with covalent bonds. If monomers form chains with cross-links, it is called a thermoset. Polymers with a linear chain structure are called thermoplastics. The characteristics of the chain structure determine the properties and behavior of polymers. Thermosets have high-temperature resistance, high strength, and rigidity than thermoplastics, but they are non-recyclable and brittle. On the contrary, thermoplastics are recyclable and flexible but are a bit costly and have low heat resistance.

Thermoplastic matrix composite material manufacturing methods can be categorized as

- Film stacking
- Thermoplastic tape-laying
- Commingled fibers
- Injection molding
- Other manufacturing methods (Balasubramanian, 2014).

Textile Reinforcements in Composite Materials Manufacturing

Textile reinforcements are the main cause of the superior mechanical properties of composite materials, which provide structural integrity, high flexibility, toughness, strength, rigidity, etc. Textile reinforcements provide, as mentioned above, the properties to composite materials in the desired directions depending on their configuration and the fiber material from which they are produced. In order to get a better understanding, it is necessary to classify textile reinforcements first.

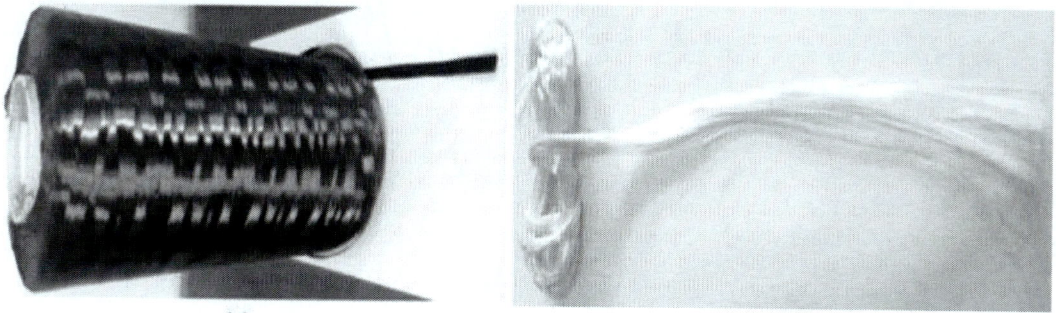

(a) (b)

Figure 4. 1D reinforcement elements (a) Carbon fiber roving, (b) Glass fiber roving.

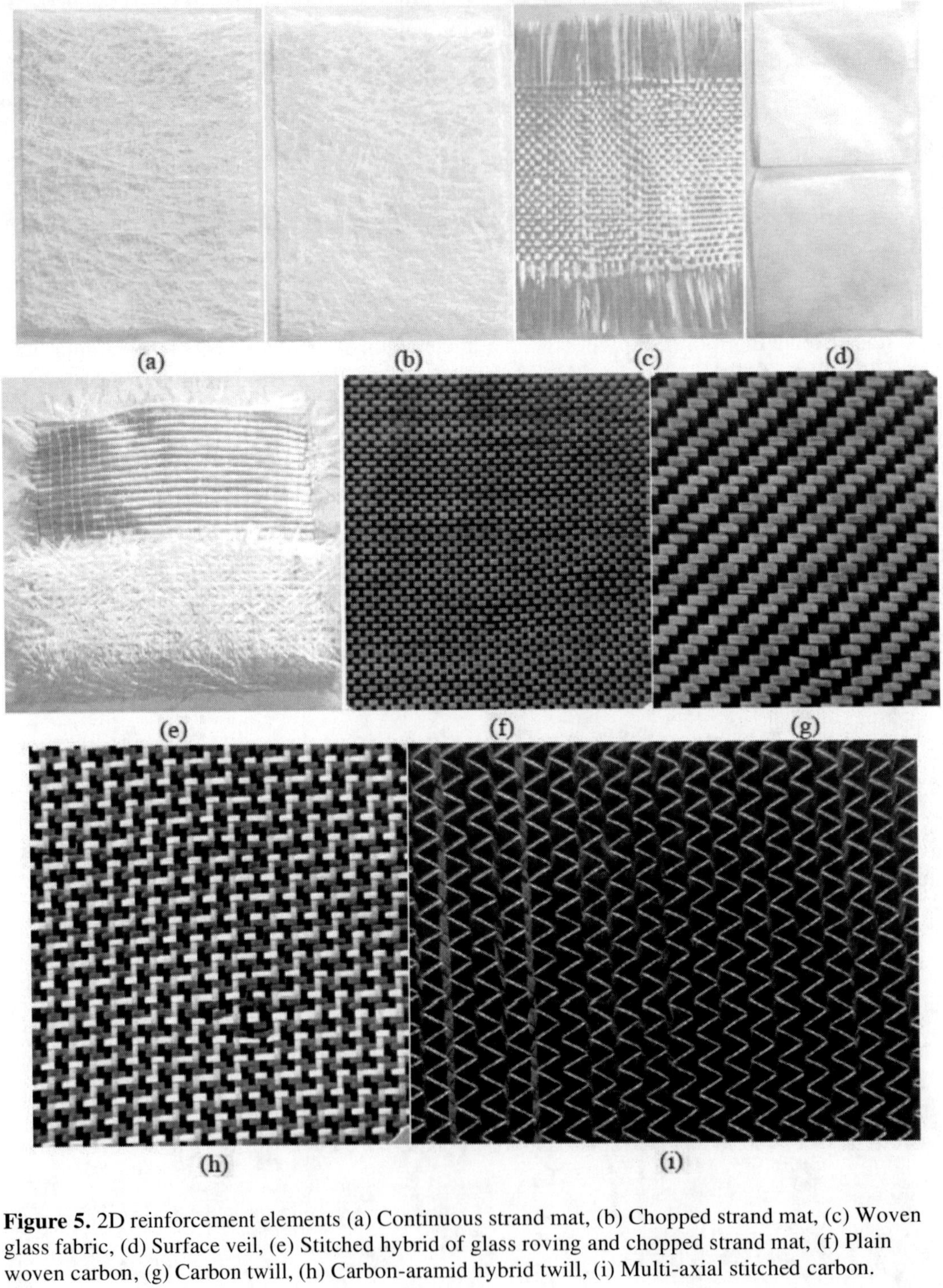

Figure 5. 2D reinforcement elements (a) Continuous strand mat, (b) Chopped strand mat, (c) Woven glass fabric, (d) Surface veil, (e) Stitched hybrid of glass roving and chopped strand mat, (f) Plain woven carbon, (g) Carbon twill, (h) Carbon-aramid hybrid twill, (i) Multi-axial stitched carbon.

Classification of Textile Materials

Textile products have been an important part of human life since ancient times. Textile technologies have continued to develop depending on this situation. Textile compounds are derived from fibers, yarns, and filaments by different production methods. Fibers used in composite manufacturing (glass, carbon, aramid, etc.) have diameter values of 5 μm – 50 μm. Fibers can be classified as continuous and discontinuous and continuous fibers are named filaments. Discontinuous fibers that have a certain length are named as short, discontinuous, or chopped depending on their lengths (Lomov et al., 2005).

Rovings are 1D reinforcements that are used to manufacture mono-axial reinforced composite materials. Chopped fibers, prepregs, weaves, and stitched textiles are used as 2D reinforcements. Chopped fibers are used in randomly oriented non-axial composite material manufacturing. Prepregs are used for mono-axial reinforcement. Weaves, braided fabrics, and stitched textiles are used as 2D reinforcement from mono-axial to multi-axial applications of composite materials. 3D reinforcements can be made by using linear elements or plane elements. 3D structures developed by using linear elements are from non-axial to multi-axial; continuous strand mats, 3D braid, multiply or 3D weaves, and stitched performs, respectively. Plane elements are used to produce honeycombs, laminates, and beam structures. Figure 4 shows the carbon fiber roving and the glass fiber roving. Some examples of 2D reinforcement elements are given in Figure 5. Figure 6 shows the 3D reinforcement such as the honeycomb structure and sandwich panel.

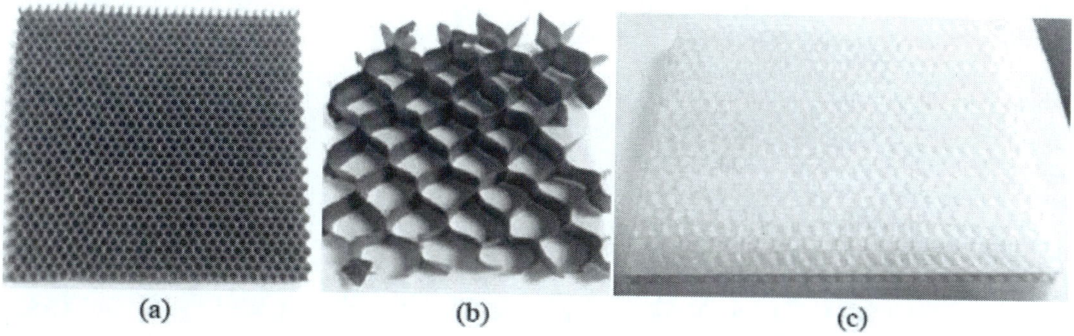

Figure 6. 3D reinforcement elements (a) Aramid honeycomb, (b) Polyethylene honeycomb, (c) Polypropylene sandwich structure.

Conclusion

The use of composite materials in industrial applications continues at an ever-increasing pace and is becoming increasingly widespread in many new application areas. The main reason for this situation is the manufacturing of composite materials that are suitable for the desired properties and preferences. However, the desired requirements are gradually changing and increasing. In order to be a solution to these requirements, the composite material manufacturing industry has to continue to develop at the same pace, which depends mainly on the developments of textile reinforcements.

Improvements in the curing processes of resins and smart composite materials are current developments. In addition, the use area of hybrid fiber-reinforced composite materials is expanding. Current developments in the chemical industry will allow the production of resins that are more resistant to environmental conditions, and the application of suitable RF or laser curing methods on these resins will allow the occurrence of new production methods. The preforms that will be produced by using high-strength fibers with the appropriate geometry utilizing textile technologies will allow the production of advanced composite materials with superior properties.

Disclaimer

None.

References

Bai, J. (2013). *Advanced Fibre-Reinforced Polymer (FRP) Composites for Structural Applications*. Woodhead Publishing. https://doi.org/10.1533/9780857098641.1.

Balasubramanian, M. (2014). *Composite Materials and Processing*. CRC Press.

Baran, İ. (2015). *Pultrusion: State of the art Process Models*. Smithers Rapra.

Bhagwan, D. A., Broutman, L. J., & Chandrasekhara, K. (2015). *Analysis and Performance of Fiber Composites*, John Wiley & Sons Inc.

Callister, W. D., & Rethwisch, D. G. (2013). *Material Science and Engineering*. Nobel.

Cripps, D., Searle, T. J., & Summerscales, J. (2000). *Comprehensive Composite Materials*. Elsevier. https://doi.org/10.1016/B0-08-042993-9/00188-1.

Fitzer, E., & Manocha, L. M. (1998). *Carbon Reinforcements and Carbon/Carbon Composites*. Springer. https://doi.org/10.1007/978-3-642-58745-0.

Işık, A. (2008). *Numerical and experimental analysis of a reinforcement element made of composite material under bending and torsional load*. [Master Thesis]. Istanbul Technical University.

Lomov, S., Verpoest, I., & Robitaille, F. (2005). *Design and manufacture of textile composites* (pp 1-61). Woodhead Publishing Limited. https://doi.org/10.1533/9781845690823.1.

Mazumdar, S. K. (2002). *Composites Manufacturing: Materials, Product and Process Engineering*. CRC Press. https://doi.org/10.1201/9781420041989.

Meyer, R. W. (1985). *Handbook of Pultrusion Technology*. Chapman and Hall. https://doi.org/10.1007/978-1-4684-7764-1_1.

Nagavally, R. R. (2017). Composite materials - history, types, fabrication techniques, advantages, and applications. *International Journal of Mechanical and Production Engineering, 5*(9), 82-87.

Park, C. H., & Lee, W. I. (2012). Compression molding in polymer matrix composites. In S. G. Advani & K. T. Hsiao (Eds.), *Manufacturing techniques for polymer matrix composites* (pp. 47-94). Woodhead Publishing. https://doi.org/10.1533/9780857096258.1.47.

Park, S. J., & Seo, M. K. (2011). Element and processing. In A. T. Hubbard (Ed), *Interface science and technology* (pp. 431-499). Elsevier. https://doi.org/10.1016/B978-0-12-375049-5.00006-2.

Starr, T. F. (2000). *Pultrusion for Engineers*. Woodhead Publishing Limited. https://doi.org/10.1533/978185 5738881.1.

Vinson, J. R., & Sierakowski, R. L. (2002). *The behavior of structures composed of composite materials* (2nd ed., pp. 1-37). Kluwer Academic Publishers. https://doi.org/10.1007/0-306-48414-5.

Yeşil, Ö. (2019). *Manufacturing of hybrid composite profiles by pultrusion method*. [Doctoral dissertation]. Uşak University.

Chapter 17

Antibacterial Electrospun Nanofibers

Seniha Morsümbül, PhD
and E. Perrin Akçakoca Kumbasar*, PhD
Department of Textile Engineering, Faculty of Engineering, Ege University, Izmir, Türkey

Abstract

Nanofibrous surfaces, which have many application areas due to their wide surface area and high porosity, are produced by the methods such as bicomponent fiber production, electrospinning, phase separation, template synthesis, and meltblown. Among these methods, electrospinning, which offers the opportunity to study with many different polymers, stands out as a relatively easier and economical method, especially in laboratory conditions. Electrospun nanofibrous surfaces can gain functional properties due to the type of polymer used or the active ingredients added to the polymer solution. Antibacterial nanofibers can be obtained by using polymers such as chitosan and bamboo, or by adding active substances such as silver, zinc, essential oil, etc., into the polymer matrix. Antibacterial nanofibrous surfaces can be used in many areas, such as medical textiles, which require antibacterial properties. This chapter provides an overview of the fabrication of antibacterial nanofibrous surfaces by electrospinning.

Keywords: antibacterial activity, electrospinning, medical textiles, nanofibrous surfaces

Introduction

Various nanotechnology applications have become remarkable in many different materials such as composite materials, filtration, medical textiles, sensors, etc. The use of nanofibrous materials in the fabrication of functional textile materials is becoming increasingly important, especially by means of their properties such as high porosity, large surface area, and the usability of different polymer types (Akduman, 2021; Başal & Oral, 2022; Şafak et al., 2020; Bostanci et al., 2020; Çay et al., 2015).

There are many different production methods for nanofibrous materials, such as template synthesis, electrospinning, phase separation, self-assembly, and drawing. Electrospinning may

* Corresponding Author's Email: perrin.akcakoca@ege.edu.tr.

In: Textile Insights and Future Aspects
Editors: Gamze D. Tetik, Gizem Celep and Fulya Yilmaz
ISBN: 979-8-89113-289-4

be the most preferred technique for nanofiber fabrication among all these methods since it has the potential to scale up for industry processing and many advantages, such as low cost, easy to apply, and usability of many different polymers (Alghoraibi & Alomari, 2019; Akduman et al., 2021a; Gao et al., 2014; Li et al., 2021).

The functionality of nanofibrous materials, and thus their industrial application possibilities, can be improved by the use of specific polymers or the incorporation of different substances such as medicines, dyes, cosmetics, etc., into the matrix (Purnamasari et al., 2022; Nazari et al., 2022; Tahir et al., 2023; Akduman et al., 2021b). Antibacterial polymers or compounds, which inhibit bacterial growth, are also materials used in the functionalization of nanofibrous surfaces. This review chapter represents the functionalization of electrospun nanofibers with antibacterial polymers or compounds.

Electrospinning

Electrospinning is a fiber spinning method that is able to generate continuous ultrafine fibers in nanometer diameter. In this method a syringe pump, a high-voltage power supply that is attached to the needle of the syringe (nozzle), and a metal collector plate positioned opposite the nozzle are used (Figure 1). A spinning solution composed of a polymer solution with/without additives is first loaded into the syringe and is pumped to the needle tip by the syringe pump at a certain rate. The polymer solution at the needle tip is exposed to a high potential voltage and the charges within the solution are induced by the electrical field provided by the high-voltage power supply. Then, a polymer jet is formed, and it flies towards the collection plate, which is charged with the reverse polarity. During this flow, the solvent in the spinning solution evaporates, and the jet is scattered as nanofibers.

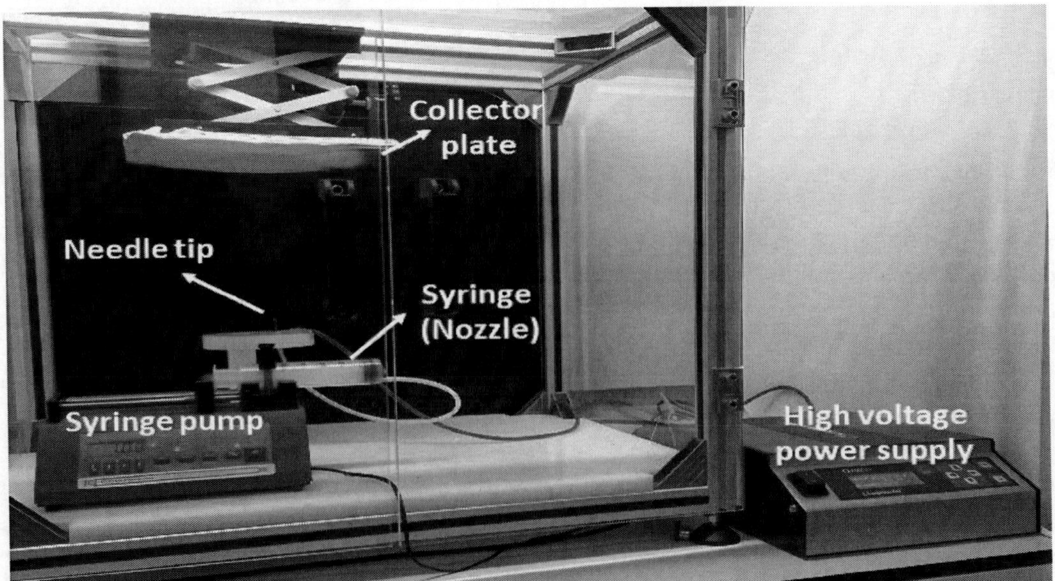

Figure 1. The electrospinning device.

Almost any natural and synthetic polymers or their blends can be used in nanofibers fabrication by the electrospinning method, if these polymers have a solubility in a solvent and sufficient molecular weight for the electrospinning process. The characteristics of the spinning solution are not the only parameter that affects the electrospinning; the process and ambient conditions also affect the properties of the obtained nanofibrous surfaces (Table 1) (Akduman & Akçakoca Kumbasar, 2017; Akçakoca Kumbasar et al., 2018).

Table 1. The factors affecting electrospinning process (Ramakrishna, et al., 2005)

Process parameters	Solution parameters	Ambient parameters
Voltage	Viscosity	Moisture
Flow rate	Concentration	Temperature
Collector type	Surface tension	Atmospheric pressure
Distance from the nozzle to collector	Electrical conductivity	
	The molecular weight of the polymer	

The electrospun nanofibrous surface can be functionalized by using specific polymers or incorporating some complexes such as medicines, chromophores, metals, etc., into the fiber matrix. Antibacterial nanofibrous surfaces, one of these types of functional nanofibrous materials, can be obtained by loading antibacterial substances into the spinning solution or using antibacterial polymers. Thus, these nanofibrous surfaces can be used in different application areas that requires antibacterial properties.

Antibacterial Nanofibrous Materials

Bacteria cause many infectious diseases and deaths worldwide. In order to reduce or even prevent these diseases and deaths, living conditions should be as sterile as possible, and antibacterial materials should be used. With the growing public health awareness to pathogenic effects, antibacterial materials that inhibit bacteria growth, have become even more remarkable in many different fields such as packaging, medical textiles, filtration, etc., (Shahidi & Wiener, 2012).

Antibacterial nanofibers, as a relatively new class of antibacterial materials, have some advantages such as large specific surface area and porosity. Compared to other antibacterial materials, specifically, the large surface area of electrospun nanofibers enables efficient loading of antibacterial agents (Wei et al., 2020).

There are various possibilities to obtain electrospun antibacterial nanofibrous materials, such as using antibacterial polymers or the incorporation of antibacterial agents into the polymer matrix during the electrospinning process (Gul et al., 2021). After the production, the antibacterial activity degree of the nanofibrous material must be determined to decide the suitability of the material for the application field. For this purpose, different standards and test methods have been developed.

Antibacterial activity tests are divided into two classes as qualitative and quantitative. In the qualitative method, the antibacterial degree of the material is determined by examining the zone formed around the material. In the quantitative method, the antibacterial degree of the material is determined by calculating the percentage of bacteria in the test environment at the beginning and end of the test period. While the most preferred methods among qualitative methods are agar disk diffusion and AATCC 147 parallel line methods, the AATCC 100 standard is the most widely used method in quantitative analysis (Pinho et al., 2011).

Different types of bacteria can be used for antibacterial tests. These bacteria are divided into two classes, gram-negative and gram-positive. Examples of gram-positive bacteria include *Bacillus subtilis, Listeria monocytogenes, Staphylococcus Aerus, Streptococcus mutans,* and *Clostridium perfringens. Escherichia Coli, Pseudomonas aeruginosa, Moraxella catarrhalis, Helicobacter pylori, Legionella pneumophila* bacteria can be given as examples of gram-negatives (Kim et al., 2015).

While the material is very effective against any type of bacteria, it may not show antibacterial activity against another bacterium (Li & Zhuang, 2020). Therefore, target bacterial species should be determined, and the appropriate polymers/substances should be used in the production of antibacterial nanofibrous materials.

There have been many different studies in the last few years about the production of antibacterial nanofibers. In this review, the recent development in the fabrication of antibacterial nanofibers by electrospinning is examined.

Antibacterial Electrospun Nanofibers Production

A large number of authors have studied the production of antibacterial electrospun nanofibers. These studies are discussed in detail below and the summary of the studies is listed in Table 2.

Shadman-Manesh et al. (2023) have reported the preparation of antibacterial electrospun polycaprolactone (PCL)-polyethylene glycol (PEG) nanofibers by incorporating egg yolk oil (EYO) into the polymer solution. They evaluated the samples by antibacterial activity, in vitro cell culturing, and in vivo trials. Their results have shown PCL-PEG-EYO nanofibrous surfaces do not have cytotoxicity up to 7 days after the cell culturing and have antibacterial activity against Staphylococcus aureus (S. aureus) and Pseudomonas aeruginosa (P. aeruginosa) bacteria strain. The authors have attributed the antibacterial potency in PCL-EG-EYO nanofibers to the presence of phosphonic acid groups and OH groups in EYO. They have specified that the higher antibacterial activity of the samples against gram-positive bacteria than that of gram-negative bacteria may be related to the more permeable layer of gram-positive bacteria cell wall than a gram-negative bacteria cell wall. Furthermore, it has been stated that enhancement in re-epithelialization, angiogenesis, collagen synthesis, and wound closure was observed after 21-day treatment period of using these antibacterial electrospun nanofibrous surfaces. The authors have concluded that the PCL-PEG-EYO nanofibrous surfaces have a major potential in the treatment of burn wounds in vivo.

Yavari Maroufi et al. (2023) have investigated the antibacterial properties of gelatin/oxidized xanthan gum nanofibers containing propolis at concentrations of 5, 12.5, and 20%. The authors have reported that the nanofibers contained in propolis have a significant inhibition ($P < 0.05$) against the S. aureus and Escherichia coli (E. coli) bacteria and this inhibition rate increased with the increasing of propolis content. They have indicated that the antibacterial activity of these nanofibers is owing to the presence of phenolic compounds such as quercetin, rutin, and naringenin within propolis. They also stated that the electrospun nanofibers contained in propolis had lower efficiency against gram-negative bacteria than gram-positive bacteria due to the outer protective membrane of gram-negative bacteria. The authors have suggested that these antibacterial electrospun nanofibers could be used as a matrix for active food packaging.

A nanofibre-based material composed of polylactic acid (PLA) electrospun fibers loaded with lemon essential oil (LEO) and covered with poly (N-isopropylacrylamide) (PNIPAAm) was developed by Xia et al. (2023) for blackberry preservation. They demonstrated that the nanofibrous surfaces incorporated with LEO have antimicrobial activities against S. aureus and E. coli bacteria compared to neat nanofibers, and the nanofibrous mats containing both LEO and PNIPAAm have the slowest bacterial growth, and the longest-lasting antibacterial activity. Moreover, the authors have reported that the electrospun nanofibers containing LEO have a higher antibacterial activity against S. aureus than that against E. coli.

Jie et al. (2023) have prepared a chitosan-urushiol nanofiber membrane by the electrospinning method and investigated the antibacterial efficiency of this nanofibrous membrane against S. aureus and E. coli bacteria. They have indicated that the neat-chitosan nanofibers showed a good antibacterial activity against S. aureus at the rate of 98% although its antibacterial efficiency on E. coli was at the rate of 11%. They have also stated that urushiol had the antibacterial rate against S. aureus and E. coli of 81.9% and 61.3%, respectively. The authors have reported that the antibacterial efficiency of the chitosan nanofibrous surface contained urushiol against both S. aureus and E. coli reached more than 94% and chitosan-urushiol exhibited a synergistic antibacterial activity against the mentioned bacteria.

In the study of Li et al. (2023), sulfonated polyarylene ether nitrile nanofibers as oil/water separation membranes have been obtained by the electrospinning method. They have stated that the antibacterial property is important for long-term storage and operation of the separation membranes. Thus, the authors in-situ grown silver nanoparticles on the nanofibrous surfaces and evaluated the antibacterial activity of these surfaces against S. aureus and E. coli bacteria. The authors have concluded that nanofibrous membranes modified with silver nanoparticles inhibit bacterial growth and can be used as high-performance nanofibrous membranes for practical wastewater treatment.

Jiao et al. (2023) have developed an antibacterial indicator absorbent pad with Janus structure that had two layers, polyurethane (PU) nanofibers as the top hydrophobic layer and polyvinyl alcohol (PVA) nanofibers as the bottom absorbent layer, for meat preservation. They have added polyhexamethylenebiguanide hydrochloride (PHMB) to the top layer as a cationic antimicrobial agent and anthocyanin to the bottom layer as an indicator to detect spoilage of meat. They have investigated the antibacterial activity and antifungal properties of the pads against S. aureus and E. coli bacteria and Aspergillus niger (A. niger) fungal. The authors have observed that the nanofibrous surfaces obtained in the study can effectively inhibit bacterial growth, prolong the shelf life of fresh meat to 7 days when stored at 4 °C, and indicate meat deterioration.

Oskouei et al. (2023) have fabricated Cu-Al layered double hydroxide/polyvinyl alcohol/chitosan composite nanofibers using the electrospinning method. They tested the antibacterial property of prepared materials against Bacillus subtilis (B. subtilis) and E. coli bacteria and observed that the composite nanofibers effectively inhibited bacterial growth. They have also stated that these nanofibers have a stronger inhibitory effect on B. subtilis (96.94%) than on E. coli (94.26%). The authors have concluded that Cu-Al layered double hydroxide can enhance the antibacterial effect of the PVA/CS nanofibers by producing reactive oxygen species and inducing oxidative.

Tajfiroozeh et al. (2023) have investigated the production of electrospun core/shell chitosan/polyethylene oxide nanofibrous materials incorporated with gallic-acid/nisin with antimicrobial activity for food packing applications. They used six bacteria (P. aeruginosa, E. coli, L. cytogenesis, S. aureus, S. typhi, and B. subtilis) for the antibacterial activity tests of the nanofibrous materials. They have indicated that neat chitosan/polyethylene oxide nanofibers exhibited no antimicrobial activity, while gallic acid (10%)/nisin (0.05% and 0.07%) incorporated nanofibrous materials exhibited the highest antimicrobial and antioxidant effects for 60 and 20 days, respectively.

Lee et al. (2023) have conjugated glycyrrhizic acid (a licorice root extract) with β-cyclodextrin to prevent aggregation of glycyrrhizic acid and then incorporate glycyrrhizic acid-conjugated β-Cyclodextrin into electrospun polyurethane nanofibers. They tested the antibacterial properties of nanofibers loaded with glycyrrhizic acid or glycyrrhizic acid-conjugated β-Cyclodextrin against S. aureus and E. coli bacteria. They indicated that glycyrrhizic acid-loaded nanofibers have higher antibacterial activity against E. coli than against S. aureus, and its antibacterial activity was dose-dependent against both bacteria specified. The authors also reported that glycyrrhizic acid-conjugated β-Cyclodextrin loaded nanofibers have higher antibacterial activity than glycyrrhizic acid-loaded nanofibers due to their dispersion without aggregation in the fiber compared to the nanofibers prepared by adding glycyrrhizic acid alone.

Elsherbiny et al. (2023) have synthesized a series of α-aminophosphonate derivatives and investigated their antibacterial activity against B. subtilis, S. aureus, E. coli, and Klebsiella pneumonia bacteria. They observed that the tolyl derivative of the aminophosphonate molecule exhibited the highest antibacterial activity against the four tested bacteria and then, they incorporated this derivative into the thermoresponsive eicosane/cellulose nanocrystals o/w emulsion. They coated the electrospun polybutylene succinate nanofibers with the thermoresponsive o/w emulsion and investigated the antibacterial activity of these coated nanofibrous surfaces. They modified the antibacterial test method (agar disk diffusion) to activate the thermoresponsive system. They preincubated the coated nanofibrous materials at 39 °C for 15 min instead of room temperature, to mimic the infected body temperature and thus ensured the melting of eicosan and the release of the active agent. They observed that samples preincubated at 39 °C had antibacterial activity against the bacteria tested while samples preincubated at room temperature had no inhibition zone. The authors have specified that this results from releasing the active agent from the thermoresponsive emulsion formulation at higher temperatures like 39 °C. The authors have stated that the coating process improves not only the antibacterial properties of the nanofibrous material but also its hydrophilicity. They have concluded that the fabricated nanofibrous materials display considerable potential for wound dressings.

Chen et al. (2023) fabricated poly(caprolactone) loaded with tannic acid and examined its antibacterial activity against S. aureus. They have observed that tannic acid-loaded nanofibers have dose-dependent antibacterial activity, while neat nanofibers have no antibacterial activity against S. aureus. They also specified that tannic acid improved the hydrophilicity and water uptake of nanofibrous materials. The authors have concluded that tannic acid-loaded nanofibrous materials have the potential for antibacterial wound dressing.

Table 2. Summary of recent studies on antibacterial electrospun nanofiber production

Polymers	Antibacterial compounds	Antibacterial tests	References
Polycaprolactone and polyethylene glycol	Egg yolk oil	Agar well diffusion method	Shadman-Manesh et al., 2023
Gelatin and oxidized xanthan gum	Propolis	Agar disk diffusion method	Yavari Maroufi et al., 2023
Polylactic acid	Lemon essential oil	Turbidimetric method	Xia et al., 2023
Chitosan	Urushiol	Plate counting method	Jie et al., 2023
Polyarylene ether nitrile	Silver	Agar disk diffusion method	Li et al., 2023
Polyurethane and polyvinyl alcohol	Polyhexamethylenebiguanide hydrochloride	Plate counting method	Jiao et al., 2023
Polyvinyl alcohol and chitosan	Cu-Al layered double hydroxides	Agar disk diffusion method	Oskouei et al., 2023
Polyethylene oxide and chitosan	Gallic-acid/nisin	Agar disk diffusion method	Tajfiroozeh et al., 2023
Polyurethane	Glycyrrhizic acid	Agar disk diffusion method	Lee et al., 2023
Polybutylene succinate	α-aminophosphonate derivatives	Agar disk diffusion method	Elsherbiny et al., 2023
Poly(caprolactone)	Tannic acid	Agar disk diffusion method	Chen et al., 2023
Polyacrylonitrile	Ce-based metal organic framework	Plate counting method	Zhang et al., 2023
Polycaprolactone and polyvinylpyrrolidone	Curcumin and silver nanoparticles	Plate counting method	Wang et al., 2022
Polyvinyl alcohol and gelatin	Peppermint oil and amoxicillin	Agar disk diffusion method, plate counting method, bacterial morphology observation, fluorescence staining and protein leakage assay	He et al., 2022
Polyurethane and gelatin	Honey and zinc-oxide nanoparticles	Agar disk diffusion method	Abolhassani et al., 2022
Polylactic acid and chitosan	Aloin	Plate counting method	Yin et al., 2022
Polycaprolactone, chitosan, polyethylene oxide	Arnebia euchroma	Dynamic contact assay and the agar plate culture method, SEM	Asghari et al., 2022
Polyethylene oxide	Basil essential oil	Agar disk diffusion method	Li et al., 2022
Polylactic acid	Geraniol and carvacrol	Agar disk diffusion method, fluorescent microscopy, SEM, antibacterial kinetic growth assays	Hemmatian et al., 2021
Polyvinyl chloride	N-halamine (7,7,9,9-tetramethyl-1,3,8-triazaspiro [4.5]-decane-2,4-dione)	Plate counting method	Çobanoğlu et al., 2021

Zhang et al. (2023) have stated that the adsorption performance of the adsorbents can be affected by the contamination of bacteria, which occurs during the removal process. Therefore, they developed an antibacterial nanofibrous adsorbent to remove the herbicide named 2,4-dichlorophenoxyacetic acid (2,4-D). They have produced a polyacrylonitrile nanofibrous membrane by electrospinning and then in-situ growth of a Ce-based metal organic framework with the UiO-66 structure on the surface of electrospun nanofibrous membranes by the solvothermal method. The authors have investigated the antibacterial activity of nanofibrous materials against E. coli and observed an antibacterial activity rate of greater than 90% with the developed nanofibrous membranes.

The core/shell structure nanofibrous materials were produced by Wang et al. (2022) using the coaxial electrospinning method. They have used polycaprolactone (PCL) and polyvinylpyrrolidone (PVP) solutions with curcumin as the core layer and the PVP solution with silver nanoparticles as the shell layer. They have indicated that compared to single-loaded nanofibrous materials with silver or curcumin, the loaded nanofibrous materials with silver and curcumin have better antibacterial activity against *S. aureus* and *E. coli* with antibacterial rates reaching 93.04% and 92.82%, respectively. The authors have also stated that the singlet oxygen generation of curcumin was improved by silver nanoparticles, and thereby a synergistic antibacterial effect was achieved combined with the antibacterial effect of silver nanoparticles.

He et al. (2022) investigated the antibacterial activity of polyvinyl alcohol/gelatin nanofibers loaded with peppermint oil and amoxicillin. They tested the nanofibrous materials against *E. coli, C. albicans,* and *S. aureus* bacteria. The authors stated that amoxicillin-loaded nanofibers had higher antibacterial activity compared to the peppermint oil-loaded nanofibers, while the nanofibers loaded with both peppermint oil and amoxicillin had the highest antibacterial activity.

Abolhassani et al. (2022) developed electrospun polyurethane-gelatin nanofibers loaded with honey and zinc oxide nanoparticles as a potential wound dressing. The antibacterial activity against *S. aureus*, *E. coli*, and *B. subtilis*, mechanical properties, and cell proliferation of the fabricated nanofibrous materials were investigated by the authors. They stated that honey-loaded nanofiber surfaces had no antibacterial effect against *S. aureus* and *B. subtilis* but were effective against *E. coli*. They also observed that zinc oxide-loaded nanofibers had antibacterial activity in all bacteria tested, and both zinc oxide- and honey-loaded nanofibrous samples had the biggest inhibition zones in all bacteria tested. Moreover, they stated that nanofibrous materials loaded with honey and zinc oxide had proper biocompatibility. They concluded that the fabricated nanofibrous materials could be a promising candidate for bandage materials.

Electrospun polylactic acid/chitosan nanofibrous materials loaded with aloin were fabricated and characterized by Yin et al. (2022). They investigated the antibacterial activity of the nanofibrous mats against *S. aureus* and *E. coli* bacteremia. They observed that the addition of aloin into the polylactic acid/chitosan nanofiber matrix improved its antibacterial activity. The authors have stated that aloin-loaded polylactic acid/chitosan nanofibrous mats exhibited good antibacterial, blood coagulative, and biocompatible properties and had the possibility of being utilized as an antibacterial wound dressing.

Asghari et al. (2022) developed a hybrid nanofibrous scaffold composed of polycaprolactone/chitosan-polyethylene oxide loaded with *Arnebia euchroma* extract for skin tissue engineering. They performed an antibacterial activity test, drug release profile, and cytotoxicity study to investigate the potential of the scaffolds in burn wound healing and treatment applications. They have observed that incorporating *Arnebia euchroma* into the nanofibrous scaffolds increases the antibacterial activity against *S. aureus* and *E. coli* bacteria. Moreover, they have stated that there is no dose-dependent antibacterial activity. The authors also evaluated the antibacterial activity of the nanofibrous scaffolds by SEM analysis. They stated that the morphology of bacteria was changed and destroyed after the application of the nanofibrous scaffold. They have indicated that the *Arnebia euchroma-loaded* nanofibrous scaffold exhibited cytocompatibility, improved cell proliferation, and cell attachment on nanofibrous scaffolds. They concluded that the nanofibrous scaffolds show promising properties for tissue engineering applications, particularly in the case of skin burns.

Li et al. (2022) produced electrospun polyethylene oxide nanofibers loaded with basil essential oil encapsulated with cationic liposomes and tested their antibacterial activity against *Listeria monocytogenes (L. monocytogenes)*. They also loaded soybean lecithin into nanofibers to investigate the effect on the antibacterial activity of basil essential oil. The authors have observed that the polyethylene oxide nanofibers have antibacterial activity against *L. monocytogenes* by basil essential oil. They have also indicated that soybean lecithin improves the antibacterial activity of nanofibers by accelerating the release of antibacterial agents from the nanofibers with the mechanism of responsiveness to bacterial phospholipase. They examined the release performance of basil essential oil with exposure to different bacterial stimulation. They used *Salmonella Typhimurium* (*S. Typhimurium*) as a model bacterium without the ability to produce phospholipase. They have observed that the antibacterial activity of the nanofibers against *L. monocytogenes* is higher than that of *S. Typhimurium*. They have stated that this results from the decomposition of soybean lecithin by phospholipase secreted by *L. monocytogenes* and, thus, more antibacterial agents are released from the nanofibers.

Hemmatian et al. (2021) have produced geraniol or carvacrol-incorporated polylactic acid nanofibers and evaluated the antibacterial activity of these nanofibers against *Staphylococcus epidermidis* (*S. epidermidis*) and *E. coli* bacteria. They have stated that the carvacrol-loaded nanofibers have better antibacterial performance than the geraniol-loaded nanofibers and the concentration increase of these compounds in the nanofibers has a positive effect on antibacterial activity. The authors have observed that the nanofibers exhibited antibacterial performance against both *E. coli* and *S. epidermidis* with over 80% of inhibition within 72h according to the antibacterial kinetic growth assays.

Çobanoğlu et al. (2021) have produced electrospun polyvinyl chloride nanofibrous materials with rechargeable antibacterial properties. They have stated that N-halamines have the ability to be repeatedly regenerated in a chlorine solution, and thus rechargeable antibacterial characteristics. They have synthesized an antibacterial agent-based N-halamine (7,7,9,9-tetramethyl-1,3,8-triazaspiro [4.5]-decane-2,4-dione (TTDD)) and incorporated it into the electrospun polyvinyl chloride nanofibers. Then, they investigated the antibacterial activity against *S. aureus* and *E. coli* bacteria and rechargeable chlorination capability of the nanofibrous materials. The authors have observed that nanofibers have rechargeable chlorination capabilities in acidic media with a minimum loss during 6 h. They have specified that chlorinated nanofibrous surfaces have dose-dependent antibacterial activity against both *S. aureus* and *E. coli* bacteria.

Conclusion

Electrospun nanofibrous materials that exhibit antibacterial properties may have a wide variety of potential applications in medical textiles, packages, or filters. In this chapter, the fabrication of antibacterial nanofibers by electrospinning, functionalization of the nanofibrous surfaces with antibacterial agents, and recent research in this field have been reported. Although studies have been performed on antibacterial nanofibrous surfaces, more detailed research should be done on the fabrication of the antibacterial electrospun nanofibers and alternative applications can be created for these fibers. Moreover, large-scale production of electrospun nanofibers needs to be developed and is expected to become more widespread for industrial applications.

Disclaimer

None.

References

Abolhassani, S., Alipour, H., Alizadeh, A., Nemati, M. M., Najafi, H., & Alavi, O. (2022). Antibacterial effect of electrospun polyurethane-gelatin loaded with honey and ZnO nanoparticles as potential wound dressing. *Journal of Industrial Textiles, 51*(1S), 954S-968S. https://doi.org/10.1177/15280837211069871.

Akçakoca Kumbasar, E. P., Morsunbul, S., & Alır, S. (2018). Photochromic nanofibers. In T. Lin (Ed.), *Novel aspects of nanofibers* (pp. 69-85). IntechOpen. https://doi.org/10.5772/intechopen.74663.

Akduman, C. (2021). Cellulose acetate and polyvinylidene fluoride nanofiber mats for N95 respirators. *Journal of Industrial Textiles, 50*(8), 1239-1261. https://doi.org/10.1177/1528083719858760.

Akduman, C., & Akçakoca Kumbasar, E. P. (2017). Electrospun polyurethane nanofibers. In F. Yılmaz (Ed.), *Aspects of polyurethanes* (pp. 17-52). IntechOpen. https://doi.org/10.5772/intechopen.69937.

Akduman, C., Demirel, V., & Tezcan, F. (2021a). Filter life comparison of different levels of nanofiber coated cleanable-surface filter for gas turbine. *Journal of Applied Polymer Science, 138*(33), 1-17. https://doi.org/10.1002/app.50820.

Akduman, Ç., Morsümbül, S., & Akçakoca Kumbasar, E. P. (2021b). The removal of reactive red 141 from wastewater: A study of dye adsorption capability of water-stable electrospun polyvinyl alcohol nanofibers. *Autex Research Journal, 21*(1), 20-31. https://doi.org/10.2478/aut-2019-0040.

Alghoraibi, I., & Alomari, S. (2019). Different methods for nanofiber design and fabrication. In A. Barhoum, M. Bechelany & A. S. H. Makhlouf (Eds.), *Handbook of nanofibers* (pp. 79-124). Springer. https://doi.org/10.1007/978-3-319-53655-2_11.

Asghari, F., Faradonbeh, D. R., Malekshahi, Z. V., Nekounam, H., Ghaemi, B., Yousefpoor, Y., Ghanbari, H., & Faridi-Majidi, R. (2022). Hybrid PCL/chitosan-PEO nanofibrous scaffolds incorporated with A. euchroma extract for skin tissue engineering application. *Carbohydrate Polymers, 278*, 1-14. https://doi.org/10.1016/j.carbpol.2021.118926.

Başal, G., & Oral, N. (2022). A Hydrophilic/hydrophobic composite structure for water harvesting from the air. *Textile and Apparel, 32*(4), 384-389. https://doi.org/10.32710/tekstilvekonfeksiyon.1074935.

Bostanci, A., Tanik, N. A., & Aykut, Y. (2020). Cellulose monoacetate/nafion (CMA/N) hybrid nanofibers as interface for electrochemical DNA biosensors. *Textile and Apparel, 29(3)*, 228-236. https://doi.org/10.32710/tekstilvekonfeksiyon.501435.

Çay, A., Akçakoca Kumbasar, E. P., & Akduman. Ç. (2015). Effects of solvent mixtures on the morphology of electrospun thermoplastic polyurethane nanofibres. *Textile and Apparel, 25*(1), 38-46.

Chen, X., Zhang, Q., Wang, Y., Meng, J., Wu, M., Xu, H., Du, L., & Yang, X. (2023). Fabrication and characterization of electrospun poly (caprolactone)/tannic acid scaffold as an antibacterial wound dressing. *Polymers, 15*(593), 1-12. https://doi.org/10.3390/polym15030593.

Çobanoğlu, B., Parin, F. N., & Yildirim, K. (2021). Production and characterization of n-halamine based polyvinyl chloride (PVC) nanowebs. *Textile and Apparel, 31*(3), 147-155. https://doi.org/10.32710/tekstilvekonfeksiyon.717601.

Elsherbiny, D. A., Abdelgawad, A. M., Shaheen, T. I., Abdelwahed, N. A. M., Jockenhoevel, S., & Ghazanfari, S. (2023). Thermoresponsive nanofibers loaded with antimicrobial α-aminophosphonate-o/w emulsion supported by cellulose nanocrystals for smart wound care patches. *International Journal of Biological Macromolecules, 233*, 1-12. https://doi.org/10.1016/j.ijbiomac.2023.123655.

Gao, Y., Bach Truong, Y., Zhu, Y., & Kyratsiz, I. L. (2014). Electrospun antibacterial nanofibers: Production, activity, and in vivo applications. *Journal of Applied Polymer Science, 131*(18), 1-13. https://doi.org/10.1002/app.40797.

Gul, A., Gallus, I., Tegginamath, A., Maryska, J., & Yalcinkaya, F. (2021). Electrospun antibacterial nanomaterials for wound dressings applications. *Membranes, 11*(12), 1-21. https://doi.org/10.3390/membranes11120908.

He, S., Jiang, L., Liu, J., Zhang, J., & Shao, W. (2022). Electrospun PVA/gelatin based nanofiber membranes with synergistic antibacterial performance. *Colloids and Surfaces A: Physicochemical and Engineering Aspects, 637*, 1-8. https://doi.org/10.1016/j.colsurfa.2021.128196.

Hemmatian, T., Seo, K. H., Yanilmaz, M., & Kim, J. (2021). The bacterial control of poly (lactic acid) nanofibers loaded with plant-derived monoterpenoids via emulsion electrospinning. *Polymers, 13*(19), 1-15. https://doi.org/10.3390/polym13193405.

Jiao, X., Xie, J., Du, H., Bian, X., Wang, C., Zhou, L., & Wen, Y. (2023). Antibacterial smart absorbent pad with Janus structure for meat preservation. *Food Packaging and Shelf Life, 37*, 1-8. https://doi.org/10.1016/j.fpsl.2023.101066.

Jie, X., Shiu, B. C., Zhang, Y., Wu, H., Ye, Y., & Fang, R. (2023). Chitosan-Urushiol nanofiber membrane with enhanced acid resistance and broad-spectrum antibacterial activity. *Carbohydrate Polymers, 312*, 1-11. https://doi.org/10.1016/j.carbpol.2023.120792.

Kim, J. H., Lee, J., Park, J., & Gho, Y. S. (2015). Gram-negative and gram-positive bacterial extracellular vesicles. *Seminars in Cell & Developmental Biology, 40*, 97-104. https://doi.org/10.1016/j.semcdb.2015.02.006.

Lee, J. C., Seksama, L. H., Park, C. H., & Kim, C. S. (2023). Enhancing the anti-bacterial activity of nanofibrous polyurethane membranes by incorporating glycyrrhizic acid-conjugated β-Cyclodextrin. *Materials Letters, 338*, 1-4. https://doi.org/10.1016/j.matlet.2023.134030.

Li, C., Bai, M., Chen, X., Hu, W., Cui, H., & Lin, L. (2022). Controlled release and antibacterial activity of nanofibers loaded with basil essential oil-encapsulated cationic liposomes against Listeria monocytogenes. *Food Bioscience, 46*, 1-9. https://doi.org/10.1016/j.fbio.2022.101578.

Li, J., & Zhuang, S. (2020). Antibacterial activity of chitosan and its derivatives and their interaction mechanism with bacteria: Current state and perspectives. *European Polymer Journal, 138*, 1-12. https://doi.org/10.1016/j.eurpolymj.2020.109984.

Li, X., He, X., Ling, Y., Bai, Z., Liu, C., Liu, X., & Jia, K. (2023). In-situ growth of silver nanoparticles on sulfonated polyarylene ether nitrile nanofibers as super-wetting antibacterial oil/water separation membranes. *Journal of Membrane Science, 675*, 1-11. https://doi.org/10.1016/j.memsci.2023.121539.

Li, Y., Zhu, J., Cheng, H., Li, G., Cho, H., Jiang, M., Gao, Q., & Zhang X. (2021). Developments of advanced electrospinning techniques: A critical review. *Advanced Materials Technologies*, *6*(11), 1-29. https://doi.org/10.1002/admt.202100410.

Nazari, A., Raeesi, M., Salehi-Mobarakeh, H., & Mahdavian, A. R. (2022). Ready-to-use optical H_2O_2 sensor based on stimuli-responsive polyacrylic film and nanofibers containing spiropyran. *Dyes and Pigments, 204*, 1-10. https://doi.org/10.1016/j.dyepig.2022.110399.

Oskouei, B., Dehghan, G., Mahdavi, M., & Feizi, M. A. H. (2023). Preparation and characterization of novel antimicrobial Cu-Al layered double hydroxide filled polyvinyl alcohol/chitosan nanofibers with peroxidase-mimic activity. *Applied Clay Science, 237*, 1-9. https://doi.org/10.1016/j.clay.2023.106914.

Pinho, E., Magalhães, L., Henriques, M., & Oliveira, R. (2011). Antimicrobial activity assessment of textiles: Standard methods comparison. *Annals of Microbiology, 61*, 493-498. https://doi.org/10.1007/s13213-010-0163-8.

Purnamasari, W., Budiastanti, T. A., Aminatun, A., Rahmah, U., Sumarsih, S., Chang, J. Y., & Fahmi, M. Z. (2022). Naproxen release behaviour from graphene oxide/cellulose acetate composite nanofibers. *RSC Advances, 12*(13), 8019-8029. https://doi.org/10.1039/d1ra09293f.

Ramakrishna, S., Fujihara, K., Teo, W., Lim, T., & Ma, Z. (2005). *An introduction to electrospinning and nanofibers*. World Scientific Publishing Co. Pte. Ltd. https://doi.org/10.1142/5894.

Şafak, Ş., Vatan, Ö., Cinkiliç. N., & Karaca E. (2020). In vitro evaluation of electrospun polysaccharide based nanofibrous mats as surgical adhesion barriers. *Textile and Apparel, 30*(2), 99-107. https://doi.org/10.32710/tekstilvekonfeksiyon.548460.

Shadman-Manesh, V., Gholipour-Kanani, A., Najmoddin, N., & Rabbani, S. (2023). Preclinical evaluation of the polycaprolactone-polyethylene glycol electrospun nanofibers containing egg-yolk oil for acceleration of full thickness burns healing. *Scientific Reports, 13*(1), 1-12. https://doi.org/10.1038/s41598-023-28065-6.

Shahidi, S. & Wiener, J. (2012). Antibacterial agents in textile industry. In V. Bobbarala (Ed.), *Antimicrobial agents* (pp. 387-406). InTech Open.

Tahir, R., Albargi, H. B., Ahmad, A., Qadir, M. B., Khaliq, Z., Nazir, A., Khalid, T., Batool, M., Arshad, S. N., Jalalah, M., Alsareii, S. A., & Harraz, F. A. (2023). Development of sustainable hydrophilic azadirachta indica loaded PVA nanomembranes for cosmetic facemask applications. *Membranes, 13*(2), 1-18. https://doi.org/10.3390/membranes13020156.

Tajfiroozeh, F., Moradi, A., Shahidi, F., Movaffagh, J., Kamali, H., Roshanak, S., & Shahroodi A. (2023). Fabrication and characterization of gallic-acid/nisin loaded electrospun core/shell chitosan/polyethylene oxide nanofiberous membranes with free radical scavenging capacity and antimicrobial activity for food packing applications. *Food Bioscience, 53*, 1-9. https://doi.org/10.1016/j.fbio.2023.102529.

Wang, Q., Liu, S., Lu, W., & Zhang, P. (2022). Fabrication of curcumin@ Ag loaded core/shell nanofiber membrane and its synergistic antibacterial properties. *Frontiers in Chemistry, 10*, 1-10. https://doi.org/10.3389/fchem.2022.870666.

Wei, Z., Wang, L., Zhang, S., Chen, T., Yang, J., Long, S., & Wang, X. (2020). Electrospun antibacterial nanofibers for wound dressings and tissue medicinal fields: A review. *Journal of Innovative Optical Health Sciences, 13*(05), 1-10. https://doi.org/10.1142/S1793545820300128.

Xia, S., Fang, D., Shi, C., Wang, J., Lyu, L., Wu, W., Lu, T., Song, Y., Guo, Y., Huang, C., & Li, W. (2023). Preparation of a thermosensitive nanofibre membrane for blackberry preservation. *Food Chemistry, 415*, 1-11. https://doi.org/10.1016/j.foodchem.2023.135752.

Yavari Maroufi, L., Norouzi, R., Ramezani, S., & Ghorbani, M. (2023). Novel electrospun nanofibers based on gelatin/oxidized xanthan gum containing propolis reinforced by Schiff base cross-linking for food packaging. *Food Chemistry, 416*, 1-10. https://doi.org/10.1016/j.foodchem.2023.135806.

Yin, J., Xu, L., & Ahmed, A. (2022). Batch preparation and characterization of electrospun porous polylactic acid-based nanofiber membranes for antibacterial wound dressing. *Advanced Fiber Materials, 4*(4), 832-844. https://doi.org/10.1007/s42765-022-00141-y.

Zhang, H., Wang, J., Teng, Y., Jia, S., Huang, H., Li, Y., & Wang, C. (2023). Ce-MOF composite electrospinning as antibacterial adsorbent for the removal of 2, 4-dichlorophenoxyacetic acid. *Chemical Engineering Journal, 462*, 1-12. https://doi.org/10.1016/j.cej.2023.142195.

Chapter 18

Hemostatic Nanofibers: A Comprehensive Review about Design and Production Methods

Mahmed Sari Njjar[1,*], MD
and Çiğdem Akduman[2], PhD
[1]Pamukkale University, Faculty of Technology, Department of Biomedical Engineering, Denizli, Türkey
[2]Pamukkale University, Denizli Vocational School of Technical Sciences,
Department of Textile Technology, Denizli, Türkey

Abstract

Nanofibers produced through the electrospinning process have been extensively studied in the biomedical field due to their high porosity, large surface area-to-volume ratio, and potential for functionalization. Nanofibers properties enable them to mimic the natural extracellular matrix, creating an optimal physiological microenvironment for hemostasis by enhancing blood coagulation and concentration. Various approaches have been explored to develop advanced nanofiber composites for effective hemostatic dressing, including the use of natural and synthetic materials, the modification of the fiber orientation, and the incorporation of hemostatic agents. This review provides a summary of recent research progress in the design of electrospun hemostatic nanofibers using a wide range of polymers for biomedical applications. The review also discusses the advantages of different structural forms of electrospun nanofibers, as well as the challenges and future directions for developing customized nanofibers for hemostasis.

Keywords: electrospinning, hemostasis, modification, nanofibers, structure

Introduction

Uncontrolled bleeding and severe complications resulting from traumatic injuries are major causes of mortality among both military troops and civilians. Traumatic injuries are responsible for more than half of all military casualties and 30% of all global deaths. Hemorrhages from complex surgeries and severe injuries frequently lead to infection, acidosis, coagulopathy,

* Corresponding Author's Email: sari.njjar@hotmail.com.

In: Textile Insights and Future Aspects
Editors: Gamze D. Tetik, Gizem Celep and Fulya Yilmaz
ISBN: 979-8-89113-289-4

multiorgan failure, extension in operation time, morbidity, and healthcare expenses (Qian et al., 2017). Effective management of life-threatening bleeding can save crucial time for subsequent treatments and reduce deaths by over 90% (Wang et al., 2021). Hemostasis is the process of preventing blood loss from a tissue or organ of the body, which occurs when platelets and blood clotting factors work together to create a blood clot and halt bleeding (Atalan, 2014). Surgeons and anesthesiologists frequently assist patients in maintaining normal hemostasis in order to reduce blood loss after trauma or severe surgery using hemostatic agents.

Tourniquets are traditionally considered the last resort for severe extremity hemorrhage. While tourniquets are undoubtedly useful in controlling distal extremity hemorrhage in some extreme situations, in rare cases, such as vein injury where a tourniquet cannot be used to control bleeding, it is appealing to use a hemostat agent to stop the bleeding arteries. To help with irregular and complex wound bleeding, commercial hemostatic products have been developed (Hu et al., 2018; Yu & Zhong, 2021). However, some of the current hemostatic agents have significant limitations, including ineffectiveness in controlling severe hemorrhage, rapid thrombogenic effects, and thermal damage.

Nanofibers have become a popular choice for achieving fast hemostasis due to their distinctive characteristics. These include their high aspect ratio, large surface area, fine flexibility, and structural similarity to natural fibrin fibers, due to their very thin fiber diameters that make them effective in capturing blood cells, platelets, and other coagulation factors. By adhering to the bleeding area, nanofibers can promote platelet adhesion and activation, accelerate local blood agglutination, and induce blood clot formation (Zhao et al., 2016). Nanofibers are generally described as fibers with a diameter of less than one micron (Süpüren et al., 2007). Nanofibers have become increasingly important research topics due to the nanostructural properties of materials with specific applications. Due to their special properties, nanofibers can be used in several biomedical applications such as tissue engineering (Han et al., 2019; Owida et al., 2022), controlled drug release (Zelkó et al., 2019), medical prostheses (Palumbo et al., 2019), hemostatic products (Lu et al., 2022), as well as other applications such as filtration (Akduman & Akçakoca Kumbasar, 2018; Akduman, 2021), composites (Zhuang et al., 2019), electrical and optical, protective clothing, cosmetics, and agricultural applications (Gorji et al., 2017).

Electrospinning is a versatile method that produces nanofibers with diameters varying from nanometers to micrometers (Zhang et al., 2019). Electrospun nanofibers have been shown to increase platelet adhesion and activation, resulting in clot formation (Nakielski et al., 2019). This review provides an overview of the latest research advances in both the structural and functional design of hemostatic nanofibers, which are produced through the electrospinning of different biomedical polymers. This article examines the advantages of using various techniques to manufacture electrospun nanofibers, as well as the primary obstacles and prospects for developing customized electrospun nanofibers that can serve as effective hemostatic agents.

Bleeding Stopping Agents (Hemostatics)

Various hemostatic agents have been developed, including powders, films, hydrogels, sponges, patches, woven and nonwoven materials, as well as knitted covers (Baggio et al., 2019; Biranje et al., 2020; Chakoli et al., 2018; Chou et al., 2017; Edwards et al., 2014; Hsu et al., 2015a; Murakami et al., 2008; Rajangam et al., 2011; Quan et al., 2015). However, only a few of these materials are easily portable and efficient when used at the site of injury (Hsu et al., 2015b). Ideal hemostatic agents should rapidly coagulate blood, be non-toxic, and biodegradable, pose minimal risk of bacterial or viral contamination, and help to enhance wound healing. Moreover, effective *in vitro* hemostats should stop bleeding within a few minutes of application to the wound area, require no mixing or additional preparation, and be easy to apply to injured areas (Biranje et al., 2021). Figure 1 shows the classification of some of the hemostats based on their sources.

Hemostatic agents are applied to control bleeding through topical or systemic administration methods. Topical agents are available as a granular powder or impregnated dressings (Sitzler, 2017). To induce hemostasis, these agents employ two strategies. In the first mechanism, the agent adheres to damaged tissue to prevent additional blood loss, thus promoting blood clotting and closing ruptured arteries. The second mechanism involves chemical reactions that activate clotting elements and platelets in injured tissue to promote clot formation. The effectiveness of topical agents depends on the clotting ability of the patient (Payne et al., 2014). Systemic agents, administered intravenously, improve primary hemostasis by encouraging fibrin formation or inhibiting fibrinolysis. They are often given in situations involving significant blood loss, such as major surgical trauma, or in individuals with hemostatic abnormalities (Fraser et al., 2008).

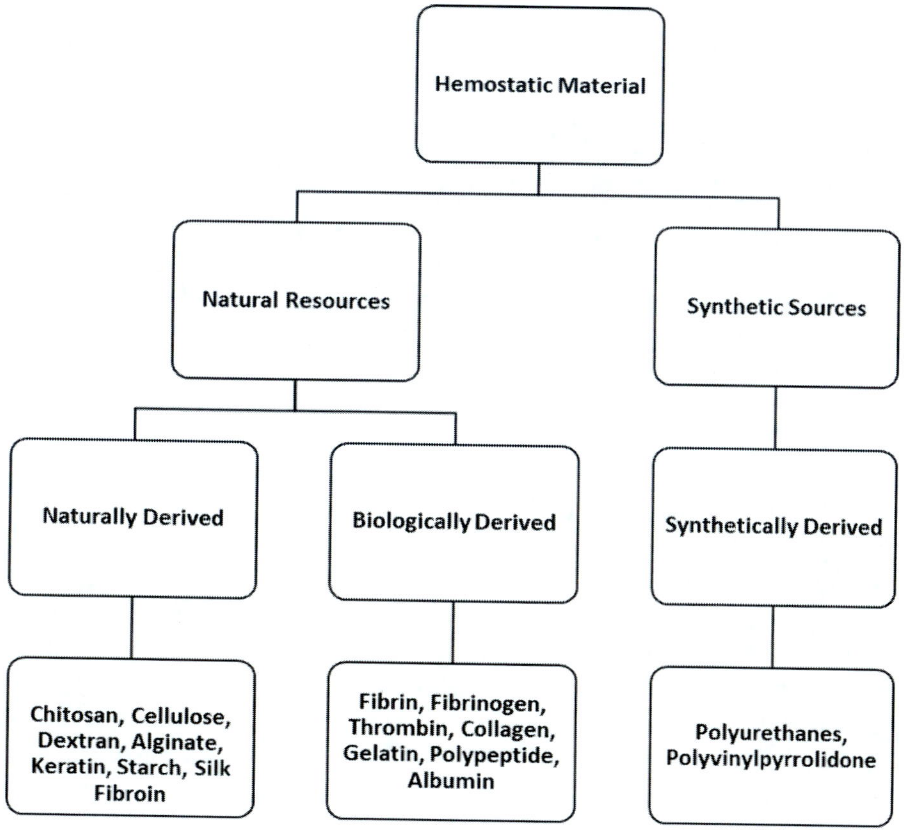

Figure 1. Classification of some of the hemostats based on natural and synthetic sources.

Commercial Hemostatic Agents

Chitosan is a copolymer composed of N-acetyl-D-glucosamine and D-glucosamine, which are linked by b-1,4-glycosidic bonds formed by alkaline chitin deacetylation (Younes & Rinaudo, 2015). This biopolymer possesses special properties such as biocompatibility and biodegradability, making it a non-toxic substance. Due to its various medicinal applications, including antibacterial activity (Rasaee et al., 2016), antioxidant (Guo et al., 2015), hemocompatibility (Ong et al., 2008), and anticancer capabilities (Ravi et al., 2018), chitosan has a significant interest in the medical field. Chitosan's ability to promote hemostatic action by attracting negatively charged erythrocytes, red blood cells, plasma, activation and aggregation of platelets and blood clotting (Howe & Cherpelis, 2013) makes it ideal for use in endoscopic sinus surgery, parenchymal hemorrhage, and urinary extravasation (Zhou et al., 2017). Two essential characteristics of chitosan that affect its hemostatic capabilities are the degree of deacetylation and the molecular weight (Sundaram et al., 2016). Commercially, chitosan is marketed in the United States under the brand names CELOXTM (Medtrade, Newport, OR) and HemCon® (HemCon Medical Technologies, Inc., Portland, OR).

Cellulose is a polymer composed of poly-b (1, 4)-D-glucose units arranged in a linear structure that is biodegradable and renewable. It is widely available in nature and can be derived from various sources, including plants, as well as certain bacteria (McCarthy et al., 2019). Cellulose is non-toxic and biocompatible, and has excellent absorption and swelling properties, which make it suitable for use in wound dressings, tissue engineering, drug delivery, and hemostats. Cellulose-based products are available in various forms, including sponges, gauzes, hydrogels, powders, and films (Wang et al., 2019). Oxidized cellulose is a widely used hemostatic agent marketed under brand names like Surgicel and Nu-Knit (Ethicon, Inc., Cincinnati, OH).

Ankaferd Blood Stopper (ABS) is an herbal extract that has traditionally been used in Türkiye and Bosnia and Herzegovina to treat bleeding. ABS is available in different forms, including tampons, sprays, and ampoules. The extract contains a standardized blend of herbs, such as *Thymus vulgaris, Glycyrrhiza glabra, Vitis vinifera, Alpinia officinarum, and Urtica dioica.* ABS works by forming an encapsulated protein network that facilitates rapid erythrocyte aggregation, known as the hemostatic ABS network (Goker et al., 2008). Additionally, ABS has antibacterial, antiseptic, and antimicrobial properties. *In vitro* studies have shown that ABS is effective against both Gram-positive and Gram-negative bacteria (Akkoc et al., 2009), foodborne pathogens (Koluman et al., 2017), resistant nosocomial pathogens (Saribas et al., 2010) and resistant strains of tuberculosis. Antimicrobial effects of Ankaferd can be attributed to its chemical ingredients.

Synthetic polymers are increasingly being used as hemostatic agents in wound healing due to their advantages over natural polymers. Polyurethanes (PU) are among the most widely used polymers in the biomedical field because of their segmental polymer structure and mechanical properties. PU can be used as hemostatic agents, vascular tissue graft adhesives, bone fixation materials, and scaffolds (Zumbardo-Bacelis et al., 2021). Synthetic PUs have been utilized as a hemostatic material in foam, film, and hydrocolloid forms because they can remain functional in the absence of hemolysis.

Poly(acrylic acid) (PAA) is a synthetic polymer that is water-soluble and possesses bioadhesive properties. It has been utilized in the development of various medical products, such as drug delivery systems and hemostatic agents (Ito et al., 2022). Feracryl®, which contains up to 2.5% iron (III) salts, was created in the early 1980s. The material exhibited robust hemostatic properties due to the formation of "artificial" clots composed of Feracryl®/plasma protein adducts. The hemostatic activity and polymer solubility were affected by the iron content. *In vivo* tests have shown that Feracryl® is non-toxic when used topically and internally during surgery (Annenkova et al., 1997).

Electrospinning Technology

Electrospinning is a manufacturing method that spins nanosized fibers from a polymer solution using a high voltage controlled by a unique electrohydrodynamic phenomenon. Figure 2 provides a simple setup of the electrospinning process. The electrospinning setup includes a high voltage power supply, an injection needle, a syringe pump, and a grounded collection area.

To extract fibers using electrospinning, an electric field is created by applying a high voltage between the nozzle and the collection area. As the pump maintains a steady flow rate, charges start to accumulate on the surface of the liquid. When the electrostatic force exceeds the surface tension, the liquid forms a conical shape called a Taylor cone (Nanoscience Instruments, 2022). Nanofibers with different structures such as beaded, smooth, porous, oriented, spiderweb, and tree-like fibers can be created by using electrospinning. The structure of the fibers is influenced by three main factors: polymer solution characteristics, processing conditions, and environmental parameters (Zhang, 2020). Table 1 lists the parameters that affect electrospinning.

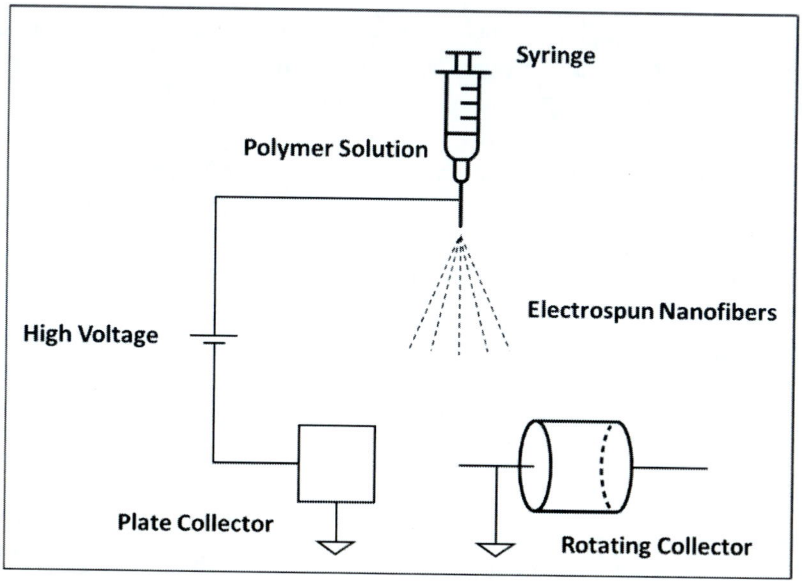

Figure 2. Representation of the setup of the electrospinning process.

Table 1. Electrospinning parameters

Solution Parameters	Processes Parameters	Environmental Parameters
Concentration	Applied voltage	Temperature
Viscosity	Flow rate	Humidity
Molecular weight	Distance between the nozzle and the collection area	Atmosphere
Surface tension	Collector types	
Solvent selection		

To prepare an electrospinning solution, a high molecular weight polymer and a suitable solvent are required to dissolve it. The molecular weight of the polymer affects the solution's rheological and electrical properties, with lower molecular weights resulting in the formation of beads rather than fibers. The solubility of the solvent influences the homogeneity of the solution, but a high solubility does not surely produce an appropriate electrospinning solution (Luo et al., 2012). The volatility of the solvent determines the evaporation speed and solidification rate of the jet, with both high and low volatility resulting in unsuitable fibers. The dielectric constant of the solvent also affects the electrostatic force between surface charges on the jet, with higher dielectric constants requiring higher voltages to maintain a steady jet (Luo et al., 2012).

The concentration and electrical conductivity of the polymer solution have also an influence. The minimum polymer concentration and polymer chain entanglement, which are responsible for the high strengths and toughness of polymeric materials, are required for the transition of electrospray to electrospinning (Husain et al., 2016). A high polymer concentration can make the electrospinning solution difficult to pump because of the high viscoelastic force, which prevents the jet from forming. Lowering viscosity and surface tension can enhance fiber formation, but a surfactant can also achieve this without lowering the polymer concentration (Hu et al., 2011).

The applied voltage, feeding rate, and the needle tip to collector distance as processing parameters have a significant impact on the fiber diameter. Higher voltages typically result in finer fibers but can also lead to bigger diameters (Demir et al., 2002). A considerable needle tip-to-collector distance is necessary to complete the elongation and solidification of the jet (Xue et al., 2019). Humidity and temperature are related to the speed of solvent evaporation and can produce structural changes in the nanofibers. Lower humidity encourages finer fibers, while higher temperatures promote their formation but accelerate solvent evaporation, limiting jet elongation (Wang et al., 2011). A balance of solution, process, and environmental parameters is required to produce smooth, uniform, and reproducible nanofibers.

The Direct Approach to Electrospinning Hemostatic Nanofibers

Hemostatic nanofibers are a promising research field in hemostasis. The high porosity, significantly high aspect ratio, variety in morphology, and flexibility in the polymer composition of hemostatic nanofibers make them a better choice for commercial hemostatic products. By controlling the solution contents and collectors, the composition and morphology of the nanofibers can be adjusted to suit specific needs (Deineka et al., 2021). Additionally, hemostatic drugs or nanoparticles can be incorporated into the nanofibers to enhance their hemostatic effect. Physical and chemical approaches can also be used to add active hemostatic components to fibers (Chen et al., 2021; Varshosaz et al., 2021).

One advantage of hemostatic nanofibers is that they can be collected directly on the bleeding site *in situ,* making them suitable for intracavitary hemostasis (Gao et al., 2020). Various natural and synthetic polymers could be used to produce hemostatic nanofibers. These include chitosan (Padalhin & Lee, 2019), collagen (Zhao et al., 2018), gelatin (Norouzi et al., 2015), and keratin (Ye et al., 2022), and synthetic polymers including polycaprolactone (PCL) (Hu & Lin, 2022), polyacrylonitrile (PAN) (Wan & Xu, 2009), polyvinyl pyrrolidone (PVP) (Li et al., 2017), polylactide (Bi et al., 2020), and polyurethane (PU) (Tan et al., 2015).

For example, Gu et al., (2013) developed a pure chitosan nanofiber membrane as a hemostatic agent. 97.2% of high porosity was measured for the developed membrane, allowing rapid blood absorption. In hemostasis, blood clotting efficiency was evaluated, and found to be 1.35 times that of commercial Surgicel®. Natural polymer-based hemostatic products can rapidly stop bleeding; however, they still face significant challenges such as high cost, limited spinnability, and potential allergenicity. Therefore, there has been an increase in the use of synthetic polymers for the creation of electrospun hemostatic membranes. Wan and Xu (2009) were the first to investigate the behaviors of blood platelets on the surface of a synthetic electrospun nanofiber.

Electrospinning polymer blends is a straightforward and effective technique for producing composite nanofibers that can enhance properties such as antibacterial efficacy, angiogenesis, and tissue regeneration, particularly in materials with poor spinnability. For example, the molecular structure chitosan and the sensitivity of the electrospinning to humidity make it challenging to achieve uniform nanofibers. To address this issue, Kim and Lee (2011) developed a gelatin/chitosan nanofiber that demonstrated superior wettability and blood clotting effectiveness compared to pure chitosan nanofibers. Additionally, multilayered electrospinning is a simple approach that can mix various polymers to create nanofibers with diverse properties (Hu et al., 2019). Another study conducted by Gu et al. (2016) also supports the effectiveness of the gelatin-blended-chitosan nanofibrous membrane for improved properties.

Multilayer nanofiber mats are known for their exceptional mechanical properties and high-volume expansion, making them more effective than single-layer structures in promoting blood clot formation (Liu et al., 2017). These nanofibers are often used to create flexible fiber sponges with high liquid absorption ratios. Compared to single-layer structures, these sponges can absorb blood more quickly, increasing the concentration of coagulation factors at the wound site, which ultimately activates the coagulation cascade reaction pathway.

Several studies in the literature explore the use of multilayered nanofibers for hemostasis and wound healing. Lee et al. (2017) conducted a study on PLGA/collagen nanofibers that release lidocaine and epinephrine in multiple layers, evaluating their effectiveness in providing analgesia and hemostasis in palatal ulcers of rabbits. The nanofibers were found to have good biocompatibility in the early stages of wound healing and released appropriate levels of lidocaine and epinephrine. Compared to the control group, the rabbits in the test group experienced faster hemostasis after surgery and faster recovery of their food and water intake. Based on the study's findings, it can be concluded that multilayer biodegradable nanofiber membranes can effectively promote hemostasis and alleviate pain during the initial healing of palatal ulcers.

Fatahian et al. (2020) conducted a study to investigate the coagulation properties of PVA nanofibers that contained transamine. They also loaded ceftriaxone into PVA nanofibers to evaluate their antibacterial properties and to compare the effects of drug concentration on antibacterial activity. Hemoglobin absorption was measured at 540 nm, and according to the results, PVA/transamine (20 mg/ml) showed lower absorption when compared to PVA/transamine (10 mg/ml). They found that PVA/transamine formulations showed higher blood coagulation ability compared to pure PVA as a control. Clots formed at the high concentration of transamine were greater than those at a low concentration of transamine (10 mg/ml), while no clots were observed in PVA nanofibers without transamines. The study demonstrated that PVA nanofibers containing transamine have the potential as hemostatic materials and that the addition of ceftriaxone can provide antibacterial properties.

In 2008, Ong et al., studied a chitosan dressing with procoagulant (polyphosphate) and antimicrobial (silver) properties. They produced different chitosan-polyphosphate formulations and evaluated their hemostatic activities *in vitro*. Their results showed that the polyphosphate ratio in chitosan was of great importance in the hemostatic characteristics. Chitosan containing 6.7% w/w and 10% w/w Polyphosphate accelerated whole blood coagulation compared to chitosan, while at higher concentrations (15% w/w) the effect regressed. Furthermore, *in vitro* studies showed that the silver-loaded chitosan-polyphosphate formulation exhibited significantly great bactericidal activity.

Sasmal and Datta (2019) developed transamine-loaded hemostatic and antibacterial PVA/chitosan nanofiber membranes for use as hemostatic patches. They evaluated the physicochemical properties and efficacy of the system, including blood coagulation, platelet adhesion, and drug release. Their results showed that the chitosan-based system with transamine (40 mg/ml) had a shorter clotting time compared to the control sample, indicating the efficacy of chitosan in hemostasis and the complementary role of transamine.

Li et al. (2019) studied carbon nanofibers, and their superhydrophobic properties as a hemostatic material. They observed that the superhydrophobicity of the nanofiber surface limits wetting with blood, preventing blood loss and bacterial attachment. When blood droplets meet the surface of the material, fibrin formation occurs, which contributes to the hemostatic effect.

Ehterami et al. (2021) produced polycaprolactone/gelatin/epinephrine scaffolds to promote bone healing. They evaluated the interaction between the developed scaffolds and blood and cells using hemolysis and MTT (3-[4,5-dimethylthiazol-2-yl]-2,5 diphenyl tetrazolium bromide) assay. Their results showed a positive contribution to cell proliferation and bone healing, but the mechanical strength of the scaffolds decreased with the addition of epinephrine.

Varshosaz et al. (2021) produced porous nanofibers with transamine-loaded polyglycerol sebacate and polyhydroxyethyl methacrylate. Their best formulation contained 35% polymer concentration, 20% polyglycerol sebacate, and 10% transamine over the total amount of polymer. It has been observed that transamine-loaded nanofibers can reduce bleeding time and volume and were effective in bleeding control. nanofibers also showed good biocompatibility with no cytotoxicity, a high porosity of 64%, low water absorption of 8.59%, and weight loss of 1.47% in L929 cells after 28 days.

Surface Modifications of Hemostatic Nanofibers

In order to improve the hemostatic properties of nanofibers, various physical and chemical approaches, and opportunities have been utilized for surface modification. One popular method involves physically coating the surface of the nanofibers with active components to improve their hemostatic efficacy through direct contact with blood or rapid release. For example, Cheng et al. (2018) immobilized thrombin onto poly(butylene succinate) electrospun nanofiber membranes using a multilayer coating method. This resulted in increased early thrombin activity in a rat liver injury model compared to commercial hemostatic gelatin sponges, leading to less blood loss and a shorter hemostatic period.

Another surface modification approach involves dip coating of prepolymer solution followed by cross-linking. Zhang et al. (2018) created a fiber/sponge composite by immersing nanofibrous membranes made of gelatin and poly(lactic-co-glycolic acid) containing cisplatin in a solution of chitosan and gelatin, the produced membrane chemically *in situ* cross-linked by genipin. The resulting sandwich-like porous fiber/sponge composites have demonstrated improved hemostatic properties.

Spraying is another physical coating approach, in which coagulation-activating chemicals are sprayed onto the surface of nanofibers. Park et al. (2017) created $CaCO_3$-containing PCL nanofibers and modified their hydrophilicity by ultrasonically spraying chitosan onto their surface. This modification allowed blood to readily penetrate the nanofibers, increasing their contact area with blood and enhancing their hemostatic properties.

Conclusion

Nanofibers have shown great promise in the field of hemostasis due to their structural similarity to fibrin clots. Extensive research has been conducted over the past two decades to develop various manufacturing processes for the creation of nanofibrous materials that can be used for both external and intracavitary hemostatic control. However, despite the numerous benefits of nanofibers in the treatment of bleeding, there are still significant limitations in research and practical implementation. In future research, particular attention should be paid to three important aspects: mechanical properties, production efficiency, and safety. Moreover, the development of functionalized nanofibers that possess diagnostic and therapeutic capabilities in addition to hemostatic efficacy should be a primary focus. With these advancements, nanofibrous hemostatic materials can be expected to have a broader range of useful and convenient applications in hemorrhage control and wound healing.

Acknowledgments

The authors would like to thank co-supervisor DVM-PhD Ahmet Koluman for the guidance and support given during the MSc project of Mahmed Sari Njjar.

Disclaimer

None.

References

Akduman, C. (2021). Cellulose acetate and polyvinylidene fluoride nanofiber mats for N95 respirators. *Journal of Industrial Textiles*, *50*(8), 1239-1261. https://doi.org/10.1177/1528083719858760.

Akduman, C., & Akcakoca Kumbasar, E. P. (2018). Nanofibers in face masks and respirators to provide better protection. *IOP Conference Series: Materials Science and Engineering, 460*, 012013. https://doi.org/10.1088/1757-899X/460/1/012013.

Akkoc, N., Akcelik, M., Haznedaroglu, I. C., Goker, H., Turgut, M., Aksu, S., Kirazli, S., & Firat, H. C. (2009). *In vitro* anti-bacterial activities of Ankaferd medicinal plant extract. *Turkiye Clinics Journal of Medical Sciences*, *29*(2), 410-415.

Annenkova, V. Z., Annenkova, V. M., Kazimirovskaia, V. B., Kononchuk, G. M., Kuznetsova, E. E., & Voronkov, M. G. (1997). Experimental toxicological research on the preparation ferakril. *Eksperimental'naia i Klinicheskaia Farmakologiia*, *60*(4), 71-73.

Atalan, N. (2014). Hemostatic Drugs. *Journal of the Society of Thoracic-Cardiovascular Anesthesia and Intensive Care*, *20*(1), 1-6. https://doi.org/10.5222/GKDAD.2014.001.

Baggio, S., Laganà, A. S., Garzon, S., Scollo, M., Raffaelli, R., Tateo, S., Ghezzi, F., & Franchi, M. (2019). Efficacy of a collagen-fibrin sealant patch (TachoSil®) as adjuvant treatment in the inguinofemoral lymphadenectomy for vulvar cancer: A double-blind randomized-controlled trial. *Archives of Gynecology and Obstetrics, 299*, 1467-1474. https://doi.org/10.1007/s00404-019-05113-z.

Bi, H., Feng, T., Li, B., & Han, Y. (2020). *In vitro* and *in vivo* comparison study of electrospun PLA and PLA/PVA/SA fiber membranes for wound healing. *Polymers*, *12*(4), 839. https://doi.org/10.3390/polym12040839.

Biranje, S. S., Madiwale, P. V., Patankar, K. C., Chhabra, R., Bangde, P., Dandekar, P., & Adivarekar, R. V. (2020). Cytotoxicity and hemostatic activity of chitosan/carrageenan composite wound healing dressing for traumatic hemorrhage. *Carbohydrate Polymers*, *239*, 116106. https://doi.org/10.1016/j.carbpol.2020.116106.

Biranje, S. S., Sun, J., Shi, Y., Yu, S., Jiao, H., Zhang, M., Wang, Q., Wang, J., & Liu, J. (2021). Polysaccharide-based hemostats: Recent developments, challenges, and future perspectives. *Cellulose*, *28*, 8899-8937. https://doi.org/10.1007/s10570-021-04132-x.

Chakoli, A. N., Ahmadi, S., He, J. M., & Huang, Y. D. (2018). Aminated multiwalled carbon nanotubes/oxidized regenerated cellulose nanocomposites for haemostatic applications. *Materials Today: Proceedings*, *5*(7), 15661-15666. https://doi.org/10.1016/j.matpr.2018.04.176.

Chen, K., Pan, H., Yan, Z., Li, Y., Ji, D., Yun, K., Su, Y., Liu, D., & Pan, W. (2021). A novel alginate/gelatin sponge combined with curcumin-loaded electrospun fibers for postoperative rapid hemostasis and prevention of tumor recurrence. *International Journal of Biological Macromolecules*, *182*, 1339-1350. https://doi.org/10.1016/j.ijbiomac.2021.05.074.

Cheng, H. H., Xiong, J., Xie, Z. N., Zhu, Y. T., Liu, Y. M., Wu, Z. Y., Yu, J., & Guo, Z. X. (2018). Thrombin-loaded poly (butylene succinate)-based electrospun membranes for rapid hemostatic application. *Macromolecular Materials and Engineering*, *303*(2), 1700395. https://doi.org/10.1002/mame.201700395.

Chou, D., Tulloch, A., Cossman, D. V., Cohen, J. L., Rao, R., Barmparas, G., Mirocha, J., & Wagner, W. (2017). The influence of collagen impregnation of a knitted dacron patch used in carotid endarterectomy. *Annals of Vascular Surgery*, *39*, 209-215. https://doi.org/10.1016/j.avsg.2016.08.011.

Deineka, V., Sulaieva, O., Pernakov, M., Korniienko, V., Husak, Y., Yanovska, A., Yusupova, A., Tkachenko, Y., Kalinkevich, O., Zlatska, A., & Pogorielov, M. (2021). Hemostatic and tissue regeneration performance of novel electrospun chitosan-based materials. *Biomedicines*, *9*(6), 588. https://doi.org/10.3390/biomedicines9060588.

Demir, M. M., Yilgor, I., Yilgor, E. E. A., & Erman, B. (2002). Electrospinning of polyurethane fibers. *Polymer*, *43*(11), 3303-3309. https://doi.org/10.1016/S0032-3861(02)00136-2.

Edwards, J. V., Graves, E., Bopp, A., Prevost, N., Santiago, M., & Condon, B. (2014). Electrokinetic and hemostatic profiles of nonwoven cellulosic/synthetic fiber blends with unbleached cotton. *Journal of Functional Biomaterials*, *5*(4), 273-287. https://doi.org/10.3390/jfb5040273.

Ehterami, A., Khastar, H., Soleimannejad, M., Salehi, M., Nazarnezhad, S., Majidi Ghatar, J., Bit, A., Jafarisani, M., Abbaszadeh-Goudarzi, G., & Shariatifar, N. (2021). Bone regeneration in rat using polycaprolactone/gelatin/epinephrine scaffold. *Drug Development and Industrial Pharmacy*, *47*(12), 1915-1923. https://doi.org/10.1080/03639045.2022.2070640.

Fatahian, R., Mirjalili, M., Khajavi, R., Rahimi, M. K., & Nasirizadeh, N. (2020). Fabrication of antibacterial and hemostatic electrospun PVA nanofibers for wound healing. *SN Applied Sciences*, *2*, 1-7. https://doi.org/10.1007/s42452-020-3084-6.

Fraser, I. S., Porte, R. J., Kouides, P. A., & Lukes, A. S. (2008). A benefit-risk review of systemic haemostatic agents: Part 1: In major surgery. *Drug Safety*, *31*, 217-230. https://doi.org/10.2165/00002018-200831030-00003.

Gao, Y., Xiang, H. F., Wang, X. X., Yan, K., Liu, Q., Li, X., Liu, R. Q., Yu, M., & Long, Y. Z. (2020). A portable solution blow spinning device for minimally invasive surgery hemostasis. *Chemical Engineering Journal*, *387*, 124052. https://doi.org/10.1016/j.cej.2020.124052.

Goker, H., Haznedaroglu, I. C., Ercetin, S., Kirazli, S., Akman, U., Ozturk, Y., & Firat, H. C. (2008). Haemostatic actions of the folkloric medicinal plant extract Ankaferd Blood Stopper®. *The Journal of International Medical Research*, *36*(1), 163–170. https://doi.org/10.1177/147323000803600121.

Gorji, M., Bagherzadeh, R., & Fashandi, H. (2017). Electrospun nanofibers in protective clothing. In M. Afshari (Ed.), *Electrospun nanofibers* (pp. 571-598). Woodhead Publishing.

Gu, B. K., Park, S. J., Kim, M. S., Kang, C. M., Kim, J. I., & Kim, C. H. (2013). Fabrication of sonicated chitosan nanofiber mat with enlarged porosity for use as hemostatic materials. *Carbohydrate Polymers*, *97*(1), 65-73. https://doi.org/10.1016/j.carbpol.2013.04.060.

Gu, B. K., Park, S. J., Kim, M. S., Lee, Y. J., Kim, J. I., & Kim, C. H. (2016). Gelatin blending and sonication of chitosan nanofiber mats produce synergistic effects on hemostatic functions. *International Journal of Biological Macromolecules*, *82*, 89-96. https://doi.org/10.1016/j.ijbiomac.2015.10.009.

Guo, M., Ma, Y., Wang, C., Liu, H., Li, Q., & Fei, M. (2015). Synthesis, anti-oxidant activity, and biodegradability of a novel recombinant polysaccharide derived from chitosan and lactose. *Carbohydrate Polymers*, *118*, 218-223. https://doi.org/10.1016/j.carbpol.2014.11.027.

Han, J., Xiong, L., Jiang, X., Yuan, X., Zhao, Y., & Yang, D. (2019). Bio-functional electrospun nanomaterials: From topology design to biological applications. *Progress in Polymer Science*, *91*, 1-28. https://doi.org/10.1016/j.progpolymsci.2019.02.006.

Howe, N., & Cherpelis, B. (2013). Obtaining rapid and effective hemostasis: Part I. Update and review of topical hemostatic agents. *Journal of the American Academy of Dermatology*, *69*(5), 659.e1-659.e17. https://doi.org/10.1016/j.jaad.2013.07.014.

Hsu, B. B., Conway, W., Tschabrunn, C. M., Mehta, M., Perez-Cuevas, M. B., Zhang, S., & Hammond, P. T. (2015a). Clotting mimicry from robust hemostatic bandages based on self-assembling peptides. *ACS Nano*, *9*(9), 9394-9406. https://doi.org/10.1021/acsnano.5b02374.

Hsu, B. B., Hagerman, S. R., Jamieson, K., Castleberry, S. A., Wang, W., Holler, E., Ljubimova, J. Y., & Hammond, P. T. (2015b). Multifunctional self-assembled films for rapid hemostat and sustained anti-infective delivery. *ACS Biomaterials Science & Engineering*, *1*(3), 148-156. https://doi.org/10.1021/ab500050m.

Hu, J., Wang, X., Ding, B., Lin, J., Yu, J., & Sun, G. (2011). One-step electro-spinning/netting technique for controllably preparing polyurethane nano-fiber/net. *Macromolecular Rapid Communications*, *32*(21), 1729-1734. https://doi.org/10.1002/marc.201100343.

Hu, W. W., & Lin, Y. T. (2022). Alginate/polycaprolactone composite fibers as multifunctional wound dressings. *Carbohydrate Polymers*, *289*, 119440. https://doi.org/10.1016/j.carbpol.2022.119440.

Hu, W., Wang, Z., Xu, Y., Wang, X., Xiao, Y., Zhang, S., & Wang, J. (2019). Remodeling of inherent antimicrobial nanofiber dressings with melamine-modified fibroin into neoskin. *Journal of Materials Chemistry B*, *7*(21), 3412-3423.

Hu, Z., Zhang, D. Y., Lu, S. T., Li, P. W., & Li, S. D. (2018). Chitosan-based composite materials for prospective hemostatic applications. *Marine Drugs, 16*(8), 273. https://doi.org/10.3390/md16080273.

Husain, O., Lau, W., Edirisinghe, M., & Parhizkar, M. (2016). Investigating the particle to fibre transition threshold during electrohydrodynamic atomization of a polymer solution. *Materials Science and Engineering: C, 65*, 240-250. https://doi.org/10.1016/j.msec.2016.03.076.

Ito, T., Yamaguchi, S., Soga, D., Yoshimoto, T., & Koyama, Y. (2022). Preparation of a bioadhesive Poly (Acrylic Acid)/Polyvinylpyrrolidone complex gel and its clinical effect on dental hemostasis. *Gels, 8*(8), 462. https://doi.org/10.3390/gels8080462.

Kim, M. Y., & Lee, J. (2011). Chitosan fibrous 3D networks prepared by freeze drying. *Carbohydrate Polymers, 84*(4), 1329-1336. https://doi.org/10.1016/j.carbpol.2011.01.029.

Koluman, A., Akar, N., & Haznedaroğlu, İ. C. (2017). Antibacterial activities of ankaferd hemostat (ABS) on shiga toxin-producing escherichia coli and other pathogens significant in foodborne diseases. *Turkish Journal of Hematology, 34*, 93-98. https://doi.org/10.4274/tjh.2015.0073.

Lee, F. Y., Lee, D., Lee, T. C., Chen, J. K., Wu, R. C., Liu, K. C., & Liu, S. J. (2017). Fabrication of multi-layered lidocaine and epinephrine-eluting PLGA/collagen nanofibers: *In vitro* and *in vivo* study. *Polymers, 9*(9), 416. https://doi.org/10.3390/polym9090416.

Li, D., Nie, W., Chen, L., Miao, Y., Zhang, X., Chen, F., Yu, B., Ao, R., Yu, B., & He, C. (2017). Fabrication of curcumin-loaded mesoporous silica incorporated polyvinyl pyrrolidone nanofibers for rapid hemostasis and antibacterial treatment. *RSC advances, 7*(13), 7973-7982. https://doi.org/10.1039/C6RA27319J.

Li, Z., Milionis, A., Zheng, Y., Yee, M., Codispoti, L., Tan, F., Poulikakos, D., & Yap, C. H. (2019). Superhydrophobic hemostatic nanofiber composites for fast clotting and minimal adhesion. *Nature Communications, 10*(1), 5562. https://doi.org/10.1038/s41467-019-13512-8.

Liu, M., Duan, X. P., Li, Y. M., Yang, D. P., & Long, Y. Z. (2017). Electrospun nanofibers for wound healing. *Materials Science and Engineering: C, 76*, 1413-1423. https://doi.org/10.1016/j.msec.2017.03.034.

Lu, X., Li, X., Yu, J., & Ding, B. (2022). Nanofibrous hemostatic materials: Structural design, fabrication methods, and hemostatic mechanisms. *Acta Biomaterialia, 154*, 49-62. https://doi.org/10.1016/j.actbio.2022.10.028.

Luo, C. J., Stride, E., & Edirisinghe, M. (2012). Mapping the influence of solubility and dielectric constant on electrospinning polycaprolactone solutions. *Macromolecules, 45*(11), 4669-4680. https://doi.org/10.1021/ma300656u.

McCarthy, R. R., Ullah, M. W., Booth, P., Pei, E., & Yang, G. (2019). The use of bacterial polysaccharides in bioprinting. *Biotechnology Advances, 37*(8), 107448. https://doi.org/10.1016/j.biotechadv.2019.107448.

Murakami, Y., Yokoyama, M., Nishida, H., Tomizawa, Y., & Kurosawa, H. (2008). A simple hemostasis model for the quantitative evaluation of hydrogel-based local hemostatic biomaterials on tissue surface. *Colloids and Surfaces B: Biointerfaces, 65*(2), 186-189. https://doi.org/10.1016/j.colsurfb.2008.04.005.

Nakielski, P., & Pierini, F. (2019). Blood interactions with nano-and microfibers: Recent advances, challenges and applications in nano-and microfibrous hemostatic agents. *Acta Biomaterialia, 84*, 63-76. https://doi.org/10.1016/j.actbio.2018.11.029.

Nanoscience Instruments. (2022, October 23). *Electrospinners & sprayers newly available*. https://www.nanoscience.com/press-releases/electrospinners-sprayers-newly-available/.

Norouzi, M., Shabani, I., Ahvaz, H. H., & Soleimani, M. (2015). PLGA/gelatin hybrid nanofibrous scaffolds encapsulating EGF for skin regeneration. *Journal of Biomedical Materials Research Part A, 103*(7), 2225-2235. https://doi.org/10.1002/jbm.a.35355.

Ong, S. Y., Wu, J., Moochhala, S. M., Tan, M. H., & Lu, J. (2008). Development of a chitosan-based wound dressing with improved hemostatic and antimicrobial properties. *Biomaterials, 29*(32), 4323-4332. https://doi.org/10.1016/j.biomaterials.2008.07.034.

Owida, H. A., Al-Nabulsi, J. I., Alnaimat, F., Al-Ayyad, M., Turab, N. M., Al Sharah, A., & Shakur, M. (2022). Recent applications of electrospun nanofibrous scaffold in tissue engineering. *Applied Bionics and Biomechanics*, 1-15. https://doi.org/10.1155/2022/1953861.

Padalhin, A. R., & Lee, B. T. (2019). Hemostasis and bone regeneration using chitosan/gelatin-BCP bi-layer composite material. *ASAIO Journal, 65*(6), 620-627. https://doi.org/10.1097/MAT.0000000000000850.

Palumbo, G., Cusanno, A., Romeu, M. G., Bagudanch, I., Negrini, N. C., Villa, T., & Farè, S. (2019). Single point incremental forming and electrospinning to produce biodegradable magnesium (AZ31) biomedical prostheses coated with porous PCL. *Materials Today: Proceedings*, *7*, 394-401. https://doi.org/10.1016/j.matpr.2018.11.101.

Park, J. Y., Kyung, K. H., Tsukada, K., Kim, S. H., & Shiratori, S. (2017). Biodegradable polycaprolactone nanofibres with β-chitosan and calcium carbonate produce a hemostatic effect. *Polymer*, *123*, 194-202. https://doi.org/10.1016/j.polymer.2017.07.013.

Payne, E. K., Berry, D. C., & Seitz, S. R. (2014). Educating the educator: Use of advanced bleeding control mechanisms in athletic training: a shift in the thought process of prehospital care - Part 2: Hemostatic agents. *Athletic Training Education Journal*, *9*(4), 193-201. https://doi.org/10.4085/0904193.

Qian, Z., Wang, H., Tuo, X., Guo, H., Xu, P., Liu, D., Wei, Y., Liu, H., Fan, Y., & Guo, X. (2017). A porous sodium polyacrylate-grafted chitosan xerogel for severe hemorrhage control synthesized from one-pot reaction. *Journal of Materials Chemistry B*, *5*(25), 4845-4851.

Quan, K., Li, G., Luan, D., Yuan, Q., Tao, L., & Wang, X. (2015). Black hemostatic sponge based on facile prepared cross-linked graphene. *Colloids and Surfaces B: Biointerfaces*, *132*, 27-33. https://doi.org/10.1016/j.colsurfb.2015.04.067.

Rajangam, T., Paik, H. J., & An, S. S. A. (2011). Development of fibrinogen microspheres as a biodegradable carrier for tissue engineering. *BioChip Journal*, *5*, 175-183. https://doi.org/10.1007/s13206-011-5211-3.

Rasaee, I., Ghannadnia, M., & Honari, H. (2016). Antibacterial properties of biologically formed chitosan nanoparticles using aqueous leaf extract of Ocimum basilicum. *Nanomedicine Journal*, *3*(4), 240-247.

Ravi, H., Kurrey, N., Manabe, Y., Sugawara, T., & Baskaran, V. (2018). Polymeric chitosan-glycolipid nanocarriers for an effective delivery of marine carotenoid fucoxanthin for induction of apoptosis in human colon cancer cells (Caco-2 cells). *Materials Science and Engineering: C*, *91*, 785-795. https://doi.org/10.1016/j.msec.2018.06.018.

Saribas, Z., Sener, B., Haznedaroglu, I., Hascelik, G., Kirazli, S., & Goker, H. (2010). Antimicrobial activity of Ankaferd Blood Stopper® against nosocomial bacterial pathogens. *Open Medicine*, *5*(2), 198-202. https://doi.org/10.2478/s11536-009-0140-4.

Sasmal, P., & Datta, P. (2019). Tranexamic acid-loaded chitosan electrospun nanofibers as drug delivery system for hemorrhage control applications. *Journal of Drug Delivery Science and Technology*, *52*, 559-567. https://doi.org/10.1016/j.jddst.2019.05.018.

Sitzler, B. (2017, January 15). *Using topical hemostatic agents*. The National Athletic Trainers' Association Now Blog. https://www.nata.org/blog/beth-sitzler/using-topical-hemostatic-agents.

Sundaram, N., Mony, U., & Jayakumar, R. (2016). Chitin and chitosan as hemostatic agents. *Encyclopedia of Polymer Science and Technology, 4e. Wiley*, 1-12. https://doi.org/10.1002/0471440264.pst647.

Süpüren, G., Kanat, Z. E., Çay, A., Kirci, T., Gülümser, T., & Tarakçioğlu, I. (2007). Nano Fibres (Part 2). *Textile and Apparel*, *17*(2), 83-89.

Tan, L., Hu, J., & Zhao, H. (2015). Design of bilayered nanofibrous mats for wound dressing using an electrospinning technique. *Materials Letters*, *156*, 46-49. https://doi.org/10.1016/j.matlet.2015.04.119.

Varshosaz, J., Choopannejad, Z., Minaiyan, M., & Kharazi, A. Z. (2021). Rapid hemostasis by nanofibers of polyhydroxyethyl methacrylate/polyglycerol sebacic acid: An *in vitro/in vivo* study. *Journal of Applied Polymer Science*, *138*(5), 49785. https://doi.org/10.1002/app.49785.

Wan, L. S., & Xu, Z. K. (2009). Polymer surfaces structured with random or aligned electrospun nanofibers to promote the adhesion of blood platelets. *Journal of Biomedical Materials Research Part A*, *89A*(1), 168-175. https://doi.org/10.1002/jbm.a.31907.

Wang, Q., Yao, Q., Liu, J., Sun, J., Zhu, Q., & Chen, H. (2019). Processing nanocellulose to bulk materials: A review. *Cellulose*, *26*, 7585-7617. https://doi.org/10.1007/s10570-019-02642-3.

Wang, X., Ding, B., Yu, J., & Yang, J. (2011). Large-scale fabrication of two-dimensional spider-web-like gelatin nano-nets via electro-netting. *Colloids and Surfaces B: Biointerfaces*, *86*(2), 345-352. https://doi.org/10.1016/j.colsurfb.2011.04.018.

Wang, Y., Zhao, Y., Qiao, L., Zou, F., Xie, Y., Zheng, Y., Chao, Y., Yang, Y., He, W., & Yang, S. (2021). Cellulose fibers-reinforced self-expanding porous composite with multiple hemostatic efficacy and shape adaptability for uncontrollable massive hemorrhage treatment. *Bioactive Materials*, *6*(7), 2089-2104. https://doi.org/10.1016/j.bioactmat.2020.12.014.

Xue, J., Wu, T., Dai, Y., & Xia, Y. (2019). Electrospinning and electrospun nanofibers: Methods, materials, and applications. *Chemical Reviews*, *119*(8), 5298-5415. https://doi.org/10.1021/acs.chemrev.8b00593.

Ye, W., Qin, M., Qiu, R., & Li, J. (2022). Keratin-based wound dressings: From waste to wealth. *International Journal of Biological Macromolecules*, *211*, 183-197. https://doi.org/10.1016/j.ijbiomac.2022.04.216.

Younes, I., & Rinaudo, M. (2015). Chitin and chitosan preparation from marine sources. Structure, properties and applications. *Marine Drugs*, *13*(3), 1133-1174. https://doi.org/10.3390/md13031133.

Yu, P., & Zhong, W. (2021). Hemostatic materials in wound care. *Burns & Trauma*, *9*, tkab019, 1-17. https://doi.org/10.1093/burnst/tkab019.

Zelkó, R., Lamprou, D. A., & Sebe, I. (2019). Recent development of electrospinning for drug delivery. *Pharmaceutics, 12*(1), 5. https://doi.org/10.3390/pharmaceutics12010005.

Zhang, S., Li, J., Chen, S., Zhang, X., Ma, J., & He, J. (2020). Oxidized cellulose-based hemostatic materials. *Carbohydrate Polymers, 230*, 115585. https://doi.org/10.1016/j.carbpol.2019.115585.

Zhang, S., Liu, H., Tang, N., Ge, J., Yu, J., & Ding, B. (2019). Direct electronetting of high-performance membranes based on self-assembled 2D nanoarchitectured networks. *Nature Communications*, *10*(1), 1458. https://doi.org/10.1038/s41467-019-09444-y.

Zhang, Z., Kuang, G., Zong, S., Liu, S., Xiao, H., Chen, X., Zhou, D., & Huang, Y. (2018). Sandwich-like fibers/sponge composite combining chemotherapy and hemostasis for efficient postoperative prevention of tumor recurrence and metastasis. *Advanced Materials*, *30*(49), 1803217. https://doi.org/10.1002/adma.201803217.

Zhao, X., Gao, J., Hu, X., Guo, H., Wang, F., Qiao, Y., & Wang, L. (2018). Collagen/polyethylene oxide nanofibrous membranes with improved hemostasis and cytocompatibility for wound dressing. *Applied Sciences*, *8*(8), 1226. https://doi.org/10.3390/app8081226.

Zhao, Y., Qiu, Y., Wang, H., Chen, Y., Jin, S., & Chen, S. (2016). Preparation of nanofibers with renewable polymers and their application in wound dressing. *International Journal of Polymer Science*, *2016*, 1-17. https://doi.org/10.1155/2016/4672839.

Zhou, J. C., Zhang, J. J., Zhang, W., Ke, Z. Y., & Zhang, B. (2017). Efficacy of chitosan dressing on endoscopic sinus surgery: A systematic review and meta-analysis. *European Archives of Oto-Rhino-Laryngology*, *274*, 3269-3274. https://doi.org/10.1007/s00405-017-4584-x.

Zhuang, Y., Lin, K., & Yu, H. (2019). Advance of nano-composite electrospun fibers in periodontal regeneration. *Frontiers in Chemistry*, *7*, 495. https://doi.org/10.3389/fchem.2019.00495.

Zumbardo-Bacelis, G. A., Meza-Villegas, L. A., Pérez-Aranda, C. A., Vargas-Coronado, R., Castillo-Cruz, O., Montaño-Machado, V., Mantovani, D., & Cauich-Rodríguez, J. V. (2021). On arginine-based polyurethane-blends specific to vascular prostheses. *Journal of Applied Polymer Science*, *138*(42), 51247. https://doi.org/10.1002/app.51247.

Chapter 19

An Analysis of Consumption Tendencies in the Global Clothing Market

Lara Orhon[1], MD
and Seher Kanat[2,*], PhD
[1]Graduate School of Natural and Applied Sciences, Ege University, İzmir, Türkiye
[2]Department of Textile Engineering, Ege University, İzmir, Türkiye

Abstract

The global clothing market, which constitutes a significant part of world trade, is one of the main markets that affect consumption tendencies in the world. In this context, this research aims to analyze the consumption tendencies in the global clothing market. In accordance with the aim of the research, the alterations within the export and import values of the leading global clothing exporter (China, European Union (27), Bangladesh, Vietnam, Türkiye, India, Malaysia, Indonesia, Hong Kong, and Pakistan) and importer countries (European Union (27), United States of America, Japan, United Kingdom, China, Canada, South Korea, Russia, Switzerland, and Australia) are analyzed in detail. The leading clothing exporter and importer countries are determined according to the data of the World Trade Organization's World Trade Statistical Review 2022. The analysis is performed on a product basis and contains data from the last decade. Besides, a survey on consumption tendencies in the global clothing market is conducted with the clothing enterprises, which are involved in the largest 1000 industrial enterprises of Türkiye, which is a significant player in the global clothing trade. Afterwards, the obtained data is analyzed. Thus, consumption alterations and the prominent product groups and trends in the clothing market, which has a significant place in the world trade, are revealed by using primary data (survey study) and secondary data (international clothing trade statistics). Subsequently, suggestions are made to clothing enterprises and brands that operate in the global clothing market, based on determined consumption tendencies. According to the obtained research results, the global clothing market has maintained its importance and share in the general global export and import in the last decade. China has maintained its leadership in global clothing export, whereas the European Union (27) is the leader of in the global clothing import. Besides, consumers have demanded mainly organic and recycled raw materials and products in recent years, and the tendency towards timeless and classic products has gradually increased. These demands are also coherent with sustainability and

---

* Corresponding Author's Email: seher.kanat@ege.edu.tr.

In: Textile Insights and Future Aspects
Editors: Gamze D. Tetik, Gizem Celep and Fulya Yilmaz
ISBN: 979-8-89113-289-4

environmental protection policies, whose importance has gradually increased in recent years.

Keywords: global clothing market, clothing sector, consumption, consumption tendencies

Introduction

Nowadays, the effects of globalization have been felt in almost every field and consumers have been altered and transformed under the effects of this wind. As knowledge sharing increases and eases, the awareness level of consumers also increases. All these effects have turned consumers into global consumers and enterprises have faced consumers who are aware of their desires and search and demand according to these. The increase in the number of global consumers has caused an alteration in global consumer tendencies.

The global clothing market, where consumers take increasingly active parts, is one of the markets in which global consumption tendencies are of importance. The global clothing market, which constitutes a significant part of world trade, is one of the leading markets that affects global consumption tendencies. In this context, this research aims to analyze the consumption tendencies in the global clothing market. Thus, the consumption alterations and the prominent product groups and trends in the clothing market, which possesses a significant place in the world trade, have been revealed, and these determined consumption tendencies have provided a basis for suggestions for clothing enterprises and brands, which operate in the global clothing market.

There are many studies in the literature that analyze the clothing buying behavior of consumers and clothing consumption tendencies and preferences on the basis of different factors and countries. The consumption tendencies in the global clothing market have generally been analyzed on the basis of consumer buying behavior. Besides, there are studies that analyze clothing consumption tendencies based on enterprises. At this point, there has been no study that analyzes the consumption tendencies in the global clothing market based on leading global clothing exporters and importer countries. Besides, this research also analyzes the knowledge of clothing enterprises and their findings about consumption tendencies in the global clothing market. These clothing enterprises are listed among the largest 1000 industrial enterprises of Türkiye, which is the fifth biggest global clothing exporter, and which predominates the alterations and tendencies within the global clothing market. In this context, this research differs from other studies within literature.

Alterations That Have Occurred Within the Marketing, Consumer, and Consumption Tendencies

Marketing concepts and, dependently, the consumers and consumption tendencies have been changed and differentiated in recent years. The marketing concept has evolved to marketing 4.0 from marketing 1.0 in recent years. The consumers have also been differentiated over the course of time in parallel to the marketing concept and, therefore, the consumer 4.0 concept has appeared. These alterations within marketing and consumers have naturally caused an alteration in consumption tendencies. At this point, the alterations that have occurred within the marketing should be analyzed.

The first phase of marketing (marketing 1.0) occurred in the industrial age. Back then, marketing mainly focused on product sales, and it did not take the consumer needs and desires into consideration. It focused on products and sales increment. The second phase of marketing is named marketing 2.0. In this phase, the consumer defines the value of the product. It is characterized by defining the needs and expectations of consumers in order to find new target markets and achieve further benefits (Lukowski, 2017). The third phase of marketing is called marketing 3.0. In this phase, consumers are active, searching, and creative individuals who are aware of the humanist dimension of a given brand and its impact on the environment. The expectations and desires of the consumers should be satisfied. Consumers can have a creative impact on products (Kotler et al., 2010; Lukowski, 2017). Finally, the fourth phase of marketing is named marketing 4.0. It is the integration of four elements as brand identity, brand image, brand integrity, and brand interaction. The first three elements were the elements of marketing 3.0. However, marketing 4.0 suggests a shift toward a more inclusive, horizontal, and social approach. Global markets that are tightly coupled with information-producing, manipulating, distributing, and consuming technologies present a development that leads to new ways to reach, collaborate, and influence potential consumers (Kotler et al., 2017; Dash et al., 2021).

The alterations which have occurred within the consumers and consumption tendencies can be analyzed under four main parts. Consumers before the 1950s are defined as consumer 1.0. They benefited from the production or delivery capability of the business. They had little choice and mainly bought products that were available on the market at that time. Consumers between 1950 and 1990 are called consumer 2.0. The properties of consumer 1.0 were still alive, supported by the development of consumerism and traditional marketing. The concepts of the brand first emerged and consumer demand for products and services was strongly driven by marketing. On the other hand, consumers between 1990 and 2015 are named consumer 3.0. The properties of consumers 1.0 and 2.0 continue but were expanded by the explosion of globalization and the Internet. Consumer retention and loyalty are key features, and the focus of competitive action is on delivering the best possible consumer experience. Finally, consumers after 2015 are defined as consumer 4.0. They are the ones whose personal goals outweigh persuasive marketing, who seek to influence and be influenced by other consumers, who employ very new ways of working and thinking, and who have high expectations. These consumers are netizens, very demanding, highly informed by electronic devices, and looking for developing, competitive, and innovative approaches to different challenges in various spheres (Freeman, 2017; Wereda & Wozniak, 2019).

Method of Research

This research aims to analyze the consumption tendencies in the global clothing market using primary (survey study) and secondary (international clothing trade statistics) data sources. Besides, it aims to make suggestions to enterprises that produce and/or market clothing products on the basis of determined consumption tendencies.

In accordance with the aim of the research, international clothing trade statistics are used as secondary data sources. In this context, the alterations in the export and import amounts of the leading global clothing exporters and importers are analyzed in detail on the basis of product groups (Orhon, 2021). According to the World Trade Organization statistics; China, European Union (27), Bangladesh, Vietnam, Türkiye, India, Malaysia, Indonesia, Hong Kong, and Pakistan are the leading global clothing exporters by 2021 (World Trade Organization, 2022). On the other hand, the European Union (27), the United States of America, Japan, the United Kingdom, China, Canada, South Korea, Russia, Switzerland, and Australia are the leading global clothing importers. The alterations in the clothing export and import amounts of these 20 countries or country groups within the last decade are analyzed by graphics on the basis of 61, 62, and 63 harmonized system-coded product groups and their four-digit sub-product groups. The necessary data for analyses have been taken from the Trademap (2023) database and analyses and graphic drawings have been actualized by the Microsoft Excel program. There are 17 sub-product groups under the 61 coded product group (knitted clothing products), whereas 17 sub-product groups are listed under the 62 coded product group (woven clothing products) and 10 sub-product groups are found under the 63 coded product group (home textiles and used clothing products).

The survey study constitutes the primary data source. The survey is conducted in clothing enterprises, which are listed among the largest 1000 industrial enterprises in Türkiye (Orhon, 2021). Türkiye is chosen because it is a significant player in the global clothing trade (it is the fifth biggest clothing exporter in the world (World Trade Organization, 2022)) and therefore it can give important information about consumption tendencies. The largest 1000 industrial enterprises of Türkiye are determined by the İstanbul Chamber of Industry, and these enterprises have a significant share within the national economy and exports. These enterprises export to leading global buyers due to their exportation potential, and also some of them export with their own brands. Therefore, clothing enterprises, which are listed within these enterprises, closely follow the global clothing consumption tendencies and the changes in the clothing products. In this context, these clothing enterprises are chosen as a tool that can reflect the global clothing consumption tendencies. 36 clothing enterprises are listed among the largest 1000 industrial enterprises in Türkiye by 2018 (İstanbul Chamber of Industry, 2019). Thus, these enterprises constitute the sample of research.

The volunteerism principle is applied in survey participation. The surveys have been completed by the marketing managers of the enterprises. The surveys have been conducted between April 2020 and November 2020 using face-to-face interviews and remote access (mail and phone) techniques. A survey form, which consists of 25 main questions, has been used as a data collection tool. The survey has been ethically approved by Ege University's Ethical Board of Social and Human Sciences Scientific Research and Publication on 29 August 2019 with the number of 354 protocols. After the repatriation and evaluation of the sent-back questionnaires, 19 of them were incorporated into the research. The obtained results are analyzed via SPSS and Microsoft Excel programs.

Findings of the Research and Their Analysis

If the global clothing export and import data in the last decade are analyzed, it can be seen that the global clothing export has increased from 460 billion dollars in 2012 to 597 billion dollars in 2021 with an increase of approximately 30% (Figure 1). Similarly, global clothing imports have increased from 425 billion dollars in 2012 to 517 billion dollars in 2021 with a 21% increase. On the other hand, global clothing export takes a 2,70% share of general global export by 2021, whereas global clothing import takes a 2,31% share of general global imports.

When the leading global clothing exporters' last decade export figures are analyzed, it can be seen that China is the leader (Figure 2). China is followed by the European Union (27), Bangladesh, Vietnam, Türkiye, and India, respectively. Türkiye has risen to fifth place in 2020 by passing over India. China's clothing export has increased by 15% within the last decade (2012-2021), whereas the European Union's clothing export has increased by 45%, Bangladesh's clothing export has increased by 131%, Vietnam's clothing export has increased by 110% and Türkiye's clothing export has increased by 35%. If the increase and decrease rates of the leading global clothing exporters are analyzed annually, it can be seen that export values generally have a tendency to increase. However, most of the leading countries have faced a decline in 2020 due to the Covid-19 pandemic.

If the leading clothing importers' last decade import figures are analyzed, it can be seen that the European Union (27) is the leader (Figure 3). The European Union (27) is followed by the United States of America, Japan, the United Kingdom, China, and Canada, respectively. The European Union's clothing import has increased by 39% within the last decade (2012-2021), whereas the USA's clothing import has increased by 17%, China's clothing import has increased by 155% and Canada's clothing import has increased by 20%. On the other hand, Japan's clothing import has decreased by 21% and the United Kingdom's clothing import has decreased by 7%. If the increase and decrease rates of the leading global clothing importers are analyzed on an annual basis, it can be seen that import values generally have a tendency to increase just as the export values.

101 graphics have been obtained from the analyses, which are actualized using secondary data. All of these graphics could not be given in the chapter. Therefore, all results have been summarized in tables (Table 1-6). As can be seen in Table 1, the export amount of knitted clothing products has increased by 29% within the last decade. Within the exports of knitted clothing products, the highest increase in exports belongs to other knitted clothes, knitted women's and girls' coats, and knitted men's and boys' underpants, pajamas, and bathrobes. On the other hand, within the exports of knitted clothing products, the largest decrease in export is amounts the knitted women's and girls' blouses.

Similarly, the export amount of woven clothing products has increased by 21% within the last decade (Table 2). Within the exports of woven clothing products, the highest increase in exports is in woven men's and boys' coats and woven women's and girls' coats. On the other hand, within the exports of woven clothing products, the largest decrease in export amounts is owned by woven ties and bow ties.

Table 3 summarizes the export amounts of used home textiles and clothing products. The export amount of these products has increased by 62% in the last decade. Inside the exports of home textiles and used clothing products, the most increased export amounts belong to floor cloths, dish cloths, dusters, similar cleaning cloths, life jackets and life belts and tarpaulins, awnings, sun blinds, tents, sails, camping goods. On the other hand, within the home textiles and used clothing products' exports, the most decreased export amount is owned by used rags, twines, ropes, and cables.

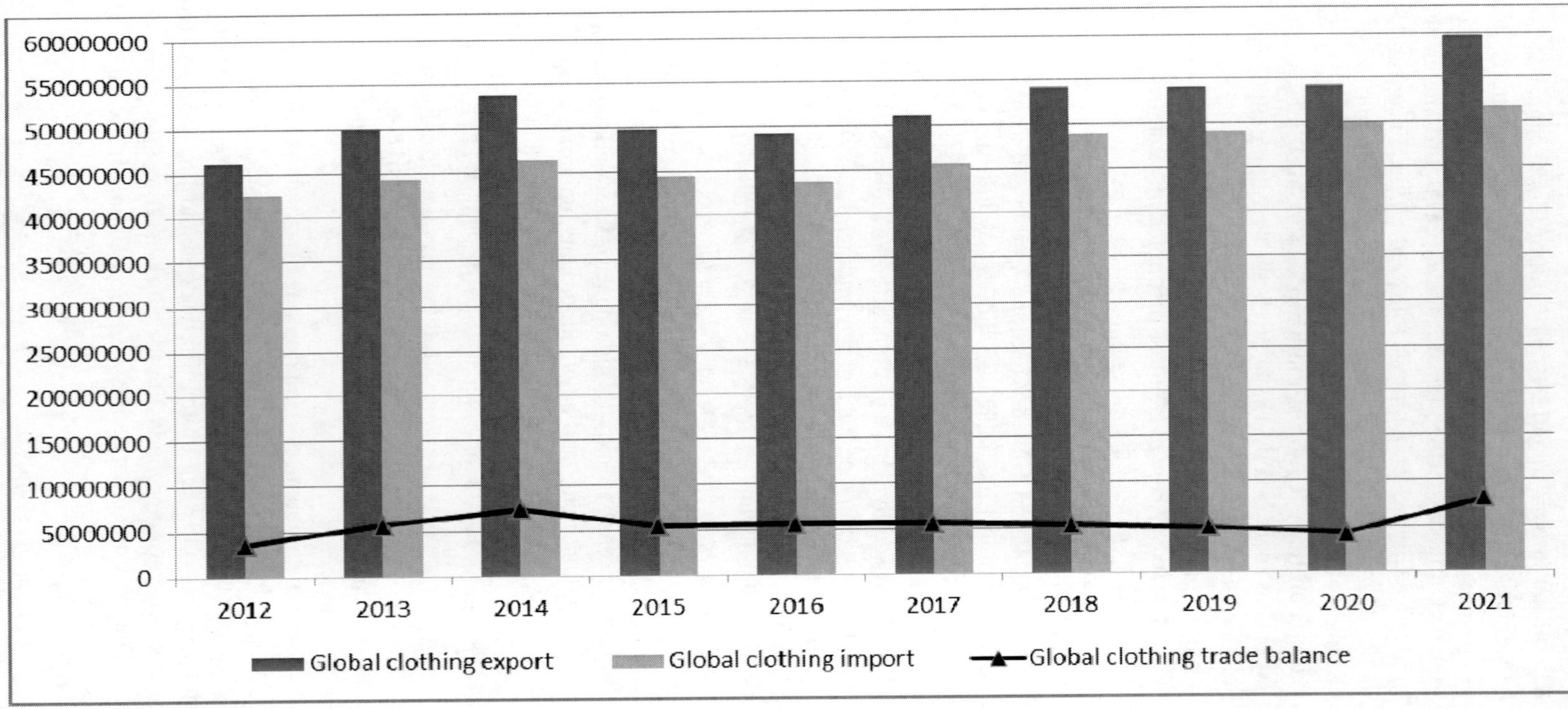

Figure 1. Global clothing export and import.

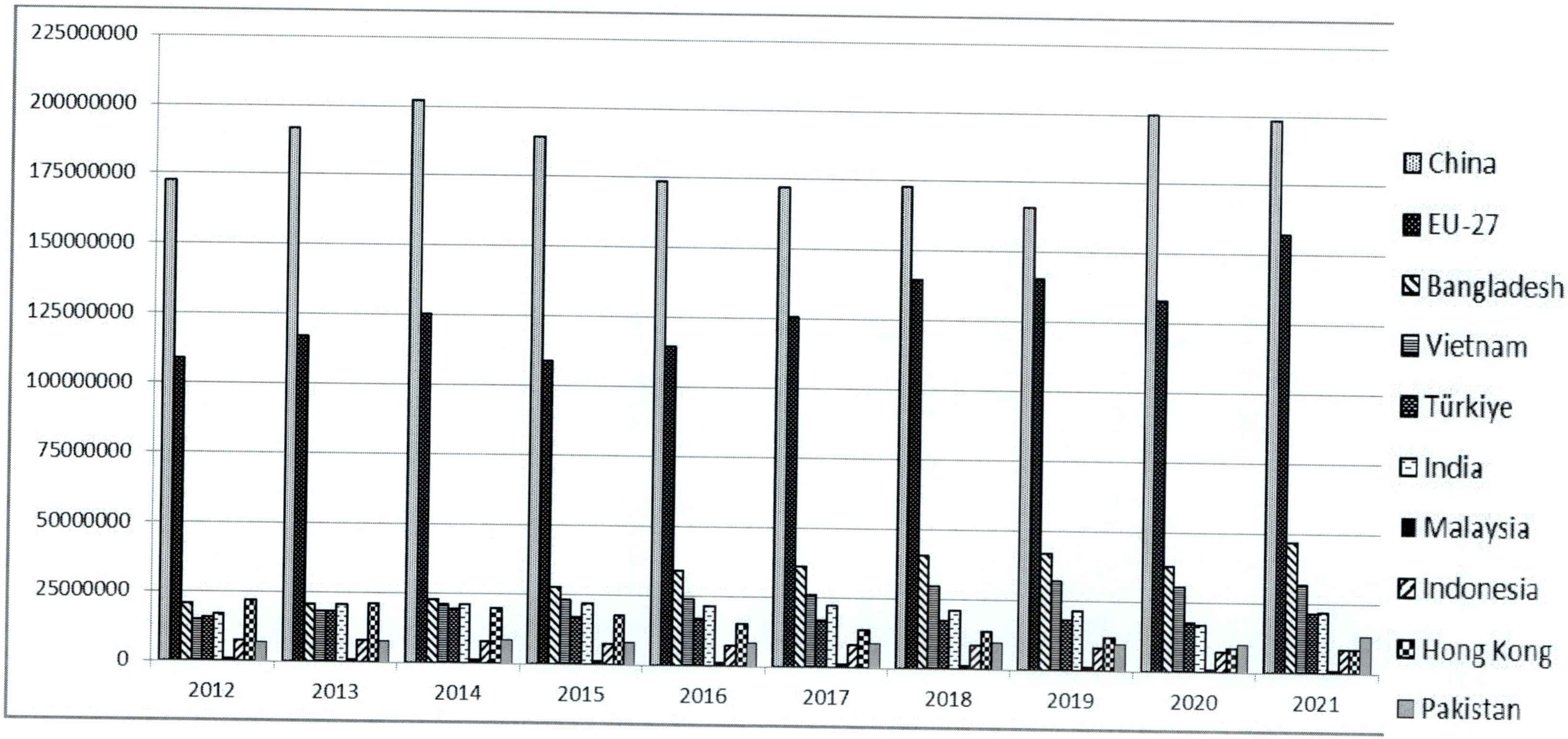

Figure 2. Leading global clothing exporters.

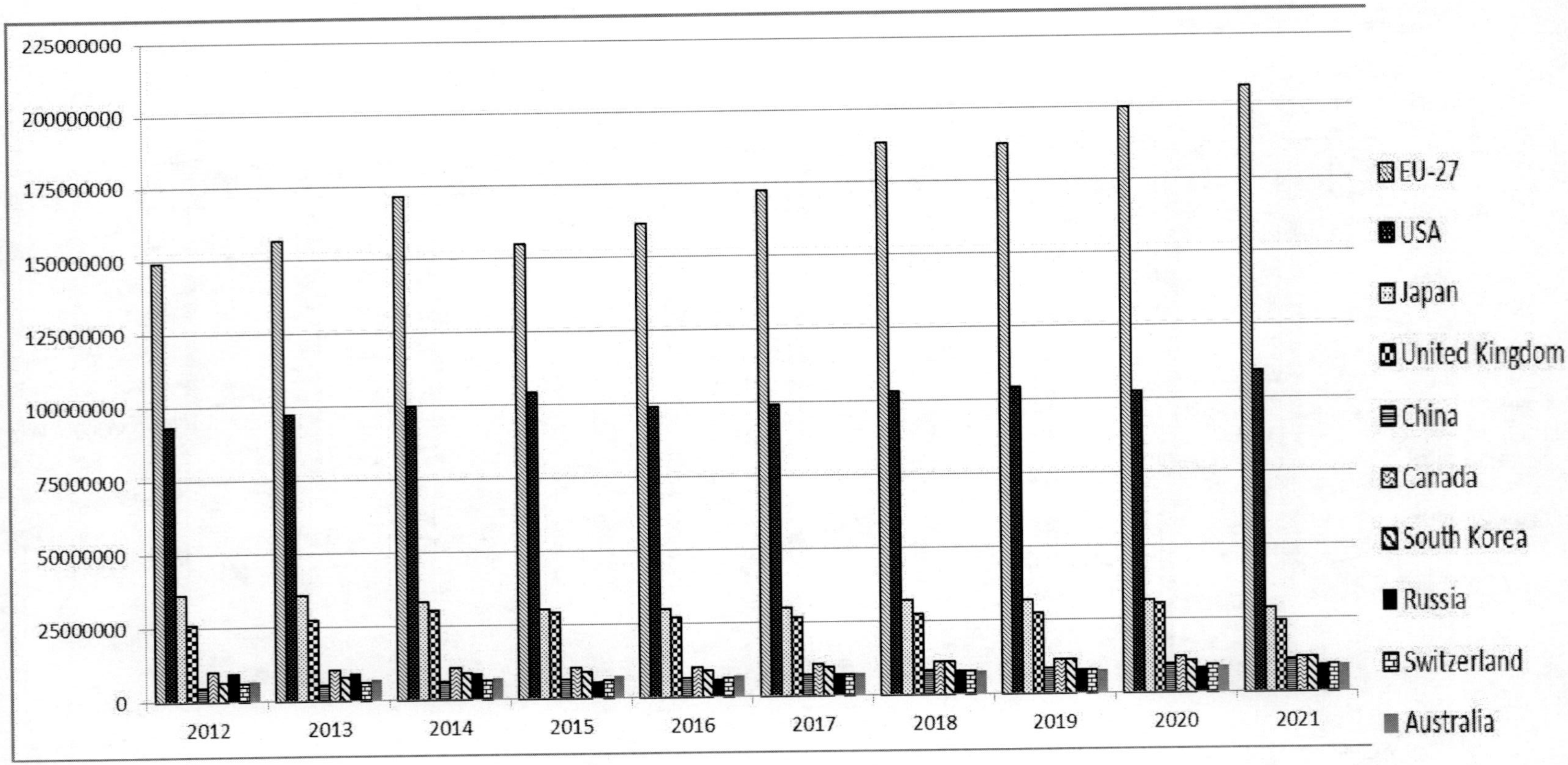

Figure 3. Leading global clothing importers.

As can be seen in Table 4, the knitted clothing products' import amount has increased by 27% within the last decade. Inside the imports of knitted clothing products, the most increased import amount belongs to knitted men's and boys' suits, jackets, trousers, and shorts. On the other hand, within the imports of knitted clothing products, the most decreased export amount is owned by knitted women's and girls' blouses.

Similarly, the import amount of woven clothing products has increased by 5% in the last decade (Table 5). Inside the imports of woven clothing products, the most increased import amount belongs to clothes made from some specific woven fabrics. On the other hand, within the imports of woven clothing products, the most decreased import amount is owned by woven ties and bow ties.

Table 6 summarizes the import amounts of home textiles and used clothing products. The import amount of these products has increased by 63% in the last decade. Inside the imports of home textiles and used clothing products, the highest number of imported products is found in floor cloths, dish cloths, dusters, similar cleaning cloths, life jackets, and life belts. On the other hand, within imports of the home textiles and used clothing products, none of the import amounts belonging to sub-product groups decreased.

Table 1. Alterations of knitted clothing products' export amounts in the last decade

Harmonized system codes	Explanation	Global increase and decrease amounts that occurred in the last decade (2012-2021)
61	**Knitted clothing products**	**29% ↑**
6101	Knitted men's and boys' coats	56% ↑
6102	Knitted women's and girls' coats	73% ↑
6103	Knitted men's and boys' suits, jackets, trousers, and shorts	14% ↑
6104	Knitted women's and girls' suits, jackets, dresses, skirts, trousers, and shorts	14% ↑
6105	Knitted men's and boys' shirts	9% ↓
6106	Knitted women's and girls' blouses	27% ↓
6107	Knitted men's and boys' underpants, pajamas, and bathrobes	71% ↑
6108	Knitted women's and girls' slips, briefs, nightdresses, pajamas, and bathrobes	59% ↑
6109	Knitted t-shirts	27% ↑
6110	Knitted pullovers and waistcoats	36% ↑
6111	Knitted babies' garments	59% ↑
6112	Knitted tracksuits, ski suits, and swimwear	38% ↑
6113	Clothes made from some specific knitted fabrics	32% ↑
6114	Other knitted clothes	98% ↑
6115	Knitted hosiery types	16% ↑
6116	Knitted gloves	51% ↑
6117	Knitted clothing accessories	29% ↑

Table 2. Alterations of woven clothing products' export amounts in the last decade

Harmonized system codes	Explanation	Global increase and decrease amounts that occurred in the last decade (2012-2021)
62	**Woven clothing products**	**21% ↑**
6201	Woven men's and boys' coats	47% ↑
6202	Woven women's and girls' coats	59% ↑
6203	Woven men's and boys' suits, jackets, trousers, and shorts	9% ↑
6204	Woven women's and girls' suits, jackets, dresses, skirts, trousers, and shorts	36% ↑
6205	Woven men's and boys' shirts	30% ↓
6206	Woven women's and girls' blouses	3% ↓
6207	Woven men's and boys' underpants, pajamas, and bathrobes	5% ↑
6208	Woven women's and girls' slips, briefs, nightdresses, pajamas, and bathrobes	39% ↑
6209	Woven babies' garments	2% ↑
6210	Clothes made from some specific woven fabrics	26% ↑
6211	Woven tracksuits and swimwear	37% ↑
6212	Woven brassieres, girdles, corsets, braces, suspenders, and garters	32% ↑
6213	Woven handkerchiefs	46% ↓
6214	Woven shawls, scarves, mufflers, mantillas, and veils	29% ↓
6215	Woven ties and bow ties	64% ↓
6216	Woven gloves	21% ↑
6217	Woven clothing accessories	20% ↑

Table 3. Alterations of home textiles and used clothing products' export amounts in the last decade

Harmonized system codes	Explanation	Global increase and decrease amounts that occurred in the last decade (2012-2021)
63	**Home textiles and used clothing products**	**62% ↑**
6301	Blankets and traveling rugs	29% ↑
6302	Bed linen, table linen, toilet linen, and kitchen linen	31% ↑
6303	Curtains, interior blinds, and curtain and bed valances	55% ↑
6304	Furnishing articles	29% ↑
6305	Sacks and bags for packaging	40% ↑
6306	Tarpaulins, awnings, sun blinds, tents, sails, camping goods	104% ↑
6307	Floor cloths, dishcloths, dusters, similar cleaning cloths, life jackets, and life belts	160% ↑
6308	Woven fabric and yarn sets that are used in the rug, tapestry, tablecloth, and serviette production	10% ↑
6309	Used clothing products	33% ↑
6310	Used rags, twines, ropes, and cables	18% ↓

According to the survey results in which clothing enterprises listed in Türkiye's largest 1000 industrial enterprises have participated, consumer preferences in the global clothing market have been changed on a medium level over the last decade. Besides, product prices have been increased in many product groups. In addition to this, the consumption amount per person in the global clothing market has also increased. Additionally, 89% of the enterprises have indicated that the awareness level of consumers on clothing products has increased in the last decade.

According to 63% of the enterprises, mostly knitted fabrics are used or demanded in clothing products on the global clothing market. According to them, the single jersey is the most demanded/used knitted fabric in the global clothing market, whereas denim and poplin are the most demanded/used woven fabrics.

According to 79% of the participating enterprises, consumers mostly demand knitted clothing products in the global clothing market. Besides, 69% of the enterprises have indicated that consumers in the global clothing market are mostly demanding basic and classical products. According to 90% of the participating enterprises, the t-shirt is the most demanded knitted clothing product. On the other hand, 79% of the enterprises have specified that a shirt/blouse is the most demanded woven clothing product.

Table 4. Alterations of knitted clothing products' import amounts in the last decade

Harmonized system codes	Explanation	Global increase and decrease amounts that occurred in the last decade (2012-2021)
61	**Knitted clothing products**	**27% ↑**
6101	Knitted men's and boys' coats	25% ↑
6102	Knitted women's and girls' coats	42% ↑
6103	Knitted men's and boys' suits, jackets, trousers, and shorts	138% ↑
6104	Knitted women's and girls' suits, jackets, dresses, skirts, trousers, and shorts	51% ↑
6105	Knitted men's and boys' shirts	17% ↓
6106	Knitted women's and girls' blouses	39% ↓
6107	Knitted men's and boys' underpants, pajamas, and bathrobes	46% ↑
6108	Knitted women's and girls' slips, briefs, nightdresses, pajamas, and bathrobes	29% ↑
6109	Knitted t-shirts	26% ↑
6110	Knitted pullovers and waistcoats	19% ↑
6111	Knitted babies' garments	23% ↑
6112	Knitted tracksuits, ski suits, and swimwear	15% ↑
6113	Clothes made from some specific knitted fabrics	74% ↑
6114	Other knitted clothes	58% ↑
6115	Knitted hosiery types	15% ↑
6116	Knitted gloves	41% ↑
6117	Knitted clothing accessories	12% ↓

Table 5. Alterations of woven clothing products' import amounts in the last decade

Harmonized system codes	Explanation	Global increase and decrease amounts that occurred in the last decade (2012-2021)
62	**Woven clothing products**	**5% ↑**
6201	Woven men's and boys' coats	38% ↑
6202	Woven women's and girls' coats	37% ↑
6203	Woven men's and boys' suits, jackets, trousers, and shorts	4% ↓
6204	Woven women's and girls' suits, jackets, dresses, skirts, trousers, and shorts	6% ↑
6205	Woven men's and boys' shirts	35% ↓
6206	Woven women's and girls' blouses	17% ↓
6207	Woven men's and boys' underpants, pajamas, and bathrobes	16% ↓
6208	Woven women's and girls' slips, briefs, nightdresses, pajamas, and bathrobes	1% ↑
6209	Woven babies' garments	24% ↓
6210	Clothes made from some specific woven fabrics	60% ↑
6211	Woven tracksuits and swimwear	28% ↑
6212	Woven brassieres, girdles, corsets, braces, suspenders, and garters	14% ↑
6213	Woven handkerchiefs	44% ↓
6214	Woven shawls, scarves, mufflers, mantillas, and veils	19% ↓
6215	Woven ties and bow ties	63% ↓
6216	Woven gloves	7% ↑
6217	Woven clothing accessories	15% ↓

Table 6. Alterations of home textiles and used clothing products' import amounts in the last decade

Harmonized system codes	Explanation	Global increase and decrease amounts that occurred in the last decade (2012-2021)
63	**Home textiles and used clothing products**	**63% ↑**
6301	Blankets and traveling rugs	38% ↑
6302	Bed linen, table linen, toilet linen, and kitchen linen	31% ↑
6303	Curtains, interior blinds, and curtain and bed valances	44% ↑
6304	Furnishing articles	37% ↑
6305	Sacks and bags for packaging	46% ↑
6306	Tarpaulins, awnings, sun blinds, tents, sails, camping goods	84% ↑
6307	Floor cloths, dishcloths, dusters, similar cleaning cloths, life jackets, and life belts	144% ↑
6308	Woven fabric and yarn sets that are used in the rug, tapestry, tablecloth, and serviette production	7% ↑
6309	Used clothing products	38% ↑
6310	Used rags, twines, ropes, and cables	20% ↑

84% of the participating enterprises have indicated that consumers mostly demand organic and/or recycled raw materials/products in the global clothing market. This finding also conforms to sustainability and environmental and human health protection policies, whose importance has increased in recent years. Besides, 95% of the participating enterprises have specified that clothing products and accessories for women are mostly demanded in the global clothing market. According to 58% of the enterprises, fast fashion products are demanded more, whereas 42% of them have indicated that the consumers incline toward classical and timeless clothing products.

Conclusion

According to the obtained research results, the global clothing market has maintained its importance and share in the general global export and import in the last decade. China has maintained its leadership in the global clothing export. Although the European Union (27) has passed China in some sub-product groups, China is the unquestionable leader in global clothing export. Besides, Türkiye is the fifth largest clothing supplier in the world. On the other hand, the European Union (27) is the leader in global clothing imports. European Union (27) is followed by the United States of America, and Japan, respectively. The European Union (27) has maintained its leadership in almost all sub-product groups, whereas the United States of America has taken over the leadership in some sub-product groups.

Global clothing export has increased by 30% in the last decade. Global export of knitted clothing products has increased by 29% in the last decade, whereas global export of woven clothing products has increased by 21% and global export of home textiles and used clothing products has increased by 62%. On the other hand, global clothing import has increased by 21% in the last decade. The global import of knitted clothing products has increased by 27% in the last decade, whereas the global import of woven clothing products has increased by 5% and the global import of home textiles and used clothing products has increased by 63%.

Both global clothing export and import have increased in the last decade on the basis of main product groups. If the sub-product groups are analyzed, there are only 8 sub-product groups in which export-based decrease has been determined. On the other hand, there are 12 sub-product groups in which import-based decrease has been determined. The significant results obtained within the scope of research and the suggestions that can be proposed on the basis of these results can be summarized follows:

1. Both global clothing export and import have increased in the last decade on the basis of main product groups. At this point, it can be said that all main product groups are demanded by consumers in increasing ratios. Therefore, enterprises can operate or continue to operate in all of these main product groups.
2. If the decadal global knitted clothing export is analyzed on the basis of sub-product groups, it can be seen that the export values are mostly increased in other knitted clothes, knitted women's and girls' coats, and knitted men's and boys' underpants, pajamas, and bathrobes. These product groups can be suggested for investment.
3. If the decadal global woven clothing export is analyzed on the basis of sub-product groups, it can be seen that the export values are mostly increased in woven men's and

boys' coats and woven women's and girls' coats. These product groups can be suggested for investment.

4. If the decadal global home textiles and used clothing products exports are analyzed on the basis of sub-product groups, it can be seen that the export values are mostly increased in floor cloths, dish cloths, dusters, similar cleaning cloths, life jackets and life belts and tarpaulins, awnings, sun blinds, tents, sails, camping goods. These product groups can be suggested for investment.
5. If the decadal global knitted clothing import is analyzed on the basis of sub-product groups, it can be seen that the import value is mostly increased in knitted men's and boys' suits, jackets, trousers, and shorts. This product group can be suggested for investment.
6. If the decadal global woven clothing import is analyzed on the basis of sub-product groups, it can be seen that, the import value is mostly increased in clothes made from some specific woven fabrics. This product group can be suggested for investment.
7. If the decadal global home textiles and used clothing product imports are analyzed on the basis of sub-product groups, it can be seen that the import value is mostly increased in other floor cloths, dish cloths, dusters, similar cleaning cloths, life jackets, and life belts. This product group can be suggested for investment.
8. Consumer awareness level about clothing products has increased in the last decade. Therefore, clothing enterprises and brands should precisely inform consumers about their products when putting their products on the global clothing market.
9. Basic and classical clothing products have been more preferred by consumers in the last decade. Consumers are inclined to classic and timeless clothing products. The pandemic process, which has been experienced since the recent year, also highlights this tendency. Thus, clothing enterprises and brands should incline to classical and timeless clothing design and production. Furthermore, they should increase the ratios of these products within their product ranges.
10. Consumers within the global clothing market have mostly demanded organic and/or recycled raw materials and/or products in the last decade. This demand also conforms with sustainability and environment and human health protection policies, whose importance has increased in recent years. Therefore, clothing enterprises and brands should incline to organic and/or recycled raw materials and/or products. They should design and produce collections consisting of these products.
11. Clothing products and accessories for women have been in more demand by consumers in the last decade because women follow fashion more closely, care about their appearance more and see clothing shopping as a relief method. Thus, clothing enterprises and brands should invest in clothing products for women, or they should increase their existing investments.

Disclaimer

None.

References

Dash, G., Kiefer, K., & Paul, J. (2021). Marketing-to-millennials: marketing 4.0, customer satisfaction and purchase intention. *Journal of Business Research, 122*, 608–620. https://doi.org/10.1016/j.jbusres.2020.10.016.

Freeman, R. W. (2020, December 10). Understanding customer 4.0 — the customer led revolution. *The Worshipful Company of Management Consultant.* https://wcomc.org/2017May31-2.

İstanbul Chamber of Industry. (2019, August 10). *Türkiye's Biggest 1000 Industrial Enterprises 2018.* Retrieved August 10, 2019, from http://www.iso500.org.tr/.

Kotler, P., Kartajaya, H., & Setiawan, I. (2010). *Marketing 3.0: From Products to Customers to the Human Spirit.* John Wiley & Sons Incorporated.

Kotler, P., Kartajaya, H., & Setiawan, I. (2017). *Marketing 4.0: Moving from Traditional to Digital.* John Wiley & Sons Incorporated.

Lukowski, W. (2017). The impact of the internet of things on value added to marketing 4.0. *Marketing Instytucji Naukowych i Badawczych, 26*(4), 187-204.

Orhon, L. (2021). *Analysis of consumption tendencies in global clothing market* [Unpublished master dissertation]. Ege University.

Trademap. (2023, April 3). *Trade Statistics for International Business Development.* Retrieved April 3, 2023, from https://www.trademap.org/.

Wereda, W., & Wozniak, J. (2019). Building relationships with customer 4.0 in the era of marketing 4.0: The case study of innovative enterprises in Poland. *Social Sciences, 8*(177), 1-27. https://doi.org/10.3390/socsci8060177.

World Trade Organization. (2023, April 3). *World Trade Statistical Review 2022.* Retrieved April 3, 2023, from https://www.wto.org/english/res_e/booksp_e/wtsr_2022_e.pdf.

Chapter 20

An Assessment of Sustainability in the Fashion and Clothing Industry in a General Framework

Derya Tama Birkocak[1,*], PhD
and Meral İşler[2], PhD
[1]Department of Textile Engineering, Ege University, Izmir, Türkey
[2]Department of Fashion Design, Selçuk University, Konya, Türkey

Abstract

The fashion and clothing industry has an important share in the world economy and also has a significant employing population worldwide due to its labor-driven nature. Consumption in this industry had spectacular growth as well, such that clothing production doubled between 2000 and 2014. Therefore, understanding the concept of sustainability and applying its requirements in this sector became essential. It refers to the ability to meet the needs of the present without compromising the ability of future generations to meet their own needs. The focus of this book chapter is to evaluate the concept of sustainability in a general framework regarding the fashion and clothing industry. To achieve this, the chapter provides an overview of sustainable development and examines the triple bottom line of sustainability, which includes economic, social, and environmental factors. Afterwards, the two outstanding movements of the fashion and clothing industry, namely fast fashion and sustainable fashion, were examined in detail; the impact of these fashion movements on the environment, humans, and economics were comparatively assessed. In addition, the three key strategies, reduce, reuse, and recycle of sustainability, were investigated towards the fashion and clothing industry.

Keywords: 3R, fashion and clothing industry, sustainability, fast fashion, slow fashion

Introduction

The fashion and clothing industry is a vast and ever-changing field that plays a significant role in the global economy. Today, the fashion industry is a global behemoth, with an estimated market worth rise to about $2,25 trillion in 2025 (Centobelli et al., 2022; Papamichael et al.,

* Corresponding Author's Email: derya.tama@ege.edu.tr.

In: Textile Insights and Future Aspects
Editors: Gamze D. Tetik, Gizem Celep and Fulya Yilmaz
ISBN: 979-8-89113-289-4

2022; Ikram, 2022). The industry includes the entire process of creating, designing, promoting, and selling clothing and accessories. From haute couture to fast fashion, significant changes have occured over the years in the industry to catch the changing consumer preferences.

The fashion and textile industry is expected to have significant growth in the coming years due to increasing consumer demand, technological advancements, and a growing focus on sustainability. Fast fashion, a term used to describe the production of low-cost clothing that is designed for short-term use, has become a dominant force in the industry in recent years (Brooksworth et al., 2022). However, this fast fashion model is facing criticism for its negative impact on the environment and for the exploitation of workers in developing countries. Conversely, there is also a growing trend in sustainable and ethical fashion; consumers demand the use of eco-friendly materials as well as transparency in all production processes with their increasing awareness (Li & Leonas, 2022; Mackie & Campbell, 2022). To meet this demand, many designers and brands are creating clothing that is both fashionable and sustainable. To meet this demand, many designers and brands are designing products that are both fashionable and sustainable.

Sustainability is a concept that has been discussed and applied in different forms throughout human history. However, the modern use of the sustainability concept first emerged in the late 20th century as a response to environmental degradation, depletion of natural resources, and concerns about climate change (Du Pisani, 2006).

During the early days of human civilization, people used natural resources in a sustainable way for food, clothing, and shelter. In the Middle Ages, the focus on sustainability changed towards long-term economic development, which led to the growth of trade and commerce. The industrial revolution, however, caused mass production, so the mass consumption of resources had a significant impact on the environment, leading to pollution, deforestation, soil degradation, etc. (Barca, 2011). The concept of sustainability as we understand it today occurred in the 1970s; in 1972 at the Stockholm Conference of the United Nations World Commission on Environment and Development the term "sustainable development" was first used. In 1984, the Brundtland Report gave a formal definition to the concept, which is about meeting present needs without compromising the ability of future generations to meet their own needs (Acar et al., 2015; Devaki et al., 2018; Li & Leonas, 2022; Papadopoulou et al., 2022). In other words, sustainability is about living within the limits of our planet's natural resources in a balance of economic growth, social progress, and environmental protection. The resources are finite and people need to use these resources conservatively and wisely; therefore, sustainability is needed to be taken as a key issue in today's world as it provides a roadmap towards a better future for all.

The concept of a sustainable economy has become a key issue for policymakers worldwide. One of the key challenges of building a sustainable economy in the world and in the nations is the need to balance economic growth with environmental protection. Sustainability helps to reduce the impact of economic activity on the environment, ensuring that natural resources are conserved for future generations (Ekins & Zenghelis, 2021). It also promotes the development of new technologies and industries, generating new economic growth opportunities. Regarding this, the European Commission (EC) announced the Green Deal in December 2019, which covers a wide range of areas, including energy, transport, agriculture, and industry. The plan aims to transform the European economy, create jobs, and improve the quality of life for citizens, and protect the environment while mitigating climate change (A European Green Deal, 2023). It involves policies and initiatives aimed at reducing greenhouse gas emissions, protecting the environment, and promoting economic growth through innovation and investment in renewable energy. The core objective of the deal is to achieve carbon neutrality by 2050, which means that the amount of carbon dioxide released into the atmosphere must be equal to the amount removed.

The Triple Bottom Line of Sustainability in Terms of the Fashion and Clothing Industry

From the production of textile fibers to the disposal of clothing, the fashion and clothing industry has a significant impact on the environment, social welfare, and economic stability. To address these challenges, sustainability has become a vital focus in the fashion industry. The three pillars of sustainability, namely economic, environmental, and social, provide a framework for creating a more sustainable fashion and clothing industry. Such a framework should be able to measure the industry's impact on the environment, society, and the economy, identify areas of improvement, and provide guidance on how to implement sustainable practices.

The three pillars of sustainability, also known as the triple bottom line, have three interconnected rings of equal size (Figure 1). Symmetrical interconnection provides balance; otherwise, different viewpoints may give one or the other greater priority (Giddings, 2002). This framework is an applied and solutions-oriented approach that includes technology and innovation; laws and governance; and economics and financial incentives (Burns, 2012; Clune et al., 2018).

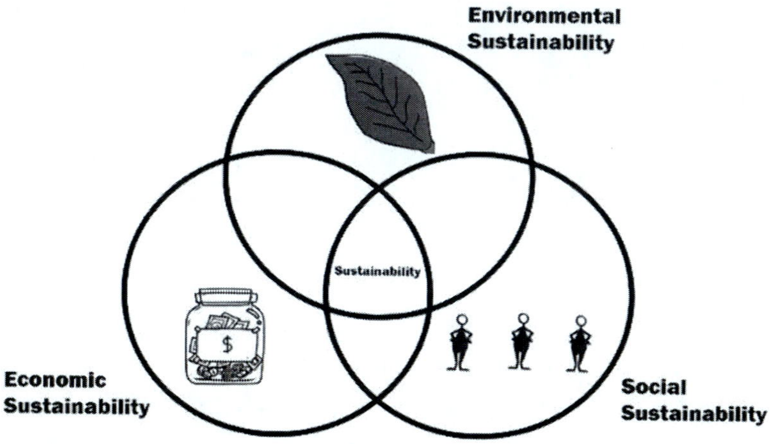

Figure 1. The triple bottom line of sustainability.

Sustainability combines the creation of a suitable and habitable space for human society within its goals, environmental protection, and ecological conservation. The economic component correctly recognizes that all of our environmental and sustainability challenges are directly linked to economic production and consumption and should be considered in conjunction with other objectives (Clune et al., 2018).

Environmental Sustainability

Environmental sustainability focuses on preserving and managing natural resources, including air, forests, water, land, biodiversity, and ecosystems. It emphasizes reducing environmental impact, protecting habitats, preventing pollution, and mitigating climate change. Environmental sustainability is achieved through actions such as reducing waste, conserving energy and using sustainable materials.

Environmental sustainability is perhaps the most critical aspect of sustainability in the fashion and clothing industry. The industry is known for its high levels of resource consumption, waste generation, and pollution, and herein, environmental sustainability refers to reducing the environmental impact of fashion production and consumption. Environmental sustainability involves looking at the entire lifecycle of a garment; from the production of raw materials such as cotton and synthetic fibers to the manufacturing, transportation, and distribution of clothing, the industry has a significant impact on the environment.

Assessing the environmental sustainability of the industry requires evaluating its carbon footprint, water footprint, waste generation, and pollution levels. The industry has been criticized for its fast fashion model, which encourages consumers to buy more and dispose of clothes quickly (Nguyen & Mogaji, 2022). To achieve environmental sustainability in the fashion and clothing industry, it is essential to focus on sustainable practices. An approach to adopting sustainable production processes that reduce the consumption of water, energy, and chemicals, including the use of renewable energy sources, the reduction of water use, and the use of non-toxic chemicals (Macchion et al., 2018). A different approach involves creating products that have the characteristics of durability, repairability, and reusability, which can extend the lifecycle of garments and reduce waste (Mackie & Campbell, 2022).

Social Sustainability

Social sustainability is the other dimension of sustainability that involves the creation of inclusive communities that are strong, equitable, and just. The initiatives promote social cohesion, enable access to education and healthcare, and ensure that individuals have reasonable employment opportunities and a good standard of living (Beyer & Arnold, 2022).

The idea behind social sustainability in terms of the fashion and clothing industry is to encourage equitability for all individuals connected to the industry, starting from farmers and factory laborers, to customers and sellers including fair wages paid to workers, safe working conditions, and treated individuals with respect and dignity (Göztaşı et al., 2022).

Focusing on several key strategies can help promote social sustainability in the fashion and clothing industry. One of these is to be transparent and accountable in the supply chain, allowing consumers to make informed decisions regarding the products they purchase (Macchion et al., 2018). Another strategy is to promote diversity and inclusion in the workforce, which can provide a more equitable and sustainable fashion industry.

The collapse of Rana Plaza and the presence of sweatshops are serious concerns about social sustainability in the fashion and clothing industry. Both examples show exploitative practices in the global supply chain that prioritize financial profits over worker safety and rights. In 2013, the Rana Plaza garment factory disaster in Bangladesh resulted in the death of more than 1,100 workers and injured thousands more. This disaster brought attention worldwide to the unsafe working conditions, low wages, and long working hours that many workers forced to experience in the global garment industry. After the tragedy, the need for improved safety regulations, better working conditions, and increased transparency and accountability in the industry is understood (Bossavie et al., 2019).

The other topic that needed to be considered is sweatshops, which are often located in developing countries with low labor standards and are characterized by poor working conditions, low wages, long working hours, and little job security (Cheek & Moore, 2003). Workers are often subjected to abuse and discrimination, with many being children, women, or migrants who are at risk of exploitation. To address these issues and promote social sustainability, it is crucial to ensure that workers are treated with dignity and respect. Moreover, companies must ensure that their suppliers follow ethical labor practices. Governments should also play a role in setting and enforcing labor standards to protect workers' rights and ensure social sustainability.

Economic Sustainability

The third crucial component of sustainability is economic sustainability. It involves creating an economic system that is fair and equitable for all stakeholders, from farmers and factory workers to consumers and retailers. This pillar also aims to create long-term economic growth without harming the environment or depleting natural resources.

In order to foster economic sustainability within the fashion and clothing industry, companies have several tasks such as applying a circular business model and incorporating sustainable materials into their product lines. The circular business model prioritizes waste minimization and the extension of product lifecycles through initiatives such as upcycling, recycling, and the creation of easily disassembled products (Choi & Li, 2015). Additionally, the use of sustainable materials such as hemp, linen, organic cotton, or recycled polyester can help to reduce the environmental pollution.

Fairtrade is an important component of economic sustainability in the fashion and clothing industry. The Fairtrade certification scheme ensures that workers are paid fair wages, have access to education and healthcare, and work under safe and gender equality conditions (Fairtrade, 2023). Fairtrade also promotes the use of sustainable farming practices that reduce environmental impacts and ensure the long-term viability of agricultural systems. This includes reducing the use of harmful chemicals and encouraging biodiversity conservation.

Fashion and Clothing Industry

The fashion and clothing industry is a multi-billion-dollar global industry that includes processes from design, production, marketing, and sales of apparel and accessories. It is a dynamic and continuously developing industry that includes changes in cultural and social trends, as well as advances in technology and manufacturing techniques. In recent years, fast fashion and slow fashion have become the most significant trends in the fashion and clothing industry. They are two opposing approaches to clothing production as well as consumption that have significantly opposite effects on the environment, people, and the economy.

Fast Fashion

Fast fashion, which is the dominating business model over the years, is characterized by the quick production of low-cost clothing that follows the last fashion trends, by creating an easily discardable behavior that can be named a "throwaway" culture (Haciola & Atilgan, 2014). Fast fashion brands constantly bring new products to their stores to keep consumers interested.

Fast fashion's 12 different seasons a year create an illusion of freshness and novelty that appeals to consumers. This is an important way for fast fashion retailers to increase their sales. The business model focuses on getting new designs from the drawing board to the shelves in a matter of weeks, which allows for a rapid response to fashion trends. This approach enables fast fashion retailers to minimize the risks associated with carrying unsold inventory from one season to the next.

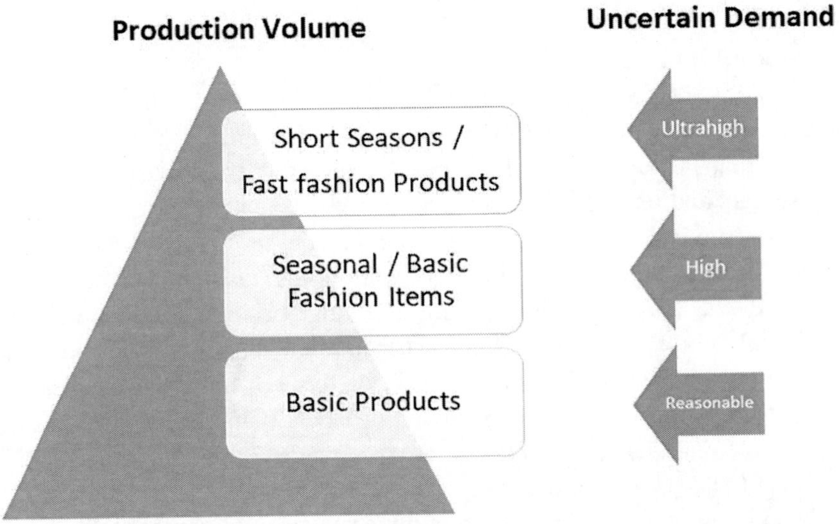

Figure 2. Demand pyramid for basic and fashion products (Regenerated; Fernie & Grant, 2015).

One of the reasons why fast fashion has been able to achieve this is the emphasis on production efficiency (Figure 2). Because fast fashion has short seasons, the supply chain must maintain a high level of efficiency in terms of minimizing production shortages and delivery times (Barnes & Lea-Greenwood, 2010). Retailers prefer production sites close to the market to procure products in a short time (Bruce & Daly, 2006). By outsourcing production to countries with lower labor costs, fast fashion retailers can manufacture products quickly and cheaply, keeping costs down and profits up.

The Impact of Fast Fashion

The fast fashion industry has come under scrutiny in recent years for its negative environmental and social impact. Especially, this business model is criticized for the exploitation of workers mainly in developing countries. The fast fashion workers are often paid low wages, work long hours in unsafe conditions, and are denied basic rights. The local communities where clothing is produced also suffer from negative impacts, including pollution and displacement. In terms of labor exploitation, workers in developing countries are often paid below the minimum wage, with long working hours and unsafe working conditions. According to the International Labour Organization, around 160 million children are engaged in child labor, with many working in the clothing industry (ILO, 2023).

In addition to social impacts, the fast fashion business model has a significant impact on the environment due to the high level of waste generated by the rapid turnover of clothing. Buying trendy and inexpensive clothing in large quantities often results in the disposal of these items after only a few wears. In the following, the environmental impact of fast fashion was collected and discussed in four groups.

Water Consumption

The production of fast fashion requires a significant amount of water. According to the report of the European Parliament, it takes approximately 2,700 liters of water to produce a cotton shirt, which is enough water for a person to drink for 2.5 years (European Parliament, 2023a). To make a pair of jeans, 3,781 liters of water is needed from the production of the cotton to the delivery of the final product to the store (World Bank, 2023). In addition, the dyeing and finishing processes of clothing production contribute to water pollution, as chemicals and dyes are often discharged into the waterways. According to the European Environment Agency (EEA), approximately 13,000 tonnes of textile microfibres, which is equivalent to 25 grams per person, enter surface water every year in Europe. Clothes are identified as the source of around 8% of microplastics that are released into the oceans in Europe. The shedding of synthetic fabrics is most significant during the first few washes, highlighting the challenges posed by the fast fashion industry, which frequently employs polyester and is commonly worn for shorter durations (EEA, 2023).

Greenhouse Gas Emissions

Fast fashion production and transportation also contribute to greenhouse gas emissions, which are a major contributor to climate change. The fashion and clothing industry is responsible for 10% of global greenhouse gas emissions, surpassing the combined emissions from international flights and maritime shipping. It is estimated that there will be a surge of more than 50% in fashion industry greenhouse gas emissions by 2030 (World Bank, 2023). Furthermore, it ranks fifth highest in terms of its impact on raw material consumption and greenhouse gas emissions (European Parliament, 2023b).

Waste Generation

Fast fashion has a significant impact on waste generation. According to estimates, every second a truckload of discarded textiles is either dumped in landfills or incinerated, where it can take hundreds of years to decompose. At the same time, it is also estimated that individuals are purchasing 60% more clothes than before and wearing them for only half as long (Unep, 2023).

Chemical Use

The production of fast fashion requires the use of chemicals such as pesticides, insecticides, and synthetic dyes. These chemicals can have severe negative impacts on both human health and the environment. For example, some pesticides used in cotton production are known to be highly toxic and can contaminate soil and water. Compared to its land use, cotton production has a significantly greater impact on pesticide and insecticide sales, accounting for 4.7% and 10% of the world's total, respectively (Better Cotton, 2023).

Slow Fashion

While fast fashion has dominated the fashion industry in recent years, there has been growing interest in slow fashion. Consumers are becoming more aware of the negative impact of fast fashion and are looking for more sustainable and ethical alternatives (McNeill & Snowdon, 2019). At that point, slow fashion comes to the fore, which is an alternative approach to the fast-paced, mass-produced fashion industry. It is a movement that aims to promote sustainable and ethical fashion practices. Slow fashion promotes mindful consumption and production of clothing items by prioritizing quality over quantity (Jung & Jin, 2014). Its aim is to add value to clothing items in a sustainable and rational way, in contrast to the fast-fashion model (Solino et al., 2020). The fundamentals of slow fashion can be categorized into four principles.

Sustainability

The use of sustainable materials and production processes is a priority in the slow fashion movement. This includes using organic or recycled materials, minimizing waste, and reducing the carbon footprint of production.

Ethical Production

Ethical in production practices is also a principle of social sustainability that includes fair labor practices, ensuring that workers are paid fair wages and work in safe and healthy conditions (McNeill & Snowdon, 2019).

Local and Hand-Crafted Production

Local and artisanal production is promoted in the slow fashion movement, instead of depending on global supply chains and mass production. This practice helps to maintain traditional crafts and techniques and supports local economies (Şener et al., 2019).

Quality over Quantity

In the slow fashion movement, a higher value on the quality of the produced garments is prioritized as they are designed to last and be worn for years. Thus, the consumers will need fewer garments to buy, leading to a reduction in waste (Pookulangara & Shephard, 2013; Jung & Jin, 2014).

As consumers become more aware of the consequences of their purchasing decisions, slow fashion has become an important movement in the fashion industry (Mangır, 2016). Slow fashion designers are also gaining more visibility, and many retailers are starting to offer slow fashion collections.

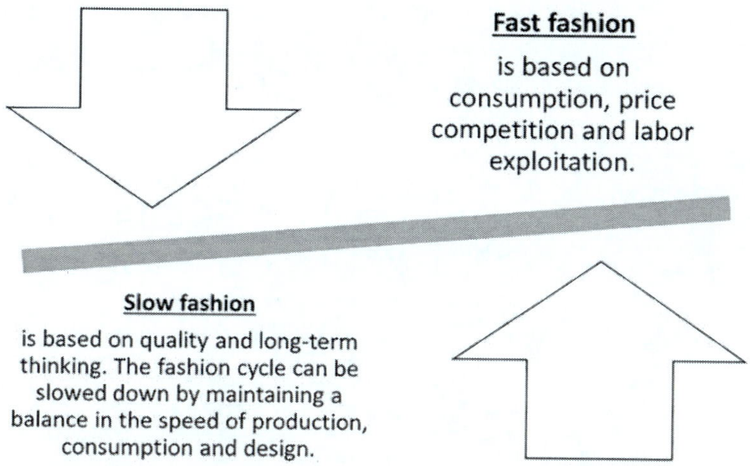

Figure 3. The fast fashion versus the slow fashion movement.

The fashion industry is complex and there are many factors to consider when it comes to fast fashion and slow fashion (Figure 3). While fast fashion has dominated the industry over the past years, there is a rising trend towards slow fashion, and customers are increasingly aware of the ecological and social implications of their fashion decisions.

Three Rs of Sustainability

As concerns about climate change and sustainable development grow, the implementation of 3R strategies in the fashion and clothing industry is becoming increasingly important. The 3R strategies of sustainability are based on three guiding principles: reduce, reuse, and recycle (Figure 4), which are widely recognized as a comprehensive approach to achieve sustainability in the fashion and clothing industry.

These principles aim to minimize waste and environmental impact by encouraging efficient use of resources, extending product lifecycles, and promoting recycling initiatives (Alhola et al., 2019).

In the circular economy model, 3R strategies contribute to achieve closed loops by preventing the generation of waste and turning waste into resources. The circular development model aims to eventually separate global economic growth from the consumption of limited resources (Zhang et al., 2022).

The waste hierarchy in 3R strategies can be visualized as an inverted pyramid where reduction waste forms the broad base at the top, followed by reuse in the middle and recycling at the bottom (Figure 5). As the amount of waste is reduced, the need for reuse decreases and as the amount of waste reused increases, the need for recycling decreases correspondingly.

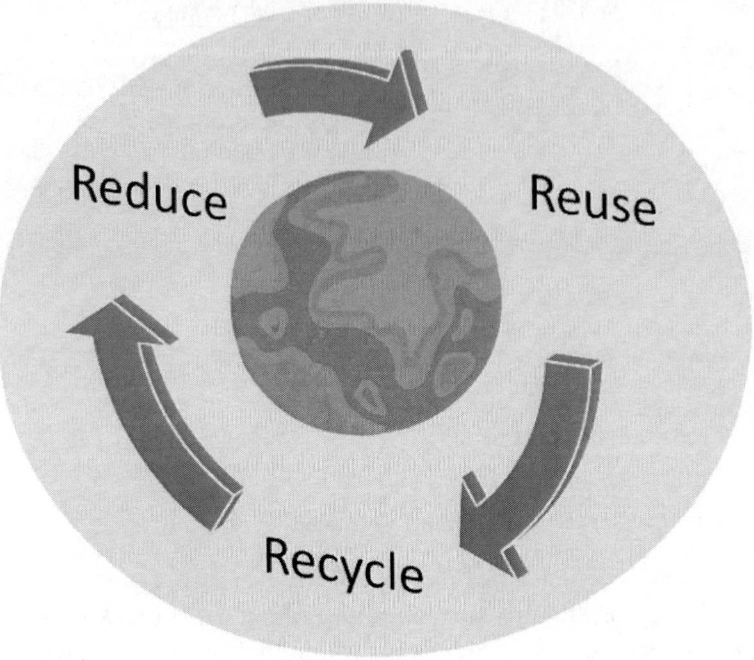

Figure 4. The 3Rs of sustainability.

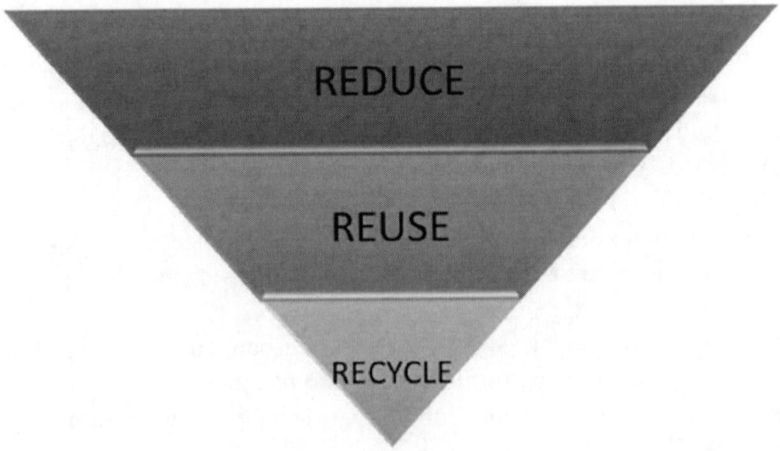

Figure 5. The waste hierarchy in 3R strategies of sustainability.

Reduce

Reducing waste in the fashion and clothing industry starts with minimizing resource consumption and waste generation during production processes (Allwood et al., 2006). The Reduce strategy is based on the principle that "the best way to manage waste is not to produce it" (Necef et al., 2020).

Strategies to achieve minimizing the use of resources include sustainable sourcing, energy-efficient manufacturing, and the use of eco-friendly materials (Allwood et al., 2006). Sustainable sourcing involves selecting raw materials that have a lower environmental impact, such as organic cotton, recycled fibers, or biodegradable alternatives (Muthu et al., 2012). To reduce waste and resource consumption can be achieved by implementing energy-efficient manufacturing processes such as optimizing energy use, minimizing water consumption, and reducing the use of harmful chemicals (Henninger et al., 2016). Finally, designing products that last longer can help reduce waste and resource consumption by extending the life cycle of garments, leading to fewer replacements and less demand for new production (Necef et al., 2020).

"Zero-waste fashion design" is a design method that can help eliminate fabric scraps and reduce textile waste in the apparel manufacturing process (Carrico & Kim, 2014; Daukantiene, 2023). Although the term zero waste has been used since 2008, traditional clothing in Europe and Asia had already incorporated the idea of zero-waste design by minimizing fabric usage during garment production (Rissanen, 2013; Paralı, 2021; Kim & Kim, 2022).

Garment production results in the generation of fabric waste, amounting to approximately 10-20% of the total fabric used (Rissanen, 2013). The concept of zero-waste fashion involves creating clothing designs that minimize or eliminate waste during the pattern-making and cutting stage (McQuillan, 2020; Daukantiene, 2023). The effectiveness of this method is heavily influenced by the width of the fabric being used (James et al., 2016) and the seasonal fashion trends, which impose strict limits on the silhouettes of garments (Daukantiene, 2023). Zero-waste fashion design requires simultaneous consideration of the technical and aesthetic elements of the garment while taking into account its two- or three-dimensional nature (Rissanen, 2013).

Reuse

Reuse emphasizes the importance of prolonging the life of products to minimize waste and resource consumption (Ekström & Salomonson, 2014). It is one of the best ways to extend the practical life of a garment by reusing the product to be used for the same purpose as when it was first produced. In the textile waste hierarchy, "reusing" refers to the process of taking discarded clothing that is still usable, repairing it, and wearing it again without making any changes to its original form (Wu et al., 2022).

Fast fashion often imposes the need to constantly wear different clothes, which leads to increased consumption and contributes to the growing problem of textile waste. However, encouraging consumers to wear the same clothes repeatedly can be an important step towards reducing the need for new clothing purchases and minimizing the negative environmental impact of textile production. Normalizing the practice of rewearing outfits by repairing them is gaining popularity among fashion-conscious individuals who are also concerned about the environment (Laitala & Klepp, 2018; McQueen, 2022). Repairing a damaged garment can return it to a condition where it is suitable for further use (Laitala & Klepp, 2018). Repairing clothes can involve a wide range of techniques, including mending holes or tears, replacing buttons or zippers, and patching up worn-out areas (McQueen, 2022). Beyond reuse, upcycling is a process that aims to transform old objects into better ones for a completely new purpose. Upcycling involves taking existing objects and creating something new and useful out of them through innovative designs and ideas (Wu et al., 2022). This approach not only promotes creativity and resourcefulness but also helps reduce waste by giving new life to objects that might otherwise have been discarded. Marques et al. (2019) participated in the "Upcycling Project by TAP (The Portuguese Airline Company)" contest by creating a collection of fashion products with high value added by transforming waste materials (Marques et al., 2019). The team used discarded fabrics from aircraft seats and metallic seat belts as their raw materials and designed a series of prototypes for their collection. By repurposing these materials, Marques et al. were able to create new products with unique designs and qualities, while also promoting sustainability and reducing waste. Overall, their project serves as an inspiring example of how upcycling can be used to create innovative, eco-friendly fashion products that add value and reduce our impact on the environment.

In recent years, the most commonly used model for clothing reuse is the second-hand fashion business model (Janigo & Wu, 2015). The rising popularity of second-hand clothing has sparked an eco-friendly and budget-conscious revolution in the fashion industry over the last few decades; however, it is still not entirely accepted by everyone (Ekström & Salomonson, 2014). The second-hand clothing has a significant positive environmental impact. As Figure 6 states, which was prepared using the data obtained from Thredup (2023), the use of second-hand clothing by consumers can significantly reduce the carbon footprint and water use.

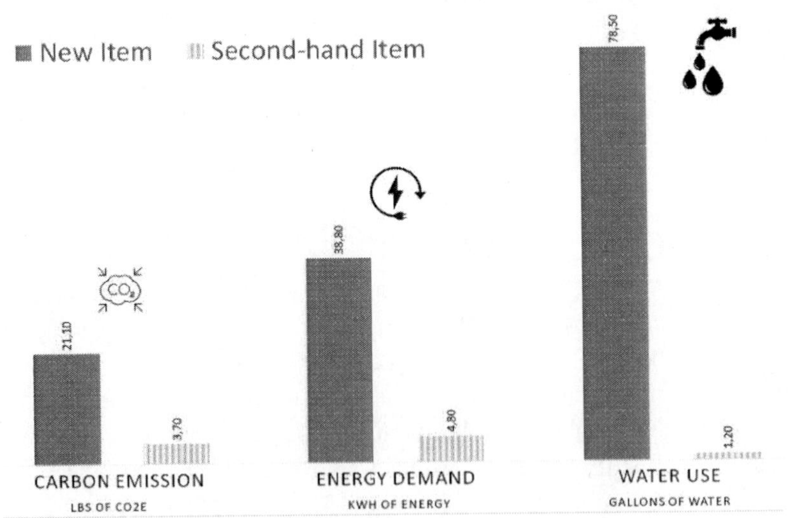

Figure 6. The footprint of new clothing and a second-hand clothing.

The benefits of using second-hand goods can be listed as below.

- *Affordability;* the products are more affordable than new ones.
- *Sustainable;* they are already pre-existing; they do not require any additional resources for production.
- *Unique style;* they are unique and rare products that are no longer in production.
- *Timeless;* due to their enduring style, these products can maintain their attractiveness over time.
- *Good Quality;* the quality of these goods is typically high, as they remain in existence.

By embracing reuse strategies such as clothing rental services and second-hand marketplaces, brands can contribute to a more sustainable future for the fashion sector. In addition, if retailers assumed the responsibility of collecting, cleaning, repairing, and reselling used clothing, it would enhance the probability of customer returns and resales, and foster trust among users (Dissanayake & Weerasinghe, 2021). With respect to these strategies, some initiatives and applications are presented below.

The North Face Renewed is a program by the outdoor clothing and gear brand The North Face that focuses on taking back gently used The North Face garments, refurbishing and repairing them to like-new condition, and reselling them at a lower cost. Customers get a $10 shopping credit by bringing back the used garments; the accepted items are subjected to a cleaning and refurbishing process, or they are either donated or recycled (The North Face, 2023).

Similar initiatives have been undertaken by Eileen Fisher and Patagonia, which have been named the Eileen Fisher Renew program (Eileen Fisher, 2023) and Patagonia Worn Wear (Patagonia, 2023), respectively. The programs aim to repair, trade-in, and resale the items; offering customers an ethical and sustainable way to dispose of their garments.

Recycling

The term "recycling" has been used since the early 1970s with increasing in environmental awareness; however, the applications were very limited; only the products with a resale value such as aluminum cans, glass bottles, and newspapers were recycled (Domina & Koch, 1997). Recycling, one of the components of the 3Rs of sustainability, as well as one of the most important concepts of waste management, is the process of converting waste materials into new products or raw materials that helps conserve natural resources (Necef et al., 2022).

In the fashion and clothing industry, textile wastes are categorized into two types, namely pre-consumer wastes and post-consumer waste.

- *Pre-consumer waste:* It is the waste generated during the manufacturing process of textiles and can include various forms of waste, such as off-cuts, selvages, shearings, and rejected materials. These materials are often produced during the processing of fibers, the production of finished yarns and textiles, technical textiles, nonwovens, garments, and footwear. Pre-consumer textile waste can also include B-grade garments or offcuts of saleable size that do not meet the required quality standards. The amount of pre-consumer waste generated can vary depending on the production process and the efficiency of waste management practices within the industry (Chavan, 2014). On the other hand, in the clothing industry, the fabric scraps generated during the cutting and sewing processes are one of the main sources of waste (Necef et al., 2013).
- *Post-consumer waste:* It is the waste generated after the product has been used and discarded by the consumer. This waste includes garments or household textiles that are no longer needed, used, and do not have the possibility to be reused. Clothing that is damaged, stained, or otherwise unlikely to be worn again may be shredded into fiber

to be used in products similar in nature to those manufactured from pre-consumer textile waste (Chavan, 2014; Celep et al., 2023).

One of the challenges in recycling textile waste is that almost all textile products are made from a blend of different fibers, which can make it difficult to separate and recycle the materials (Daukantiene, 2023). According to Kessler et al. (2021), the current amount of textile products recovered for potential reuse and recycling is limited (Kessler et al., 2021). This is evidenced by the fact that in 2020, the global cotton recycling rate was only 0.96%, while the polyester recycling rate was 14.7% (European Parliament, 2023b). However, it should be noted that a significant portion of recycled polyester (99%) is produced from plastic bottles rather than textile waste (Şener et al., 2022), as reported by the European Parliament Briefing in 2022.

Recycling technologies for textile waste can be broadly divided into two categories: mechanical recycling and chemical recycling.

- *Mechanical recycling: It* involves the use of physical processes such as shredding, cutting, and grinding to break down textile waste into smaller fibers or particles that can be used to create new products. This method is limited to fabrics made of the same fiber type, as the blending of fibers can make the process difficult.
- *Chemical recycling:* Chemical recycling typically involves the depolymerization of materials, which breaks down the long chains of polymers in textiles into their constituent monomers. These monomers can then be purified and repolymerized back into new fibers, which can be used to create new textiles or other products (Dissanayake & Weerasinghe, 2021; Celep et al., 2023). Chemical recycling can handle mixed fiber fabrics, and it can be a more sustainable alternative to mechanical recycling.

Recycled textile materials can be used in a wide range of applications such as home textiles, lingerie (Ünal & Acar, 2022), upholstery (Esi & Baykal, 2022), auxiliary (El-Sayed et al., 2021), carpets, blankets (Chen et al., 2021), etc. Other potential applications for recycled textiles include acoustic insulation, building materials, and packaging materials.

Conclusion

The fashion and clothing industry plays a significant role in the global economy, generating employment and contributing to economic growth. However, it also has a great impact on the environment and society in a negative way. In response, there is a growing interest in sustainable and ethical practices within the industry, with many businesses and organizations making efforts to address these issues.

Sustainability requires a collaborative effort of individuals, communities, businesses, and policymakers. Thus, it is necessary to make changes both in production processes and consumption attitudes (Niinimäki et al., 2020). However, this will require the participation, and cooperation of all stakeholders, including the textile industry, fashion businesses, consumers, and policymakers. The textile industry will need to invest in clean technology, while fashion businesses will need to develop new business models that prioritize sustainability. Consumers will also need to modify their consumption habits to reduce waste and support sustainable fashion choices. Policymakers can contribute to these efforts by enacting legislation and global business rules that promote sustainability and reduce the negative impact of the fashion industry. Overall, collective efforts are needed from all stakeholders to achieve a more sustainable and ethical fashion industry.

Disclaimer

None.

References

A European Green Deal. (2023, February 12). *Striving to be the first climate-neutral continent.* https://commission.europa.eu/strategy-and-policy/priorities-2019-2024/european-green-deal_en.

Acar, E., Kilic, M., & Güner, M. (2015). Measurement of sustainability performance in textile industry by using a multi-criteria decision making method. *Textile and Apparel, 25*(1), 3-9.

Alhola, K., Ryding, S. O., Salmenperä, H., & Busch, N. J. (2019). Exploiting the potential of public procurement: Opportunities for circular economy. *Journal of Industrial Ecology, 23*(1), 96-109. https://doi.org/10.1111/jiec.12770.

Allwood, J. M., Laursen, S. E., De Rodríguez, C. M., & Bocken, N. M. (2006). Well dressed. The present and future sustainability of clothing and textiles in the United Kingdom. *Technical Annex, 1*.

Barca, S. (2011). Energy, property, and the industrial revolution narrative. *Ecological Economics, 70*(7), 1309-1315. https://doi.org/10.1016/j.ecolecon.2010.03.012.

Barnes, L., & Lea-Greenwood, G. (2010). Fast fashion in the retail store environment. *International Journal of Retail & Distribution Management, 38*(10), 760-772. https://doi.org/10.1108/09590551011076533.

Better Cotton. (2023, February 18). *Pesticides and crop protection.* https://bettercotton.org/field-level-results-impact/key-sustainability-issues/pesticides-and-crop-protection-in-cotton-farming/.

Beyer, K., & Arnold, M. G. (2022). Social sustainability in an evolving circular fashion industry: Identifying and triangulating concepts across different publication groups. *NachhaltigkeitsManagementForum 30,* 29–54. https://doi.org/10.1007/s00550-022-00527-x.

Bossavie, L., Cho, Y., & Heath, R. (2019). The effects of international scrutiny on manufacturing workers: Evidence from the Rana Plaza collapse in Bangladesh. *World Bank Policy Research Working Paper,* (9065).

Brooksworth, F., Mogaji, E., & Bosah, G. (2022). Brand, consumer and sustainability perspectives in fashion marketing: Conclusion and research agenda. In F. Brooksworth, E. Mogaji & G. Bosah (Eds.), *Fashion marketing in emerging economies volume I: Brand, consumer and sustainability perspectives* (pp. 267-278). Cham: Springer International Publishing.

Bruce, M., & Daly, L. (2006). Buyer behaviour for fast fashion. *Journal of Fashion Marketing and Management: An International Journ*al, *10*(3), 329-344. https://doi.org/10.1108/13612020610679303.

Burns, T. R. (2012). The sustainability revolution: A societal paradigm shift. *Sustainability, 4*(6), 1118-1134. https://doi.org/10.3390/su4061118.

Carrico, M., & Kim, V. (2014). Expanding zero-waste design practices: A discussion paper. *International Journal of Fashion Design, Technology and Educatio*n, *7*(1), 58-64. https://doi.org/10.1080/17543266.2013.837967.

Celep, G., Tetik, G., & Yilmaz, F. (2023). Limitations of textile recycling: The reason behind the development of alternative sustainable fibres. In H. Ibrahim (Ed), *Next-generation textiles* (pp.1-26). IntechOpen. https://doi.org/10.5772/intechopen.105118.

Centobelli, P., Abbate, S., Nadeem, S. P., & Garza-Reyes, J. A. (2022). Slowing the fast fashion industry: An all-round perspective. *Current Opinion in Green and Sustainable Chemistry, 38,* 100684, 1-9. https://doi.org/10.1016/j.cogsc.2022.100684.

Chavan, R. B. (2014). Environmental sustainability through textile recycling. *Journal of Textile Science and Engineering, 2*(2), 1-5.

Cheek, W. K., & Moore, C. E. (2003). Apparel sweatshops at home and abroad: Global and ethical issues. *Journal of Family and Consumer Sciences, 95*(1), 9-19.

Chen, X., Memon, H. A., Wang, Y., Marriam, I., & Tebyetekerwa, M. (2021). Circular economy and sustainability of the clothing and textile industry. *Materials Circular Economy, 3*(1), 1-9. https://doi.org/10.1007/s42824-021-00026-2.

Choi, T. M., & Li, Y. (2015). Sustainability in fashion business operations. *Sustainability, 7*(11), 15400-15406. https://doi.org/10.3390/su71115400.

Clune, W. H., & Zehnder, A. J. (2018). The three pillars of sustainability framework: Approaches for laws and governance. *Journal of Environmental Protection, 9*(3), 211-240. https://doi.org/10.4236/jep. 2018.93015.

Daukantiene, V. (2023). Analysis of the sustainability aspects of fashion: A literature review. *Textile Research Journal, 93*(3-4), 991-1002. https://doi.org/10.1177/00405175221124971.

Devaki, E., Suganya, M., & Sreelakshmi S. (2018). Sustainable fashion – a review. In M. Parthiban, M. R. Srikrishnan & P. Kandhavadivu (Eds.), *Sustainability in fashion and apparels: Challenges and solutions* (pp.73-82). CRC Press.

Dissanayake, D. G. K., & Weerasinghe, D. (2022). Towards circular economy in fashion: Review of strategies, barriers and enablers. *Circular Economy and Sustainability, 2,* 25-45. https://doi.org/10.1007/s43615-021-00090-5.

Domina, T., & Koch, K. (1997). The textile waste lifecycle. *Clothing and Textiles Research Journal, 15*(2), 96-102. https://doi.org/10.1177/0887302X9701500204.

Du Pisani, J. A. (2006). Sustainable development–historical roots of the concept. *Environmental Sciences, 3*(2), 83-96. https://doi.org/10.1080/15693430600688831.

EEA. (2023, February 18). *Microplastics from textiles: Towards a circular economy for textiles in Europe.* https://www.eea.europa.eu/publications/microplastics-from-textiles-towards-a.

Eileen Fisher. (2023, March 12). *A casual friday state of mind—all week long.* https://www.eileenfisher.com/circular-by-design/journal/sustainability/renew-program-reaches-2-million-garments.html?loc=US.

Ekins, P., & Zenghelis, D. (2021). The costs and benefits of environmental sustainability. *Sustainability Science, 16,* 949-965. https://doi.org/10.1007/s11625-021-00910-5.

Ekström, K. M., & Salomonson, N. (2014). Reuse and recycling of clothing and textiles—A network approach. *Journal of Macromarketing, 34*(3), 383-399. https://doi.org/10.1177/0276146714529658.

El-Sayed, H., AbouTaleb, M., & Mowafi, S. (2021). Potential applications of textile wastes and by-products in preparation of textile auxiliaries. *Egyptian Journal of Chemistry, 64*(8), 4433-4447. https://doi.org/10.21608/EJCHEM.2021.79398.3899.

Esi, B., & Baykal, P. D. (2022). Optimization of the yarn slippage at the seam in chenille upholstery fabrics including recycled and virgin polyester yarns. *Textile and Apparel, 32*(1), 1-8. https://doi.org/10.32710/tekstilvekonfeksiyon.928820.

European Parliament. (2023a, February 18). *The impact of textile production and waste on the environment (infographics).* https://www.europarl.europa.eu/news/en/headlines/society/20201208STO93327/the-impact-of-textile-production-and-waste-on-the-environment-infographic.

European Parliament. (2023b, February 18). *Textiles and the environment.* https://www.europarl.europa.eu/RegData/etudes/BRIE/2022/729405/EPRS_BRI(2022)729405_EN.pdf.

Fairtrade. (2023, February 15). https://www.fairtrade.net/.

Fernie, J., & Grant, D. B. (2015). *Fashion logistics: Insights into the fashion retail supply chain.* Kogan Page Publishers.

Giddings, B., Hopwood, B., & O'brien, G. (2002). Environment, economy and society: fitting them together into sustainable development. *Sustainable Development, 10*(4), 187-196. https://doi.org/10.1002/sd.199.

Göztaşı E. T., Ünal Z. B., & Acar, E. (2022). Evaluation of selected apparel subcontractors in terms of social compliance. *Osmaniye Korkut Ata University Journal of The Institute of Science and Technology, 5*(2), 653-667.

Haciola, Y., & Atilgan, T. (2014). Research on fast fashion and fast fashion retailing in Turkish apparel sector. *Textile and Apparel, 24*(2), 143-150.

Henninger, C. E., Alevizou, P. J., & Oates, C. J. (2016). What is sustainable fashion?. *Journal of Fashion Marketing and Management: An International Journal, 20*(4), 400-416. https://doi.org/10.1108/JFMM-07-2015-0052.

Ikram, M. (2022). Transition toward green economy: Technological innovation's role in the fashion industry. *Current Opinion in Green and Sustainable Chemistry*, 37, 100657, 1-10. https://doi.org/10.1016/j.cogsc.2022.100657.

ILO. (2023, February 18). *What is child labour?*. https://www.ilo.org/ipec/facts/lang--en/index.htm.

James, A. M., Roberts, B. M., & Kuznia, A. (2016). Transforming the sequential process of fashion production: Where zero-waste pattern cutting takes the lead in creative design. *International Journal of Fashion Design, Technology and Education, 9*(2), 142-152. https://doi.org/10.1080/17543266.2016.1167253.

Janigo, K. A, & Wu, J. (2015). Collaborative redesign of used clothes as a sustainable fashion solution and potential business opportunity. *Fashion Practice, 7*(1), 75-97.

Jung, S., & Jin, B. (2014). A theoretical investigation of slow fashion: Sustainable future of the apparel industry. *International Journal of Consumer Studies, 38*(5), 510-519. https://doi.org/10.1111/ijcs.12127.

Kessler, L., Matlin, S. A., & Kuemmerer, K. (2021). The contribution of material circularity to sustainability—Recycling and reuse of textiles. *Current Opinion in Green and Sustainable Chemistry, 32*, 100535. https://doi.org/10.1016/j.cogsc.2021.100535.

Kim, S., & Kim, H. Y. (2022). Creative exploration: Zero-waste fashion design practices with traditional Korean clothing. *International Journal of Fashion Design, Technology and Education,* 1-16. https://doi.org/10.1080/17543266.2022.2148293.

Laitala, K., & Klepp, I. G. (2018). Care and production of clothing in Norwegian homes: Environmental implications of mending and making practices. *Sustainability, 10*(8), 2899. https://doi.org/10.3390/su10082899.

Li, J., & Leonas, K. K. (2022). Sustainability topic trends in the textile and apparel industry: A text mining-based magazine article analysis. *Journal of Fashion Marketing and Management: An International Journal, 26*(1), 67-87. https://doi.org/10.1108/JFMM-07-2020-0139.

Macchion, L., Da Giau, A., Caniato, F., Caridi, M., Danese, P., Rinaldi, R. & Vinelli, A. (2018). Strategic approaches to sustainability in fashion supply chain management. *Production Planning & Control, 29*(1), 9-28. https://doi.org/10.1080/09537287.2017.1374485.

Mackie, L., & Campbell, D. (2022). Sustainability and the supply chain. In F. Brooksworth, E. Mogaji & G. Bosah (Eds.), *Fashion marketing in emerging economies volume I: Brand, consumer and sustainability perspectives* (pp. 163-220). Cham: Springer International Publishing.

Mangır, A. F. (2016). Fast and slow fashion for sustainable development. *Journal of Selçuk University Social Sciences Vocational School, 19*, 143-154.

Marques, A. D., Moreira, B., Cunha, J. & Moreira, S. (2019). From waste to fashion–a fashion upcycling contest. *Procedia CIRP, 84*, 1063-1068. https://doi.org/10.1016/j.procir.2019.04.217.

McNeill, L. S., & Snowdon, J. (2019). Slow fashion–Balancing the conscious retail model within the fashion marketplace. *Australasian Marketing Journal, 27*(4), 215-223. https://doi.org/10.1016/j.ausmj.2019.07.005.

McQueen, R. H., McNeill, L. S., Kozlowski, A., & Jain, A. (2022). Frugality, style longevity and garment repair–environmental attitudes and consumption behaviour amongst young Canadian fashion consumers. *International Journal of Fashion Design, Technology and Education, 15*(3), 371-384. https://doi.org/10.1080/17543266.2022.2072958.

McQuillan, H. (2020). Digital 3D design as a tool for augmenting zero-waste fashion design practice. *International Journal of Fashion Design, Technology and Education, 13*(1), 89-100. https://doi.org/10.1080/17543266.2020.1737248.

Muthu, S. S., Li, Y., Hu, J. Y., & Ze, L. (2012). Carbon footprint reduction in the textile process chain: Recycling of textile materials. *Fibers and Polymers, 13*, 1065-1070. https://doi.org/10.1007/s12221-012-1065-0.

Necef, Ö. K., Birkocak, D. T., Kılıç, A., Boz, S., Abreu, M. J., & Öndoğan, Z. (2022). Investigating the effect of recycled cotton included fabrics on the thermal behaviour by using a female thermal manikin. *Autex Research Journal*. https://doi.org/10.2478/aut-2022-0020.

Necef, Ö. K., Seventekin, N., & Pamuk, M. (2013). A study on recycling the fabric scraps in apparel manufacturing industry. *Textile and Apparel, 23*(3), 286-289.

Necef, Ö. K., Tama, D., & Boz, S. (2020). Examples of sustainability strategies applied in fashion industry evaluation. *Turkish Journal of Fashion Design and Management, 2*(2), 67-78.

Nguyen, N. P., & Mogaji, E. (2022). Emerging economies in fashion global value chains: brand positioning and managerial implications. In F. Brooksworth, E. Mogaji & G. Bosah (Eds.), *Fashion marketing in emerging economies volume I: Brand, consumer and sustainability Perspectives* (pp. 109-133). Cham: Springer International Publishing.

Niinimäki, K., Peters, G., Dahlbo, H., Perry, P., Rissanen, T., & Gwilt, A. (2020). The environmental price of fast fashion. *Nature Reviews Earth & Environment, 1*(4), 189-200. https://doi.org/10.1038/s43017-020-0039-9.

Papadopoulou, M., Papasolomou, I., & Thrassou, A. (2022). Exploring the level of sustainability awareness among consumers within the fast-fashion clothing industry: A dual business and consumer perspective. *Competitiveness Review: An International Business Journal, 32*(3), 350-375. https://doi.org/10.1108/CR-04-2021-0061.

Papamichael, I., Chatziparaskeva, G., Pedreno, J. N., Voukkali, I., Candel, M. B. A., & Zorpas, A. A. (2022). Building a new mind set in tomorrow fashion development through circular strategy models in the framework of waste management. *Current Opinion in Green and Sustainable Chemistry, 36,* 100638. https://doi.org/10.1016/j.cogsc.2022.100638.

Paralı, A. (2021). Investigation of zero waste pattern preparation techniques within the scope of sustainable fashion: The example of mosaic method. *Ulakbilge - Journal of Social Sciences, 9*(67), 1457-1474.

Patagonia. (2023, March 12). https://www.patagonia.com/stories/worn-wear/.

Pookulangara, S., & Shephard, A. (2013). Slow fashion movement: Understanding consumer perceptions—An exploratory study. *Journal of Retailing and Consumer Services, 20*(2), 200-206. https://doi.org/10.1016/j.jretconser.2012.12.002.

Rissanen, T. I. (2013). *Zero-waste fashion design: a study at the intersection of cloth, fashion design and pattern cutting*. [Doctoral dissertation, University of Technology, Sydney].

Solino, L. J. S., de Lima Teixeira, B. M., & de Medeiros Dantas, I. J. (2020). The sustainability in fashion: A systematic literature review on slow fashion. *International Journal for Innovation Education and Research, 8*(10), 164-202.

Şener, T., Bişkin, F., & Dündar, N. (2022). The effects of perceived value, environmental concern and attitude on recycled fashion consumption. *Journal of Fashion Marketing and Management: An International Journal*, *27*(4), 595-611. https://doi.org/10.1108/JFMM-01-2021-0003.

Şener, T., Bişkin, F., & Kılınç, N. (2019). Sustainable dressing: consumers' value perceptions towards slow fashion. *Business Strategy and the Environment, 28*(8), 1548-1557. https://doi.org/10.1002/bse.2330.

The North Face. (2023, March 12). https://www.thenorthfacerenewed.com/.

Thredup. (2023, February 25). https://www.thredup.com/resale/2021/#size-and-impact.

UNEP. (2023, February 18). https://www.unep.org/news-and-stories/story/environmental-costs-fast-fashion.

Ünal, Z. B., & Acar, E. (2022). *Women's underwear and current trends*. Iksad Publishing.

World Bank. (2023, February 18). *How much do our wardrobes cost to the environment?*. https://www.worldbank.org/en/news/feature/2019/09/23/costo-moda-medio-ambiente.

Wu, D., Zhuang, M., Zhang, X., & Zhao, Y. (2022). Towards circular fashion: Design for community-based clothing reuse and upcycling services under a social innovation perspective. *Sustainability, 15*(1), 262. https://doi.org/10.3390/su15010262.

Zhang, C., Hu, M., Di Maio, F., Sprecher, B., Yang, X., & Tukker, A. (2022). An overview of the waste hierarchy framework for analyzing the circularity in construction and demolition waste management in Europe. *Science of the Total Environment, 803*, 149892. https://doi.org/10.1016/j.scitotenv.2021.149892.

Chapter 21

Innovative Joining Methods in the Apparel Industry

Serkan Boz[1], PhD
and Mehmet Küçük[2,*], PhD
[1]Department of Fashion Design, Faculty of Fashion and Design, Ege University, Izmir, Türkey
[2]Department of Textile Engineering, Faculty of Engineering, Ege University, Izmir, Türkey

Abstract

In the apparel industry, due to their two-dimensional structure, fabric surfaces must be combined within the framework of certain rules with various joining methods in order to cover three-dimensional objects. There are different methods as an alternative to traditional joining methods (lock stitch, chain stitch, etc.), also called sewing. These are thermal, radiational, and frictional (ultrasonic) joining. The sewing method is the process of combining fabric surfaces with threads by considering sufficient strength, flexibility, and aesthetic properties. The joining process in sewing is carried out with a sewing needle that creates holes in the fabric surfaces. While alternative joining methods provide seamless joining, they do not create any holes in the fabric surfaces, since there is no sewing needle in these methods. Joining processes with thermal, laser, and ultrasound energy are based on the melting, fusing, and cooling of the joined surfaces. During the ultrasonic (frictional) joining process, the high-frequency mechanical vibration energy is focused and transmitted to the surfaces to be joined. Surfaces rub against each other due to high-frequency mechanical vibration energy, releasing heat. The resulting heat causes the synthetic fabric or joining material to melt. Adhesion occurs as a result of melting in the part exposed to vibration and pressure. When the fabric moves away from the vibration energy; with a sudden loss of heat, this adhesion becomes permanent and ultrasonic joining is obtained. During the thermal joining process, the surfaces to be joined are melted separately as a result of contact with the hot element/air. The surfaces are cooled and solidified under controlled pressure. In this way, thermal joining is achieved. During laser (radiational) joining, laser beams are passed over the fabric to create the joint. The laser is an intense energy source that creates a thermal effect without physical contact. When the materials heated by the laser reach the joining temperature, they absorb the generated energy. In this way, flexible, waterproof, high-strength joints can be obtained. The basis of these three methods is based on thermal joining, but they differ from each other by heat generation sources. Thermoplastic fabrics and/or adhesive materials, which are being used intensively in today's garment operations, have also led to the spread of thermal joining methods. In

* Corresponding Author's Email: mehmet.kucuk@ege.edu.tr.

In: Textile Insights and Future Aspects
Editors: Gamze D. Tetik, Gizem Celep and Fulya Yilmaz
ISBN: 979-8-89113-289-4

this study, innovative alternative joining techniques used in the apparel industry, their use areas, the advantages and disadvantages of these techniques are explained in detail.

Keywords: alternative joining methods, ultrasonic joining, thermal joining, laser joining, innovative sewing

Introduction

The shape of the fabric, which is the most common form of textile materials, can be changed by folding, cutting, stretching, and pulling methods. Although the fabric has the ability to cover simple 3-dimensional (3D) objects, it is possible to cover the complex 3D structure of the human body; only by combining 2D panels. In the garment industry, many different methods such as sewing, ultrasonic joining, thermal joining, and laser joining can be applied with their advantages and disadvantages.

Sewing is the process of joining panels with thread, with sufficient strength, flexibility, and aesthetic properties. However, intermittent joining and needle holes occurs during sewing. It is necessary to use other joining methods for continuous joining and sewing without needle holes. Joining processes with thermal, laser, and ultrasound; they are all based on the melting and cooling of the combined thermoplastic surfaces. In the thermal joining process; as a result of contact with the hot element, the surfaces to be joined are melted separately. The surfaces are cooled and solidified under controlled pressure. The main disadvantage of this method is; extremely hot contact causes the fibers to deteriorate (Shi & Little, 2000).

Light Technology Group and Union Special Company have developed the LEB method for high-speed welding of thermoplastics using the non-contact technique (Whitwell & Fraser, 1971). In this method, the laser is moved over the fabric to create a joining. The laser offers an intense energy source that creates a thermal effect without physical contact. Thermoplastic materials absorb sufficient energy at that wavelength when the joining temperature is reached. Properly controlled LEB provides flexible, waterproof, high-strength joints (Black, 1995), but highly focused, extremely intense laser radiation causes degradation of some thermoplastics (Potente & Uebbing, 1997).

In our modern age, technology is developing rapidly and manufacturing industries are trying to make the most of this situation. One of these technologies is the use of ultrasonic energy in the textile and apparel industry. The term ultrasonic is a term that means faster than sound and refers to sounds that cannot be heard by the human ear. In this study, the types of joining machines using ultrasonic energy, their functions, the products in which they are used, and the advantages they provide will be explained, respectively.

Ultrasonic joining has become the preferred method in industries for joining thermoplastic materials. In general, ultrasonic joining is a method used to combine thermoplastic or non-thermoplastic materials with thermoplastic materials.

It has been observed that the term “sewing” or “welding” is also used instead of the term “joining” in different references. The term “joining” will be used in other parts of the text.

All three methods are applied in different areas of the industry. However, during apparel production, ultrasonic joining comes to the fore due to its advantages, and its use is becoming more common day by day.

Joining Mechanisms

Innovative joining techniques, which have been developed and used as an alternative to traditional joining methods, have many advantages. These methods can be grouped in different ways. According to the most common classification; these methods can be grouped into three categories: thermal (by conduction or convection); radiational (e.g., infrared, laser, or induction), and frictional (e.g., rotary, vibrational, or ultrasonic). Although the joining techniques are different in these methods, the basis of the process is; melting, joining, and cooling the thermoplastic surfaces to be joined (Boles, 2012; Kayar et al., 2015; Hussen et al., 2022) (Figure 1). These techniques, it enables surfaces to be joined without any bonding material (such as needle, thread, or adhesive) (Appleby, 2009; Boles, 2012; Seram & Cabon, 2013; Hussen et al., 2022). In addition, since no chemicals are used, they are considered environmentally friendly techniques (Kayar, 2014).

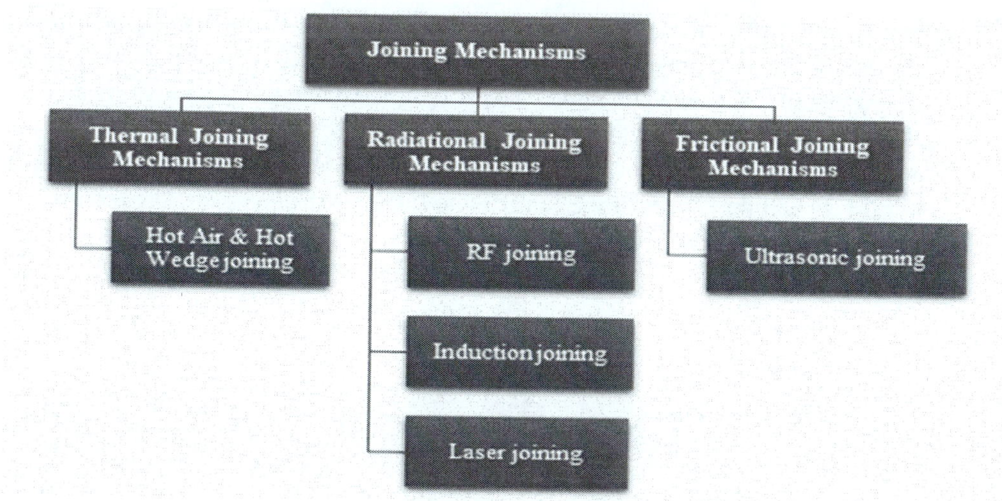

Figure 1. Joining mechanisms.

The joining behaviors of different polymers according to the methods are shown in Table 1.

Table 1. Polymers and the joining mechanisms' behaviors (Pierlot, 2011)

Polymer	Hot Air & Hot Wedge Joining	Ultrasonic Joining	RF Joining (Dielectric)	Laser Joining (Through Transmission)	Induction Joining
Polyethylene	****	****	*	The top fabric layer must be transparent to IR radiation, while the second layer or an intermediate bonding layer absorbs the radiation	A conduction or ferromagnetic susceptor gasket or tape is necessary generally molded from the same polymer. The method allows flexibility in joining dissimilar materials
Polypropylene	****	****	*		
Polyester	b***	****	***		
Polyamide	b***	***	***		
Polyacrylonitrile	c*	*	*		
Polyvinylchloride	****	**	****		
Polyurethane	****	****	****		
Polyphenylenesulfide	**	**	*		

Natural Fibers	Possible for fabrics with more than 50% thermoplastic fiber or use thermoplastic bonding tape: ***	
**** Easiest; *Difficult; Related to ease of joining not bond strength, polymer fillers, and additives may impact joining ease and performance		

Thermal Joining Mechanism

Hot Air & Hot Wedge Joining

The hot air & hot wedge joining is used mostly today that is in underwear and outdoor sports products. The joints obtained by this technique are thin, soft, flexible, waterproof, and easy to care. In addition, a wide variety of textile surfaces, such as natural, synthetic, and blends, can be joined with this technique (Jevsnik et al., 2017).

An external heat unit is used to join thermoplastic surfaces according to the convection or conduction technique. The heat obtained in this unit is transmitted to the surfaces to be joined with hot wedges or hot air. These two methods have similar characteristics. However, there are some different features between them.

For instance, the hot wedge has no noise or fumes compared to the hot air. The heated wedge is inserted between the layers of the fabric. The fabric surface is softened and pressure is used to form the seal (Figure 2).

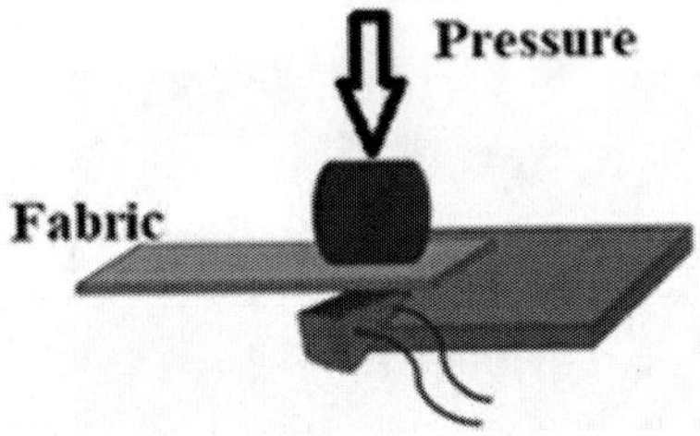

Figure 2. Hot wedge.

The hot air can weld irregular shape seams, e.g., circles. In this technique, heat is provided by blowing air across electrical heating elements and is injected at the joining point.

Wedge and nozzles can be structured in the desired joint form (Figure 3).

Hot air tape joining technology is also used in apparel, especially for covering the needle holes created by classical sewing methods. A coated adhesive tape (usually PUR coated) can be used over or just behind the sewing line to make the seam water and air permeable in this method (Figure 4). It is frequently preferred during the production of outdoor clothing and products (e.g., coats, tents, boots, awnings, etc.).

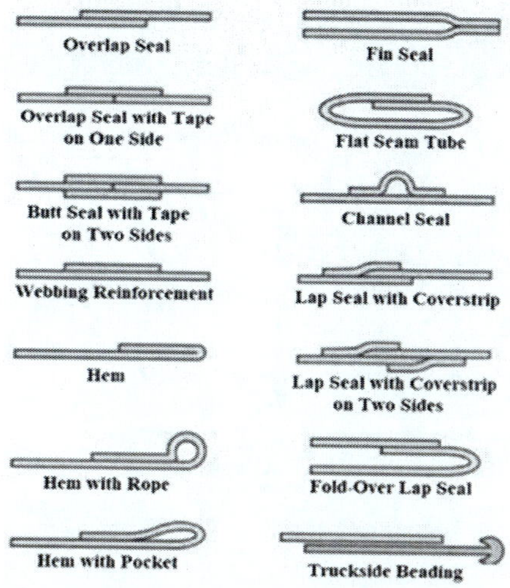

Figure 3. The seal types.

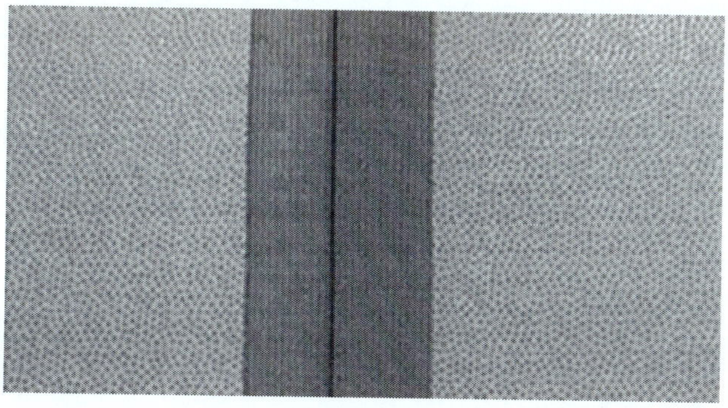

Figure 4. Hot air tape joining.

Radiational Joining Mechanisms

One of the joining methods for thermoplastic surfaces is based on internal heating. RF, induction, and laser generators are most commonly used for internal heating joining methods.

Radio Frequency (RF) Joining

RF joining technology uses high-frequency alternating electromagnetic energy for internal heating (Figure 5).

RF joining is a fast method. It is possible to weld multiple layers simultaneously and creates a good and uniform joint appearance.

This method has some disadvantages, such as material restriction and limited application depending on the thickness of the materials.

RF joining technology is suitable for polymers such as PVC, PUR, PA (Nylon), PES, special grades PP, PE, and cellulose acetate, while it is unsuitable for PTFE and PS.

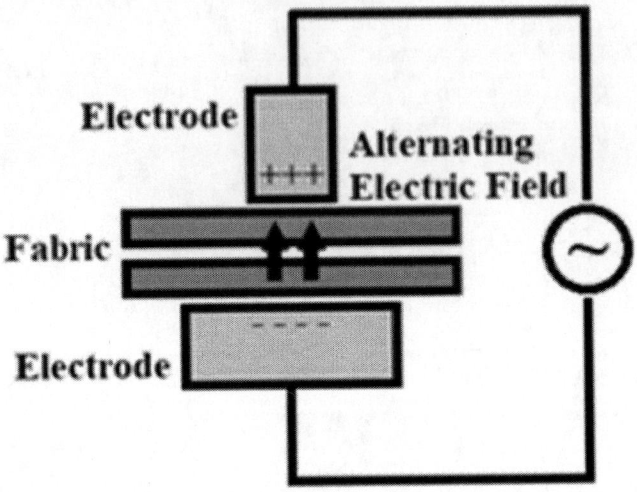

Figure 5. RF joining mechanism.

Some RF joining applications are listed below.

- Chair upholstery
- Quilting
- Table mats
- Polymer Ease
- Tarpaulins
- Awnings
- Tents
- Pool liners,
- Transportation (truck) covers
- Airbags
- Bonding vinyl pads to automotive
- Carpets, etc.

Induction Joining

Induction joining is also called electromagnetic joining. The induction joining method has the following advantages:

- The joint area can be heated remotely.
- The power level of heating and cooling can be easily controlled.
- Complex joints with multiple curvatures can be welded.

- It is possible to make joints with dissimilar materials, e.g., thermoplastics to ceramics, aluminum.
- This method is less dependent on the properties of the materials being welded.
- Induction joining is a rapid method in addition.

Besides all these advantages, the high cost appears as a disadvantage.

Laser Joining

Laser joining is an effective method for joining surfaces. It is preferred in many industrial areas. Due to the flexibility and quality of the joint obtained, it is frequently used in the field of automotive textiles. Two plastic parts with similar chemical properties (the same polymer matrix) are required to have an efficient joining zone (Binetruy et al., 2010).

Infrared (IR) heating using a laser has two approaches. These are direct joining and Through Transmission IR Laser Joining. The applications of laser joining are listed as follows.

- Bed assemblies
- Medical furniture
- Coating
- Clothing
- Airbags, etc.

Laser joining can be done in different ways. One of these methods can be preferred depending on the material properties and the shape of the joint line (Table 2).

Table 2. Laser joining types

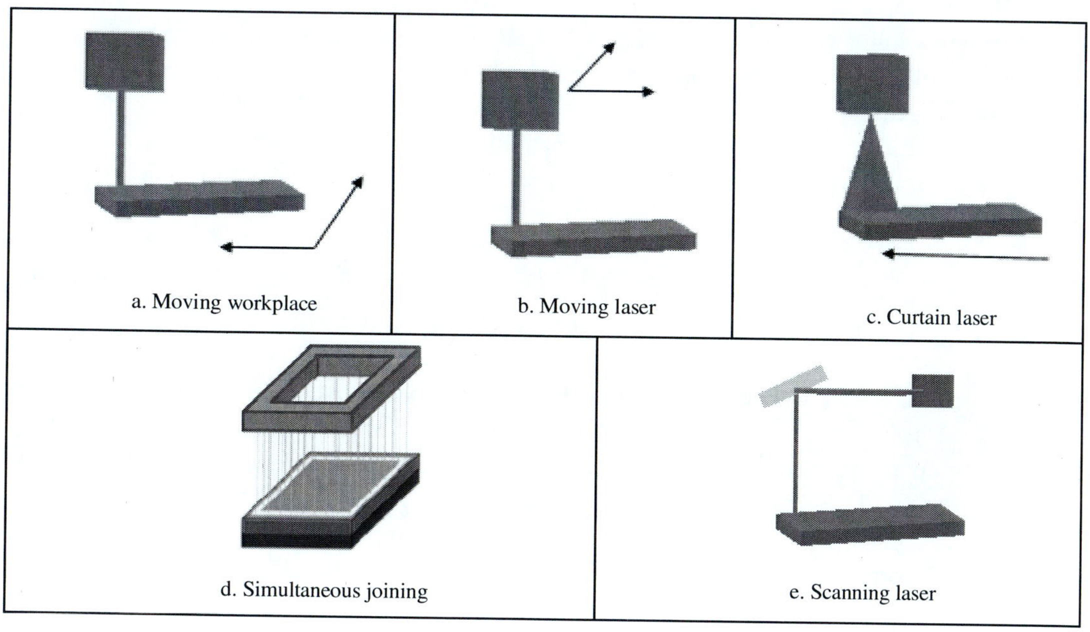

a. Moving workplace

b. Moving laser

c. Curtain laser

d. Simultaneous joining

e. Scanning laser

Frictional Joining Mechanisms

Ultrasonic Joining

In addition to the above-mentioned methods, the ultrasonic joining method is especially preferred for apparel production (clothing). The fact that the hardening after melting is at a minimum level and that the changes in the physical properties such as elasticity, drape, and strength of the joined areas are minimum make this method preferred during the production of clothing.

Ultrasonic energy has vibrational energy beyond the audible frequency of sound. Ultrasonic joining can be used to join surfaces containing at least 60% thermoplastic material. In this technique, there is no need for a preparation process such as bobbin preparation or yarn bobbin replacement (Kayar et al., 2015; Ghosh & Reddy, 2009; Mao & Goswami, 2001). The technique is also preferred by many designers in new generation designs (Seram and Cabon, 2014). The technique has a wide range of applications, including the automotive industry, medical and hygiene products, sports, work and protective clothing, coating and packaging, underwear, filters, and technical textiles (Hussen et al., 2022).

Ultrasonic devices have been used in processes such as cleaning, degradation, metal and plastic joining, drying, extraction, atomization, and reducing chemical reactions for years. However, the use of ultrasonic systems in textiles and related industries has become widespread in recent years, in the mid-90s.

The apparel industry uses ultrasonic energy in a wide variety of processes. The first major application of this technology in fabric and fiber joining dates back to the 1970s (it was first used to create mattress pads and bedspreads). Today, there are ultrasonic devices suitable for trimming, cutting, buttoning, product shaping and assembly, joining, and some other uses in the textile industry. Ultrasonic energy is an important tool used to join, cut, and form textiles, nonwovens, and film surfaces and transform them into a convenient form that can be used by many industries or for the end user.

Ultrasonic joining has become the preferred method in industries for joining thermoplastic materials. In general terms, ultrasonic joining is a method used to combine thermoplastic or non-thermoplastic materials with thermoplastics (Boz, 2008). Ultrasonic joining is an advanced technique of synthetic materials to obtain waterproof joining. Fabrics can be 100% synthetic or up to 40% natural fiber content. It is a fast, clean (no thread), and economical method for garment manufacturers.

Energy Conversion Mechanismin Ultrasonic Joining

Ultrasonic joining technology uses high-frequency vibration to join two or more thermoplastic materials together. Vibrations create a rapid increase in temperature within the material. This increase in temperature causes the material to melt, coalesce, and form a bond. Similar melting temperatures are preferred when joining two different materials such that each layer will arrive at the coalescence stage at the same time. Other relevant material variables include thickness, density, and uniformity across the surface. Ultrasonically joined edges are not abraded (Boz, 2008).

The material in the ultrasonic joining process; it is fed between the Horn, which transmits ultrasonic vibrations, and a stationary anvil (or rotating wheel). High-frequency mechanical vibrations (20-40 kHz) are transmitted to the thermoplastic materials, causing frictional heat to form on the inner surface and thus reaching sufficient temperature for melting and joining (Figure 6). Heat is required on the inner surface for joining. The ultrasonic joining process takes place strictly on the inner part (Kuttruff, 1991). A heat-activated material should be placed between fabrics without synthetic content or with a natural fiber content of over 40%. Ultrasonic vibration and pressure allow the heat-activated material to melt and penetrate (Branson Ultrasonic Corp., 1996).

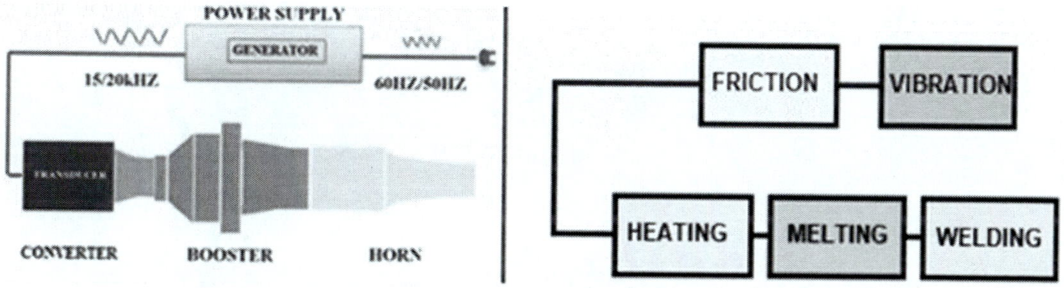

Figure 6. Energy conversion.

Transducer (Converter): converts high-frequency electrical energy into high-frequency mechanical energy (vibration amplitude) (Figure 7).

Booster: Modifies the vibration amplitude coming from the converter.

Horn: Adjusts the vibration amplitude coming from the booster and applies (transmits) it to the parts to be joined.

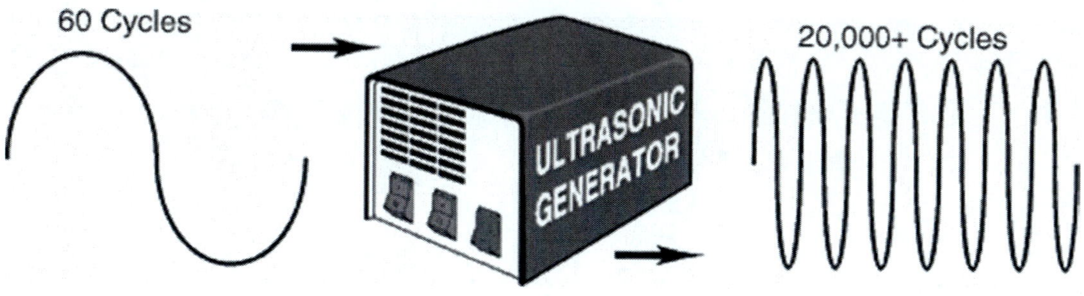

Figure 7. Transducer energy change (Boz, 2008).

Ultrasonic Joining Machines

Ultrasonic joining machines are similar in structure to classical sewing machines, but they do not have parts to carry needles and sewing threads. Since the joining process takes place in the form of connecting the fabrics to each other; "horn" and "anvil" have the most important task. The horn is connected to the ultrasonic energy generator. It focuses the high-frequency mechanical vibration energy and transmits it to fabrics. Fabrics trapped between the horn and the anvil create heat by friction against each other due to the high-frequency mechanical vibration energy. Heat causes the synthetic fabric or bonding material to melt. Adhesion occurs as a result of melting in the part exposed to vibration and pressure. When the fabric moves away from the vibration energy; with a sudden loss of heat, this union becomes permanent and ultrasonic joining is achieved.

The Parts of the Ultrasonic Joining Machine

In ultrasonic joining machines, as in classical sewing machines, alternatives are offered in order to provide ease of sewing with different placements of the body (Figures 8 and 9).

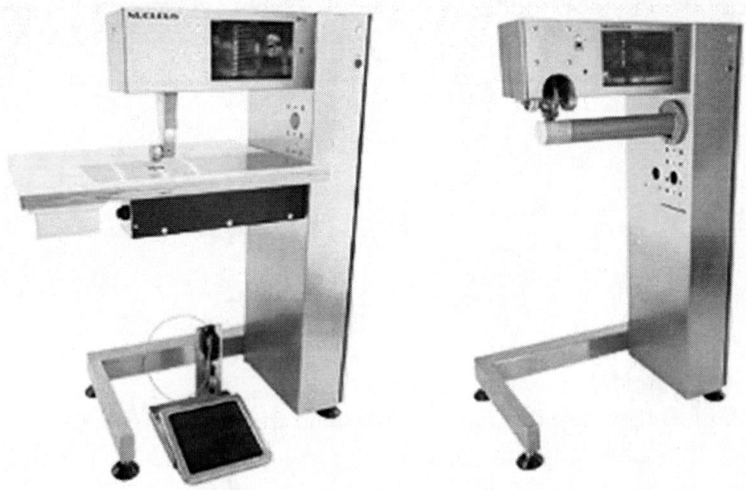

Figure 8. The flatbed and cylindrical arm ultrasonic joining machine (www.nucleusultrasonics. com).

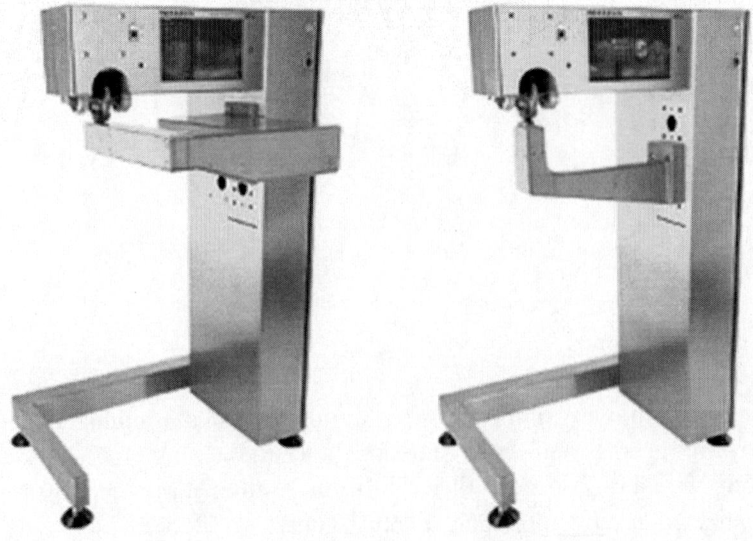

Figure 9. Free arm ultrasonic joining machines (www.nucleusultrasonics.com).

Horn and Anvil

Ultrasonic vibrations apply pulses perpendicular to the surface and provide periodic, tangential stress of the surfaces to each other. Significant stress on the micro-rough surface reveals its plastic deformation. The plasticized micro-rough surface causes an increase in sound absorption and the desired temperature. Therefore, the increase in a temperature at these points shows melting behavior (Kuttruff, 1991; Abramov, 1994). The roughness on the inner surface of the sample at the beginning of the joining cycle is effective in heating. These asperities distribute heat around the ultrasonic horn and anvil. Thermal energy causes an increase in the temperature of the inner surface. For this reason, the displacement of the polymer chains increases, and the formation of strong polymer chains is ensured. When vibration is displaced, the inner surface cools and the long polymers cling to each other, resulting in a joining suture (Shi & Little, 2000).

The horns and anvil can appear in different forms on different machines. Horn types to the fabric to be joined;

- Vertically positioned and linearly moving,
- Vertically positioned and circularly moving
- Horizontally positioned and circular moving ones.

The horn is connected to the ultrasonic energy transducers; while it is located at the bottom of the machine in most machines (Figure 10), it can be located at the head of the machine in some machines (Figure 11).

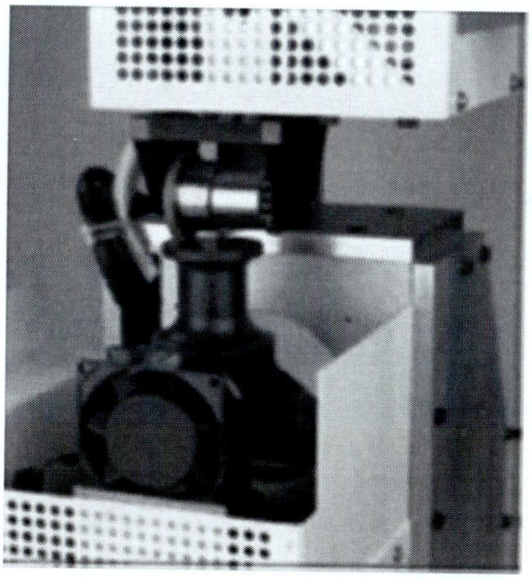

Figure 10. Horizontally positioned and circular moving horn (Boz, 2008).

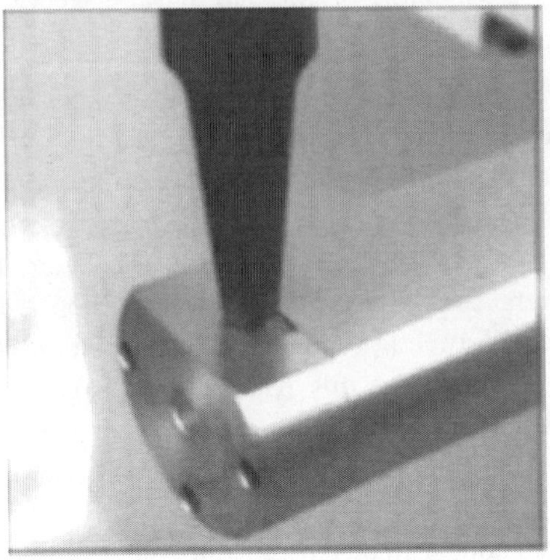

Figure 11. Vertically positioned and linearly moving horn (Boz, 2008).

The anvils are positioned in the opposite direction as the horns. They are usually in the form of wheels. The pattern created during joining is transmitted to the fabric through the teeth on the roller surface. Due to these roles, very different patterns can be transferred to the fabric (Figure 12).

Figure 12. Roller types (www.nucleusultrasonics.com).

During ultrasonic joining, wheels that can be cut and sealed at the same time can be used instead of patterned rollers. When anvils with a narrow surface area (pointed) instead of large ones are used, the joined materials can be cut right next to the point where they are joined by increasing the joining pressure. Thus, a clean joint is created as in overlock sewing machines (Figure 13).

Different Applications with Ultrasonic Joining

The joining machines using ultrasonic energy can be used in different applications apart from garment production. The most common ultrasonic joining machines are presented below.

- Quilting machines,
- Surgical mask machines
- Sanitary pad production

Ultrasonic Quilting Joining Machines

In classical quilting machines, quilting takes place by moving the needles (during sewing) to the right and left on the fabric layers transmitted to the machine. In the ultrasonic quilting joining machine, the pattern is embroidered on the fabrics with a roller that extends along the width of the fabric (Figure 14).

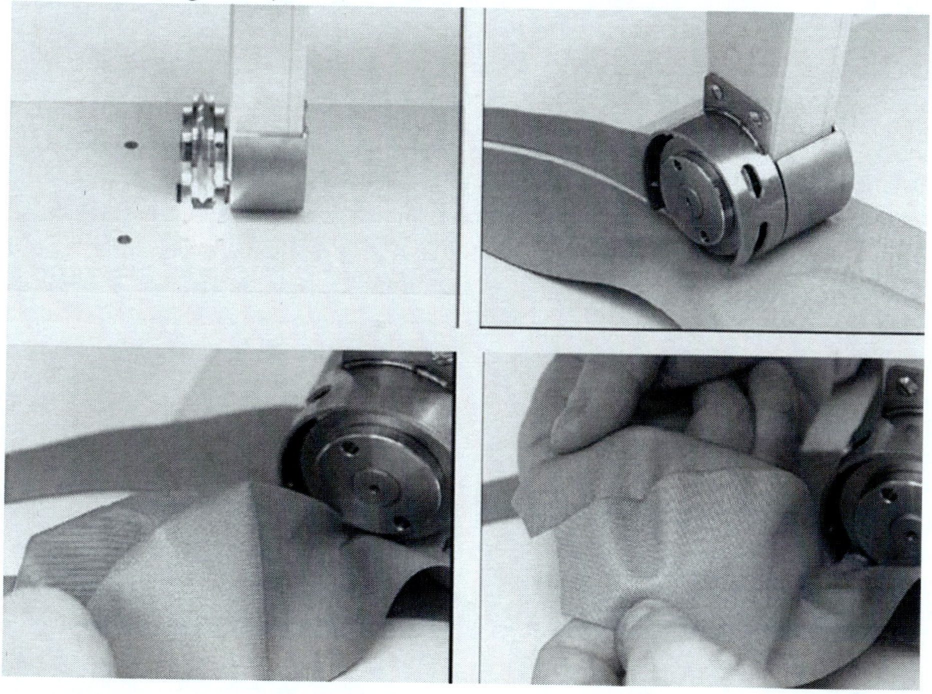

Figure 13. Cut seal roller type (Boz, 2008).

Figure 14. Ultrasonic quilting joining machines.

Ultrasonic quilting machines have a wide range of uses, including bedspreads, blankets, mattresses, coats, sleeping bags, auto covers, snow jackets, etc. It is suitable for 1.60–2.40 m fabric width. In addition to classical quilting patterns, it is possible to obtain patterned quilts by transferring different patterned rollers.

Figure 15. Joining image obtained by ultrasonic quilting machine.

Surgical Mask Ultrasonic Joining Machines

Automation systems are also being developed in the field of ultrasonic joining machines. The most common example of these is surgical mask machines. The fabrics placed in the machine as strips are folded and cut into desired sizes and taken to the ultrasonic joining process. The fabrics entering the machine as strips come out as ready-to-use surgical masks (Figure 16).

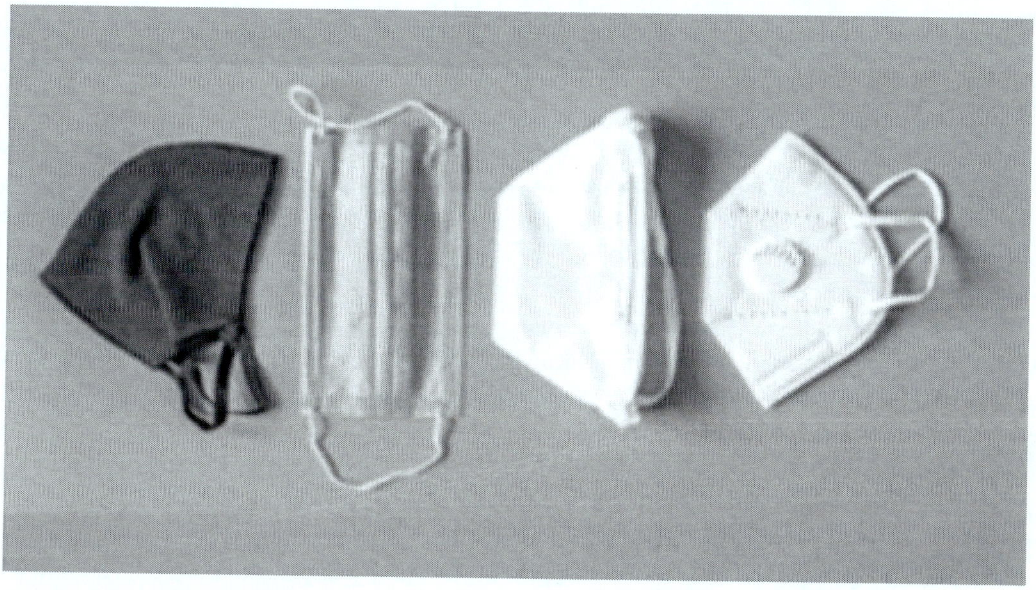

Figure 16. Surgical mask samples.

Sanitary Pad Production

Another common usage area of ultrasonic joining automation systems is the production of sanitary pads. As in mask automatons, different layers are brought together and combined. The roller used for the pad performs both the joining and cutting processes at the same time. Similar production methods are used in diapers.

Factors Affecting Ultrasonic Joining Performance

Two different parameters emerge during the formation of ultrasonic welding. These

- Properties of the material to be joined
- Settings used on the machine during assembly.

When these two factors are not applied in an optimum form, unification does not occur. As a result, the joining process may be of low strength, or material problems (e.g., thinning) may occur.

Material Properties Affecting Ultrasonic Joining Performance

The ability of textiles and films to be ultrasonically bonded is generally dependent on the thermoplastic material content. Plastic materials can be divided into three groups such as amorphous, semi-crystalline, and crystalline plastics. Amorphous thermoplastics are characterized by their irregular molecular structures and wide melting ranges. When heated, they slowly soften and show viscoelastic deformation, and slowly solidify during cooling. Ultrasonic energy can easily travel through amorphous thermoplastics. Crystalline thermoplastics, on the other hand, have a regular molecular structure and melt in a narrow temperature range. Since such plastic materials absorb ultrasonic energy, high-amplitude ultrasonic energies are needed to transmit ultrasonic energy from their structures, and therefore their joining properties are lower than those of amorphous plastics. Soft thermoplastics with a low modulus of elasticity reduce ultrasonic vibrations, making joining difficult. Generally, as the plastic material softens, ultrasonic shaping or point joining becomes easier.

The energy transmission feature of the plastic determines which method will be chosen, directly or indirectly, from the ultrasonic joining method. As the distance between the sonotrode contact surface and the joining area increases, the ability of the plastic to transmit ultrasonic energy gains importance. Thermoplastic types have important effects on the joining ability. For example, thermoplastics produced by molding have high molecular weights and high melting temperatures. Therefore, there are brittle and hard surface layers. Therefore, joining thermoplastics produced by molding is more difficult than joining plastics produced by injection/extrusion. As a result, thermoplastics produced by injection/extrusion should be joined only with thermoplastics produced by the same method, and thermoplastics produced by molding should be joined only with thermoplastics produced by molding.

The ultrasonic joining of plastics produced by injection is very common. The ultrasonic joining ability of these plastics varies depending on their melting temperatures, elasticity modules, friction coefficient, and thermal conductivity. Generally, if the rigidity of the plastics increases, the joining ability also increases. The horn is held closer to the joining surface when welding plastics with low modulus of elasticity such as polyethylene and polypropylene. However, with the correct selection of the joining pressure and amplitude, good results can be obtained from the source of most plastics (Erkul, 2001).

Table 3. The ability of materials to be combined in comparison to each other (Branson Ultrasonic Welding Machine Catalogs, 2007)

Material	Woven	Nonwoven	Knitting	Coating	Laminate	Film
Acrylic	4	X	4	X	X	(c)
Eva	X	X	X	2	X	1(e)
Nylon (Pa)	2	2	2	2	2	2
Polyester	2	1	2	1	1	1
Polyethylene	X	1	X	1	1	4-5(a)
Polypropylene	1	1	2	1	1	1
PVC	3-5	X	X	3-5	3-5	3-5(b)
Saran	X	X	X	1	X	1(e)
Surlin	X	X	X	1	X	X
Urethane	X	X	X	1	X	1(d)
Natural Fibers with Joining Agents	2	2	2	2	2	X

Codes: 1 = the easiest; 5 = the hardest.

The codes in Table 3 show the ease of combining the most commonly used thermoplastic materials. These are just proportional representations. They do not express the achievable adhesion strength.

Note: The parts with lines (–) (X) indicate that these fibers are not used for those material groups.

a) The thin polyethylene film (less than 0.003"-0.076 mm) is generally considered weak for ultrasonic bonding because it has a low coefficient of friction and is prone to stall during bonding.
b) It is difficult to predict in advance due to the large number of additives used in the manufacture of PVC sheets or fibers. Plasticizers are often added to PVC to increase flexibility. However, when the amount of plasticizer is increased, the ultrasonic bonding ability of PVC may be inhibited.
c) Acrylics can be attached or cut by ultrasonically. Continuous bonding is often unsatisfactory due to its frangibility and low strength.
d) Thermoplastic urethane coated materials (ester-based) are easily joined, but good strength may not be achieved.
e) Cutting and sewing can be easily done.

Welding of the same types of plastic materials to each other does not pose any problem in terms of joining. However, due to the different melting temperatures, stiffness, friction coefficients, elasticity modules, and thermal conductivity of plastics, the desired result cannot be obtained from welding different types of plastic materials.

Machine Parameters Affecting Ultrasonic Joining Performance

In the ultrasonic joining process, heating occurs as a result of friction on the inner surface. Ultrasonic joining strength is determined according to the degree of bonding, which depends on the inner surface temperature. Thus, the joining strength is associated with the thermal energy on the inner surface. The parameters shown in Figure 18 affect ultrasonic welding.

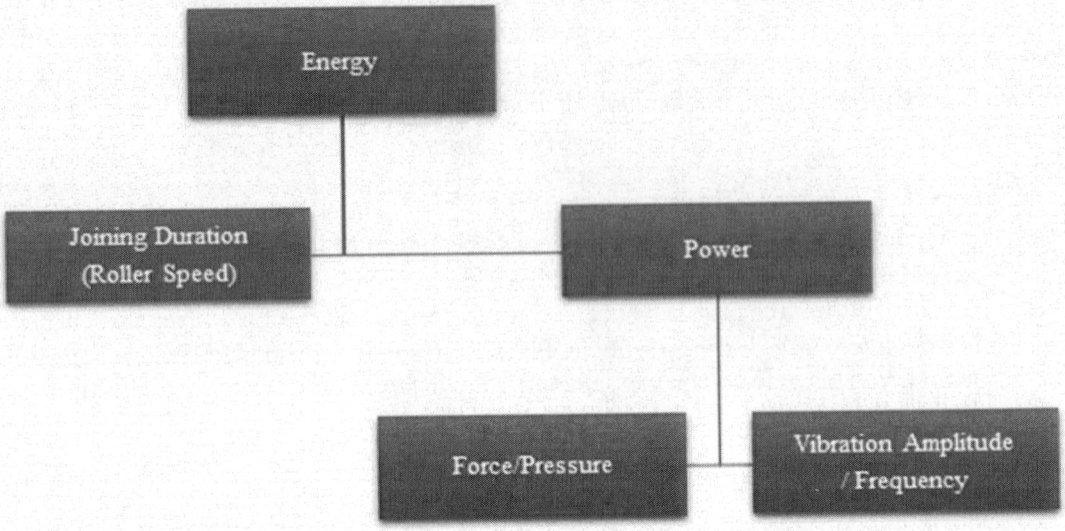

Figure 17. Machine parameters affecting ultrasonic welding.

From the control panel shown above, the ultrasound power (Watt) to be transferred to the layers to be joined and the pressure (Newton) to be applied to the layers are adjusted (Figure 18). The roller type can be selected from the control panel screen below (Figure 19).

Each of the parameters used in ultrasonic joining applications is directly related to the quality of joining. The vibration amplitude, the joining pressure, and the joining time should be selected appropriately for each joining application. While determining each variable, other variables should be kept constant, and different numbers of joining pieces should be studied separately from each other. The result of each trial should be measured and recorded, and the optimum value should be found. There is various joining efficiency or joining quality measurements used to determine the best joining conditions. Depending on the use area of the product with the selected measurement method, it can be the penetration of the joining, the breaking strength, or the impermeability test (Erkul, 2001).

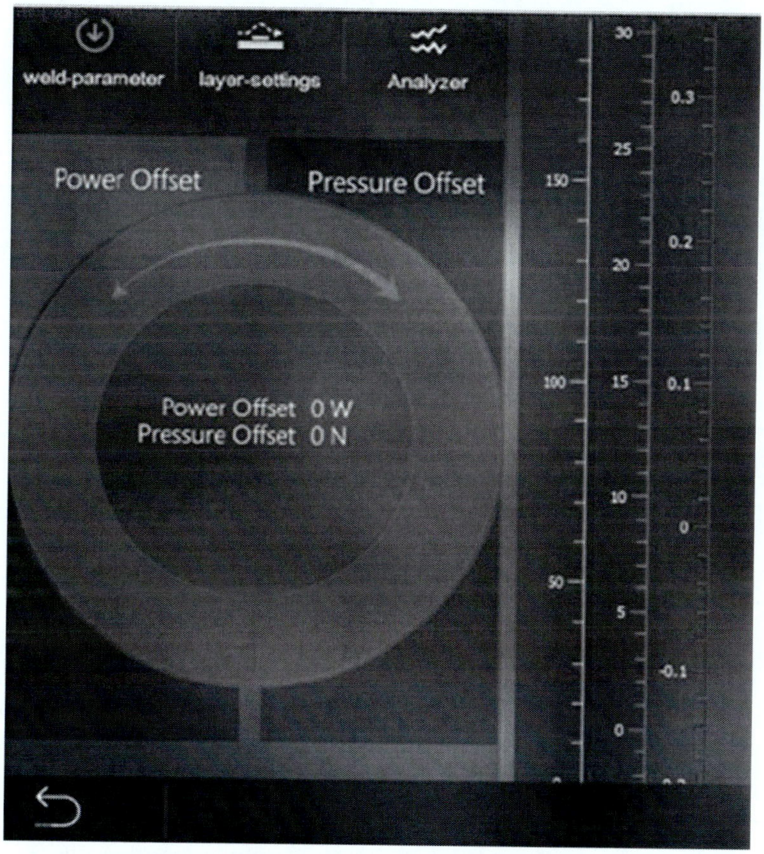

Figure 18. Control panel-1 (Nucleus DX1 Ultrasonic Welding Machine Catalogue).

Figure 19. Control panel-2 (Nucleus DX1 Ultrasonic Welding Machine Catalogue).

Joining Pressure

It is important to determine the optimum pressure value in order to produce a sufficient weld. The purpose of the jointing pressure is to ensure the vibration transition between the material and the horn and the bonding of the heated surfaces to each other. Low pressure causes poor contact, and thus poor energy transfer. The increase in the amount of pressure initially increases the joining strength, but high pressures change the direction of the polymer chains during ultrasonic joining and cause a decrease in the joining strength (Benatar, 1989). Excessively increased pressure damages the fibers and causes a decrease in the strength of the fabric.

Joining Time/Roller Rotation Speed

Joining time is the process in which only mechanical vibrations are applied to fabric surfaces. It is very important not to exceed the joining period, which is calculated with the result of the trials. If quality joining is desired, the optimum value of the joining time should be determined. Excessive joining time causes the formation of joining burrs on the inner and outer edges, deterioration of the quality of the joining, and cracks in the areas requiring sealing (Erkul, 2001). The effect of the joining time on the strength of the joining shows a similar trend to the effect of the joining pressure. This is because; The increase in joining time causes an increase in energy transfer. Thus, the joining strength increases until the maximum strength is reached. If the joining time is too long, the fibers may be damaged (charring, burning, melting) and thus the joining strength decreases (Shi & Little, 2000).

While the value of the joining time is set in static ultrasonic joining machines, the rotational speed of the wheel determines the interaction time between the fabric and the horn in ultrasonic sewing machines. In order to prolong the joining time in ultrasonic sewing machines, the sewing speed should be reduced, and to reduce it, the sewing speed should be increased. In addition, after the fabrics are combined in ultrasonic sewing machines, pullers are placed behind the rollers to keep the fabrics together for a while, so that the heated thermoplastic fabrics are kept together for a while. In addition, cooling devices (i.e., wind/air blower) can be added to remove the heat as soon as possible and solidify the weld.

Ultrasound Power

The mechanical vibration effect applied to the fabric surfaces to be joined can be adjusted by ultrasound power. Ultrasound power is the most important variable factor in the joining process. High ultrasound power is needed to make fast and effective joining of crystalline plastic parts with high melting points. It is necessary to bring the ultrasound power to the optimum value such as the joining time and the joining pressure. Adjusting the ultrasound power too much or too little has a similar effect on the joining time and the joining pressure. Ultrasound power is determined by entering a percentage (%) in some machines and wattage in some machines (Figure 19).

Maximum joining strength; long joining time, low joining pressure, and high energy containing high vibration amplitude parameters are provided (Wei et al., 1985). Machine parameters are generally determined by preliminary trials according to variable fabric parameters. These parameters can be saved in new-generation machines and later called from among the saved settings when working with the same fabric/fabric group.

Usage Areas of Ultrasonic Joining

Today, ultrasonic sewing is preferred in clothing, medical and hygienic products, workwear and protective clothing, outdoor products, automotive and motorcycle materials, covers, packaging, etc., technical textile products (Figures 20-25).

Clothing

Figure 20. Use of ultrasonic joining in clothing.

- Bras
- Panties and other underwear
- Active sportswear

Medical and Hygienic Products

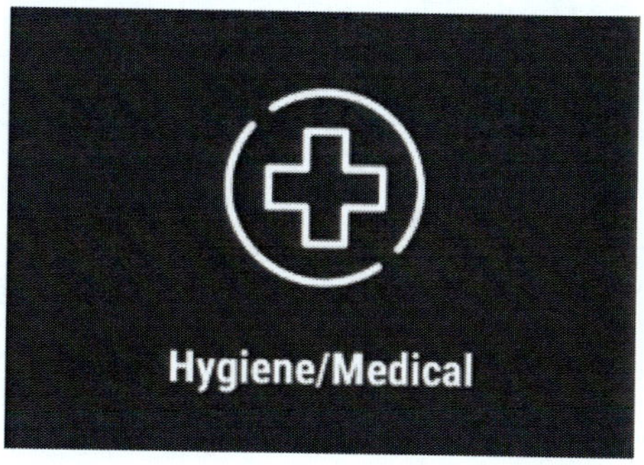

Figure 21. Use of ultrasonic joining in medical and hygienic products.

- Orthopedic bandages
- Medical pillows and mattress covers
- Hygienic products
- Surgical clothing and fabrics

Workwear and Protective Clothing

Figure 22. Use of ultrasonic joining in workwear and protective clothing.

- Bulletproof clothing
- All work clothes
- Police, military, firefighter, and other security personnel attire

Outdoor Products

Figure 23. The use of ultrasonic joining in sportswear and outdoor products.

- Waterproof jacket and trousers
- Waterproof shoes
- Backpacks, carry bags, and sleeping bags
- Tents
- T-shirts

- Jump mats
- life jackets
- Inflatable boats
- Beach clothes

Automotive and Motorcycle Products

Figure 24. Ultrasonic joining use in automotive and motorcycle.

- Seat arm, head, and all upholstery
- Convertible car head
- Cover (e.g., car cover)
- Motorcycle seat
- Tar tarps

Covering and Packaging, Foil, Filter, and Technical Textiles

Figure 25. Use of ultrasonic joining in cover and packaging.

- Covers
- Tar tarp and boat covers
- Packaging materials
- Pool covers
- Water tanks and containers
- Awnings
- Big tents
- Big tents
- Flat roof and bunkers
- Pool metal leaves
- Pool coating materials
- Balloons
- Bathroom curtains
- Air and dust filters
- Billboards, flags, and pennants
- Sewage and wastewater places
- Inflatable structures

Conclusion

When comparing the classical methods with the innovative joining methods developed in today's technology, the following points can be highlighted.

Since the use of the ultrasonic joining methods in the clothing industry is not very common yet, the production of ultrasonic joining machines is limited. Although limited production increases the cost of parts and machines, the fact that they save material and time due to the absence of needles and threads reduces the cost of production.

The increase in the use of thermoplastics and the search for alternative joining methods will continue to increase the use of ultrasonic welding.

It can be said that ultrasonic joining machines are an indispensable joining tool for materials with high thermoplastic fiber content. Due to the wide variety of roller types, different pattern alternatives can be offered.

Disclaimer

None.

References

Abramov, O. V. (1994). *Ultrasound in liquid and solid metals*. CRC Press.

Appleby, C. K. (2009). Development of fabric seaming for clothing using ultrasonic sealing technique [*Senior Honors Theses & Projects*]. Eastern Michigan University.

Benatar, A., Eswaran, R. V., & Nayar, S. K. (1989). Ultrasonic welding of thermoplastics in the near-field. *Polymer Engineering and Science*, 29(23), 1689-1698. https://doi.org/10.1002/pen.760292311.

Binetruy, C., Bueter, A., Castaing, P., Clement, S., Deleglise, M., Franz, C., Giessl, A., Knapp, W., Oumarou, M., Renard, J., & Roesner, A. (2010). *Analysis of laser welding of long fiber reinforced composites*. Laser Institute of America, 562.

Black, S. S. (1995). *Union special to debut laser-enhanced bonding*. Bobbin, 98-104.

Boles, K. (2012). Examination of alternative fabric joining techniques compared to traditional sewing. *McNair Scholars Research Journal,* 5(3), 1-17.

Boz, S. (2008). Investigation of ultrasonic energy usage in apparel industry (Publication No. 216082). [*Master dissertation*]. Ege University.

Branson Ultrasonic Welding Machine Catalogs, (2007).

Erkul, M. (2001). Ultrasonic welding of plastics (Publication no. 106198). [*Master dissertation*]. Yıldız Teknik University.

Ghosh, S., & Reddy, R. (2009). Ultrasonic sealing of polyester and spectra fabrics using thermo plastic properties. *Journal of Applied Polymer Science*, 113, 1082–1089. https://doi.org/10.1002/app.30050.

Hussen, M. S., Kyosev, Y. K., Pietsch, K., Rothe, S., & Kabish, A. K. (2022). Effect of ultrasonic welding process parameters on seam strength of PVC-coated hybrid textiles for weather protection. *Journal of Industrial Textiles,* 51(2S), 2960S-3003S. https://doi.org/10.1177/15280837211057579.

Jevsnik, S., Eryürük, S. H., Kalaoğlu, F., Karagüzel, Kayaoğlu, B., Komarkova, P., Golombikova, V., & Stjepanovic, Z. (2017). Seam properties of ultrasonic welded multilayered textile materials. *Journal of Industrial Textiles,* 46(5), 1193-1211. https://doi.org/10.1177/1528083715613632.

Kayar, M. (2014). Analysis of ultrasonic seam tensile properties of thermal bonded nonwoven fabrics. *Journal of Engineered Fibers and Fabrics*, 9(3), 8-18. https://doi.org/10.1177/155892501400900302.

Kayar, M., Mistik, S. I., & Inan, D. (2015). Analysing effect of the factors on ultrasonic seam tensile properties of nonwoven fabrics by Nested Anova Design. *International Journal of Clothing Science and Technology*, 27(6), 803-817. https://doi.org/10.1108/IJCST-05-2014-0066.

Kuttruff, H. (1991). Ultrasonics fundamentals and applications, Elsevier Science.

Mao, Z., & Goswami, B. C. (2001). Studies on the process of ultrasonic bonding of nonwovens: Part 2: Experiments and results. *Journal of Engineered Fibers and Fabrics*, os-10(3), 17-28. https://doi.org/10.1177/1558925001OS-01000307.

Nucleus DX1 Ultrasonic Welding Machine Catalogue, (2023).

Pierlot, T. (2011). *Welding technology for joining textile media* [PowerPoint slides].

Potente, H., & Uebbing, M. (1997). Friction welding of polyamides. *Polymer Engineering and Science,* 37(4), 726-736. https://doi.org/10.1002/pen.11716.

Seram, N., & Cabon, D. (2013). Investigating the possibility of constructing different seam types for clothing using ultrasonic. *International Journal of Clothing Science and Technology*, 25(2), 90-98.

Seram, N., & Cabon, D. (2014). Threadless seaming - application of ultrasonic technology in fashion. *Research Journal of Textile and Apparel,* 18(4), 76-81.

Shi, W., & Little, T. (2000). Mechanisms of ultrasonic joining of textile materials. *International Journal of Clothing Science and Technology*, 12(5), 331-350. http://dx.doi.org/10.1108/09556220010377869.

Technical Report from Branson Ultrasonic Corp., (1996).

Wei, K. Y., Vigo, T. L., & Goswami, B. C. (1985). Structure-property relationships of thermally bonded polypropylene non-wovens. *Journal of Applied Polymer Science*, 30, 1523-1534. https://doi.org/10.1002/app.1985.070300417.

Whitwell, J. C., & Fraser, W. A. (1971). Infrared laser bonding of thermoplastic monofilaments. *Textile Research Journal*, 41(12), 1003-1005. https://doi.org/10.1177/004051757104101210.

www.nucleusultrasonics.com. (2023).

Chapter 22

CAD Systems for the Clothing Industry: Significance, Functions and Applications

Meral İşler[1,*], PhD
and Derya Tama Birkocak[2], PhD
[1]Department of Fashion Design, Selçuk University, Konya, Türkey
[2]Department of Textile Engineering, Ege University, Izmir, Türkey

Abstract

The fashion industry, which has a significant role in the world economy, is a wide sector that covers the production of clothing and its accessories, as well as the marketing of these products. As consumer expectations and industry competition continue to evolve, it has become essential to respond to consumer requirements quickly and at the lowest cost. Adapting to the requirements of the industry will be possible with efficient production and the inclusion of the latest technology. In addition to machine technologies used in clothing production, CAD technologies are indispensable. CAD systems, which perform different functions, are used in many stages of the production processes, starting from the design stage of clothing production to the final product. The CAD systems were categorized into five categories, namely body measuring, design, garment pattern design, virtual fitting, and marker design. In this chapter, CAD systems used in garment production processes are discussed from a broad perspective, including the characteristics of the production stages in which they are used, the functions of the systems, and the types frequently used in the market. The final part of the study discusses future trends and industry expectations of CAD systems used in garment production.

Keywords: Computer-Aided Design (CAD) Systems, 2D CAD, 3D CAD, garment pattern, virtual fitting

Introduction

Procedures involving Computer-Aided Design (CAD) and Computer-Aided Manufacturing (CAM) play a crucial role across multiple industries. In recent decades, there has been an

* Corresponding Author's Email: meralisler@selcuk.edu.tr.

In: Textile Insights and Future Aspects
Editors: Gamze D. Tetik, Gizem Celep and Fulya Yilmaz
ISBN: 979-8-89113-289-4

increasing implementation of CAD/CAM systems across all domains of the textile and apparel industry (Dwivedi & Dwivedi, 2013). The transition to computerization has supplanted the majority of sluggish, laborious, manpower-heavy, and less-efficient manual tasks. Computer-Aided Design (CAD) systems have become crucial in the clothing sector, offering a platform for designers to efficiently create, modify, and optimize designs (Volino et al., 2005). CAD systems have had a transformative impact on the clothing sector, fostering the development of innovative designs and streamlining production processes (Yan & Fiorito, 2002; Baral, 2011; Uddiin et al., 2021; Hossain et al., 2023). By leveraging advanced algorithms and visualization techniques, CAD systems have revolutionized the way clothing is designed and manufactured (Nayak & Padhye, 2015).

CAD systems have a storied and transformative past in the clothing industry. The journey from the initial adoption of digital design tools to the incorporation of cutting-edge technologies such as 3D modeling demonstrates the continuous evolution of CAD systems in response to the dynamic demands of the industry (Yan & Fiorito, 2002; Baral, 2011; Spahiu et al., 2014; Uddiin et al., 2021; Hossain et al., 2023;). Incorporation of CAD systems into the clothing sector started in the late 20th century, signifying a marked deviation from conventional manual methods. Initially, these systems were used primarily for rudimentary design tasks and pattern generation (Jhanji, 2018). During the 1980s and 1990s, 2D CAD systems underwent enhancements to offer more advanced tools for pattern classification, marking, and plotting. This era also witnessed the birth of CAD systems explicitly devised for the clothing sector, featuring functions tailored to the specific needs of the industry (Zhang et al., 2008; Nayak & Padhye, 2015). With the dawn of the 21st century, 3D CAD systems were introduced into the clothing industry. These systems facilitated the three-dimensional visualization of designs, providing a more authentic depiction of garments and fostering more complex design methodologies (Moritz & Youn, 2022). The past decade has marked the integration of state-of-the-art technologies such as Virtual Reality (VR), Artificial Intelligence (AI), and 3D printing with CAD systems. These technologies have enriched the capabilities of CAD systems, fostering more creative design processes and enhancing production efficiency (Thomassey & Zeng, 2018). The chronicle of CAD systems in the clothing industry is one of consistent growth and innovation. As CAD systems persist in their development and amalgamation with upcoming technologies, their impact on the clothing industry is expected to continue to evolve and broaden. With this expansion, the positive and negative impacts of these systems on the industry become clearer.

Advantages of (CAD) Systems

CAD systems provide an abundant set of design tools and features, empowering designers to explore creative designs and methods (Yan & Fiorito, 2002; Fontana, 2005; Sayem et al., 2010; Baral, 2011; Uddiin et al., 2021; Hossain et al., 2023). These tools equip designers with the capacity to envision designs in 3D, experiment with diverse materials and color schemes, and implement changes promptly and effortlessly (Collier & Collier, 1990; Nayak & Padhye, 2015). CAD systems simplify the design process, minimizing lead times and production expenses. They facilitate designers in rapidly creating and altering designs, reducing the time and cost related to conventional design approaches (Stjepanovic, 1995; Spelic, 2020). Moreover, CAD systems can automatically produce patterns and classification, thereby accelerating the production process even further (Hu, 2011). CAD systems provide an exceptional level of precision and accuracy in design and production. They enable exact measurements and intricate designs, decreasing the likelihood of mistakes and waste during the production process (Oppong et al., 2013; Nayak & Padhye, 2015).

CAD systems promote more sustainable practices in the clothing industry by optimizing material utilization and reducing waste. The ability of CAD systems to optimize pattern layouts to minimize fabric waste contributes to more sustainable and cost-efficient production practices (Hu, 2011). CAD systems can be combined with other technologies such as Virtual Reality (VR), Artificial Intelligence (AI), and 3D printing, boosting their capabilities and use in the clothing industry (Thomassey & Zeng, 2018).

Disadvantages of (CAD) Systems

A considerable hurdle for implementing CAD systems is the substantial initial investment that they require. The expenses associated with CAD software, hardware, and related infrastructure can be prohibitive, particularly for small to medium-sized businesses (Yan & Fiorito, 2002; Baral, 2011; Uddiin et al., 2021; Hossain et al., 2023). CAD systems are intricate tools that require specialized skills and training for efficient use. This training can be lengthy and expensive, which could pose obstacles to adoption, especially for smaller firms with limited resources (Nayak & Padhye, 2015).

As CAD systems are computer-based, they rely on a dependable power supply and operational hardware and software. Any complications with these components can lead to disruptions in the design and production process (Hu, 2011). Despite CAD systems offering cutting-edge visualization techniques, they are unable to offer the tactile experience that is often vital in the clothing design process. Designers are unable to feel the texture or drape of fabric in a digital setting, which could potentially restrict the design process (Nayak & Padhye, 2015). Even with the advanced features of CAD systems, they have technical constraints. Certain intricate design elements or details may not be fully represented or reproduced in a digital environment (Yan & Fiorito, 2002; Baral, 2011; Uddiin et al., 2021; Hossain et al., 2023).

CAD systems, in conjunction with CAM systems, are utilized at many stages of the production process, starting from the design phase of the garment. CAD/CAM systems used in the clothing industry often include the following components (Jhanji, 2018):

- A graphics package for artistic design tasks
- Input hardware, such as a digitizer or scanner
- An interactive hardware workstation
- Pattern Design Software (PDS) system
- Software for pattern grading
- Lay planning or marker-making software
- Output devices like plotters and printers
- Computer-controlled cutters for single, medium, and high-ply bulk cutting
- Production and management control software
- Computerized sewing and pressing machinery
- Overhead transport system for material handling.

This chapter deals with CAD systems used in the clothing industry.

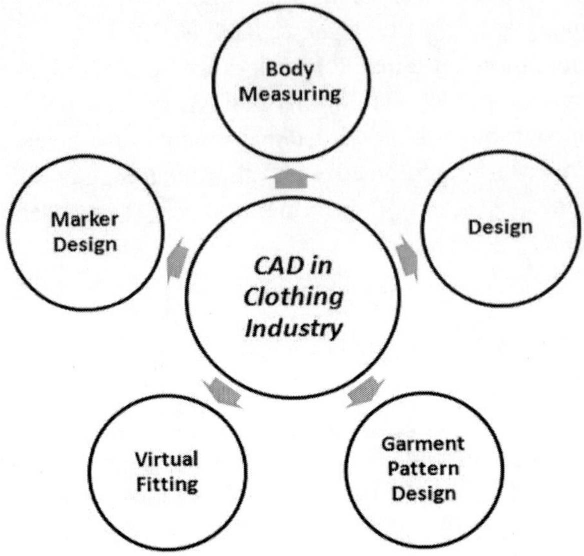

Figure 1. Application areas of CAD systems in the clothing industry.

CAD Systems in the Clothing Industry

CAD systems, which are frequently used in many sectors, offer the opportunity to apply from the simplest to the most complex processes. CAD systems used in the clothing industry can be examined in five groups: body measurement, design, garment pattern design, marker design, and virtual fitting (Figure 1).

Body Measuring

In the universe of customized clothing, measurements derived from 3D body models can be employed to craft patterns that accommodate an individual's unique fit. For size studies, collective data from numerous individuals can help establish a broader range of standard sizes that more accurately cater to the demographic. In the case of virtual fittings, 3D body models assist in visualizing the fit of various garments on a person's physique. CAD systems have gained widespread acceptance for body measurements, providing precise and uniform measurements. Some salient features of CAD systems used for this purpose include:

With adjustable measurement parameters, CAD systems allow customization of measurement parameters, facilitating the customization of specific body measurements such as chest, waist, and hip circumference, as well as various lengths and widths according to individual requirements (Qiu, 2014; Tama & Öndoğan, 2020). With non-touch measurement, employ non-touch methods of measurement, ensuring a more comfortable experience for the person being measured and averting potential measurement inaccuracies stemming from physical contact (Simmons & Istook, 2003). CAD systems can merge body measurement data with design software, enabling the production of personally fitted clothing designs directly derived from body measurements (Ashdown, 2007). These systems have the ability to generate databases of body measurements, which prove beneficial for mass customization, sizing studies, ergonomic investigations, and demographic research (Apeagyei, 2010). In these systems, body measurement data can be used to construct parametric models, which are mathematical models capable of adapting to changes in certain parameters, like varying body sizes or shapes (Apeagyei et al., 2007; Chen et al., 2019). CAD systems, utilizing the captured body measurements, can emulate the fit of a garment on a specific body, aiding in the evaluation of design and fit prior to the creation of a physical prototype (Ashdown, 2007).

Body scanners in the fashion industry play a variety of roles, such as assisting in the creation of custom clothing, conducting sizing investigations and allowing virtual fittings. These devices efficiently obtain body measurements from customers, a vital procedure for manufacturing custom attire. A proficient body scanning system should fulfill the following requirements (Yu, 2004; Uhm et al., 2015):

- It should be capable of quickly obtaining measurements.
- Measurements should be taken in a non-contact manner without requiring the customer to remove their clothing.
- Regardless of the lighting conditions or the color of the customer's clothing, it should produce accurate measurements.
- It should be able to calculate body circumferences without relying on body landmark information.
- The collected data should be ready for immediate integration into the production processes.
- The system should be easily transportable, simple to set up, and compact in size.

These body scanners work primarily based on two key technologies; 3D scanning and infrared depth sensing:

3D scanning technology in the clothing industry involves the use of multiple sensors that capture the shape of a person's body in three dimensions. The data is then processed by software to create a digital 3D model of the person's body (D'Apuzzo, 2007). The scanner typically uses structured light or laser technology to capture the body's shape. Structured light scanners project a pattern of light onto the body, and sensors capture the deformation of the pattern on the body. This information is used to calculate the depth and surface information (Daanen & Psikuta, 2018).

In infrared depth sensing technology, body scanners use infrared sensors to capture depth information. These scanners project infrared light onto the body. The light bounces back to the sensor and the time it takes for the light to return is used to calculate depth information (Uhm et al., 2015).

2D CAD System for Design

2D and 3D CAD systems used in the clothing industry are used to perform design, pattern and marker preparation, and simulation processes. The systems used in the model design phase are designed to design clothes on a silhouette, apply texture and color to the garment, design a fabric surface, create the technical drawing of the designed garment, apply sewing types, etc. providing facilities (Çınar, 2019). These systems are also frequently used in the preparation of print and embroidery patterns in order to give the garment an ornamental feature. In addition to the systems such as CorelDRAW, and Adobe Illustrator, which are also used in different design disciplines, there are systems such as Tex-Design, Weavelt directly for fashion and textile design (Figure 2).

2D CAD systems hold a significant position in garment illustration due to their key features that empower designers to effectively visualize and convey their designs. In its 2D CAD systems' drawing and sketch tools module, it presents a compilation of drawing tools for designers to generate digital sketches of apparel. These tools enable the creation of exact lines and curves and often facilitate the application of colors, textures, and patterns (Wang, 2004).

Designers have the capacity to create illustrations on distinct layers with 2D CAD systems, permitting separate work on various design elements. This proves especially useful when crafting intricate garments with multiple components or exploring different design elements (Jhanji, 2018). Such systems enable the application of diverse colors, textures, and patterns to illustrations, aiding in visualization of potential fabric and material appearances in a finished piece (Sayem et al., 2010; Dai, 2011).

These systems also possess the ability to scale and modify illustrations, which is beneficial for adjusting design proportions or developing variant styles (Carr &Lutham, 2014). Detailed annotations, essential for communicating specific design aspects, construction details, or manufacturing directives, are made possible by CAD systems. Integration with other software used in the garment design process, such as 3D modeling tools, PLM (Product Lifecycle Management) systems, and CAM (Computer-Aided Manufacturing) systems, is often feasible with CAD systems, promoting a streamlined and efficient design process (Chaudhary et al., 2020). Facilitating effortless file sharing and collaboration by CAD systems allows designers, manufacturers, and other stakeholders to remotely view and discuss designs (Hu et al., 2008). CAD systems also allow designers to create and save templates for various garment types, which can greatly expedite the design process for similar future styles (Sayem et al., 2010; Jhanji, 2018).

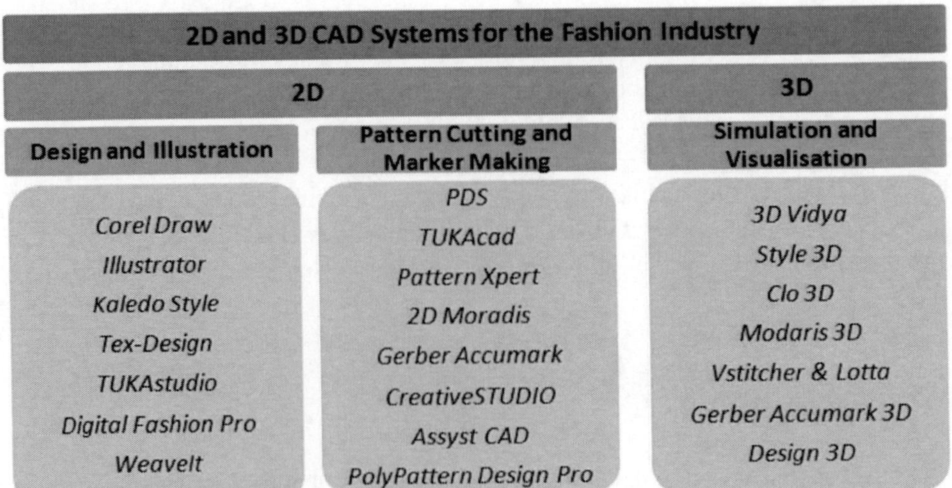

Figure 2. 2D and 3D CAD software for the fashion industry (Regenerated; Sayem et al., 2023).

2D CAD System for Making Pattern and Marker Design

2D CAD systems play a vital role in the apparel industry, as they facilitate the design process and perform pattern and marker preparation in a short time and at high quality. These systems allow users to create and modify patterns digitally, facilitating quick adjustments and alterations (Papachristou et al., 2019). The system also offers tools to duplicate, mirror, rotate, and scale patterns. Grading, the process of increasing or decreasing a base size pattern to create additional sizes, is an essential feature of 2D CAD systems. It enables precise and rapid size scaling while maintaining the shape, fit, balance, and proportions of the original pattern (Meng et al., 2012; Lee, 2014). These systems can also generate markers, which are efficient layouts of pattern pieces in the fabric to minimize waste. CAD systems can calculate the most fabric-efficient arrangement, reducing material costs (Kılıç, 2019). Moreover, 2D CAD systems often include a digital catalog or library, where designers can store and retrieve patterns, style variations, and size ranges (Kadolph & Marcketti, 2013).

CAD systems can also provide estimates of fabric usage and costs, helping budgeting and planning (Kunz et al., 2016). CAD systems can often be integrated with other software used in garment production, allowing for a smoother and more efficient production process (Papachristou et al., 2019). 2D CAD systems can help standardize the pattern creation process, leading to greater consistency and accuracy across all patterns and sizes (Kadolph & Marcketti, 2013). Additionally, CAD systems allow for easy sharing and collaboration, making it possible for designers, pattern makers, and other stakeholders to work together from different locations (Glock & Kunz, 2004).

3D for CAD System Virtual Fitting

The fit of the garment refers to the degree to which a product accommodates the ease of style and movement ease. As inferred from its definition, fit encompasses three key components: anthropometry, comfort, and clothing production. Each of these primary components is connected to several subcomponents (Figure 3), and some of them are interlinked (Sayem, 2023).

The complexity of assessing the fit of the garment arises from its dependence on numerous factors. Therefore, fitting evaluation is a crucial aspect of the garment development and production process.

Since clothing fit is affected by so many factors, making it difficult to adapt to the body, many subjective and objective physical fit assessment methods have emerged. With the development of technology, software has also been developed that evaluates clothing fit in a virtual environment. Today, virtual fit assessments have become as useful as physical fit assessments. When the physical and virtual compliance evaluation process is examined in Figure 4 is examined, there are differences regarding the process (Morlock et al., 2023).

In conventional production processes, the creation and assessment of physical prototypes for fit and design accuracy are required, which demands substantial input from the production team. On the other hand, virtual fitting eliminates the need for physical prototypes by creating and adjusting virtual models that are displayed on 3D digital mannequins within the software. Physical fitting entails fitting a physical prototype on a live model or a mannequin, while virtual fitting uses a 3D simulation on a digital model (Istook & Hwang, 2001). Virtual fitting offers greater flexibility, as alterations can be quickly performed within the software. However, each modification in physical fitting requires a new prototype, making the process more time-consuming and resource-heavy (Daanen & Ter Haar, 2013). Virtual fitting is generally more cost-effective and quicker as it bypasses the need for physical prototypes and models (Tama et al., 2016). However, physical fitting can be expensive and time-intensive due to the multiple rounds of sampling and fitting. The physical fitting allows for a tactile evaluation of the garment, including aspects such as fabric texture, drape, and weight, which is a component that virtual fitting lacks (Istook & Hwang, 2001). The core principles of fit assessment remain consistent, whether in physical or virtual try-ons. However, significant differences emerge when applying these methods, particularly concerning the production or simulation stage, the skills required from the workforce, and the final result (Glock & Kunz, 2004). The transition from physical to virtual trials also involves a change in the skill set of the workforce. Traditional methods require expertise in pattern making, sewing, and physical garment evaluation, whereas virtual fitting requires skills in 3D modeling software, digital pattern making, and virtual fit evaluation (Liu et al., 2017; Hwang Shin & Lee, 2020). Finally, the results of these processes differ. Physical try-ons result in a tangible prototype that can be physically tried on and felt by people, while virtual try-ons generate a digital model that can be viewed and assessed on a screen. Despite the convenience of virtual fitting, it lacks the tactile experience provided by physical try-ons (Daanen & Ter Haar, 2013).

There are numerous virtual application software applications available for garment fit assessment, including Clo 3D, Lectra 3D Prototype, OptiTex, and V-Stitcher 3D (Sayem et al., 2023). These virtual fitting assessments simulate the human body, based on the mechanical properties of fabrics and their interaction.

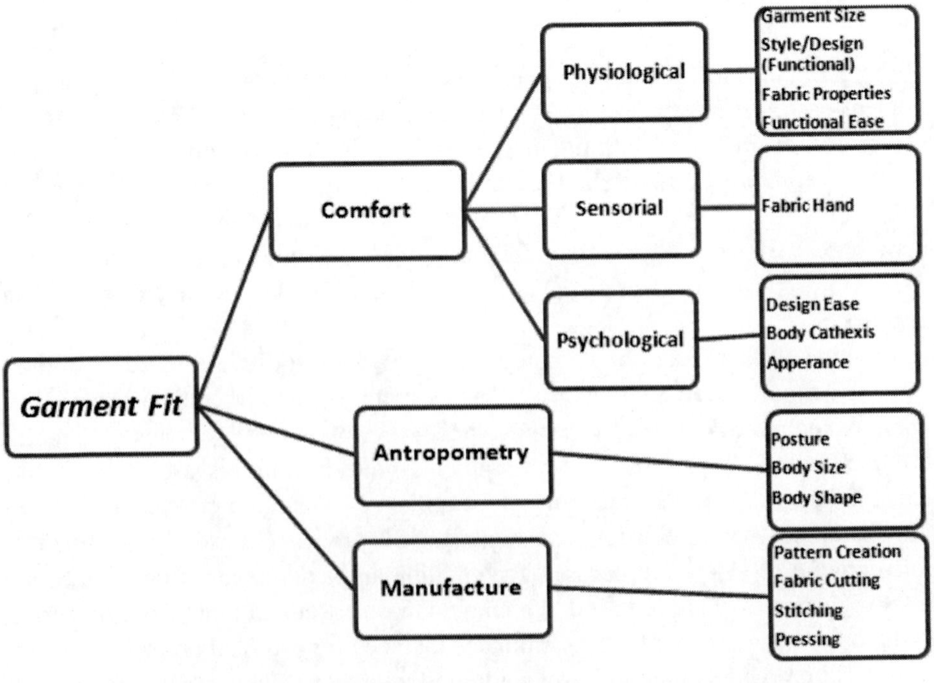

Figure 3. Garment fit components and sub-components (Regenerated; Sayem, 2023).

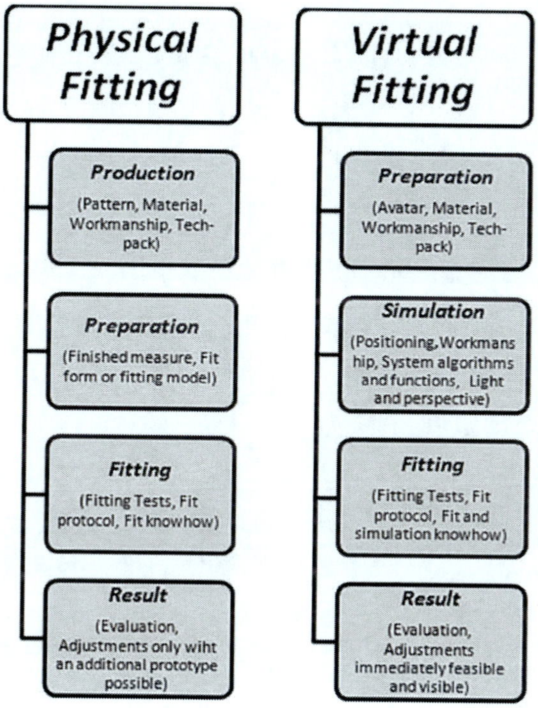

Figure 4. Comparison of physical fitting and virtual fitting (Regenerated; Morlock et al., 2023).

These softwares generally consist of three main parts (Sayem et al., 2010); 3D avatar module, fabric module, and virtual pattern sewing module.

- In the avatar module, you can create a body of any gender and size. Critical body measurements such as height, waist, and hip circumference (Lee & Park, 2017) determine the dimensions of the parametric manikin. Desired changes in the dimensions can be made quickly.
- The fabric section, which is prepared with the technical parameters of the fabrics, aims to increase realism by making use of the various mechanical properties of the fabrics. The software has the opportunity to introduce new fabrics as well as the fabrics defined in the system (Wang, 2022).
- The virtual pattern sewing section, on the other hand, allows one to evaluate the fit of the garment obtained by dressing the predefined or created clothes on the 3D human body. During a virtual 3D experiment, both consumers and designers can visualize the static and dynamic behavior of selected fabrics and the effect of simulated pressures between the human body and fabrics, as well as the effect of garment fit on comfort, as demonstrated by fashion styles (Tama et al., 2016; Liu et al., 2018).

In these virtual try-ons software, a garment pattern can be prepared in the system or in a different system. The prepared garment is dressed in the virtual mannequin, and the compatibility of different fabrics is examined by changing the fabric preferences on the mannequin. In addition, after the dressing process is completed, the dynamic appearance of the garment is evaluated by running the virtual mannequin (Figure 5).

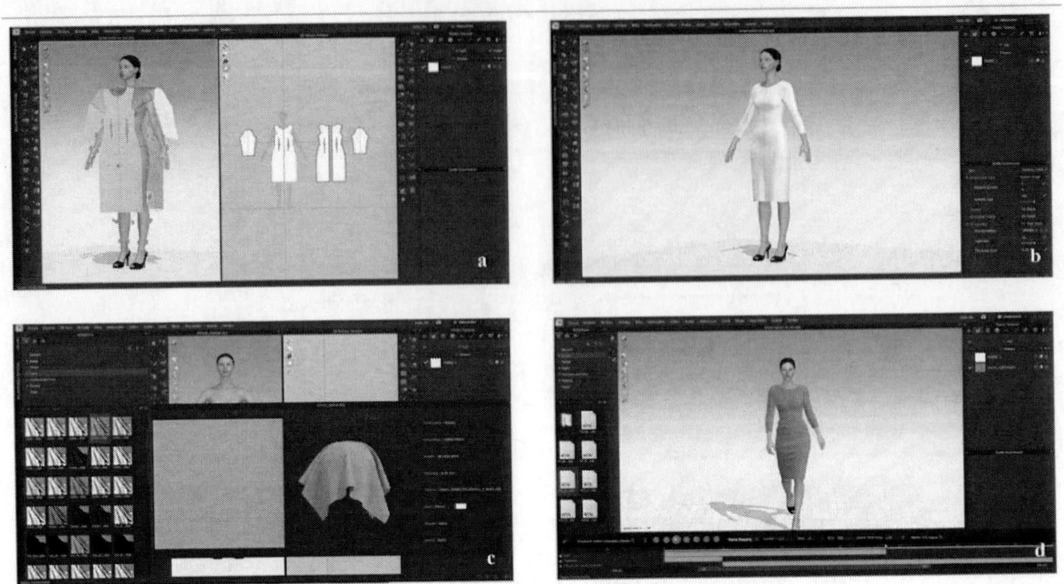

Figure 5. CLO 3D fitting (a. Preparing the clothing pattern in accordance with the avatar, b. Dressing the cut on the avatar, c. Fabric selection, d. Evaluation of compatibility by executing the avatar).

Visual evaluation can be made after the garment is put on the mannequin, as well as the systems creating pressure and strain maps and giving an idea about how incompatible which region is. Abundance, potty, and sewing problems in the garment are also seen in the mannequin (Figure 6).

Figure 6. V-Stitcher 3D fitting (a. Virtual mannequin b. Dressing the suit on the avatar, c. Strain map, d. Pressure Map).

The 3D software, which is used more and more for the simulation, visualization, and visual harmony of clothes, saves time and resources compared to traditional methods. It is known that they run on similar systems; however, there are slight differences from each other. These similarities and differences are as follows (Gerber, 2023; V-Stitcher, 2023; Human Solutions, 2023; CLO 3D, 2023; Modaris 3D, 2023).

- All these software platforms support 3D modeling and visualization of clothing. They allow designers to create a virtual prototype of a garment and visualize it on a 3D avatar or mannequin.
- Offers the ability to simulate different types of fabrics and draping behaviors, allowing designers to see how a garment would look with different materials.
- These platforms also include tools for grading (resizing) patterns, as well as tools for creating and adjusting pattern pieces to create different sizes.
- Most of these software solutions are designed to work seamlessly with other CAD software and can import and export data in a variety of formats.
- The user interfaces and the level of user-friendliness may differ between these programs. Some may be more intuitive or easier to learn than others.
- Some of these software packages may offer special features that others do not.
- While many of these software solutions offer integration with other systems, the level of integration and the specific systems with which they work may vary.
- The cost of these software solutions can vary significantly, as can the terms of license agreements. Some can be purchased as a one-time purchase, while others may require a monthly or annual subscription.

Conclusion

Digitalization in the fashion industry is impacting competition among companies by accelerating the design process. Digital integration in the fashion industry encompasses 3D garment designs, pattern creation for custom-designed garments, pattern evaluation, style editing, and virtual try-ons. Alongside this, digital advancements such as smart textiles and wearable technologies, virtual reality, and the like are creating the potential for progression in an entirely new and sustainable direction, far beyond the traditional structure of the fashion industry (Paralı, 2022). Industry 5.0, also known as the fifth industrial revolution, shifts the focus from automation and technology to collaboration between humans and machines. It aims to integrate digital, physical, and biological systems in a way that enhances human creativity and problem-solving capabilities. It is predicted that the effect of CAD/CAM systems on the sector will increase in the coming years. The use of robots in processes such as cutting and sewing in the sector, the use of artificial intelligence and machine learning techniques in production, supply chain, and customer experiences, the dissemination of 3D printing productions, the use of virtual and augmented reality technologies in display and presentation emphasize the importance of this technology. All these developments appear as potential of CAD/CAM systems, which will take place depending on the needs of industry manufacturers and consumers.

Disclaimer

None.

References

3D Design Solutions. (n.d.). *Gerber Technology*. Retrieved March 12, 2023, from https://www.gerbertechnology.com/fashion-software/3d-design-solutions/.

3D Vidya. (n.d.). *Human Solutions*. Retrieved March 12, 2023, from https://www.human-solutions.com/fashion/3d-simulation-vidya/.

Apeagyei, P. R., (2010). Application of 3D body scanning technology to human measurement for clothing Fit. *International Journal of Digital Content Technology and its Applications*, *4*(7), 58-68.

Apeagyei, P. R., Otieno, R., & Tyler, D. (2007). Ethical practice and methodological considerations in researching body cathexis for fashion products. *Journal of Fashion Marketing and Management: An International Journal, 11*(3), 332-348. https://doi.org/10.1108/13612020710763092.

Ashdown, S. P. (2007). An investigation of the structure of sizing systems: A comparison of three multidimensional optimized sizing systems and four existing sizing systems. *International Journal of Clothing Science and Technology, 19*(5), 349-362.

Baral, L. M. (2011). Feasibility study of CAD and CAM technology in garments industries of Bangladesh. *Advanced Materials Research, 264*, 1563-1567. https://doi.org/10.4028/www.scientific.net/AMR.264-265.1563.

Carr, H., & Lutham, B. (2014). *The technology of clothing manufacture*. Blackwell Publishing.

Chaudhary, S., Kumar, P., & Johri, P. (2020). Maximizing performance of apparel manufacturing industry through CAD adoption. *International Journal of Engineering Business Management, 12*, 1-12. https://doi.org/10.1177/1847979020975528.

Chen, Y., Song, Z., Xu, W., Martin, R. R., & Cheng, Z. Q. (2019). Parametric 3D modeling of a symmetric human body. *Computers & Graphics, 81*, 52-60. https://doi.org/10.1016/j.cag.2019.03.013.

CLO. (n.d.). (2023, March 12). *CLO virtual fashion*. https://www.clo3d.com/.

Collier, B. J., & Collier, J. R., (1990). CAD/CAM in the textile and apparel industry. *Clothing and Textiles Research Journal, 8*(3), 7-13. https://doi.org/10.1177/0887302X9000800302.

Çınar, P. (2019, November 5-7). Fashion industry evolving under the influence of new technologies in the 21st century. *Bilge Kagan 2nd International Science Congress Barcelona/SPAIN*, 88-95.

Daanen, H. A. M., & Psikuta, A. (2018). 3D body scanning. In R. Nayak & R. Padhye (Eds.), *Automation in garment manufacturing* (pp. 237-252). Woodhead Publishing. https://doi.org/10.1016/B978-0-08-101211-6.00010-0.

Daanen, H. A. M., & Ter Haar, F. B. (2013). 3D whole body scanners revisited. *Displays, 34*(4), 270-275. https://doi.org/10.1016/j.displa.2013.08.011.

Dai, L. (2011). The 3D digital technology of fashion design. *2011 International Symposium on Computer Science and Society* (pp. 178-180). IEEE. https://doi.org/10.1109/ISCCS.2011.56.

D'Apuzzo, N. (2007). *3D body scanning technology for fashion and apparel industry.* In Videometrics IX Vol. 6491, pp. 203-214. SPIE.

Dwivedi, A., & Dwivedi, A. (2013). Role of computer and automation in design and manufacturing for mechanical and textile industries: CAD/CAM. *International Journal of Innovative Technology and Exploring Engineering (IJITEE), 3*(3), 174-181.

Fontana, M., Rizzi, C., & Cugini, U. (2005). 3D virtual apparel design for industrial applications. *Computer-Aided Design, 37*(6), 609-622. https://doi.org/10.1016/j.cad.2004.09.004.

Glock, R. E., & Kunz, G. I. (2004). *Apparel manufacturing: Sewn product analysis.* Pearson Publishing.

Hossain, M. S., Samad, M. A., Ali, M. H., Sutradhar, P., Islam, J., Ali, M. H., Al-Amin, M., Babar, U. Z. M., Karim, R. M., & Ullah, M. (2023). An observation toward computer aided processes in garments production. Comparison and analysis of CAD/CAM Software in Bangladesh. *Innovación y Software, 4*(1), 6-23.

Hu, J. (2011). *Computer technology for textiles and apparel*. Elsevier.

Hu, Z. H., Ding, Y. S., Zhang, W. B., & Yan, Q. (2008). An interactive co-evolutionary CAD system for garment pattern design. *Computer-Aided Design, 40*(12), 1094-1104. https://doi.org/10.1016/j.cad.2008.10.010.

Hwang Shin, S. J., & Lee, H. (2020). The use of 3D virtual fitting technology: comparison between sourcing agents contractors and domestic suppliers in the apparel industry. *International Journal of Fashion Design, Technology and Education, 13*(3), 300-307.

Istook, C. L., & Hwang, S. J. (2001). 3D body scanning systems with application to the apparel industry. *Journal of Fashion Marketing and Management: An International Journal, 5*(2), 120-132. https://doi.org/10.1108/EUM0000000007283.

Jhanji, Y. (2018). Computer-aided design - garment designing and patternmaking. In R. Nayak & R. Padhye (Eds.), *Automation in garment manufacturing* (pp. 253-290). Woodhead Publishing. https://doi.org/10.1016/B978-0-08-101211-6.00011-2.

Kadolph, S. J., & Marcketti, S. B. (2013). *Textiles*. Pearson Education.

Kılıç, A. (2019). Comparison of CAD and manual system efficiency in pre-production preparation process. *Textile and Apparel, 29*(1), 73-84. https://doi.org/10.32710/tekstilvekonfeksiyon.437799.

Kunz, G. I., Karpova E., & Garner, M. B. (2016). *Going global: The textile and apparel industry*. Fairchild Books.

Lee, E., & Park, H. (2017). 3D Virtual fit simulation technology: strengths and areas of improvement for increased industry adoption. *International Journal of Fashion Design, Technology and Education, 10*(1), 59-70.

Lee, Y., A. (2014). Computer design and digital fit of clothing. In D. Gupta & N. Zakaria (Eds.), *Anthropometry, apparel sizing and design* (pp. 305-319). Woodhead Publishing. https://doi.org/10.1533/9780857096890.2.305.

Liu, K., Zeng, X., Bruniaux, P., Tao, X., Kamalha, E., & Wang, J. (2018). Garment fit evaluation using machine learning technology. In S. Thomassey & X. Zeng (Eds.), *Artificial intelligence for fashion industry in the big data era* (pp. 273-288). Springer. https://doi.org/10.1007/978-981-13-0080-6_14.

Liu, K., Zeng, X., Bruniaux, P., Wang, J., Kamalha, E., & Tao, X. (2017). Fit evaluation of virtual garment try-on by learning from digital pressure data. *Knowledge-Based Systems, 133*, 174-182. https://doi.org/10.1016/j.knosys.2017.07.007.

Meng, Y., Mok, P., Y., & Jin, X. (2012). Computer aided clothing pattern design with 3D editing and pattern alteration. *Computer-Aided Design, 44*(8), 721-734. https://doi.org/10.1016/j.cad.2012.03.006.

Modaris® 3D. (n.d.). *Lectra*. Retrieved March 12, 2023, from https://www.lectra.com/en/products/software-solutions/modaris-3d.

Moritz, A., & Youn, S. Y. (2022). Spatial ability of transitioning 2D to 3D designs in virtual environment: Understanding spatial ability in apparel design education. *Fashion and Textiles, 9*(1), 29. https://doi.org/10.1186/s40691-022-00293-w.

Morlock, S., Pirch, C., & Klepser, A. (2023). The virtual fitting process - How precisely does 3D simulation represent physical reality? In A. S. M., Sayem (Ed.), *Digital fashion innovations* (pp. 39-61). CRC Press.

Nayak, R., Padhye, R., Wang, L., Chatterjee, K., & Gupta, S. (2015). The role of mass customisation in the apparel industry. *International Journal of Fashion Design, Technology and Education, 8*(2), 162-172.

Oppong, J. A., Biney-Aidoo, V., & Antiaye, E. (2013). Evaluating the benefits of computer aided design (CAD) in fashion education, the case of Accra polytechnic. *Journal of Education and Practice, 4*(21), 1-10.

Papachristou, E., Kyratsis, P., & Bilalis, N. (2019). A comparative study of open-source and licensed CAD software to support garment development learning. *Machines, 7*(2), 30. http://dx.doi.org/10.3390/machines7020030.

Paralı A. (2022). Digital technologies and their implications in the textile and fashion industry. In V. Ö. Akgün & M. Diktaş (Eds.), *Opening the doors of the digital world- Current articles on digitization from different perspectives* (pp.485 -511). Nobel Publishing.

Qiu, Y. (2014). Body scanning and 3D CAD software in the garment industry. *In proceedings of the international conference on education technology and information system* (ICETIS 2014).

Sayem, A. S. M. (2023). Clothing fit evaluation: From physical to virtual. In A. S. M. Sayem (Ed.), *Digital fashion innovations* (pp. 17-38). CRC Press.

Sayem, A. S. M., Chakraborty, S., Hoque, S. A., Saha, K., Mica, M. T., & Ahsan, M. (2023). Defining digital fashion and tracking the developments in relevant technologies. In A. S. M. Sayem (Ed.), *Digital fashion innovations* (pp. 3-14). CRC Press.

Sayem, A. S. M., Kennon, R., & Clarke, N. (2010). 3D CAD systems for the clothing industry. *International Journal of Fashion Design, Technology and Education, 3*(2), 45-53.

Simmons, K. P., & Istook, C. L. (2003). Body measurement techniques: Comparing 3D bodyscanning and anthropometric methods for apparel applications, *Journal of Fashion Marketing and Management, 7*(3), 306-332. https://doi.org/10.1108/13612020310484852.

Spahiu, T., Shehi, E., & Piperi, E. (2014). *Advanced CAD/CAM systems for garment design and simulation.* In 6th International Conference of Textile 1-6.

Spelic, I. (2020). The current status on 3D scanning and CAD/CAM applications in textile research. *International Journal of Clothing Science and Technology, 32*(6), 891-907. https://doi.org/10.1108/IJCST-07-2018-0094.

Stjepanovic, Z. (1995). Computer-aided processes in garment production: features of CAD/CAM hardware. *International Journal of Clothing Science and Technology, 7*(2/3), 81-88. https://doi.org/10.1108/09556229510087236.

Tama D., & Öndoğan, Z. (2020). Calculating the percentage of body measurement changes in dynamic postures in order to provide fit in skiwear. *Journal of Textiles and Engineer, 27*(120), 271-282.

Tama D., Kılıç, A. Ş., Öndoğan, Z., & Nizamoğlu, S. (2016). The usability of 3D flattening in design and pattern preparation of tight-fit garments. *Çukurova University Journal of the Faculty of Engineering and Architecture, 31*(SI 2), 169-174.

Thomassey, S., & Zeng, X. (2018). Introduction: artificial intelligence for fashion industry in the big data era. In S. Thomassey & X. Zeng (Eds.), *Artificial intelligence for fashion industry in the big data era* (pp. 1-6). Springer Singapore. https://doi.org/10.1007/978-981-13-0080-6_1.

Uddiin, M. M., Rahman, M. M., & Helal, M. M. (2021). The benefits of using bomputerized system over manual system in apparel production. *GUB Journal of Science and Engineering (GUBJSE), 8*(01), 36-40.

Uhm, T., Park, H., & Park, J. I. (2015). Fully vision-based automatic human body measurement system for apparel application. *Measurement, 61*, 169-179. https://doi.org/10.1016/j.measurement.2014.10.044.

Volino, P., Cordier, F., & Magnenat-Thalmann, N. (2005). From early virtual garment simulation to interactive fashion design. *Computer-Aided Design, 37*(6), 593-608. https://doi.org/10.1016/j.cad.2004.09.003.

VStitcher: 3D Design Software. (n.d.). *Browzwear*. Retrieved March 12, 2023, from https://browzwear.com/products/v-stitcher.

Wang, C. C. (2004). CAD tools in fashion/garment design. *Computer-Aided Design and Applications, 1*(1-4), 53-62.

Wang, T. (2022). Intelligent analysis of computer-aided garment design CAD based on virtual reality model for surplus fabric removal and reuse without segmentation of cutting pieces, *Advances in Multimedia, 2022,* 1-13. https://doi.org/10.1155/2022/7974639.

Yan, H., & Fiorito, S. S. (2002). Communication: CAD/CAM adoption in US textile and apparel industries. *International Journal of Clothing Science and Technology, 14*(2), 132-140. https://doi.org/10.1108/09556220210424233.

Yu, W. (2004). 3D body scanning. In J. Fan, W. Yu & L. Hunter (Eds.), *Clothing appearance and fit: Science and technology* (pp. 135-168) Woodhead Publishing.

Zhang, L., Zhang H., & Li, Y. (2008). The new development of CAD in clothing. *Computer and Information Science, 1*(3), 104-106.

Chapter 23

Virtual Garment Design and Sample Production Systems Used in the Apparel Industry

Özlem Topbaş[1,*], PhD
and Gülşah Pamuk[2], PhD
[1]Mosi Textile R&D Center, Izmir, Türkey
[2]Emel Akın Vocational School, Ege University, Izmir, Türkey

Abstract

The apparel industry, which is one of the sectors where international competition is most intense, is also considered one of the highest income-generating sectors. It is possible to specify factors such as quality, qualified workforce, fast delivery opportunities, and advanced technological infrastructure as the most important factors that determine the competitiveness of companies operating in the field of apparel. In addition to these factors, companies have to manufacture uniquely designed products. In today's apparel industry, smaller orders are organized rather than bulk orders. Since the preferences of consumers are constantly changing, companies have to respond to consumer requests as soon as possible. For this reason, companies have to take advantage of the opportunities offered by technology in all processes, from product design to production. There are various systems used in garment design and garment pattern preparation in the apparel industry. These systems are called Computer-Aided Design (CAD) systems. By means of these systems, it is possible to design a garment and perform various editing operations on the designed garment. Once a garment is designed and its patterns are created with CAD systems in a 2D platform, a digital prototype is produced in a 3D environment. In the apparel industry, the use of CAD systems, which are used to reduce sample costs, is becoming increasingly common. By using these systems, it is possible to design clothes in the virtual environment, to dress the designed clothes on avatars, work with different color variants, and send the garment to the customer online. In this chapter, it is aimed to introduce the 3D garment design and sample preparation systems used in the apparel industry. The features of these systems, which are widely used in the apparel industry, and the advantages they provide to companies are mentioned. In addition, the stages of virtual sample production from a dress model are explained using the V-Stitcher program.

Keywords: 3D design, digital sample, garment design, sustainable fashion, apparel industry

[*] Corresponding Author's Email: ozlem.topbas@mositekstil.com.

In: Textile Insights and Future Aspects
Editors: Gamze D. Tetik, Gizem Celep and Fulya Yilmaz
ISBN: 979-8-89113-289-4

Introduction

Technological developments in all areas of our lives also affect the apparel industry. High technology studies are involved in all processes from the design to production of a ready-made garment. The textile and ready-made garment industry has the feature of being the locomotive sector in the economy of developing countries and plays an active role in the transition to the industrialization process. In addition, the fact that this sector is a labor-intensive sector also solves the employment problem in these countries. Due to the labor-intensive nature of ready-made garment production, it is concentrated in countries where labor costs are low. However, global brands come mainly from developed countries that are known for their fashion and design capabilities. In other words, the production of ready-made garments has shifted to countries with cheap labor over the years due to the rapidly increasing competition. Textile, starting from fiber, includes processes such as yarn, weaving, knitting, dyeing, and printing, and ready-made garment covers the processes that will turn this process into a usable item. The part from fiber to yarn and the finished fabric is evaluated within the textile sector, and the process from fabric to clothing is evaluated within the ready-made clothing sector.

Computer-aided design and manufacturing systems offer significant advantages to companies in the apparel industry. Computer-Aided Design (CAD) is a term used for the creation, development, and analysis of any design work in a computer environment (Groover & Zimmers, 1983). CAD systems can perform labor-intensive and time-consuming processes such as model design, garment pattern preparation, grading, and cutting plan preparation in a short time and with minimum error. It also increases the creativity of designers. In addition, it offers advantages such as reducing production costs, increasing the prestige of the company, and responding to consumer demands in a short time. Computer-Aided Manufacturing (CAM) systems are widely used in the apparel industry. These systems ensure that labor-intensive processes in apparel processes are carried out by computer control. For example, fully automatic spreading systems are used in the fabric spreading process. These machines are designed to spread both knitted and woven fabrics. One of the labor-intensive processes in the apparel industry is the cutting process. The computer-aided production system used for the cutting process in the clothing industry is called Cutter. It consists of different parts, such as the main control panel, and the vacuum unit. The vacuum level can be adjusted parametrically according to the fabric to be cut and the height of the marker. Fully automatic sewing machines and embroidery machines, which are widely used in the apparel industry, are also included in CAM systems.

Studies on the use of CAD systems in the apparel industry started in the 1970s. The main purpose of these studies was the preparation of cutting plans in the computer environment. Today, many companies are working in this field, and studies on the realization of clothing design in the computer environment are increasing. CAD software is used in the preparation of garment patterns and markers. However, there is an increasing interest in software used for garment design and garment dressing on virtual mannequins. Thus, many researchers had the opportunity to conduct research on these systems (Elnashar et al., 2011; Lim & Istook, 2010; Bilgic & Turemez, 2019; Cichocka et al., 2014; Liu & Jang, 2013; Baytar, 2018; Tama et al., 2014).

CAD systems offer a wide range of advantages to apparel companies. The most important of these advantages is the reduction of production costs. As it is well known, the most important cost item in a garment is the expense of the fabric. Approximately 50-60% of the total cost of a garment is fabric expense. The preparation of cutting plans in CAD systems reduces the fabric expenses of apparel companies. Another important advantage of these systems is that they allow apparel companies to respond to consumer demands in a short time. In addition, advantages such as minimizing the errors caused by garment patterns, increasing the prestige of the company, and saving labor and time are also very important.

A CAD system used for garment pattern and marker preparation consists of three main parts. The main unit of the system is the Workstation, where many operations are performed. It is possible to prepare a garment pattern at the Workstation. In addition, edits can be made on the garment patterns transferred to the system using the Digitizer.

The Digitizer is the second main unit of a CAD system, and it is used to transfer the manually prepared garment patterns to the Workstation. To perform the digitization process accurately and completely, garment patterns must be placed on the digit table in the correct position. By moving an apparatus called "Cursor" on the garment patterns, the image of each pattern is transferred to the Workstation with x-y coordinates. The cursor is a very sensitive instrument, and it is very important to hold it with both hands while working. All operations in the menu section of the digit table are performed by pressing the A key of the cursor. The digitization operation starts by selecting "start piece" in the menu and ends by selecting "end input." In the digit process, it is possible to obtain the garment patterns of the other sizes in the set of sizes using the garment patterns of the main size. However, first, it is necessary to perform the pattern-grading process. Today, many apparel companies use computer-aided design systems for the pattern grading process. The purpose of the pattern grading process is to obtain the garment patterns of 'll the sizes in the set of sizes by using the measurement differences between the sizes in the size table. There are different grading methods in the apparel industry.

The last unit of a CAD system is the plotter. The plotter is used to draw garment patterns and markers. It is possible to perform the drawing process with a pen or inkjet. If the drawing process is performed with a pencil, these plotters are called horizontal plotters. If the drawing process is performed by inkjet, such plotters are called vertical plotters.

Garment Pattern Preparation in the CAD System: Preliminary

Various files are used in the Workstation for the preparation of garment patterns. It is possible to examine these files in two sections, parameter files, and marker preparation files. However, before data entry into these files, a main folder must be created on the Workstation. All the garment patterns and marker plans prepared in the system should be included in this main folder. To briefly mention the parameter files, it is possible to say that these files contain some information for notch type, marker, and garment pattern drawing.

Marker preparation files are the files in which the pattern properties of the garment and the technical properties of the fabric to be used are transferred to the system. For example, information such as the name of the garment pattern and the size of the garment to be written on the garment patterns are defined in the annotation file. The most common information in the annotation file is as follows: marker name, marker description, order number, order description, model name, length, marker width, utilization, size, bundle, and piece name. The information can be placed side by side or one below the other on the garment pattern.

Another part of the marker preparation files is the Lay Limits File. This file contains the technical specifications of the fabric to be used in the production of the designed garment. Fabric spread is one of the most important processes in the apparel industry. There are various options for fabric spreading such as single-ply, face-to-face, book fold, and tubular. Generally, woven fabrics are used in the production of clothing products such as shirts, jackets, and blouses. The single-ply option is selected for woven fabrics. In the planned productions for knitted fabrics, the tubular option is chosen.

The technical features of the garment patterns to be included in the marker are specified in the Model File. The main headings in the model file are as follows: Piece name, fabric, flips, add piece, x shrink stretch, y shrink stretch.

The last file in the marker preparation files is the Order File and it must be prepared for each new marker. The order file consists of various contents such as marker name, order number, target utilization, target length, block buffer, etc.

Marker Making

In the apparel industry, production is generally made from a garment model in different colors and in different size distributions, according to order and sales forecasts. This color and size distribution, which is determined before production starts, is called the assortment of that model. Sometimes there may be only color assortment or only size assortment. In the color assortment, more colors are produced for the colors that are thought to be sold more depending on the fashion, while less production is made for the colors that are thought to be sold less. In size assortment, the numerical majority in the order is usually collected in medium sizes. Order less from the smallest and largest sizes. The reason is that the number of potential buyers in medium sizes is higher than the number of buyers in large and small sizes. Color and size assortment is also considered when calculating the cost in the order.

Marker preparation is the most important stage of the whole cutting process. Because 50-60% of the production cost of a garment is the fabric cost. The way the garment patterns are placed on the marker affects the quality, material use, and production time of the product. The marker is a generally rectangular surface, used to determine the use of materials according to technological possibilities and comply with quality requirements.

The last thing to do before starting to place the markers in the CAD system is to prepare the marker. After all marker preparation files have been prepared completely and without errors, the Order File must be converted to a marker. If there is a previously prepared marker in the system for the relevant Order File, the system gives a warning message. From this message, it is understood that the marker has already been prepared in the system. The system asks if the user currently wants to record on this marker. The marker display consists of four main sections: Menu Bar, Icon Menu, Marker Area, and Status Bar. There is also a box called "Toolbar" which contains shortcuts to various functions (Pamuk, 2017).

The area where the marker will be placed is in the form of a rectangle at the bottom of the computer screen. The width of the marker area is the width of the fabric to be worked with. In the marker area, there is a line that appears as a dashed line. Placing all garment patterns on the marker, to the left of this line, which can also be called the target productivity line, is an indication that the marker placement process was carried out successfully in accordance with the target efficiency value entered in the Order File.

It should not be forgotten that there will be a decrease in marker efficiency when moving to the right of this line. It is possible to create this line in two ways:

- *Target Length:* It expresses the maximum marker length targeted by placing all patterns in the marker area;
- *Target Efficiency:* It expresses the maximum marker efficiency targeted by placing all garment patterns in the marker area (Pamuk, 2017).
- A toolbar is used to provide convenience to the user in the marker-making process. All the functions in the Toolbar are applied by right-clicking on the garment patterns. Some of the most important features of the toolbar are listed as follows:
- *Auto Slide:* The purpose of this menu is to place all the clothing patterns in the icon on the cake with the rectangle to be created by pressing the right button continuously after selecting one of the "Area, Length, Width" functions. In the process of placing the garment patterns one by one in the marker area, the patterns on the marker screen are dragged to the marker area with the left button of the mouse. Making such an application for each garment pattern can be time-consuming for some models. Instead, it is more accurate to make automatic marker placement first and then take corrective actions to increase productivity (Pamuk, 2017).
- *Group Slide:* It has a similar feature to the Auto Slide function. However, it preserves the positions of the established garment patterns. It is the function where the movement of some selected garment patterns is performed as a group.
- *Overlap:* It is used to place the garment patterns on top of each other. The places of the garment patterns to be placed on top of each other are determined, and the dragged garment pattern is placed on top of the other by pressing the right button of the mouse.
- *Align:* It positions two garment patterns so that their ends are aligned. The garment patterns can only move horizontally and vertically. In general, it gives better results on flat items such as sheets and duvet covers.
- *Flip:* It allows the garment patterns to be flipped on the X and Y axes. The operation is performed by right-clicking on the garment pattern to be flipped.
- *Rotate:* It is used to rotate the garment patterns at to the desired degree. For example, 45 CW: Rotates the garment patterns 45° clockwise; 45 CCW: Rotates the garment patterns 45° counter clockwise.
- *Place:* It is the function used to place the garment patterns in the desired area of the marker. The garment pattern is brought to the desired place with the left button and placed with the right button of the mouse.
- *Fold:* In tubular fabrics, it is used to fold the garment pattern and place the fabric layer or to open the pattern with the fabric layer folded.
- *Center:* It is the function used to place the selected garment patterns in the middle of the empty area on the marker or in the middle of the two garment patterns where they will be placed.
- *ToolBox Override:* It is used to remove all the controls for a single garment pattern.

Virtual Garment Design and Sample Production Stages

There are various three-dimensional dressing systems used in the apparel industry. In this research, the V-Stitcher system was studied. There are seven main files used in the design process with the V-Stitcher program. The names of these files are as follows: File, Edit, Wiev, Tools, 3D Environment, Garment, and Help. The main stages of the model creation by using V-Stitcher are as follows:

Avatar Selection

The V-Stitcher is where the patterns are imported, stitched together, and arranged. Avatars are edited and imported into V-Stitcher as well, and garments are paired with avatars in V-Stitcher. The two-dimensional garment patterns prepared on the CAD system are transferred into three-dimensional modeling system before selecting the avatar (Bilgic & Turemez, 2019).

The designer can arrange avatar's skin tone, body resolution (original, high, max), hair (bald, beach boy, coach, cold spark, cowboy, delta force, explorer, havana, natural fling, short hair, stardust, top gun, urban soldier), final poses (arms down, arms forward, arms spread, balanced, basketball, bicycle, casual, driving, jump shot, fold over, football, etc.), pose angle (hands at 15, hands at 30, hands at 45, hands at 60, hands at 75, hands at 90). The designer can also arrange lights, environment, and 3D view (front, left, back, right, top, bottom, etc.)

The desired size can be selected from the avatars on the 3D screen, or an avatar can be created from scratch by entering different sizes. Avatar options are available for women, men, and children. Avatar has the option for skin texture. It is also possible to work with the desired measurements. Parametric avatars can move from pose to pose. Scanned avatars do not come with poses unless they are scanned in multiple poses. Imported avatars remain locally within the software in the avatars list and can be used between files (V-Stitcher Training Manual, 2018). The avatar selection process can be seen in Figure 1 (a). and (b).

Virtual Sewing

In this process, the garment patterns are stitched (V-Stitcher Training Manual, 2018). The sewing parts are checked on the three-dimensional screen. The desired stitch images are selected from the "Stitches" window. It is possible to work with different stitch options, such as lock stitch and chain stitch. If some pieces were not sewn or sewn incorrectly, the user can pause the simulation and can make the corrections. After this, the garment is simulated again for a few seconds. When the "Garment" menu is selected, the "Garment Info" can be seen on the screen. In this section, there is much information like company ID, garment ID, season, style ID, style name, designer, etc. The images of the visual sewing process are shown in Figure 2 (a) and (b).

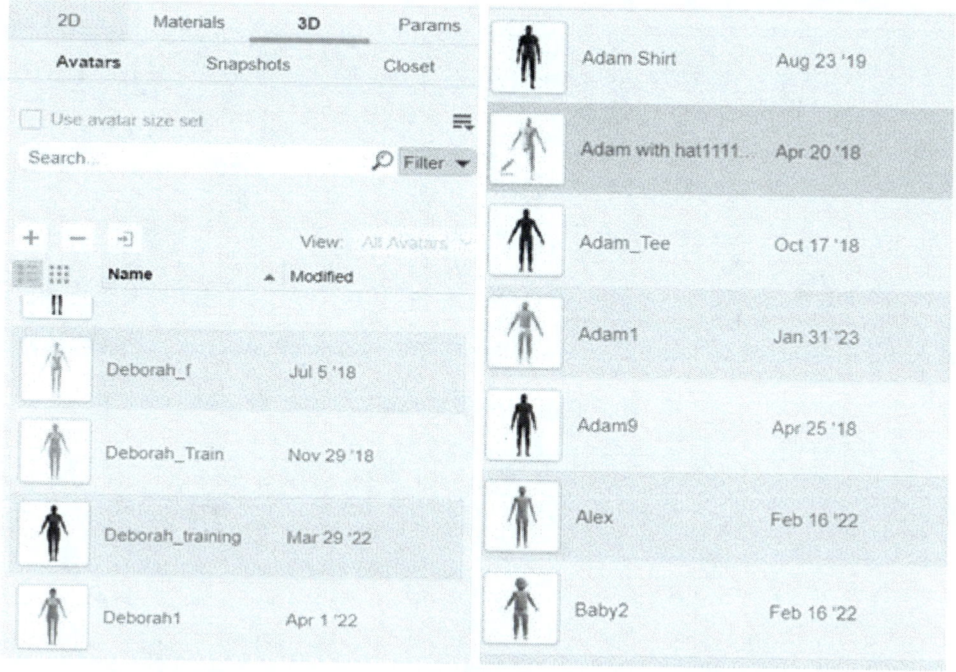

(a)

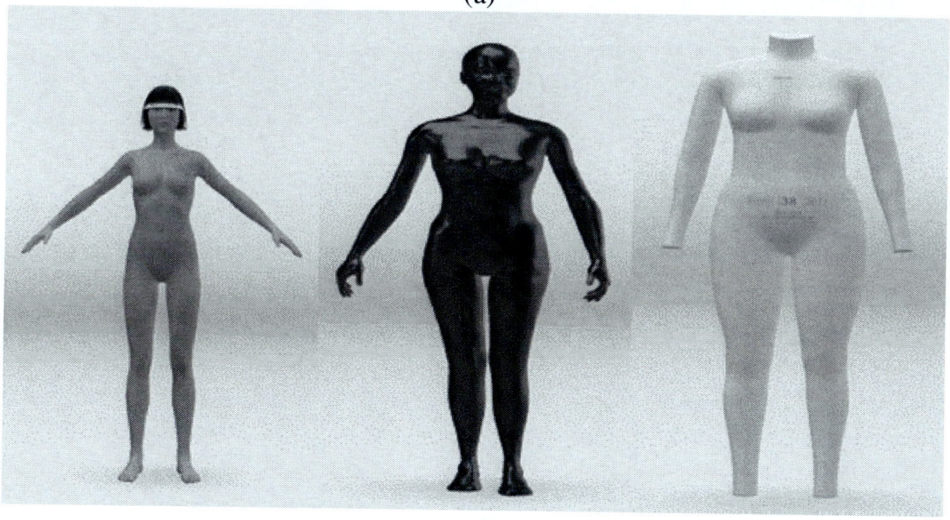

(b)

Figure 1. (a) and (b) The avatar selection process.

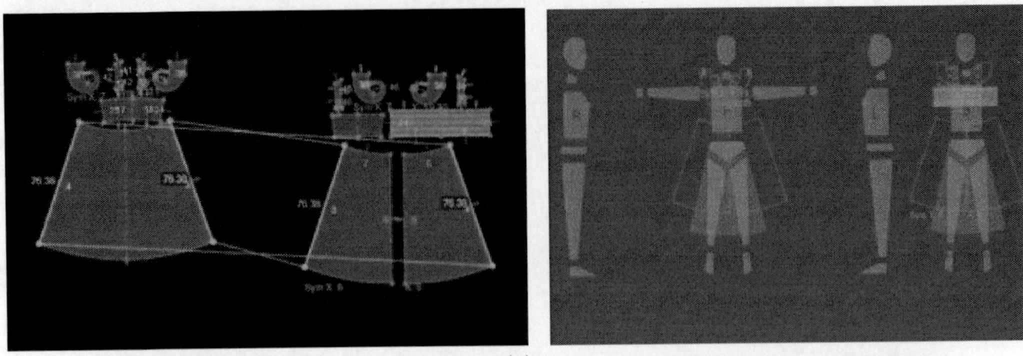

(a)

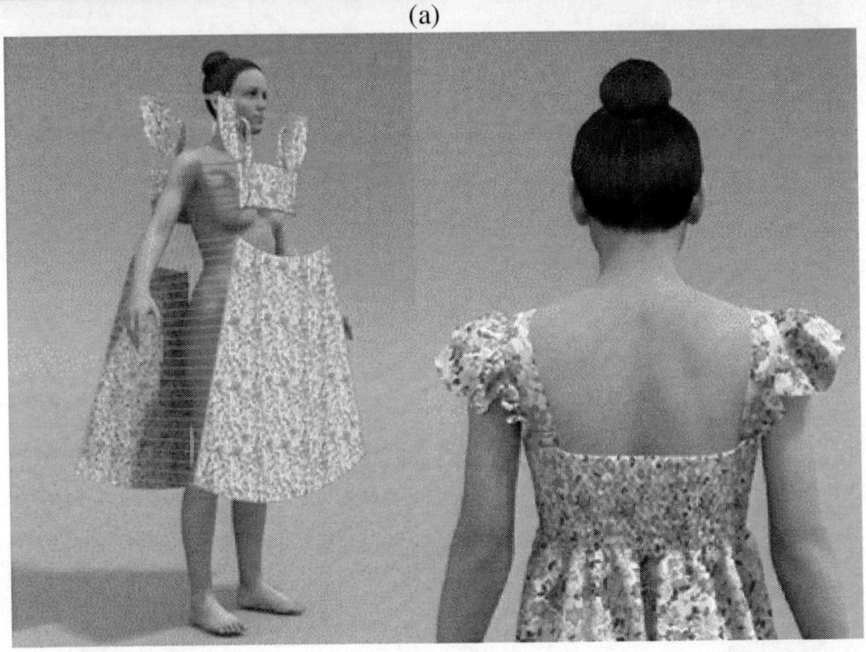

(b)

Figure 2. (a) and (b) Virtual sewing process.

Materials Selection

The fabric type is selected according to the fabric properties of the model. Then, the fabric selection is made in accordance with the weight and mixture. When choosing the "Materials" section, the designer can select the sources for the fabric type in the "Libraries" section. If "Knits" is chosen, "knit jersey, knit stretch jersey, knit rib, knit mesh, knit French terry, knit fleece, knit miscellaneous" options are seen. If "woven" is chosen, "natural fibers, natural polymers, synthetic polymers, and refined" options are seen. In this section, the different types of accessories such as buttons, zippers, etc., can also be seen. The texture tool can be used on all over prints and images placed in material map channels, just as the texture tool can be used on fabrics (V-Stitcher Training Manual, 2018). Images that are applied to the garment in one instance at a time are called Artwork & Trims. Images can be replaced with any other image file and will be placed on the garment where they were replaced. It is possible to work with different fabrics and accessories. The user can store variations of the garment with changes that can be made in the fabric, color, artwork, trims, and seams (Figure 3 (a) and (b)). The materials section of the system for fabric and accessories can be seen in Figure 4 and Figure 5, respectively.

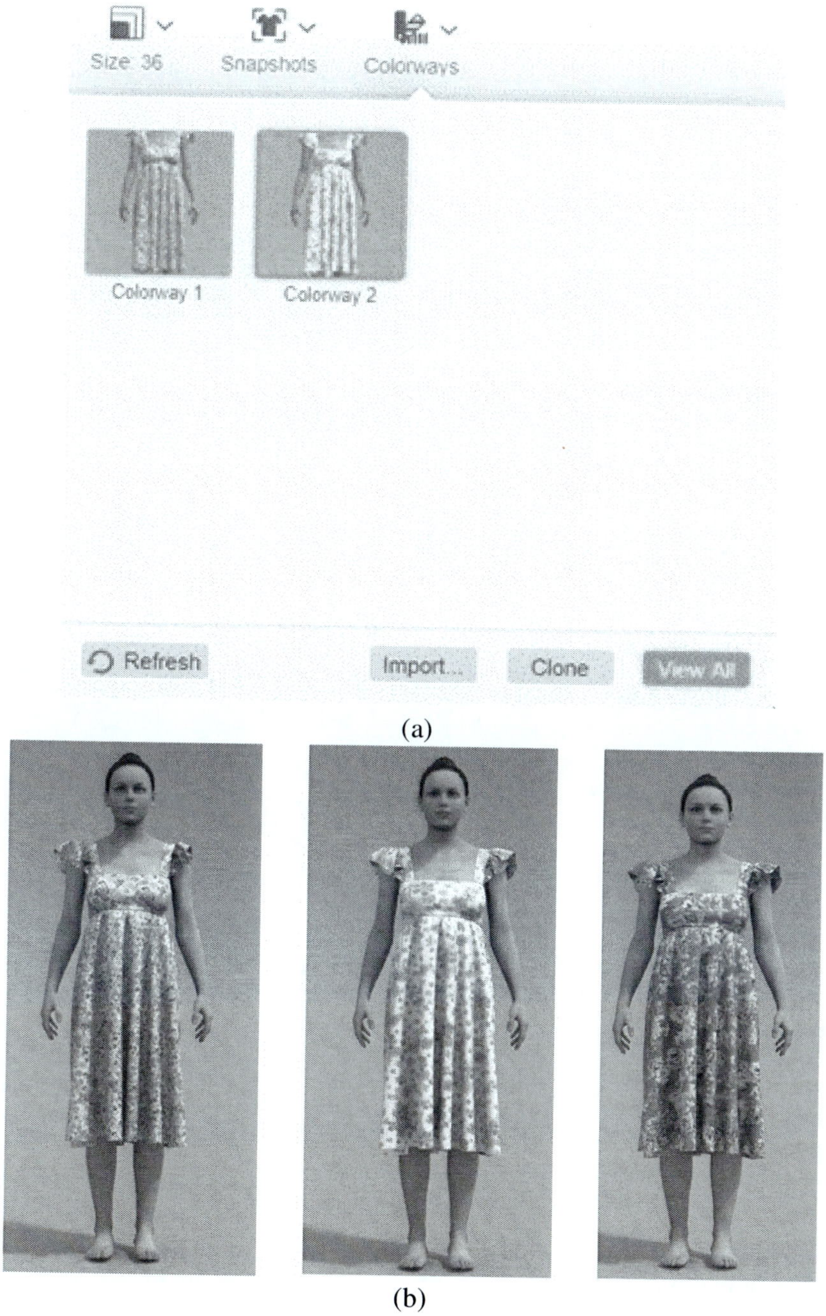

Figure 3. (a) and (b) Creating variants.

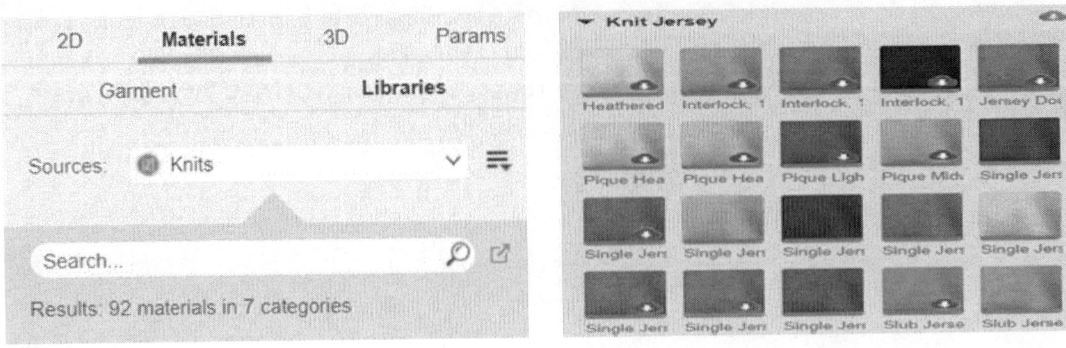

Figure 4. Fabric selection.

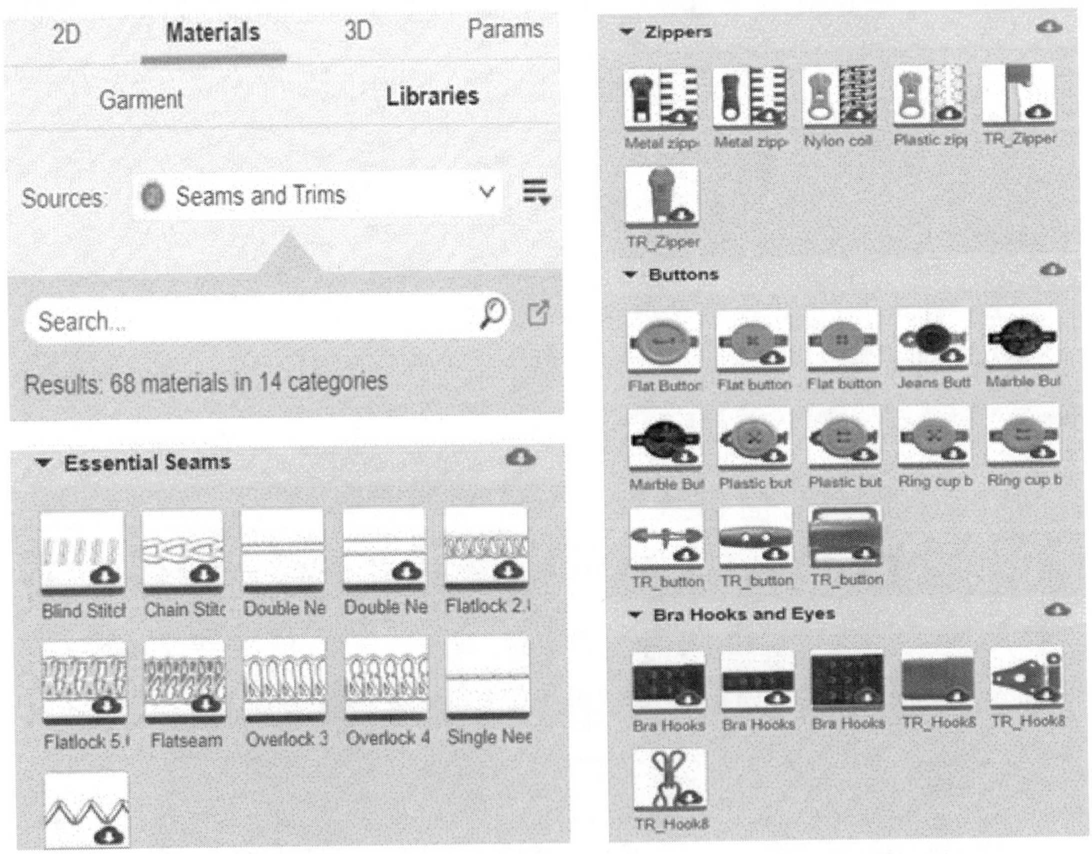

Figure 5. Accessory selection.

Grading

The grading process is easily performed and the desired dimensions are entered. It can create as many sizes of gradation as desired. An example of the pattern grading process can be seen in Figure 6.

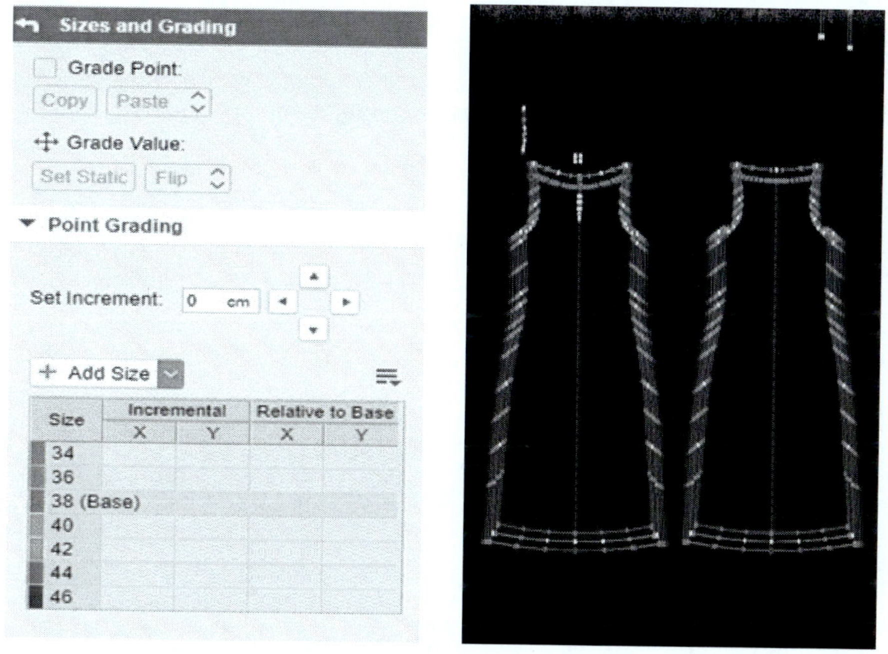

Figure 6. Pattern grading process.

Creating Multiple Avatars

Size control is done by bringing avatars side by side. The variants can be dressed as desired (Figure 7). The bodies of avatars can be changed.

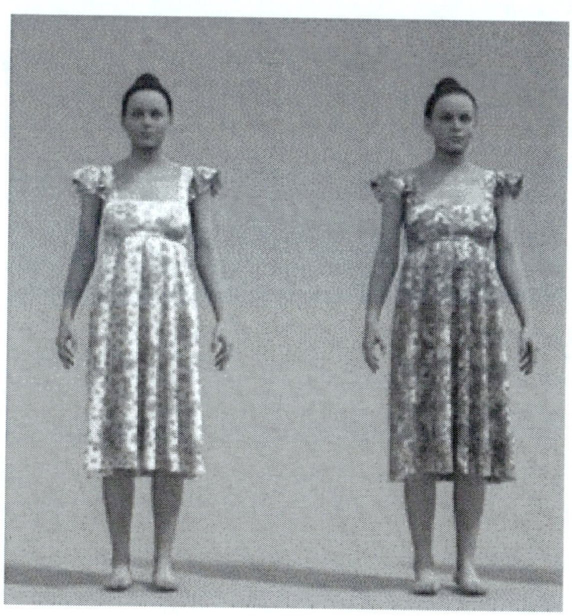

Figure 7. Creating multiple avatars process.

Annotate

It allows designers to take notes by marking that area on the model. The notes can be viewed by clicking on the model (Figure 8).

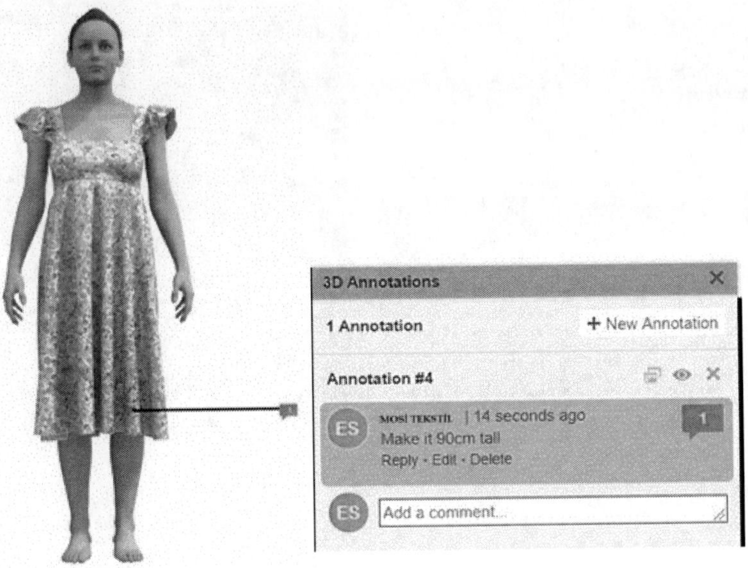

Figure 8. Annotation process.

Rendering

In its most basic definition, rendering; it is the process of processing a drawing, video, or a different design object created, drawn, made, or developed on the computer, taking its final form and extracting it from the program in which it was created in usable condition. In other words, the rendering process ensures that a raw product produced on the computer becomes usable. The light groups that emerge because of the interaction of lights, reflections, and objects in a digital design area are calculated during the rendering process. The rendering software starts by calculating all these objects, objects, lights, effects, and other details. It reveals the final version of the developed design by optimizing this propagation through shadow and light calculation.

There are 3 types of rendering: a. Normal (basic image capture), b. Schematic (black and white image capture), and c. Raytracing (photorealistic image) (V-Stitcher Training Manual, 2018). The designed garment is ready to be sent to the customer with the rendering processes.

Production Times of a Digital Garment Sample

A digital garment sample of a dress (Figure 8), the main steps of which are shown above, was manufactured using V-Stitcher in the Mosi Textile Pattern Department. Production times of the digital sample are given in Table 1.

Table 1. The production times of a digital garment sample

Importing 2D patterns (min)	Editing patterns on avatars (min)	3D stitching (min)	Pre-simulation (min)	Fabric image/draping (min)	Snapshots (min)	Color variants (min)	Rendering (min)	Total (min)
00.27	01.11	20.18	01.30	01.36	00.21	00.33	13.07	39.03

Then, these garment patterns are exported to V-Stitcher. The total time for this process was 00.27 min. The second process was to edit the garment patterns on the avatar. For this study, a woman avatar was chosen. The total time to edit garment patterns on avatar was 01.11 min. The three-dimensional stitching process is one of the most time-consuming operations in the system. As seen in Table 1, three-dimensional stitching and rendering processes take the longest time in digital sample production. The duration of the visual sewing process may vary according to the user's experience and the model properties. The shortest process is to save snapshots; snapshots can be used to save styling options and finalized draping results. The correct determination of the stitch types requires the user to have knowledge of the apparel. It is also interesting that the total time for the color variants was just 00.33 min. This process, which would normally take a long time under real production conditions, took place in a very short time in V-Stitcher. It is clear that it will take quite a long time to obtain this dress with traditional manufacturing methods. However, as can be seen in this case study, the time required to design the dress, create color variants, and have it ready for shipping to the customer is set at just 39 minutes.

Conclusion

The apparel industry is a labor-intensive industry and plays an important role in the economic development of many countries. The apparel companies aim to create a successful product in all processes from product design to production, from marketing to sales.

A ready-to-wear product that is produced according to consumer preferences, needs to be presented to the ready-made clothing market quickly. It takes an average of six weeks from the receipt of the order of a garment to the delivery of it to the customer. This period may be longer depending on various factors such as company capacity, model features, and the customer's overseas location. The prolongation of the time required to organize the order creates dissatisfaction among the customers. In the face of this situation, apparel companies experience customer losses and have difficulty in reaching new customers.

The shortening of collection preparation time intervals during the season, the changes in model selection and application processes, and the shortening of the product production cycle require garment companies to provide faster feedback than before.

With the development of technologies, the use of computer-controlled systems is becoming widespread at every stage of garment production. Computer-aided design systems are used in processes such as garment pattern and marker preparation, which are activities carried out before production. These systems prevent companies from losing time and effort. CAD systems also help reduce production costs. In today's competitive conditions, ready-made clothing companies have to adapt technological developments to their production processes. It is possible to say that the use of three-dimensional clothing design systems is very popular today. First of all, these systems increase the creativity of designers. The fabric and accessory libraries in these systems offer great convenience to designers. Designers can create designs with different types of fabric and different accessories. They can see the different color combinations of the products they have designed in seconds and arrange their designs in a very short time.

In the apparel industry, the preparation of digital samples of garments in a computer environment is becoming more common day by day. The systems used offer many advantages to companies. Digital samples are prepared in a very short time and sent to customers in the digital environment.

Sample preparation is a very important process in apparel production. The correct and complete preparation of the sample is very important. A product prepared according to customer requests primarily creates customer satisfaction. However, most of the time, customers want to see different colors and designs of ready-made garments. For example, for a customer who wants to see a different color of the prepared sample, it is necessary to obtain a different color of the fabric used in a very short time. This situation is quite challenging for apparel companies. First, the fabric manufacturers are contacted and the requested fabric is supplied. After the fabric of different colors is supplied, the sample production process begins. The new sample produced is sent to the customer. The product is revised in line with the customer's criticisms. The process proceeds this way until the customer makes the final decision. It is possible to optimize all these time-consuming and costly processes by using three-dimensional dressing systems and to respond to customer requests and demands in a very short time.

In traditional apparel production, after the garment patterns are prepared, the fabric is supplied and after the spreading and cutting processes, the sewing phase of the product begins. Quality control is carried out at all these stages. After the product is sewn, the final operations such as ironing and packaging are done. These stages are carried out for both sample production and mass production. There are different cost factors at each of these stages in traditional apparel manufacturing. The use of three-dimensional dressing systems greatly reduces the costs of sample preparation. Processes such as garment design, sewing, fit checks, etc., are carried out virtually. The way the designed product will appear to users can be seen instantly by using avatars. In addition, working with different color variants is quite simple. It is possible to make changes to digital samples according to the demands of the customer. Digital samples are sent to customers online. Thus, transportation costs are also eliminated.

As can be seen, CAD systems used in the apparel industry have many advantages. These systems provide convenience to designers in subjects such as garment pattern preparation, garment design, marker making, and digital sample preparation. To actively use these systems in apparel companies, there is a need for technical personnel specialized in this field. Designers who want to actively use computer-aided design and three-dimensional dressing systems need to have knowledge about preparing basic garment patterns, model application, pattern grading, garment pattern preparation and application, model definition, and model analysis.

Disclaimer

None.

References

Baytar, F. (2018). Apparel CAD patternmaking with 3D simulations: Impact of recurrent use of virtual prototypes on students' skill development. *International Journal of Fashion Design, Technology and Education, 11*(2), 187-195.

Bilgic, H., & Turemez, F. (2019). Determining production fabric with three-dimensional modeling system (V-Stitcher). *International Journal of Scientific and Technological Research, 5*(1), 63-70.

Cichocka, A., Bruniaux, P., & Frydrych, I. (2014). 3D garment modelling-creation of a virtual mannequin of the human body. *Fibres & Textiles in Eastern Europe, 22*(6), 123-131.

Elnashar, E. A., Kuzmichev, V., & Sakharova, N. (2011, November 22-24). Fitting stretch model of three-dimensional for simulating apparel surfaces. *1st SMARTEX-Egypt 2011 (World Textiles Conference).*

Groover, M. P., & Zimmers, E. W. (1983). *CAD/CAM: Computer-aided design and manufacturing*. Pearson Education.

Lim, H. S., & Istook, C. (2010). A comparative study on virtual try-on systems using body measurement input. *The International Journal of Costume Culture, 13*(2), 118-129.

Liu, Y. M., & Jang, H. K. (2013). A study on the functional characteristics of apparel 3D CAD system. *Advanced Materials Research, 627,* 501-505. https://doi.org/10.4028/www.scientific.net/amr.627.501.

Pamuk, O. (2017). *Gerber AccuMark V10: Basic Concepts and Marker Making*. Ege University Publications.

Tama, D., Encan, B. C., & Ondogan, Z. (2014, April 2-5). Future Skills of Cad Systems in Apparel Industry. *XIIIth International Izmir Textile and Apparel Symposium, Izmir, Turkey*, 158-162. https://www.textotex.com/application/static/data/file/eng/Derya%20TAMA.pdf.

V-Stitcher Training Manual, 2018.

Index

C

D

N

O

P

Q

R

S

W

X

Y

Z